Physical-Layer Security

From Information Theory to Security Engineering

This complete guide to physical-layer security presents the theoretical foundations, practical implementation, challenges, and benefits of a groundbreaking new model for secure communication. Using a bottom-up approach from the link level all the way to end-to-end architectures, it provides essential practical tools that enable graduate students, industry professionals, and researchers to build more secure systems by exploiting the noise inherent to communication channels.

The book begins with a self-contained explanation of the information-theoretic limits of secure communications at the physical layer. It then goes on to develop practical coding schemes, building on the theoretical insights and enabling readers to understand the challenges and opportunities related to the design of physical-layer security schemes. Finally, applications to multi-user communications and network coding are also included.

Matthieu Bloch is an Assistant Professor in the School of Electrical Engineering of the Georgia Institute of Technology. He received a Ph.D. in Engineering Science from the Université de Franche-Comté, Besançon, France, in 2006, and a Ph.D. in Electrical Engineering from the Georgia Institute of Technology in 2008. His research interests are in the areas of information theory, error-control coding, wireless communications, and quantum cryptography.

João Barros is an Associate Professor in the Department of Electrical and Computer Engineering of the Faculdade de Engenharia da Universidade do Porto, the Head of the Porto Delegation of the Instituto de Telecomunicações, Portugal, and a Visiting Professor at the Massachusetts Institute of Technology. He received his Ph.D. in Electrical Engineering and Information Technology from the Technische Universität München (TUM), Germany, in 2004 and has since published extensively in the general areas of information theory, communication networks, and security. He has taught short courses and tutorials at various institutions and received a Best Teaching Award from the Bavarian State Ministry of Sciences and the Arts, as well as the 2010 IEEE ComSoc Young Researcher Award for Europe, the Middle East, and Africa.

Physical-Layer Security

From Information Theory to Security Engineering

MATTHIEU BLOCH
Georgia Institute of Technology

JOÃO BARROS
University of Porto

CAMBRIDGE
UNIVERSITY PRESS

University Printing House, Cambridge CB2 8BS, United Kingdom

Cambridge University Press is part of the University of Cambridge.

It furthers the University's mission by disseminating knowledge in the pursuit of education, learning and research at the highest international levels of excellence.

www.cambridge.org
Information on this title: www.cambridge.org/9780521516501

First published 2011

A catalogue record for this publication is available from the British Library

ISBN 978-0-521-51650-1 Hardback

To our families

Contents

Preface

This book is the result of more than five years of intensive research in collaboration with a large number of people. Since the beginning, our goal has been to understand at a deeper level how information-theoretic security ideas can help build more secure networks and communication systems. Back in 2008, the actual plan was to finish the manuscript within one year, which for some reason seemed a fairly reasonable proposition at that time. Needless to say, we were thoroughly mistaken. The pace at which physical-layer security topics have found their way into the main journals and conferences in communications and information theory is simply staggering. In fact, there is now a vibrant scientific community uncovering the benefits of looking at the physical layer from a security point of view and producing new results every day. Writing a book on physical-layer security thus felt like shooting at not one but multiple moving targets.

To preserve our sanity we decided to go back to basics and focus on how to bridge the gap between theory and practice. It did not take long to realize that the book would have to appeal simultaneously to information theorists, cryptographers, and network-security specialists. More precisely, the material could and should provide a common ground for fruitful interactions between those who speak the language of security and those who for a very long time focused mostly on the challenges of communicating over noisy channels. Therefore, we opted for a mathematical treatment that addresses the fundamental aspects of information-theoretic security, while providing enough background on cryptographic protocols to allow an eclectic and synergistic approach to the design of security systems.

The book is intended for several different groups: (a) communication engineers and security specialists who wish to understand the fundamentals of physical-layer security and apply them in the development of real-life systems, (b) scientists who aim at creating new knowledge in information-theoretic security and applications, (c) graduate students who wish to be trained in the fundamental techniques, and (d) decision makers who seek to evaluate the potential benefits of physical-layer security. If this book leads to many exciting discussions at the white board among diverse groups of people, then our goal will have been achieved.

Finally, we would like to acknowledge all our colleagues, students, and friends who encouraged us and supported us during the course of this project. First and foremost, we are deeply grateful to Steve McLaughlin, who initiated the project and let us run with it. Special thanks are also due to Phil Meyer and Sarah Matthews from Cambridge University Press for their endless patience as we postponed the delivery of the manuscript countless times. We express our sincere gratitude to Demijan Klinc and Alexandre

Pierrot, who proofread the entire book in detail many times and relentlessly asked for clarification, simplification, and consistent notation. We would like to thank Glenn Bradford, Michael Dickens, Brian Dunn, Jing Huang, Utsaw Kumar, Ebrahim Molavian-Jazi, and Zhanwei Sun for attending EE 87023 at the University of Notre Dame when the book was still a set of immature lecture notes. The organization and presentation of the book have greatly benefited from their candid comments. Thanks are also due to Nick Laneman, who provided invaluable support. Willie Harrison, Xiang He, Mari Kobayashi, Ashish Khisti, Francesco Renna, Osvaldo Simeone, Andrew Thangaraj, and Aylin Yener offered very constructive comments. The book also benefited greatly from many discussions with Prakash Narayan, Imre Csiszár, Muriel Médard, Ralf Koetter, and Pedro Pinto, who generously shared their knowledge with us. Insights from research by Miguel Rodrigues, Luísa Lima, João Paulo Vilela, Paulo Oliveira, Gerhard Maierbacher, Tiago Vinhoza, and João Almeida at the University of Porto also helped shape the views expressed in this volume.

Matthieu Bloch, Georgia Institute of Technology
João Barros, University of Porto

Notation

$\mathrm{GF}(q)$	Galois field with q elements
$\mathbb{R}$	field of real numbers
$\mathbb{C}$	field of complex numbers
$\mathbb{N}$	set of natural numbers ($\mathbb{N}^*$ excludes 0)
$\mathcal{X}$	alphabet or set
$\lvert\mathcal{X}\rvert$	cardinality of $\mathcal{X}$
$\mathrm{cl}(\mathcal{X})$	closure of set $\mathcal{X}$
$\mathrm{co}(\mathcal{X})$	convex hull of set $\mathcal{X}$
$\mathbb{1}$	indicator function
$\{x_i\}_n$	ensemble with n elements $\{x_1, \dots, x_n\}$
x	generic element of alphabet $\mathcal{X}$
$\lvert x\rvert$	absolute value of x
$\lceil x\rceil$	unique integer n such that $x \leqslant n < x + 1$
$\lfloor x\rfloor$	unique integer n such that $x - 1 \leqslant n \leqslant x$
$[\![x, y]\!]$	sequence of integers between $\lfloor x\rfloor$ and $\lceil y\rceil$
x^+	positive part of x, that is $x^+ = \max(x, 0)$
$\mathrm{sign}(x)$	$+1$ if $x \geqslant 0$, -1 otherwise
x^n	sequence $x_1, \dots, x_n$
$\bar{x}^n$	sequence with n repetitions of the same element x
ϵ	usually, a "small" positive real number
$\delta(\epsilon)$	a function of ϵ such that $\lim_{\epsilon\to 0} \delta(\epsilon) = 0$
$\delta_\epsilon(n)$	a function of ϵ and n such that $\lim_{n\to\infty} \delta_\epsilon(n) = 0$
$\delta(n)$	a function of n such that $\lim_{n\to\infty} \delta(n) = 0$
$\mathbf{x}$	column vector containing the n elements $x_1, x_2, \dots, x_n$
$\mathbf{x}^\mathsf{T}$	transpose of $\mathbf{x}$
$\mathbf{x}^\dagger$	Hermitian transpose of $\mathbf{x}$
$\mathbf{H}$	matrix
$(h_{ij})_{m,n}$	$m \times n$ matrix whose elements are h_{ij}, with $i \in [\![1, m]\!]$ and $j \in [\![1, n]\!]$
$\lvert\mathbf{H}\rvert$	determinant of matrix $\mathbf{H}$
$\mathrm{tr}(\mathbf{H})$	trace of matrix $\mathbf{H}$
$\mathrm{rk}(\mathbf{H})$	rank of matrix $\mathbf{H}$
$\mathrm{Ker}(\mathbf{H})$	kernel of matrix $\mathbf{H}$

X	random variable implicitly defined on alphabet $\mathcal{X}$
p_{X}	probability distribution of random variable X
$\mathsf{X} \sim p_{\mathsf{X}}$	random variable X with distribution p_{X}
$\mathcal{N}(\mu, \sigma^2)$	Gaussian distribution with mean μ and variance σ^2
$\mathcal{B}(p)$	Bernoulli distribution with parameter p
$p_{\mathsf{X}\vert\mathsf{Y}}$	conditional probability distribution of X given Y
$\mathcal{T}_\epsilon^n(\mathsf{X})$	strong typical set with respect to p_{X}
$\mathcal{T}_\epsilon^n(\mathsf{XY})$	strong joint-typical set with respect to p_{XY}
$\mathcal{T}_\epsilon^n(\mathsf{XY}\vert x^n)$	conditional strong typical set with respect to p_{XY} and x^n
$\mathcal{A}_\epsilon^n(\mathsf{X})$	weak typical set with respect to p_{X}
$\mathcal{A}_\epsilon^n(\mathsf{XY})$	joint weak typical set with respect to p_{XY}
$\mathbb{E}_{\mathsf{X}}$	expected value over random variable X
$\mathrm{Var}(\mathsf{X})$	variance of random variable X
$\mathbb{P}_{\mathsf{X}}$	probability of an event over X
$\mathbb{H}(\mathsf{X})$	Shannon entropy of discrete random variable X
$\mathbb{H}_b$	binary entropy function
$\mathbb{H}_c(\mathsf{X})$	collision entropy of discrete random variable X
$\mathbb{H}_\infty(\mathsf{X})$	min-entropy of discrete random variable X
$\mathbb{h}(\mathsf{X})$	differential entropy of continuous random variable X
$\mathbb{I}(\mathsf{X};\mathsf{Y})$	mutual information between random variables X and Y
$\mathbf{P}_e(\mathcal{C})$	probability of error of a code $\mathcal{C}$
$\mathbf{E}(\mathcal{C})$	equivocation of a code $\mathcal{C}$
$\mathbf{L}(\mathcal{C})$	information leakage of a code $\mathcal{C}$
$\mathbf{U}(\mathcal{S})$	uniformity of keys guaranteed by key-distillation strategy $\mathcal{S}$
$\underline{\lim}_{x\to c} f(x)$	limit inferior of $f(x)$ as x goes to c
$\overline{\lim}_{x\to c} f(x)$	limit superior of $f(x)$ as x goes to c
$f(x) = O(g(x))$	If g is non-zero for large enough values of x, $f(x) = O(g(x))$ as $x \to a$ if and only if $\overline{\lim}_{x\to\infty} \vert f(x)/g(x)\vert < \infty$.

Abbreviations

AES	Advanced Encryption Standard
AWGN	additive white Gaussian noise
BC	broadcast channel
BCC	broadcast channel with confidential messages
BEC	binary erasure channel
BSC	binary symmetric channel
CA	certification authority
DES	Data Encryption Standard
DMC	discrete memoryless channel
DMS	discrete memoryless source
DSRC	Dedicated Short-Range Communication
DSS	direct sequence spreading
DWTC	degraded wiretap channel
EAP	Extensible Authentication Protocol
EPC	Electronic Product Code
ESP	Encapsulating Security Payload
FH	frequency hopping
GPRS	General Packet Radio Service
GSM	Global System for Mobile Communications
IETF	Internet Engineering Task Force
IP	Internet Protocol
LDPC	low-density parity-check
LLC	logical link control
LLR	log-likelihood ratio
LPI	low probability of intercept
LS	least square
LTE	Long Term Evolution
MAC	multiple-access channel
MIMO	multiple-input multiple-output
NFC	near-field communication
NIST	National Institute of Standards and Technology, USA
OSI	open system interconnection
PKI	public key infrastructure
RFID	radio-frequency identification

RSA	Rivest–Shamir–Adleman
SIM	subscriber identity module
SSL	Secure Socket Layer
TCP	Transmission Control Protocol
TDD	time-division duplex
TLS	transport layer security
TWWTC	two-way wiretap channel
UMTS	Universal Mobile Telecommunication System
WTC	Wiretap channel
XOR	exclusive OR

Part I

Preliminaries

1 An information-theoretic approach to physical-layer security

A simple look at today's information and communication infrastructure is sufficient for one to appreciate the elegance of the layered networking architecture. As networks flourish worldwide, the fundamental problems of transmission, routing, resource allocation, end-to-end reliability, and congestion control are assigned to different layers of protocols, each with its own specific tools and network abstractions. However, the conceptual beauty of the layered protocol stack is not easily found when we turn our attention to the issue of network security. In the early days of the Internet, possibly because network access was very limited and tightly controlled, network security was not yet viewed as a primary concern for computer users and system administrators. This perception changed with the increase in network connections. Technical solutions, such as personnel access controls, password protection, and end-to-end encryption, were developed soon after. The steady growth in connectivity, fostered by the advent of electronic-commerce applications and the ubiquity of wireless communications, remains unhindered and has resulted in an unprecedented awareness of the importance of network security in all its guises.

The standard practice of adding authentication and encryption to the existing protocols at the various communication layers has led to what could be rightly classified as a patchwork of security mechanisms. Given that data security is so critically important, it is reasonable to argue that security measures should be implemented at *all* layers where this can be done in a cost-effective manner. Interestingly, one layer has remained almost ignored in this shift towards secure communication: the physical layer, which lies at the lowest end of the protocol stack and converts bits of information into modulated signals. The state of affairs described is all the more striking since randomness, generally perceived as a key element of secrecy systems, is abundantly available in the stochastic nature of the noise that is intrinsic to the physical communication channel. On account of this observation, this book is entirely devoted to an emerging paradigm: security technologies that are embedded at the *physical layer* of the protocol architecture, a segment of the system where little security exists today.

The absence of a comprehensive physical-layer security approach may be partly explained by invoking the way security issues are taught. A typical graduate course in cryptography and security often starts with a discussion of Shannon's information-theoretic notion of perfect secrecy, but information-theoretic security is quickly discarded and regarded as no more than a beautiful, yet unfeasible, theoretical construct. Such an exposition is designed to motivate the use of state-of-the-art encryption

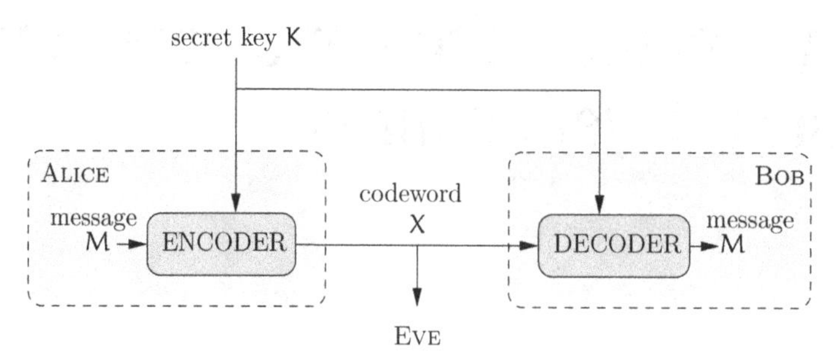

Figure 1.1 Shannon's model of a secrecy system.

algorithms, which are insensitive to the characteristics of the communication channel and rely on mathematical operations assumed to be hard to compute, such as prime factorization.

In this introductory chapter, we approach the subject in a different way. First, we give a bird's-eye view of the basic concepts of information-theoretic security and how they differ from classical cryptography. Then, we discuss in general terms some of the major achievements of information-theoretic security and give some examples of its potential to strengthen the security of the physical layer. The main idea is to exploit the randomness of noisy communication channels to guarantee that a malicious eavesdropper obtains no information about the sent messages: security is ensured not relative to a hard mathematical problem but by the physical uncertainty inherent to the noisy channel.

1.1 Shannon's perfect secrecy

Roughly speaking, the objective of secure communication is twofold; upon transmission of a message, the intended receivers should recover the message without errors while nobody else should acquire any information. This fundamental principle was formalized by Shannon in his 1949 paper [1], using the model of a secrecy system illustrated in Figure 1.1. A transmitter attempts to send a message M to a legitimate receiver by encoding it into a *codeword* X.[1] During transmission, the codeword is observed by an *eavesdropper* (called the enemy cryptanalyst in Shannon's original model) without any degradation, which corresponds to a worst-case scenario in which the communication channel is error-free. In real systems, where some form of noise is almost always present, this theoretical assumption corresponds to the existence of powerful error-correction mechanisms, which ensure that the message can be recovered with arbitrarily small probability of error. As is customary in cryptography, we often refer to the transmitter as "Alice," to the legitimate receiver as "Bob," and to the eavesdropper as "Eve."

In this worst-case scenario, the legitimate receiver must have some advantage over the eavesdropper, otherwise the latter would be able to recover the message M as well. The solution to this problem lies in the use of a secret key K, known only to the transmitter

[1] In cryptography, X is also called a cryptogram or ciphertext.

Table 1.1 Example of a one-time pad

Message	M	0	1	0	1	0	0	0	1	1	0	1
Key	K	1	0	0	1	1	0	0	0	1	0	1
Cryptogram	$X = M \oplus K$	1	1	0	0	1	0	0	1	0	0	0

and the legitimate receiver. The codeword X is then obtained by computing a function of the message M and the secret key K.

Shannon formalized the notion of secrecy by quantifying the average uncertainty of the eavesdropper. In information-theoretic terms, messages and codewords are treated as random variables, and secrecy is measured in terms of the conditional entropy of the message given the codeword, denoted as $\mathbb{H}(M|X)$. The quantity $\mathbb{H}(M|X)$ is also called the eavesdropper's *equivocation*; *perfect secrecy* is achieved if the eavesdropper's equivocation equals the a-priori uncertainty one could have about the message, that is

$$\mathbb{H}(M|X) = \mathbb{H}(M).$$

This equation implies that the codeword X is statistically independent of the message M. The absence of correlation ensures that there exists no algorithm that would allow the cryptanalyst to extract information about the message. We will see in Chapter 3 that perfect secrecy can be achieved only if $\mathbb{H}(K) \geqslant \mathbb{H}(M)$; that is, the uncertainty about the key must be at least as large as the uncertainty about the message. In other words, we must have at least one secret bit for every bit of information contained in the message.

From an algorithmic perspective, perfect secrecy can be achieved by means of a simple procedure called a one-time pad (or Vernam's cipher), an example of which is shown in Table 1.1 for the case of a binary message and a binary key. The codeword is formed by computing the binary addition (XOR) of each message bit with a separate key bit. If the key bits are independent and uniformly distributed, it can be shown that the codeword is statistically independent of the message. To recover the message, the legitimate receiver need only add the codeword and the secret key. On the other hand, the eavesdropper does not have access to the key; therefore, from her perspective, every message is equally likely and she cannot do better than randomly guessing the message bits.

Although the one-time pad can achieve perfect secrecy with low complexity, its applicability is limited by the following requirements:

- the legitimate partners must generate and store long keys consisting of random bits;
- each key can be used only once (otherwise the cryptanalyst has a fair chance of discovering the key);
- the key must be shared over a secure channel.

To solve the problem of distributing long keys in a secure manner, we could be tempted to generate long pseudo-random sequences using a smaller seed. However, information theory shows that the uncertainty of the eavesdropper is upper bounded by the number of random key bits used. The smaller the key the greater the probability that the eavesdropper will succeed in extracting some information from the codeword. In this case,

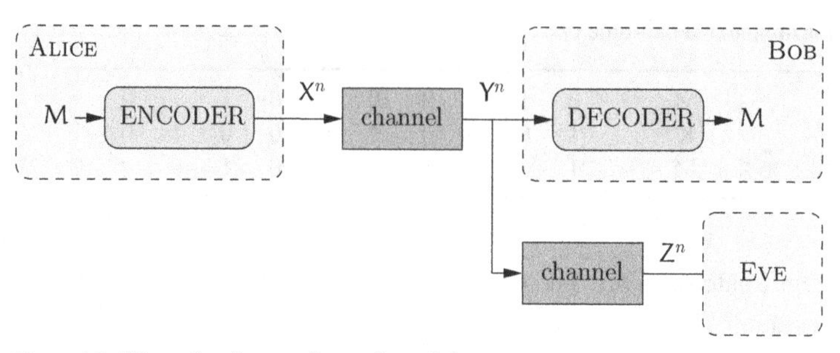

Figure 1.2 Wyner's wiretap channel model.

the only obstacle faced by the eavesdropper is computational complexity, which leads directly to the concept of computational security.

The aforementioned caveats regarding the one-time pad are arguably responsible for the skepticism with which security practitioners dismiss the usefulness of information-theoretic security. We shall now see that a closer look at the underlying communications model may actually yield the solution towards wider applicability.

1.2 Secure communication over noisy channels

As mentioned before, random noise is an intrinsic element of almost all physical communication channels. In an effort to understand the role of noise in the context of secure communications, Wyner introduced the *wiretap channel* model illustrated in Figure 1.2. The main differences between this approach and Shannon's original secrecy system are that

- the legitimate transmitter encodes a message M into a codeword X^n consisting of n symbols, which is sent over a noisy channel to the legitimate receiver;
- the eavesdropper observes a noisy version, denoted by Z^n, of the signal Y^n available at the receiver.

In addition, Wyner suggested a new definition for the secrecy condition. Instead of requiring the eavesdropper's equivocation to be exactly equal to the entropy of the message, we now ask for the *equivocation rate* $(1/n)\mathbb{H}(\mathsf{M}|\mathsf{Z}^n)$ to be arbitrarily close to the entropy rate of the message $(1/n)\mathbb{H}(\mathsf{M})$ for sufficiently large codeword length n. With this relaxed security constraint, it can be shown that there exist channel codes that *asymptotically* guarantee both an arbitrarily small probability of error at the intended receiver and secrecy. Such codes are colloquially known as *wiretap codes*. The maximum transmission rate that is achievable under these premises is called the *secrecy capacity*, and can be shown to be strictly positive whenever the eavesdropper's observation Z^n is "noisier" than Y^n.

In the seventies and eighties, the impact of Wyner's results was limited due to several important obstacles. First, practical code constructions for the wiretap channel were not available. Second, the wiretap channel model restricts the eavesdropper by assuming that

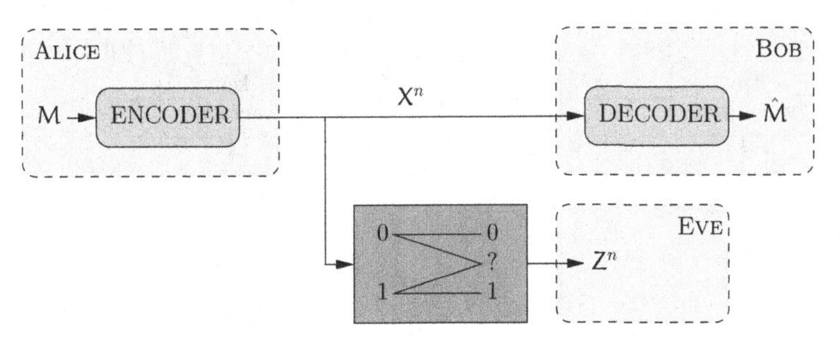

Figure 1.3 Communication over a binary erasure wiretap channel.

she suffers from more noise than is experienced by the legitimate receiver. In addition, soon after the notion of secrecy capacity appeared, information-theoretic security was overshadowed by Diffie and Hellman's seminal work on public-key cryptography, which relies on mathematical functions believed hard to compute and has dominated security research since then.

1.3 Channel coding for secrecy

Although the previous results on the secrecy capacity prove the existence of codes capable of guaranteeing reliable communication while satisfying a secrecy condition, it is not immediately clear how such codes can be constructed in practice. Consider the channel model illustrated in Figure 1.3, in which Alice wants to send one bit of information to Bob over an error-free channel while knowing that Eve's channel is a binary erasure channel, which erases an input symbol with probability ϵ. If Alice sends an uncoded bit, then Eve is able to obtain it correctly with probability $1-\epsilon$, leading to an equivocation equal to ϵ. It follows that, unless $\epsilon = 1$, Eve is able to obtain a non-trivial amount of information.

Alternatively, Alice could use an encoder that assigns one or more codewords to each of the two possible messages, 0 and 1. Suppose she takes all the binary sequences of length n and maps them in such a way that those with even parity correspond to $\mathsf{M} = 0$ and those with odd parity are assigned to $\mathsf{M} = 1$. If Bob receives one of these codewords over the error-free channel, he can obtain the correct message value by determining the parity of the received codeword. Eve, on the other hand, is left with an average of $n\epsilon$ erasures. As soon as one or more bits are erased, Eve loses her ability to estimate the parity of the binary sequence transmitted. This event happens with probability $1-(1-\epsilon)^n$ and it can be shown that

$$\mathbb{H}(\mathsf{M}|\mathsf{Z}^n) \geqslant 1-(1-\epsilon)^n,$$

which goes to unity as n tends to infinity. In others words, for sufficiently large codeword length, we get an equivocation that is arbitrarily close to the entropy of a message. The drawback of this coding scheme is that the transmission rate of $1/n$ goes to zero asymptotically with n as well. Alice and Bob can communicate securely by assigning

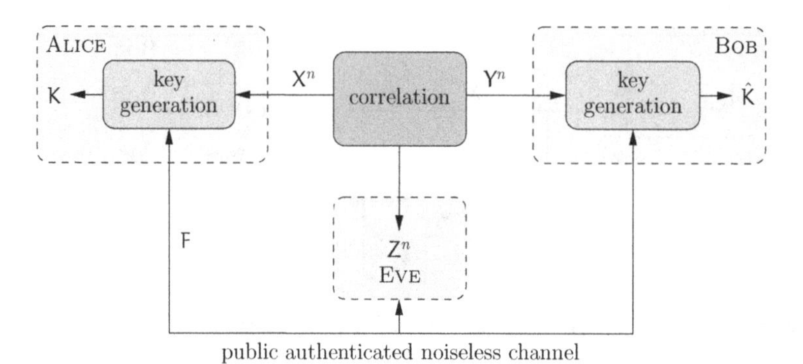

Figure 1.4 Secret-key agreement from correlated observations.

multiple codewords to the same message, but the secrecy achieved bears a price in terms of rate.

The intuition developed for the binary erasure wiretap channel should carry over to more general models. If Bob's channel induces fewer errors than Eve's, Bob should still be able to recover messages using a channel code; in contrast, Eve should be left with a list of possible codewords and messages. Asymptotic perfect secrecy can be achieved if this list covers the entire set of messages and their probability given that the received noisy codeword is roughly uniform. Unfortunately, to this day, practical wiretap code constructions are known for only a few specific channels.

1.4 Secret-key agreement from noisy observations

If Alice and Bob are willing to settle for generating a secret key instead of communicating a secret message straight away, then they can use the noisy channel to generate correlated random sequences and subsequently use an error-free communication channel to agree on a secret key. Such a situation is illustrated in Figure 1.4, in which Alice, Bob, and Eve obtain correlated observations X^n, Y^n, and Z^n, respectively; Alice and Bob then generate a key K on the basis of their respective observations and a set of messages F exchanged over the error-free channel. In the early nineties, Maurer and Ahlswede and Csiszár showed that, even if messages F are made available to the eavesdropper, Alice and Bob can generate a key oblivious to the eavesdropper such that $\mathbb{H}(K|Z^nF)$ is arbitrarily close to $\mathbb{H}(K)$. Provided that authentication is in place, granting Eve access rights to all feedback messages does not compromise security.

To gain some intuition for why public feedback is useful, consider an instance of Wyner's setup, in which the main channel and the eavesdropper's channel are binary symmetric channels. When Alice transmits a random symbol X over the main channel, Bob obtains $Y = X \oplus D$ and Eve observes $Z = X \oplus E$, where D and E are Bernoulli random variables that correspond to the noise added by the main channel and the eavesdropper's channel, respectively. Assume further that Bob's channel is noisier than Eve's, in the sense that $\mathbb{P}[D = 1] \geqslant \mathbb{P}[E = 1]$. Bob now uses the feedback channel in the following manner. To send a symbol V, he adds the noisy observation received from the channel and sends $V \oplus Y = V \oplus X \oplus D$ over the public channel. Since Alice knows X, she can perform a simple binary addition in order to obtain $V \oplus D$. Eve, on the other hand, has only a noisy observation Z, and it can be shown that her optimal estimation of V is $V \oplus Y \oplus Z = V \oplus D \oplus E$. Thus, Alice and Bob effectively transform a wiretap scenario that is advantageous to Eve into a channel in which she suffers from more errors than do Alice and Bob.

From a practical perspective, the design of key-agreement schemes from correlated observations turns out to be a simpler problem than the construction of codes for the wiretap channel. In fact, a wiretap code needs to guarantee *simultaneously* reliable communication to the legitimate receiver and security against the eavesdropper. On the other hand, since a key does not carry any information in itself, the reliability and security constraints can be handled *separately*. For instance, Alice would first send error-correction data to Bob, in the form of parity bits, which would allow him to revert the bit flips caused by the noise in the channel. Even if the error-correcting bits are sent over the public channel, the fact that Eve's observation contains more errors than Bob's is sufficient to guarantee that she is unable to arrive at the same sequence as Alice and Bob. Alice and Bob would then use a well-chosen hash function to transform their common sequence of symbols into a much shorter key and, because of her errors, Eve is unable to predict the output of the hash. Finally, the key would be used as a one-time pad to ensure information-theoretic security.

1.5 Active attacks

Thus far, we have assumed that Eve is a passive eavesdropper, who wishes to extract as much information as possible from the signals traversing the channel. However, if she can afford the risk of being detected by the legitimate partners, she has a wide range of active attacks at her disposal. Eve could impersonate Alice or Bob to cause further confusion, intercept and forge messages that are sent over the noisy channels and the error-free public channel, or simply send jamming signals to perturb the communication.

Sender authentication is a tacit assumption in most contributions in the area of information-theoretic security. Except in special and rare instances of the wiretap scenario, a shared secret in the form of a small key is necessary to authenticate the first transmissions. Subsequent messages can be authenticated using new keys that can be generated at the physical layer using some of the methods in this book. Alternatively, if Alice and Bob are communicating over a wireless channel, then they can sometimes exploit the reciprocity of the channel to their advantage. The receiver can associate a certain channel impulse response with a certain transmitter and it is practically impossible for an adversary located at a different position to be able to generate a similar impulse response at the receiver.

With authentication in place, it is impossible for the attacker to impersonate the legitimate partners and to forge messages. However, the attacker may decide to obstruct the communication by means of jamming. This can be done in a blind manner by transmitting noise, or in a more elaborate fashion exploiting all the available information on the codes and the signals used by the legitimate partners. It is worth pointing out that the use of jamming is not restricted to the active attackers. Cooperative jamming techniques, by which one or more legitimate transmitters send coded jamming signals to increase the confusion of the attacker, can be used effectively to increase the secrecy capacity in multi-user channels. Sophisticated signal processing, most notably through the use of multiple antennas, can also further enhance the aforementioned security benefits.

1.6 Physical-layer security and classical cryptography

There are many fundamental differences between the classical cryptographic primitives used at higher layers of the protocol stack and physical-layer security based on information-theoretic principles. It is therefore important to understand what these differences are and how they affect the choice of technology in a practical scenario.

Classical computational security uses public-key cryptography for authentication and secret-key distribution and symmetric encryption for the protection of transmitted data. The combination of state-of-the-art algorithms like RSA and the Advanced Encryption Standard (AES) is deemed secure for a large number of applications because so far no efficient attacks on public-key systems are publicly known. Many symmetric ciphers were broken in the past, but those that were compromised were consistently replaced by new algorithms, whose cryptanalysis is more difficult and requires more computational effort. Under the assumption that the attacker cannot break hard cryptographic primitives, it is possible to design systems that are secure with probability one. The technology is readily available and inexpensive.

However, there are also disadvantages to the computational model. The security of public-key cryptography is based on the conjecture that certain one-way functions are hard to invert, which remains unproven from a mathematical point of view. Computing power continues to increase at a very fast pace, such that brute-force attacks that were once deemed unfeasible are now within reach. Moreover, there are no precise metrics to compare the strengths of different ciphers in a rigorous way. In general, the security of a cryptographic protocol is measured by whether it survives a set of attacks or not. From the works of Shannon and Wyner, one concludes that the ruling cryptographic paradigm can never provide information-theoretic security, because the communication channel between the friendly parties and the eavesdropper is noiseless and the secrecy capacity is zero. Moreover, existing key-distribution schemes based on the computational model require a trusted third party as well as complex protocols and system architectures. If multiple keys are to be generated, it is usually possible to do so only from a single shared secret and at the price of reduced data protection.

The main advantages of physical-layer security under the information-theoretic security model come from the facts that no computational restrictions are placed on the eavesdropper and that very precise statements can be made about the information that is leaked to the eavesdropper as a function of the channel quality. Physical-layer security has already been realized in practice through quantum key distribution and, in theory, suitably long codes can come exponentially close to perfect secrecy. The system architecture for security is basically the same as the one for communication. Instead of distributing keys, it is possible to generate on-the-fly as many secret keys as desired.

However, we must accept some disadvantages as well. First and foremost, information-theoretic security relies on average information measures. The system can be designed and tuned for a specific level of security, claiming for instance that with very high probability a block will be secure; however, it might not be possible to guarantee confidentiality with probability one. We are also forced to make assumptions about the communication channels that might not be accurate in practice. In most cases, one would make very conservative assumptions about the channels, which is likely to result in low secrecy capacities or low secret-key or message exchange rates. A few systems have been deployed, most notably for optical communication, but the technology is not very widely available and is still expensive.

In light of the brief comparisons above, it is likely that any deployment of a physical-layer security protocol in a classical system would be part of a layered security solution whereby confidentiality and authentication are provided at a number of different layers, each with a specific goal in mind. This modular approach is how virtually all systems are designed, so, in this context, physical-layer security provides an additional layer of security that does not yet exist in communication networks.

1.7 Outline of the rest of the book

The main objective of this book is to lay out the theoretical foundations of physical-layer security and to provide practical tools for implementing it in real systems. The different chapters cover essential theory and mathematical models for assessing physical-layer security and characterizing its fundamental limits, coding schemes for data security at the physical layer, and system aspects of physical-layer security.

Chapter 2 summarizes fundamental notions of information theory required in order to understand subsequent chapters. Our presentation emphasizes the mathematical tools and notions of particular relevance to physical-layer security.

Chapter 3 introduces the seminal results regarding secrecy capacity for communication channels, highlighting the mathematical techniques used in the derivations.

Chapter 4 focuses on the fundamental limits and methodologies of secret-key agreement, including the reconciliation of correlated sequences and how privacy amplification allows strong secrecy.

Chapter 5 discusses the fundamental limits of secure communication over Gaussian and wireless channels.

Chapter 6 covers some of the techniques used to achieve physical-layer security in practice, including the design of codes for wiretap channels as well as the construction of codes for secret-key agreement.

Chapter 7 addresses system issues related to the integration of physical-layer security in contemporary communications architectures and gives examples of practical applications.

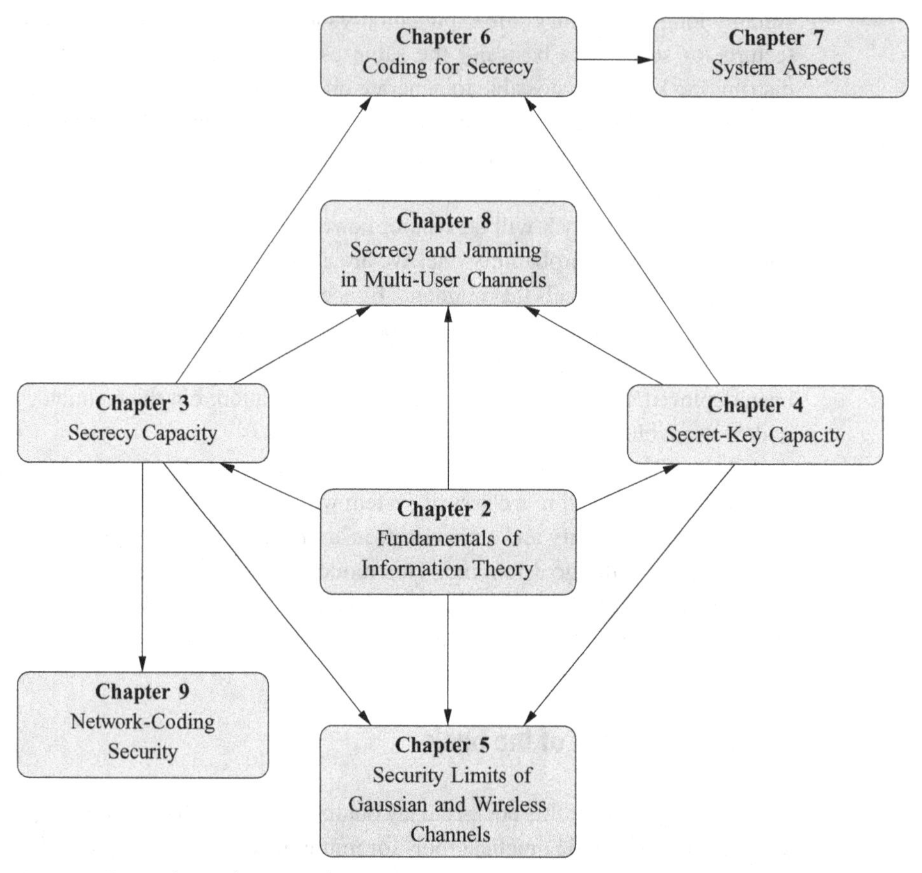

Figure 1.5 Dependences between chapters.

Chapter 8 discusses physical-layer security in multi-user systems and shows how the secrecy rates of multi-terminal networks can be increased through the appropriate use of feedback, cooperation, and jamming.

Chapter 9 deals with network-coding security. Although it is not necessarily implemented at the physical layer, network coding combines aspects of information and coding theory with close connections to information-theoretic security. By allowing intermediate nodes to mix different information flows through non-trivial operations, network coding offers a number of security challenges and opportunities.

The dependences between the chapters of the book are illustrated in Figure 1.5. The reader familiar with the tools and techniques of information theory can probably skip Chapter 2 and start at Chapter 3. The fundamental concepts and results of information-theoretic security are presented in Chapter 3 and Chapter 4 and are leveraged in Chapter 5 and Chapter 8 to study specific applications. Chapter 6, Chapter 7, and Chapter 9 rely on the notions introduced in earlier chapters but can be read independently.

2 Fundamentals of information theory

We begin with a brief overview of some of the fundamental concepts and mathematical tools of information theory. This allows us to establish notation and to set the stage for the results presented in subsequent chapters. For a comprehensive introduction to the fundamental concepts and methods of information theory, we refer the interested reader to the textbooks of Gallager [2], Cover and Thomas [3], Yeung [4], and Csiszár and Körner [5].

The rest of the chapter is organized as follows. Section 2.1 provides an overview of the basic mathematical tools and metrics that are relevant for subsequent chapters. Section 2.2 illustrates the fundamental proof techniques used in information theory by discussing the point-to-point communication problem and Shannon's coding theorems. Section 2.3 is entirely devoted to network information theory, with a special emphasis on distributed source coding and multi-user communications as they relate to information-theoretic security.

2.1 Mathematical tools of information theory

The following subsections describe a powerful set of metrics and tools that are useful to characterize the fundamental limits of communication systems. All results are stated without proof through a series of lemmas and theorems, and we refer the reader to standard textbooks [2, 3, 4] for details. Unless specified otherwise, all random variables and random vectors used throughout this book are real-valued random vectors.

2.1.1 Useful bounds

We start by recalling a few inequalities that are useful to bound the probabilities of rare events.

Lemma 2.1 (Markov's inequality). *Let* X *be a non-negative real-valued random variable. Then,*

$$\forall a > 0 \quad \mathbb{P}[\mathrm{X} \geqslant a] \leqslant \frac{\mathbb{E}_{\mathrm{X}}[\mathrm{X}]}{a}.$$

The following consequence of Markov's inequality is particularly useful.

Lemma 2.2 (Selection lemma). *Let $\mathrm{X}_n \in \mathcal{X}_n$ be a random variable and let $\mathcal{F}$ be a finite set of functions $f : \mathcal{X}_n \to \mathbb{R}^+$ such that $|\mathcal{F}|$ does not depend on n and*

$$\forall f \in \mathcal{F} \quad \mathbb{E}_{\mathrm{X}_n}[f(\mathrm{X}_n)] \leqslant \delta(n).$$

Then, there exists a specific realization x_n of X_n such that

$$\forall f \in \mathcal{F} \quad f(x_n) \leqslant \delta(n).$$

Proof. Let $\epsilon_n \triangleq \delta(n)$. Using the union bound and Markov's inequality, we obtain

$$\begin{aligned}
\mathbb{P}_{\mathrm{X}_n}\left[\bigcup_{f \in \mathcal{F}} \{f(\mathrm{X}_n) \geqslant (|\mathcal{F}|+1)\epsilon_n\}\right] &\leqslant \sum_{f \in \mathcal{F}} \mathbb{P}_{\mathrm{X}_n}[f(\mathrm{X}_n) \geqslant (|\mathcal{F}|+1)\epsilon_n] \\
&\leqslant \sum_{f \in \mathcal{F}} \frac{\mathbb{E}_{\mathrm{X}_n}[f(\mathrm{X}_n)]}{(|\mathcal{F}|+1)\epsilon_n} \\
&\leqslant \frac{|\mathcal{F}|}{|\mathcal{F}|+1} \\
&< 1.
\end{aligned}$$

Therefore, there exists *at least* one realization x_n of X_n such that

$$\forall f \in \mathcal{F} \quad f(x_n) \leqslant (|\mathcal{F}|+1)\epsilon_n.$$

Since $\epsilon_n = \delta(n)$ and $|\mathcal{F}|$ is finite and independent of n, we can write $(|\mathcal{F}|+1)\epsilon_n$ as $\delta(n)$. □

We call Lemma 2.2 the "selection lemma" because it tells us that, if $\mathbb{E}_{\mathrm{X}_n}[f(X_n)] \leqslant \delta(n)$ for all $f \in \mathcal{F}$, we can select a specific realization x_n such that $f(x_n) \leqslant \delta(n)$ for all $f \in \mathcal{F}$. Two other useful consequences of Markov's inequality are Chebyshev's inequality and Chernov bounds.

Lemma 2.3 (Chebyshev's inequality). *Let X be a real-valued random variable. Then,*

$$\forall a > 0 \quad \mathbb{P}[|\mathrm{X} - \mathbb{E}_{\mathrm{X}}[\mathrm{X}]| \geqslant a] \leqslant \frac{\mathrm{Var}(\mathrm{X})}{a^2}.$$

Lemma 2.4 (Chernov bounds). *Let X be a real-valued random variable. Then, for all $a > 0$,*

$$\begin{aligned}
\forall s > 0 \quad \mathbb{P}[\mathrm{X} \geqslant a] \leqslant \mathbb{E}_{\mathrm{X}}\left[\mathrm{e}^{s\mathrm{X}}\right]\mathrm{e}^{-sa}, \\
\forall s < 0 \quad \mathbb{P}[\mathrm{X} \leqslant a] \leqslant \mathbb{E}_{\mathrm{X}}\left[\mathrm{e}^{s\mathrm{X}}\right]\mathrm{e}^{-sa}.
\end{aligned}$$

2.1.2 Entropy and mutual information

In this section, we define a series of useful information-theoretic quantities, whose operational significance will become clear in the next section.

Definition 2.1. *Let* X *and* X′ *be two discrete random variables defined on the same alphabet* $\mathcal{X}$*. The variational distance between* X *and* X′ *is*

$$\mathbb{V}(\mathsf{X},\mathsf{X}') \triangleq \sum_{x\in\mathcal{X}} |p_\mathsf{X}(x) - p_{\mathsf{X}'}(x)|.$$

Definition 2.2. *Let* $\mathsf{X} \in \mathcal{X}$ *be a discrete random variable with distribution* p_X*. The Shannon entropy (or entropy for short) of* X *is defined as*

$$\mathbb{H}(\mathsf{X}) \triangleq -\sum_{x\in\mathcal{X}} p_\mathsf{X}(x)\log p_\mathsf{X}(x),$$

with the convention that $0\log 0 \triangleq 0$*. Unless specified otherwise, all logarithms are taken to the base two and the unit for entropy is called a bit.*

If $\mathcal{X} = \{0, 1\}$, then X is a binary random variable and its entropy depends solely on the parameter $p = \mathbb{P}[\mathsf{X} = 0]$. The *binary entropy function* is defined as

$$\mathbb{H}_b(p) \triangleq -p\log p - (1-p)\log(1-p).$$

Lemma 2.5. *For any discrete random variable* $\mathsf{X} \in \mathcal{X}$

$$0 \leqslant \mathbb{H}(\mathsf{X}) \leqslant \log|\mathcal{X}|.$$

The equality $\mathbb{H}(\mathsf{X}) = 0$ *holds if and only if* X *is a constant while the equality* $\mathbb{H}(\mathsf{X}) = \log|\mathcal{X}|$ *holds if and only if* X *is uniform on* $\mathcal{X}$*.*

Conceptually, $\mathbb{H}(\mathsf{X})$ can be viewed as a measure of the average amount of information contained in X or, equivalently, the amount of uncertainty that subsists until the outcome of X is revealed.

Proposition 2.1 (Csiszár and Körner). *Let* X *and* X′ *be two discrete random variables defined on the same alphabet* $\mathcal{X}$*. Then,*

$$\left|\mathbb{H}(\mathsf{X}) - \mathbb{H}(\mathsf{X}')\right| \leqslant \mathbb{V}(\mathsf{X},\mathsf{X}')\log\left(\frac{|\mathcal{X}|}{\mathbb{V}(\mathsf{X},\mathsf{X}')}\right).$$

Definition 2.3. *Let* $\mathsf{X} \in \mathcal{X}$ *and* $\mathsf{Y} \in \mathcal{Y}$ *be two discrete random variables with joint distribution* p_{XY}*. The joint entropy of* X *and* Y *is defined as*

$$\mathbb{H}(\mathsf{XY}) \triangleq -\sum_{x\in\mathcal{X}}\sum_{y\in\mathcal{Y}} p_{\mathsf{XY}}(x,y)\log p_{\mathsf{XY}}(x,y).$$

Definition 2.4. *Let* $\mathsf{X} \in \mathcal{X}$ *and* $\mathsf{Y} \in \mathcal{Y}$ *be two discrete random variables with joint distribution* p_{XY}*. The conditional entropy of* X *given* Y *is defined as*

$$\mathbb{H}(\mathsf{Y}|\mathsf{X}) \triangleq -\sum_{x\in\mathcal{X}}\sum_{y\in\mathcal{Y}} p_{\mathsf{XY}}(x,y)\log p_{\mathsf{Y}|\mathsf{X}}(y|x).$$

By expanding $\mathbb{H}(\mathsf{XY})$ with Bayes' rule, one can verify that

$$\mathbb{H}(\mathsf{XY}) = \mathbb{H}(\mathsf{X}) + \mathbb{H}(\mathsf{Y}|\mathsf{X}).$$

This expansion generalizes to the entropy of a random vector $X^n = (X_1, \dots, X_n)$ as

$$\mathbb{H}(X^n) = \mathbb{H}(X_1) + \mathbb{H}(X_2|X_1) + \cdots + \mathbb{H}(X_n|X^{n-1})$$
$$= \sum_{i=1}^{n} \mathbb{H}(X_i|X^{i-1}),$$

with the convention that $\mathbb{H}(X_1|X^0) \triangleq \mathbb{H}(X_1)$. This expansion is known as the *chain rule of entropy*.

Lemma 2.6 ("Conditioning does not increase entropy"). *Let $X \in \mathcal{X}$ and $Y \in \mathcal{Y}$ be two discrete random variables with joint distribution p_{XY}. Then,*

$$\mathbb{H}(X|Y) \leqslant \mathbb{H}(X).$$

In other words, this lemma asserts that knowledge of Y cannot increase our uncertainty about X.

Definition 2.5. *Let $X \in \mathcal{X}$ and $Y \in \mathcal{Y}$ be two discrete random variables with joint distribution p_{XY}. The mutual information between X and Y is defined as*

$$\mathbb{I}(X;Y) \triangleq \mathbb{H}(X) - \mathbb{H}(X|Y).$$

Let $X \in \mathcal{X}$, $Y \in \mathcal{Y}$, and $Z \in \mathcal{Z}$ be discrete random variables with joint distribution p_{XYZ}. The conditional mutual information between X and Y given Z is

$$\mathbb{I}(X;Y|Z) \triangleq \mathbb{H}(X|Z) - \mathbb{H}(X|YZ).$$

Intuitively, $\mathbb{I}(X;Y)$ represents the uncertainty about X that is not resolved by the observation of Y. By using the chain rule of entropy, one can expand the mutual information between a random vector $X^n = (X_1, \dots, X_n)$ and a random variable Y as

$$\mathbb{I}(X^n;Y) = \sum_{i=1}^{n} \mathbb{I}(X_i;Y|X^{i-1}),$$

with the convention that $\mathbb{I}(X_1;Y|X^0) \triangleq \mathbb{I}(X_1;Y)$. This expansion is known as the *chain rule of mutual information*.

Lemma 2.7. *Let $X \in \mathcal{X}$ and $Y \in \mathcal{Y}$ be two discrete random variables with joint distribution p_{XY}. Then,*

$$0 \leqslant \mathbb{I}(X;Y) \leqslant \min(\mathbb{H}(X), \mathbb{H}(Y)).$$

The equality $\mathbb{I}(X;Y) = 0$ holds if and only if X and Y are independent. The equality $\mathbb{I}(X;Y) = \mathbb{H}(X)$ ($\mathbb{I}(X;Y) = \mathbb{H}(Y)$) holds if and only if X is a function of Y (Y is a function of X).

Lemma 2.8. *Let $X \in \mathcal{X}$, $Y \in \mathcal{Y}$, and $Z \in \mathcal{Z}$ be three discrete random variables with joint distribution p_{XYZ}. Then,*

$$0 \leqslant \mathbb{I}(X;Y|Z) \leqslant \min(\mathbb{H}(X|Z), \mathbb{H}(Y|Z)).$$

The equality $\mathbb{I}(\mathsf{X};\mathsf{Y}|\mathsf{Z}) = 0$ *holds if and only if* X *and* Y *are conditionally independent given* Z. *In this case, we say that* $\mathsf{X} \rightarrow \mathsf{Z} \rightarrow \mathsf{Y}$ *forms a Markov chain. The equality* $\mathbb{I}(\mathsf{X};\mathsf{Y}|\mathsf{Z}) = \mathbb{H}(\mathsf{X}|\mathsf{Z})$ *(*$\mathbb{I}(\mathsf{X};\mathsf{Y}|\mathsf{Z}) = \mathbb{H}(\mathsf{Y}|\mathsf{Z})$*) holds if and only if* X *is a function of* Y *and* Z *(*Y *is a function of* X *and* Z*).*

Lemma 2.9 (Data-processing inequality). *Let* $\mathsf{X} \in \mathcal{X}$, $\mathsf{Y} \in \mathcal{Y}$, *and* $\mathsf{Z} \in \mathcal{Z}$ *be three discrete random variables such that* $\mathsf{X} \rightarrow \mathsf{Y} \rightarrow \mathsf{Z}$ *forms a Markov chain. Then,*

$$\mathbb{I}(\mathsf{X};\mathsf{Y}) \geqslant \mathbb{I}(\mathsf{X};\mathsf{Z}).$$

An equivalent form of the data-processing inequality is $\mathbb{H}(\mathsf{X}|\mathsf{Y}) \leqslant \mathbb{H}(\mathsf{X}|\mathsf{Z})$, which means that, on average, processing Y can only increase our uncertainty about X.

Lemma 2.10 (Fano's inequality). *Let* $\mathsf{X} \in \mathcal{X}$ *be a discrete random variable and let* X' *be any estimate of* X *that takes values in the same alphabet* $\mathcal{X}$. *Let* $P_{\mathrm{e}} \triangleq \mathbb{P}[\mathsf{X} \neq \mathsf{X}']$ *be the probability of error obtained when estimating* X *with* X'. *Then,*

$$\mathbb{H}(\mathsf{X}|\mathsf{X}') \leqslant \mathbb{H}_{\mathrm{b}}(P_{\mathrm{e}}) + P_{\mathrm{e}} \log(|\mathcal{X}| - 1),$$

where $\mathbb{H}_{\mathrm{b}}(P_{\mathrm{e}})$ *is the binary entropy function defined earlier.*

Fano's inequality is the key ingredient of many proofs in this book because it relates an information-theoretic quantity (the conditional entropy $\mathbb{H}(\mathsf{X}|\mathsf{X}')$) to an operational quantity (the probability of error P_{e}). In what follows, we often write Fano's inequality in the form $\mathbb{H}(\mathsf{X}|\mathsf{X}') \leqslant \delta(P_{\mathrm{e}})$ to emphasize that $\mathbb{H}(\mathsf{X}|\mathsf{X}')$ goes to zero if P_{e} goes to zero.

Definition 2.6. *A function* $f : \mathcal{I} \rightarrow \mathbb{R}$ *defined on a set* $\mathcal{I}$ *is convex on* $\mathcal{I}$ *if, for all* $(x_1, x_2) \in \mathcal{I}^2$ *and for all* $\lambda \in [0, 1]$,

$$f(\lambda x_1 + (1 - \lambda)x_2) \leqslant \lambda f(x_1) + (1 - \lambda) f(x_2).$$

If the equation above holds with strict inequality, f *is strictly convex on* $\mathcal{I}$. *A function* $f : \mathcal{I} \rightarrow \mathbb{R}$ *defined on a set* $\mathcal{I}$ *is (strictly) concave on* $\mathcal{I}$ *if the function* $-f$ *is (strictly) convex on* $\mathcal{I}$.

Lemma 2.11 (Jensen's inequality). *Let* $\mathsf{X} \in \mathcal{X}$ *be a random variable and let* $f : \mathcal{X} \rightarrow \mathbb{R}$ *be a convex function. Then,*

$$\mathbb{E}[f(\mathsf{X})] \geqslant f(\mathbb{E}[\mathsf{X}]).$$

If f *is strictly convex, then equality holds if and only if* X *is a constant.*

Lemma 2.12. *Let* $\mathsf{X} \in \mathcal{X}$ *and* $\mathsf{Y} \in \mathcal{Y}$ *be two discrete random variables with joint distribution* p_{XY}. *Then,*

- $\mathbb{H}(\mathsf{X})$ *is a concave function of* p_{X}*;*
- $\mathbb{I}(\mathsf{X};\mathsf{Y})$ *is a concave function of* p_{X} *for* $p_{\mathsf{Y}|\mathsf{X}}$ *fixed;*
- $\mathbb{I}(\mathsf{X};\mathsf{Y})$ *is a convex function of* $p_{\mathsf{Y}|\mathsf{X}}$ *for* p_{X} *fixed.*

If X is a continuous random variable then, in general, $\mathbb{H}(\mathsf{X})$ is not well defined. It is convenient to define the differential entropy as follows.

Definition 2.7. *Let* $\mathsf{X} \in \mathcal{X}$ *be a continuous random variable with distribution* p_{X}. *The differential entropy of* X *is defined as*

$$\mathbb{h}(\mathsf{X}) \triangleq -\int_{x\in\mathcal{X}} p_{\mathsf{X}}(x)\log p_{\mathsf{X}}(x)\mathrm{d}x.$$

The notions of joint entropy, conditional entropy, and mutual information for continuous random variables are identical to their discrete counterparts, but with the differential entropy $\mathbb{h}$ in place of the entropy $\mathbb{H}$. With the exception of Lemma 2.5 and Fano's inequality, all of the properties of entropy and mutual information stated above hold also for continuous random variables. In addition, the following properties will be useful in the next chapters.

Lemma 2.13. *If* $\mathsf{X} \in \mathcal{X}$ *is a continuous random variable with variance* $\mathrm{Var}(\mathsf{X}) \leqslant \sigma^2$, *then*

$$\mathbb{h}(\mathsf{X}) \leqslant \frac{1}{2}\log\left(2\pi e\sigma^2\right),$$

with equality if and only if X *has a Gaussian distribution with variance* σ^2.

Lemma 2.14 (Entropy–power inequality). *Let* $\mathsf{X} \in \mathcal{X}$ *and* $\mathsf{Y} \in \mathcal{Y}$ *be independent continuous random variables with entropy* $\mathbb{h}(\mathsf{X})$ *and* $\mathbb{h}(\mathsf{Y})$, *respectively. Let* X' *and* Y' *be independent Gaussian random variables such that* $\mathbb{h}(\mathsf{X}') = \mathbb{h}(\mathsf{X})$ *and* $\mathbb{h}(\mathsf{Y}') = \mathbb{h}(\mathsf{Y})$. *Then,*

$$\mathbb{h}(\mathsf{X}+\mathsf{Y}) \geqslant \mathbb{h}\left(\mathsf{X}'+\mathsf{Y}'\right),$$

or, equivalently,

$$2^{2\mathbb{h}(\mathsf{X}+\mathsf{Y})} \geqslant 2^{2\mathbb{h}(\mathsf{X}')} + 2^{2\mathbb{h}(\mathsf{Y}')}.$$

2.1.3 Strongly typical sequences

Let $x^n \in \mathcal{X}^n$ be a sequence whose n elements are in a finite alphabet $\mathcal{X}$. The number of occurrences of a symbol $a \in \mathcal{X}$ in the sequence x^n is denoted by $N(a; x^n)$, and the *empirical distribution* (or histogram) of x^n is defined as the set $\{N(a;x^n)/n : a \in \mathcal{X}\}$.

Definition 2.8 (Strong typical set). *Let* p_{X} *be a distribution on a finite alphabet* $\mathcal{X}$ *and let* $\epsilon > 0$. *A sequence* $x^n \in \mathcal{X}^n$ *is (strongly)* ϵ*-typical with respect to* p_{X} *if*

$$\forall a \in \mathcal{X} \quad \left|\frac{1}{n}N(a;x^n) - p_{\mathsf{X}}(a)\right| \leqslant \epsilon\, p_{\mathsf{X}}(a).$$

The set of all ϵ*-typical sequences with respect to* p_{X} *is called the strong typical set and is denoted by* $\mathcal{T}_\epsilon^n(\mathsf{X})$.

In other words, the *typical set* $\mathcal{T}_\epsilon^n(\mathsf{X})$ contains all sequences x^n whose empirical distribution is "close" to p_{X}. The notion of typicality is particularly useful in information theory because of a result known as the *asymptotic equipartition property* (AEP).

Theorem 2.1 (AEP). *Let p_X be a distribution on a finite alphabet $\mathcal{X}$ and let $0 < \epsilon < \min_{x\in\mathcal{X}} p_X(x)$. Let X^n be a sequence of independent and identically distributed (i.i.d.) random variables with distribution p_X. Then,*

$$1 - \delta_\epsilon(n) \leqslant \mathbb{P}\left[X^n \in \mathcal{T}_\epsilon^n(X)\right] \leqslant 1,$$
$$(1 - \delta_\epsilon(n))2^{n(\mathbb{H}(X)-\delta(\epsilon))} \leqslant \left|\mathcal{T}_\epsilon^n(X)\right| \leqslant 2^{n(\mathbb{H}(X)+\delta(\epsilon))},$$
$$\forall x^n \in \mathcal{T}_\epsilon^n(X) \quad 2^{-n(\mathbb{H}(X)+\delta(\epsilon))} \leqslant p_{X^n}(x^n) \leqslant 2^{-n(\mathbb{H}(X)-\delta(\epsilon))}.$$

In simple terms, the AEP states that, for sufficiently large n, the probability that the realization x^n of a sequence of i.i.d. random variables belongs to the typical set is close to unity. Moreover, for practical purposes we may assume that the probability of any strongly typical sequence is about $2^{-n\mathbb{H}(X)}$ and the number of strongly typical sequences is approximately $2^{n\mathbb{H}(X)}$. In some sense, the AEP provides an operational interpretation of entropy.

Remark 2.1. *It is possible to provide explicit expressions for $\delta_\epsilon(n)$ and $\delta(\epsilon)$ [6], but the rough characterization used in Theorem 2.1 is sufficient for our purposes. In particular, it makes it easier to keep track of small terms that depend on ϵ or n because we can write equations such as $\delta(\epsilon) + \delta(\epsilon) = \delta(\epsilon)$ without worrying about the exact dependence on ϵ.*

The notion of typicality generalizes to multiple random variables. Assume that $(x^n, y^n) \in \mathcal{X}^n \times \mathcal{Y}^n$ is a pair of sequences with elements in finite alphabets $\mathcal{X}$ and $\mathcal{Y}$. The number of occurrences of a pair $(a, b) \in \mathcal{X} \times \mathcal{Y}$ in the pair of sequences (x^n, y^n) is denoted by $N(a, b; x^n, y^n)$.

Definition 2.9 (Jointly typical set). *Let p_{XY} be a joint distribution on the finite alphabets $\mathcal{X} \times \mathcal{Y}$ and let $\epsilon > 0$. Sequences $x^n \in \mathcal{X}^n$ and $y^n \in \mathcal{Y}^n$ are ϵ-jointly typical with respect to p_{XY} if*

$$\forall (a, b) \in \mathcal{X} \times \mathcal{Y} \quad \left|\frac{1}{n}N(a, b; x^n, y^n) - p_{XY}(a, b)\right| \leqslant \epsilon\, p_{XY}(a, b).$$

The set of all ϵ-jointly typical sequences with respect to p_{XY} is called the jointly typical set and is denoted by $\mathcal{T}_\epsilon^n(XY)$.

One can check that $\mathcal{T}_\epsilon^n(XY) \subseteq \mathcal{T}_\epsilon^n(X) \times \mathcal{T}_\epsilon^n(Y)$. In other words, $(x^n, y^n) \in \mathcal{T}_\epsilon^n(XY)$ implies that $x^n \in \mathcal{T}_\epsilon^n(X)$ and $y^n \in \mathcal{T}_\epsilon^n(Y)$. This property is known as the *consistency* of joint typicality. Notice that the jointly typical set $\mathcal{T}_\epsilon^n(XY)$ is the typical set $\mathcal{T}_\epsilon^n(Z)$ for the random variable $Z = (X, Y)$. Therefore, the result below follows directly from Theorem 2.1.

Corollary 2.1 (Joint AEP). *Let p_{XY} be a joint distribution on the finite alphabets $\mathcal{X} \times \mathcal{Y}$ and let $0 < \epsilon < \min_{(x,y)\in\mathcal{X}\times\mathcal{Y}} p_{XY}(x, y)$. Let (X^n, Y^n) be a sequence of i.i.d. random variables with joint distribution p_{XY}. Then,*

$$1 - \delta_\epsilon(n) \leqslant \mathbb{P}\left[(X^n, Y^n) \in \mathcal{T}_\epsilon^n(XY)\right] \leqslant 1,$$
$$(1 - \delta_\epsilon(n))2^{n(\mathbb{H}(XY)-\delta(\epsilon))} \leqslant \left|\mathcal{T}_\epsilon^n(XY)\right| \leqslant 2^{n(\mathbb{H}(XY)+\delta(\epsilon))},$$
$$\forall (x^n, y^n) \in \mathcal{T}_\epsilon^n(XY) \quad 2^{-n(\mathbb{H}(XY)+\delta(\epsilon))} \leqslant p_{X^nY^n}(x^n, y^n) \leqslant 2^{-n(\mathbb{H}(XY)-\delta(\epsilon))}.$$

It is also useful to introduce a conditional typical set, for which we can establish a conditional version of the AEP.

Definition 2.10. *Let p_{XY} be a joint distribution on the finite alphabets $\mathcal{X} \times \mathcal{Y}$ and let $\epsilon > 0$. Let $x^n \in \mathcal{T}_\epsilon^n(\mathsf{X})$. The set*

$$\mathcal{T}_\epsilon^n(\mathsf{XY}|x^n) \triangleq \left\{y^n \in \mathcal{Y}^n : (x^n, y^n) \in \mathcal{T}_\epsilon^n(\mathsf{XY})\right\}$$

is called the conditional typical set with respect to x^n.

Theorem 2.2 (Conditional AEP). *Let p_{XY} be a joint distribution on the finite alphabets $\mathcal{X} \times \mathcal{Y}$ and suppose that $0 < \epsilon' < \epsilon \leqslant \min_{(x,y)\in\mathcal{X}\times\mathcal{Y}} p_{\mathsf{XY}}(x, y)$. Let $x^n \in \mathcal{T}_{\epsilon'}^n(\mathsf{X})$ and let $\tilde{\mathsf{Y}}^n$ be a sequence of random variables such that*

$$\forall y^n \in \mathcal{Y}^n \quad p_{\tilde{\mathsf{Y}}^n}(y^n) = \prod_{i=1}^n p_{\mathsf{Y}|\mathsf{X}}(y_i|x_i).$$

Then,

$$1 - \delta_{\epsilon\epsilon'}(n) \leqslant \mathbb{P}\left[\tilde{\mathsf{Y}}^n \in \mathcal{T}_\epsilon^n(\mathsf{XY}|x^n)\right] \leqslant 1,$$
$$(1 - \delta_{\epsilon\epsilon'}(n))2^{n(\mathbb{H}(\mathsf{Y}|\mathsf{X})-\delta(\epsilon))} \leqslant \left|\mathcal{T}_\epsilon^n(\mathsf{XY}|x^n)\right| \leqslant 2^{n(\mathbb{H}(\mathsf{Y}|\mathsf{X})+\delta(\epsilon))},$$
$$\forall y^n \in \mathcal{T}_\epsilon^n(\mathsf{XY}|x^n) \quad 2^{-n(\mathbb{H}(\mathsf{Y}|\mathsf{X})+\delta(\epsilon))} \leqslant p_{\mathsf{Y}^n|\mathsf{X}^n}(y^n|x^n) \leqslant 2^{-n(\mathbb{H}(\mathsf{Y}|\mathsf{X})-\delta(\epsilon))}.$$

The conditional AEP means that, if x^n is a typical sequence and if $\tilde{\mathsf{Y}}^n$ is distributed according to $\prod_{i=1}^n p_{\mathsf{Y}|\mathsf{X}}(y_i|x_i)$, then $\tilde{\mathsf{Y}}^n$ is jointly typical with x^n with high probability for n large enough. In addition, the number of sequences y^n that are jointly typical with x^n is approximately $2^{n\mathbb{H}(\mathsf{Y}|\mathsf{X})}$, and their probability is on the order of $2^{-n\mathbb{H}(\mathsf{Y}|\mathsf{X})}$. The following corollary of the conditional AEP will be useful.

Corollary 2.2. *Let p_{XY} be a joint distribution on the finite alphabets $\mathcal{X} \times \mathcal{Y}$ and let $0 < \epsilon < \mu_{\mathsf{XY}}$ with $\mu_{\mathsf{XY}} \triangleq \min_{(x,y)\in\mathcal{X}\times\mathcal{Y}} p_{\mathsf{XY}}(x, y)$. Let $\tilde{\mathsf{Y}}^n$ be a sequence of i.i.d. random variables with distribution p_{Y}. Then,*

- *if $x^n \in \mathcal{T}_\epsilon^n(\mathsf{X})$,*

$$(1 - \delta_\epsilon(n))2^{-n(\mathbb{I}(\mathsf{X};\mathsf{Y})+\delta(\epsilon))} \leqslant \mathbb{P}\left[\tilde{\mathsf{Y}}^n \in \mathcal{T}_\epsilon^n(\mathsf{XY}|x^n)\right] \leqslant 2^{-n(\mathbb{I}(\mathsf{X};\mathsf{Y})-\delta(\epsilon))};$$

- *if $\tilde{\mathsf{X}}^n$ is a sequence of random variables independent of $\tilde{\mathsf{Y}}^n$ and with arbitrary distribution $p_{\tilde{\mathsf{X}}^n}$ on $\mathcal{X}^n$,*

$$\mathbb{P}\left[(\tilde{\mathsf{X}}^n, \tilde{\mathsf{Y}}^n) \in \mathcal{T}_\epsilon^n(\mathsf{XY})\right] \leqslant 2^{-n(\mathbb{I}(\mathsf{X};\mathsf{Y})-\delta(\epsilon))}.$$

In other words, if $\tilde{\mathsf{Y}}^n$ is generated independently of x^n, the probability that $\tilde{\mathsf{Y}}^n$ is jointly typical with x^n is small and on the order of $2^{-n\mathbb{I}(\mathsf{X};\mathsf{Y})}$. Corollary 2.2 generalizes to more than two random variables; in particular, we make extensive use of the following result.

Corollary 2.3. *Let p_{UXY} be a joint distribution on the finite alphabets $\mathcal{U} \times \mathcal{X} \times \mathcal{Y}$ and suppose $0 < \epsilon < \mu_{\mathsf{UXY}}$ with $\mu_{\mathsf{UXY}} \triangleq \min_{(u,x,y)\in\mathcal{U}\times\mathcal{X}\times\mathcal{Y}} p_{\mathsf{UXY}}(u, x, y)$. Let $(\tilde{\mathsf{U}}^n, \tilde{\mathsf{X}}^n)$ be* *a sequence of random variables with arbitrary distribution* $p_{\tilde{U}^n\tilde{X}^n}$ *on* $\mathcal{U}^n \times \mathcal{X}^n$. *Let* $\tilde{Y}^n$ *be a sequence of random variables conditionally independent of* $\tilde{X}^n$ *given* $\tilde{U}^n$ *such that*

$\forall(u^n, x^n, y^n) \in \mathcal{U}^n \times \mathcal{X}^n \times \mathcal{Y}^n$

$$p_{\tilde{U}^n\tilde{X}^n\tilde{Y}^n}(u^n, x^n, y^n) = \left(\prod_{i=1}^{n} p_{Y|U}(y_i|u_i)\right) p_{\tilde{U}^n\tilde{X}^n}(u^n, x^n).$$

Then,

$$\mathbb{P}_{\tilde{U}^n\tilde{X}^n\tilde{Y}^n}\left[(\tilde{U}^n, \tilde{X}^n, \tilde{Y}^n) \in \mathcal{T}_\epsilon^n(UXY)\right] \leqslant 2^{-n(\mathbb{I}(X;Y|U)-\delta(\epsilon))}.$$

2.1.4 Weakly typical sequences

Strong typicality requires the relative frequency of each possible symbol to be close to the corresponding probability; however, the notion of strong typicality does not apply to continuous random variables and it is sometimes convenient to use a *weaker* notion of typicality, which merely requires the empirical entropy of a sequence to be close to the true entropy of the corresponding random variable. All definitions and results in this section are stated for discrete random variables but hold also for continuous random variables on replacing the entropy $\mathbb{H}$ with the differential entropy $\mathbb{h}$.

Definition 2.11 (Weakly typical set). *Let* p_X *be a distribution on a finite alphabet* $\mathcal{X}$ *and let* $\epsilon > 0$. *A sequence* $x^n \in \mathcal{X}^n$ *is (weakly)* ϵ*-typical with respect to* p_X *if*

$$\left|-\frac{1}{n}\log p_{X^n}(x^n) - \mathbb{H}(X)\right| \leqslant \epsilon.$$

The set of all weakly ϵ*-typical sequences with respect to* p_X *is called the weakly typical set and is denoted* $\mathcal{A}_\epsilon^n(X)$.

The weak version of the AEP then follows from the weak law of large numbers.

Theorem 2.3 (AEP). *Let* p_X *be a distribution on a finite alphabet* $\mathcal{X}$ *and let* $\epsilon > 0$. *Let* X^n *be a sequence of i.i.d. random variables with distribution* p_X. *Then,*

- *for n sufficiently large,* $\mathbb{P}\left[X^n \in \mathcal{A}_\epsilon^n(X)\right] > 1 - \epsilon$;
- *if* $x^n \in \mathcal{A}_\epsilon^n(X)$, *then* $2^{-n(\mathbb{H}(X)+\epsilon)} \leqslant p_{X^n}(x^n) \leqslant 2^{-n(\mathbb{H}(X)-\epsilon)}$;
- *for n sufficiently large,* $(1-\epsilon)2^{n(\mathbb{H}(X)-\epsilon)} \leqslant \left|\mathcal{A}_\epsilon^n(X)\right| \leqslant 2^{n(\mathbb{H}(X)+\epsilon)}$.

Definition 2.12 (Jointly weak typical set). *Let* p_{XY} *be a joint distribution on the finite alphabets* $\mathcal{X} \times \mathcal{Y}$ *and let* $\epsilon > 0$. *Sequences* $x^n \in \mathcal{X}^n$ *and* $y^n \in \mathcal{Y}^n$ *are jointly (weakly)* ϵ*-typical with respect to* p_{XY} *if*

$$\left|-\frac{1}{n}\log p_{X^nY^n}(x^n, y^n) - \mathbb{H}(XY)\right| \leqslant \epsilon,$$

$$\left|-\frac{1}{n}\log p_{X^n}(x^n) - \mathbb{H}(X)\right| \leqslant \epsilon,$$

$$\left|-\frac{1}{n}\log p_{Y^n}(y^n) - \mathbb{H}(Y)\right| \leqslant \epsilon.$$

The set of all jointly weakly ϵ-typical sequences with respect to p_{XY} is called the jointly weakly typical set and is denoted $\mathcal{A}_\epsilon^n(XY)$.

Theorem 2.4 (joint AEP). *Let p_{XY} be a joint distribution on the finite alphabets $\mathcal{X} \times \mathcal{Y}$ and let $\epsilon > 0$. Let (X^n, Y^n) be a sequence of i.i.d. random variables with joint distribution p_{XY}. Then,*

- *for n sufficiently large, $\mathbb{P}\left[(X^n, Y^n) \in \mathcal{A}_\epsilon^n(XY)\right] > 1 - \epsilon$;*
- *if $(x^n, y^n) \in \mathcal{A}_\epsilon^n(XY)$, then $2^{-n(\mathbb{H}(XY)+\epsilon)} \leqslant p_{X^nY^n}(x^n, y^n) \leqslant 2^{-n(\mathbb{H}(XY)-\epsilon)}$;*
- *for n sufficiently large, $(1-\epsilon)2^{n(\mathbb{H}(XY)-\epsilon)} \leqslant \left|\mathcal{A}_\epsilon^n(XY)\right| \leqslant 2^{n(\mathbb{H}(XY)+\epsilon)}$.*

With weak typicality, there is no exact counterpart to the conditional AEP given in Corollary 2.2 but the following result holds nevertheless.

Theorem 2.5. *Let p_{XY} be a joint distribution on the finite alphabets $\mathcal{X} \times \mathcal{Y}$ and let $\epsilon > 0$. Let $\tilde{Y}^n$ be a sequence of i.i.d. random variables with distribution p_Y, and let $\tilde{X}^n$ be an independent sequence of i.i.d. random variables with distribution p_X. Then,*

$$\mathbb{P}\left[(\tilde{X}^n, \tilde{Y}^n) \in \mathcal{A}_\epsilon^n(XY)\right] \leqslant 2^{-n(\mathbb{I}(X;Y)-\delta(\epsilon))}.$$

In subsequent chapters, we use the term AEP for both strong and weak typicality; however, it will be clear from the context whether we refer to the theorems of Section 2.1.3 or those of Section 2.1.4.

2.1.5 Markov chains and functional dependence graphs

The identification of Markov chains among random variables that depend on each other via complicated relations is a recurrent problem in information theory. In principle, Markov chains can be identified by manipulating the joint probability distribution of random variables, but this is often a tedious task. In this short section, we describe a graphical yet correct method for identifying Markov chains that is based on the *functional dependence graph* of random variables.

Definition 2.13 (Functional dependence graph). *Consider m independent random variables and n functions of these variables. A functional dependence graph is a directed graph having $m + n$ vertices, and in which edges are drawn from one vertex to another if the random variable of the former vertex is an argument in the function defining the latter.*

Example 2.1. Let $M \in \mathcal{M}$ and $Z^n \in \mathbb{R}^n$ be independent random variables. Let $\{f_i\}_n$ be a set of functions from $\mathcal{M}$ to $\mathbb{R}^n$. For $i \in [\![1, n]\!]$ define the random variables $X_i = f_i(M)$ and $Y_i = X_i + Z_i$. The functional dependence graph of the random variables M, X^n, Y^n, and Z^n is shown in Figure 2.1.

Definition 2.14 (d-separation). *Let $\mathcal{X}$, $\mathcal{Y}$, and $\mathcal{Z}$ be disjoint subsets of vertices in a functional dependence graph $\mathcal{G}$. The subset $\mathcal{Z}$ is said to* d-separate *$\mathcal{X}$ from $\mathcal{Y}$ if there*

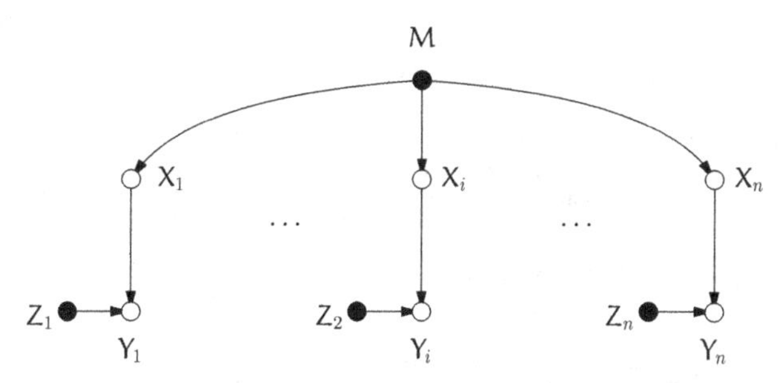

Figure 2.1 Functional dependence graph of variables in Example 2.1. For clarity, the independent random variables are indicated by filled circles (•) whereas the functions of these random variables are indicated by empty circles (∘).

exists no path between a vertex of $\mathcal{X}$ and a vertex of $\mathcal{Y}$ after the following operations have been performed:

- *construct the subgraph $\mathcal{G}'$ consisting of all vertices in $\mathcal{X}$, $\mathcal{Y}$, and $\mathcal{Z}$, as well as the edges and vertices encountered when moving backward starting from any of the vertices in $\mathcal{X}$, $\mathcal{Y}$, or $\mathcal{Z}$;*
- *in the subgraph $\mathcal{G}'$, delete all edges coming out of $\mathcal{Z}$;*
- *remove all arrows in $\mathcal{G}'$ to obtain an undirected graph.*

The usefulness of d-separation is justified by the following theorem.

Theorem 2.6. *Let $\mathcal{X}$, $\mathcal{Y}$, and $\mathcal{Z}$ be disjoint subsets of the vertices in a functional dependence graph. If $\mathcal{Z}$ d-separates $\mathcal{X}$ from $\mathcal{Y}$, and if we collect the random variables in $\mathcal{X}$, $\mathcal{Y}$, and $\mathcal{Z}$ in the random vectors* X, Y, *and* Z, *respectively, then* $\mathrm{X} \rightarrow \mathrm{Z} \rightarrow \mathrm{Y}$ *forms a Markov chain.*

Theorem 2.6 is particularly useful in the converse proofs of channel coding theorems.

Example 2.2. On the basis of the functional dependence graph of Figure 2.1, one can check that, for any $i \neq j$, $\mathrm{X}_i \rightarrow \mathrm{X}_j \rightarrow \mathrm{Y}_j$.

2.2 The point-to-point communication problem

The foundations of information theory were laid by Claude E. Shannon in his 1948 paper "A mathematical theory of communication" [7]. In his own words, *the fundamental problem of communication is that of reproducing at one point either exactly or approximately a message selected at another point.* If the message – for example, a letter from the alphabet, the gray level of a pixel or some physical quantity measured by a sensor – is to be reproduced at a remote location with a certain fidelity, some amount of information must be transmitted over a physical channel. This observation is the basis of Shannon's general model for *point-to-point* communication reproduced in Figure 2.2. It consists of the following elements.

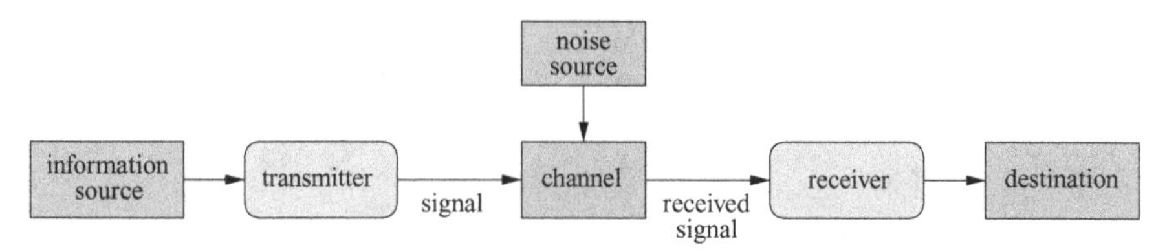

Figure 2.2 Shannon's communication model (from [7]).

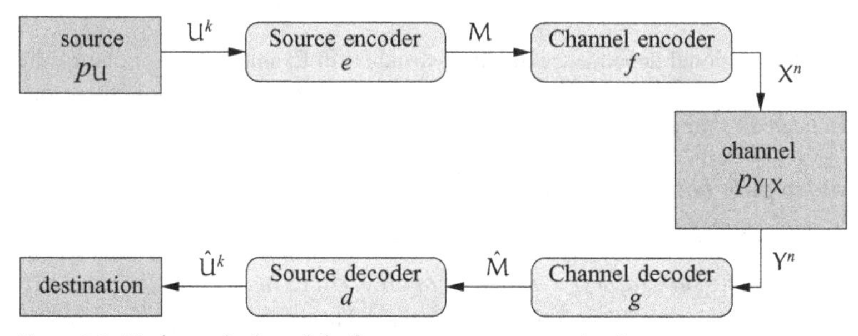

Figure 2.3 Mathematical model of a two-stage communication system.

- The *information source* generates messages according to some random process.
- The *transmitter* observes the messages and forms a signal to be sent over the channel.
- The *channel* is governed by a *noise source*, which corrupts the original input signal; this models the physical constraints of a communication system, for instance thermal noise in electronic circuits or multipath fading in a wireless medium.
- The *receiver* takes the received signal, forms a reconstructed version of the original message, and delivers the result to the *destination*.

Given the statistical properties of the information source and the noisy channel, the goal of the communication engineer is to design the transmitter and the receiver in a way that allows the information sent by the source to reach its destination in a *reliable* way. Information theory can help us achieve this goal by characterizing the fundamental mechanisms behind communication systems and providing us with precise mathematical conditions under which reliable communication is possible.

2.2.1 Point-to-point communication model

To give a precise formulation of the point-to-point communication problem, we require definitions for each of its constituent modules. We assume that the *source* and the *channel* are described by discrete-time random processes, and we determine that the receiver and the transmitter agree on a common *code*, specified by an *encoder* and *decoder* pair. As illustrated in Figure 2.3, we consider a two-stage system in which the source is compressed before being encoded for channel transmission, and channel outputs are decoded before being decompressed. The basic relationships among the components in Figure 2.3 are described in the following lines.

Definition 2.15. *A discrete memoryless source (DMS) $(\mathcal{U}, p_{\mathsf{U}})$ generates a sequence of i.i.d. symbols (or letters) from the finite alphabet $\mathcal{U}$ according to the probability distribution p_{U}. The random variable representing a source symbol is denoted by* U.

Definition 2.16. *A discrete memoryless channel (DMC) $\left(\mathcal{X}, p_{\mathsf{Y}|\mathsf{X}}, \mathcal{Y}\right)$ is described by a finite input alphabet $\mathcal{X}$, a finite output alphabet $\mathcal{Y}$, and a conditional probability distribution $p_{\mathsf{Y}|\mathsf{X}}$, such that* X *and* Y *denote the channel input and the channel output, respectively. The set of conditional probabilities (also called transition probabilities) can be represented by a channel transition probability matrix $\left(p_{\mathsf{Y}|\mathsf{X}}(y|x)\right)_{\mathcal{X},\mathcal{Y}}$.*

In what follows, we illustrate many results numerically with the following DMCs.

Example 2.3. A binary symmetric channel with cross-over probability $p \in [0, 1]$, denoted by BSC(p), is a DMC $\left(\{0, 1\}, p_{\mathsf{Y}|\mathsf{X}}, \{0, 1\}\right)$ characterized by the transition probability matrix

$$\begin{pmatrix} 1-p & p \\ p & 1-p \end{pmatrix}.$$

Example 2.4. A binary erasure channel with erasure probability $\epsilon \in [0, 1]$, denoted by BEC(ϵ), is a DMC $\left(\{0, 1\}, p_{\mathsf{Y}|\mathsf{X}}, \{0, ?, 1\}\right)$ characterized by the transition probability matrix

$$\begin{pmatrix} 1-\epsilon & \epsilon & 0 \\ 0 & \epsilon & 1-\epsilon \end{pmatrix}.$$

Definition 2.17. *A* $(2^{kR}, k)$ source code $\mathcal{C}_k$ *for a DMS $(\mathcal{U}, p_{\mathsf{U}})$ consists of*

- *a message set $\mathcal{M} = \llbracket 1, 2^{kR} \rrbracket$;*
- *an encoding function $e : \mathcal{U}^k \to \mathcal{M}$, which maps a sequence of k source symbols u^k to a message m;*
- *a decoding function $d : \mathcal{M} \to \mathcal{U}^k \cup \{?\}$, which maps a message m to a sequence of source symbols $\hat{u}^k \in \mathcal{U}^k$ or an error message* **?**.

The *compression rate* of the source code is defined as $(1/k)\log\lceil 2^{kR} \rceil$ in bits[1] per source symbol, and its probability of error is

$$\mathbf{P}_{\mathrm{e}}(\mathcal{C}_k) \triangleq \mathbb{P}\left[\hat{\mathsf{U}}^k \neq \mathsf{U}^k \mid \mathcal{C}_k\right].$$

Definition 2.18. *A rate R is an achievable compression rate for the source $(\mathcal{U}, p_{\mathsf{U}})$ if there exists a sequence of $\left(2^{kR}, k\right)$ source codes $\{\mathcal{C}_k\}_{k\geqslant 1}$, such that*

$$\lim_{k\to\infty} \mathbf{P}_{\mathrm{e}}(\mathcal{C}_k) = 0,$$

that is, the source sequences can be reconstructed with arbitrarily small probability of error with compression rates arbitrarily close to R.

[1] Unless specified otherwise, all logarithms are taken to the base two.

Definition 2.19. *A* $(2^{nR}, n)$ channel code $\mathcal{C}_n$ *for a DMC* $(\mathcal{X}, p_{Y|X}, \mathcal{Y})$ *consists of*

- *a message set* $\mathcal{M} = [\![1, 2^{nR}]\!]$*;*
- *an encoding function* $f : \mathcal{M} \to \mathcal{X}^n$*, which maps a message* m *to a* codeword x^n *with* n *symbols;*
- *a decoding function* $g : \mathcal{Y}^n \to \mathcal{M} \cup \{?\}$*, which maps a block of* n *channel outputs* y^n *to a message* $\hat{m} \in \mathcal{M}$ *or an error message* ?.

The set of codewords $\{f(m) : m \in [\![1, 2^{nR}]\!]\}$ is called the *codebook* of $\mathcal{C}_n$. With a slight abuse of notation, we denote the codebook itself by $\mathcal{C}_n$ as well. Unless specified otherwise, messages are represented by a random variable M uniformly distributed in $\mathcal{M}$, and the rate of the channel code is defined as $(1/n)\log\lceil 2^{nR}\rceil$ in bits per channel use. The average probability of error is defined as

$$\mathbf{P}_e(\mathcal{C}_n) \triangleq \mathbb{P}\left[\hat{\mathsf{M}} \neq \mathsf{M} \mid \mathcal{C}_n\right].$$

Definition 2.20. *A rate* R *is an achievable transmission rate for the DMC* $(\mathcal{X}, p_{Y|X}, \mathcal{Y})$ *if there exists a sequence of* $(2^{nR}, n)$ *codes* $\{\mathcal{C}_n\}_{n\geqslant 1}$ *such that*

$$\lim_{n\to\infty} \mathbf{P}_e(\mathcal{C}_n) = 0;$$

that is, messages can be transmitted at a rate arbitrarily close to R *and decoded with arbitrarily small probability of error. The channel capacity of the DMC is defined as*

$$C \triangleq \sup\{R : R \text{ is an achievable transmission rate}\}.$$

The typical goal of information theory is to characterize achievable rates on the basis of information-theoretic quantities that depend only on the given probability distributions and not on the block lengths k or n. A theorem that confirms the existence of codes for a class of achievable rates is often referred to as a *direct result* and the arguments that lead to this result constitute an *achievability proof*. On the other hand, when a theorem asserts that codes with certain properties do not exist, we speak of a *converse result* and a *converse proof*. A fundamental result that includes both the achievability and the converse parts is called a *coding theorem*. The mathematical tools that enable this characterization are those presented in Section 2.1, and we illustrate their use by discussing two of Shannon's fundamental coding theorems. These results form the basis of information theory and are of great use in several of the proofs developed in subsequent chapters.

Remark 2.2. *Notice that the formulation of the point-to-point communication problem does not put any constraints either on the computational complexity or on the delay of the encoding and decoding procedures. In other words, the goal is to describe the fundamental limits of communications systems irrespective of their technological limitations.*

2.2.2 The source coding theorem

The *source coding theorem* gives a complete solution (achievability and converse) for the point-to-point communication problem stated in Section 2.2.1 when the channel is noiseless, that is $Y = X$. In that case, it is not necessary to use a channel code to compensate for the impairments caused by the channel, but it is still useful to encode the messages produced by the source to achieve a more efficient representation of the source information in bits per source symbol. This procedure is called *source coding* or *data compression*. The main idea is to consider only a subset $\mathcal{A}$ of all possible source sequences $\mathcal{U}^k$, and assign a different index $i \in [\![1, |\mathcal{A}|]\!]$ to each of the sequences $u^k \in \mathcal{A}$. If the source produces a sequence $u^k \in \mathcal{A}$, then the encoder outputs the corresponding index i, otherwise it outputs some predefined constant. The decoder receives the index i and outputs the corresponding sequence in $\mathcal{A}$.

Since information theory is primarily concerned with the fundamental limits of reliable communication, it is possible to prove the existence of codes without having to search for explicit code constructions. One technique, which is particularly useful in information-theoretic problems related to source coding, consists of *throwing* sequences $u^n \in \mathcal{U}^n$ randomly into a finite set of bins, such that the sequences that land in the same bin share a common bin index. If each sequence is assigned a bin at random according to a uniform distribution, then we refer to this procedure as *random binning*. If we want to prove that there exists a code such that the error probability goes to zero, it suffices to show that the average of the probability of error taken over all possible bin assignments goes to zero and to use the selection lemma. The following theorem exploits random binning to characterize the set of achievable compression rates.

Theorem 2.7 (Source coding theorem). *For a discrete memoryless source* $(\mathcal{U}, p_{\mathsf{U}})$,

$$\inf\{R : R \text{ is an achievable compression rate}\} = \mathbb{H}(\mathsf{U}).$$

In other words, if a compression rate R satisfies $R > \mathbb{H}(\mathsf{U})$ then R is achievable and any achievable compression rate R must satisfy $R \geqslant \mathbb{H}(\mathsf{U})$.

Proof. We start with the achievability part of the proof, which is based on random binning. The idea is to randomly assign each source sequence to one of a finite number of bins; then, as long as the number of bins is larger than $2^{k\mathbb{H}(\mathsf{U})}$, the probability of finding more than one typical sequence in the same bin is very small. If each typical sequence is mapped to a different bin index, an arbitrarily small probability of error can be achieved by letting the decoder output the typical sequence that corresponds to the received index. Formally, let $\epsilon > 0$ and $k \in \mathbb{N}^*$. Let $R > 0$ be a rate to be specified later. We construct a $(2^{kR}, k)$ source code $\mathcal{C}_k$ as follows.

- *Binning.* For each sequence $u^k \in \mathcal{T}_\epsilon^k(\mathsf{U})$, draw an index uniformly at random in the set $[\![1, 2^{kR}]\!]$. The index assignment defines the encoding function

$$e : \mathcal{U}^k \to [\![1, 2^{kR}]\!],$$

  which is revealed to the encoder and decoder.
- *Encoder.* Given an observation u^k, output $m = e(u^k)$ if $u^k \in \mathcal{T}_\epsilon^k(\mathsf{U})$; otherwise output $m = 1$.
- *Decoder.* Given message m, output $\hat{u}^k$ if it is the unique sequence such that $\hat{u}^k \in \mathcal{T}_\epsilon^k(\mathsf{U})$ and $e(\hat{u}^k) = m$; otherwise output an error ?.

The random variable that represents the randomly generated encoding function e is denoted by E while the random variable that represents the randomly generated code $\mathcal{C}_k$ is denoted by C_k. We proceed to bound $\mathbb{E}[\mathbf{P}_e(\mathsf{C}_k)]$. First, note that $\mathbb{E}[\mathbf{P}_e(\mathsf{C}_k)]$ can be expressed in terms of the events

$$\mathcal{E}_0 = \{\mathsf{U}^k \notin \mathcal{T}_\epsilon^k(\mathsf{U})\},$$
$$\mathcal{E}_1 = \{\exists \hat{u}^k \neq \mathsf{U}^k : \mathsf{E}(\hat{u}^k) = \mathsf{E}(\mathsf{U}^k) \text{ and } \hat{u}^k \in \mathcal{T}_\epsilon^k(\mathsf{U})\}$$

as $\mathbb{E}[\mathbf{P}_e(\mathsf{C}_k)] = \mathbb{P}[\mathcal{E}_0 \cup \mathcal{E}_1]$. By the union bound,

$$\mathbb{E}[\mathbf{P}_e(\mathsf{C}_k)] \leqslant \mathbb{P}[\mathcal{E}_0] + \mathbb{P}[\mathcal{E}_1]. \tag{2.1}$$

By the AEP,

$$\mathbb{P}[\mathcal{E}_0] \leqslant \delta_\epsilon(k) \tag{2.2}$$

and we can upper bound $\mathbb{P}[\mathcal{E}_1]$ as

$$\begin{aligned}
\mathbb{P}[\mathcal{E}_1] &= \sum_{u^k} p_{\mathsf{U}^k}(u^k)\, \mathbb{P}\left[\exists \hat{u}^k \neq u^k : \mathsf{E}(\hat{u}^k) = \mathsf{E}(u^k) \text{ and } \hat{u}^k \in \mathcal{T}_\epsilon^k(\mathsf{U})\right] \\
&\leqslant \sum_{u^k} p_{\mathsf{U}^k}(u^k) \sum_{\substack{\hat{u}^k \in \mathcal{T}_\epsilon^k(\mathsf{U}) \\ \hat{u}^k \neq u^k}} \mathbb{P}\left[\mathsf{E}(\hat{u}^k) = \mathsf{E}(u^k)\right] \\
&= \sum_{u^k} p_{\mathsf{U}^k}(u^k) \sum_{\substack{\hat{u}^k \in \mathcal{T}_\epsilon^k(\mathsf{U}) \\ \hat{u}^k \neq u^k}} \frac{1}{\lceil 2^{kR} \rceil} \\
&\leqslant \sum_{u^k} p_{\mathsf{U}^k}(u^k) \frac{1}{\lceil 2^{kR} \rceil} \left|\mathcal{T}_\epsilon^k(\mathsf{U})\right| \\
&\leqslant \sum_{u^k} p_{\mathsf{U}^k}(u^k) \frac{1}{\lceil 2^{kR} \rceil} 2^{k(\mathbb{H}(\mathsf{U})+\delta(\epsilon))} \\
&\leqslant 2^{k(\mathbb{H}(\mathsf{U})+\delta(\epsilon)-R)}.
\end{aligned}$$

Hence, if we choose $R > \mathbb{H}(\mathsf{U}) + \delta(\epsilon)$, we have

$$\mathbb{P}[\mathcal{E}_1] \leqslant \delta_\epsilon(k). \tag{2.3}$$

On substituting (2.2) and (2.3) into (2.1), we obtain $\mathbb{E}[\mathbf{P}_e(\mathsf{C}_k)] \leqslant \delta_\epsilon(k)$. By applying the selection lemma to the random variable C_k and the function $\mathbf{P}_e$, we conclude that there exists at least one source code $\mathcal{C}_k$ such that $\mathbf{P}_e(\mathcal{C}_k) \leqslant \delta_\epsilon(k)$. Since ϵ can be chosen arbitrarily small, all rates $R > \mathbb{H}(\mathsf{U})$ are achievable.

We now establish the converse result and show that any achievable rate must satisfy $R \geqslant \mathbb{H}(\mathsf{U})$. Let R be an achievable rate and let $\epsilon > 0$. By definition, there exists a source code $\mathcal{C}_k$ such that $\mathbf{P}_e(\mathcal{C}_k) \leqslant \delta(\epsilon)$. If we let M denote the message output by the encoder, then Fano's inequality guarantees that

$$\frac{1}{k}\mathbb{H}(\mathsf{U}^k|\mathsf{M}\mathcal{C}_k) \leqslant \delta(\mathbf{P}_e(\mathcal{C}_k)) \leqslant \delta(\epsilon).$$

We drop the conditioning on $\mathcal{C}_k$ in subsequent calculations to simplify the notation. Note that

$$\begin{aligned}\mathbb{H}(\mathsf{U}) &= \frac{1}{k}\mathbb{H}(\mathsf{U}^k)\\ &= \frac{1}{k}\mathbb{I}(\mathsf{U}^k;\mathsf{M}) + \frac{1}{k}\mathbb{H}(\mathsf{U}^k|\mathsf{M})\\ &\leqslant \frac{1}{k}\mathbb{I}(\mathsf{U}^k;\mathsf{M}) + \delta(\epsilon)\\ &\leqslant \frac{1}{k}\mathbb{H}(\mathsf{M}) + \delta(\epsilon)\\ &\leqslant R + \delta(k) + \delta(\epsilon).\end{aligned}$$

Since ϵ can be chosen arbitrarily small and k can be chosen arbitrarily large, we obtain $R \geqslant \mathbb{H}(\mathsf{U})$. □

Remark 2.3. *Alternatively, the achievability part of the source coding theorem can be established on the basis of the AEP alone. In fact, for large k the AEP guarantees that any sequence u^k produced by the source $(\mathcal{U}, p_{\mathsf{U}})$ belongs with high probability to the typical set $\mathcal{T}_\epsilon^k(\mathsf{U})$; hence, we need only index the approximately $2^{k\mathbb{H}(\mathsf{U})}$ typical sequences to achieve arbitrarily small probability of error and the corresponding rate is on the order of $\mathbb{H}(\mathsf{U})$.*

2.2.3 The channel coding theorem

The *channel coding theorem* gives a complete solution (achievability and converse) for the point-to-point communication problem stated in Section 2.2.1 when the source $(\mathcal{U}, p_{\mathsf{U}})$ is uniform over $\mathcal{U}$. According to the source coding theorem, there is no need to encode the source since $\mathbb{H}(\mathsf{U}) = \log|\mathcal{U}|$ is maximal. We simply group the source symbols in sequences of length k. Letting $M \triangleq |\mathcal{U}|^k$ and $\mathcal{M} = [\![1, M]\!]$, we index each sequence of length k with an integer $m \in \mathcal{M}$. We use a channel code of rate $(1/n)\log M$ to transmit the messages produced by source U over a discrete memoryless channel $(\mathcal{X}, p_{\mathsf{Y}|\mathsf{X}}, \mathcal{Y})$.

As was done for the source coding theorem, it is possible to prove the existence of codes without having to search for explicit code constructions. The idea is to construct a random code by drawing the symbols of codewords independently at random according to a fixed probability distribution p_{X} on $\mathcal{X}$. Then, if we want to prove that there exists a code such that the error probability goes to zero for n sufficiently large, it suffices to show that the average of the probability of error taken over all possible random codebooks goes to zero for n sufficiently large and use the selection lemma. This technique is referred to as *random coding* and is used in the proof of the following theorem to characterize the set of achievable rates.

Theorem 2.8 (Channel coding theorem). *The capacity of a DMC* $(\mathcal{X}, p_{Y|X}, \mathcal{Y})$ *is* $C = \max_{p_X} \mathbb{I}(X;Y)$. *In other words, if* $R < C$ *then* R *is an achievable transmission rate and an achievable transmission rate must satisfy* $R \leqslant C$.

Proof. We begin with the achievability part based on random coding. We choose a probability distribution p_X on $\mathcal{X}$ and, without loss of generality, we assume that p_X is such that $\mathbb{I}(X;Y) > 0$. Let $0 < \epsilon < \mu_{XY}$, where $\mu_{XY} = \min_{(x,y)\in\mathcal{X}\times\mathcal{Y}} p_{XY}(x,y)$, and let $n \in \mathbb{N}^*$. Let $R > 0$ be a rate to be specified later. We construct a $(2^{nR}, n)$ code $\mathcal{C}_n$ as follows.

- *Codebook construction.* Construct a codebook with $\lceil 2^{nR} \rceil$ codewords, labeled $x^n(m)$ with $m \in [\![1, 2^{nR}]\!]$, by generating the symbols $x_i(m)$ for $i \in [\![1, n]\!]$ and $m \in [\![1, 2^{nR}]\!]$ independently according to p_X. The codebook is revealed both to the encoder and to the decoder.
- *Encoder f.* Given m, transmit $x^n(m)$.
- *Decoder g.* Given y^n, output $\hat{m}$ if it is the unique message such that $(x^n(\hat{m}), y^n) \in \mathcal{T}_\epsilon^n(XY)$; otherwise, output an error **?**.

The random variable that represents the randomly generated codebook $\mathcal{C}_n$ is denoted by C_n. We first develop an upper bound for $\mathbb{E}[\mathbf{P}_e(C_n)]$. Notice that

$$\begin{aligned}\mathbb{E}[\mathbf{P}_e(C_n)] &= \mathbb{E}_{C_n}\left[\mathbb{P}\left[M \neq \hat{M} \mid C_n\right]\right] \\ &= \sum_m \mathbb{E}_{C_n}\left[\mathbb{P}\left[M \neq \hat{M} \mid M = m, C_n\right]\right] p_M(m).\end{aligned}$$

By virtue of the symmetry of the random code construction, we have that $\mathbb{E}_{C_n}\left[\mathbb{P}\left[M \neq \hat{M} \mid M = m, C_n\right]\right]$ is independent of m. Therefore, we can assume without losing generality that message $m = 1$ was sent and write

$$\mathbb{E}[\mathbf{P}_e(C_n)] = \mathbb{E}_{C_n}\left[\mathbb{P}\left[M \neq \hat{M} \mid M = 1, C_n\right]\right].$$

Notice that $\mathbb{E}[\mathbf{P}_e(C_n)]$ can be expressed in terms of the events

$$\mathcal{E}_i = \{(X^n(i), Y^n) \in \mathcal{T}_\epsilon^n(XY)\} \quad \text{for } i \in [\![1, 2^{nR}]\!]$$

as $\mathbb{E}[\mathbf{P}_e(C_n)] = \mathbb{P}\left[\mathcal{E}_1^c \cup \bigcup_{i\neq 1} \mathcal{E}_i\right]$. By the union bound,

$$\mathbb{E}[\mathbf{P}_e(C_n)] \leqslant \mathbb{P}\left[\mathcal{E}_1^c\right] + \sum_{i\neq 1} \mathbb{P}[\mathcal{E}_i]. \tag{2.4}$$

By the AEP,

$$\mathbb{P}\left[\mathcal{E}_1^c\right] \leqslant \delta_\epsilon(n). \tag{2.5}$$

Since Y^n is the output of the channel when $X^n(1)$ is transmitted and since $X^n(1)$ is independent of $X^n(i)$ for $i \neq 1$, note that Y^n is independent of $X^n(i)$ for $i \neq 1$; hence,

Corollary 2.2 applies and

$$\mathbb{P}[\mathcal{E}_i] \leqslant 2^{-n(\mathbb{I}(X;Y)-\delta(\epsilon))} \quad \text{for } i \neq 1. \tag{2.6}$$

On substituting (2.5) and (2.6) into (2.4), we obtain

$$\mathbb{E}[\mathbf{P}_e(C_n)] \leqslant \delta_\epsilon(n) + \lceil 2^{nR} \rceil 2^{-n(\mathbb{I}(X;Y)-\delta(\epsilon))}.$$

Hence, if we choose the rate R such that $R < \mathbb{I}(X;Y) - \delta(\epsilon)$, then

$$\mathbb{E}[\mathbf{P}_e(C_n)] \leqslant \delta_\epsilon(n).$$

By applying the selection lemma to the random variable C_n and the function $\mathbf{P}_e$, we conclude that there exists a $(2^{nR}, n)$ code $\mathcal{C}_n$ such that $\mathbf{P}_e(\mathcal{C}_n) \leqslant \delta_\epsilon(n)$. Since ϵ can be chosen arbitrarily small and since the distribution p_X is arbitrary, we conclude that all rates $R < \max_{p_X} \mathbb{I}(X;Y)$ are achievable.

We now establish the converse part of the proof. Let R be an achievable rate and let $\epsilon > 0$. For n sufficiently large, there exists a $(2^{nR}, n)$ code $\mathcal{C}_n$ such that

$$\frac{1}{n}\mathbb{H}(M|\mathcal{C}_n) \geqslant R \quad \text{and} \quad \mathbf{P}_e(\mathcal{C}_n) \leqslant \delta(\epsilon).$$

In the remaining part of the proof we drop the conditioning on $\mathcal{C}_n$ to simplify the notation. By virtue of Fano's inequality, it also holds that

$$\frac{1}{n}\mathbb{H}(M|Y^n) \leqslant \delta(\mathbf{P}_e(\mathcal{C}_n)) = \delta(\epsilon).$$

Therefore,

$$\begin{aligned}
R &\leqslant \frac{1}{n}\mathbb{H}(M) \\
&\leqslant \frac{1}{n}\mathbb{I}(M;Y^n) + \frac{1}{n}\mathbb{H}(M|Y^n) \\
&\leqslant \frac{1}{n}\mathbb{I}(M;Y^n) + \delta(\epsilon) \\
&\overset{(a)}{\leqslant} \frac{1}{n}\mathbb{I}(X^n;Y^n) + \delta(\epsilon) \\
&= \frac{1}{n}\mathbb{H}(Y^n) - \frac{1}{n}\mathbb{H}(Y^n|X^n) + \delta(\epsilon) \\
&\overset{(b)}{=} \frac{1}{n}\sum_{i=1}^{n}\left(\mathbb{H}(Y_i|Y^{i-1}) - \frac{1}{n}\mathbb{H}(Y_i|X_i)\right) + \delta(\epsilon) \\
&\overset{(c)}{\leqslant} \frac{1}{n}\sum_{i=1}^{n}\left(\mathbb{H}(Y_i) - \frac{1}{n}\mathbb{H}(Y_i|X_i)\right) + \delta(\epsilon) \\
&= \frac{1}{n}\sum_{i=1}^{n}\mathbb{I}(X_i;Y_i) + \delta(\epsilon) \\
&\leqslant \max_{p_X}\mathbb{I}(X;Y) + \delta(\epsilon),
\end{aligned}$$

where (a) follows from the data-processing inequality applied to the Markov chain $M \rightarrow X^n \rightarrow Y^n$, (b) follows because the channel is memoryless, and (c) follows because conditioning does not increase entropy. Since ϵ can be chosen arbitrarily small, we obtain $R \leqslant \max_{p_X} \mathbb{I}(X;Y)$. □

The channel coding theorem shows that the channel capacity is equal to the maximum mutual information between the channel input X and the channel output Y, where the maximization is carried out over all possible input probability distributions p_X. The proof technique and structure are common to most proofs in subsequent chapters.

Example 2.5. The capacity of a binary symmetric channel BSC(p) is $1 - \mathbb{H}_b(p)$. The capacity of a binary erasure channel BEC(ϵ) is $1 - \epsilon$.

Among the many channel models, the additive white Gaussian noise (AWGN) channel (Gaussian channel for short) takes a particularly prominent role in information and communication theory, because it captures the impact of thermal noise and interference on wired and wireless communications. The channel output at each time $i \geqslant 1$ is given by $Y_i = X_i + N_i$, where X_i denotes the transmitted symbol and $\{N_i\}_{i \geqslant 1}$ are i.i.d. random variables with distribution $\mathcal{N}(0, \sigma^2)$. Since the channel capacity of the Gaussian channel can be infinite without further restrictions, we add an average power constraint in the form of

$$\frac{1}{n}\sum_{i=1}^{n} \mathbb{E}\left[X_i^2\right] \leqslant P.$$

Theorem 2.9. *The capacity of a Gaussian channel is given by*

$$C = \frac{1}{2}\log\left(1 + \frac{P}{\sigma^2}\right),$$

where P denotes the power constraint and σ^2 is the variance of the noise.

Sketch of proof. The proof developed for Theorem 2.8 does not apply directly to the Gaussian channel because of the power constraint imposed on channel inputs and the continuous nature of the channel. Nevertheless, it is possible to develop a similar proof by using weakly typical sequences and the weak AEP (see for instance [3, Chapter 9]). The power constraint can be dealt with by introducing an error event that accounts for the sequences violating the power constraint in the codebook generation. □

2.3 Network information theory

Shannon's coding theorems characterize the fundamental limits of communication between two users. However, in many communication scenarios – for example, satellite broadcasting, cellular telephony, the Internet, and wireless sensor networks – the information is sent by one or more transmitting nodes to one or more receiving nodes

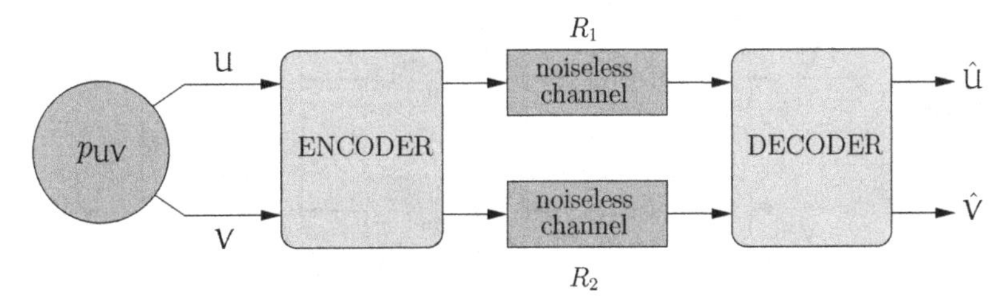

Figure 2.4 Joint encoding of correlated sources.

over more or less intricate communication networks. The interactions between the users of said networks introduce a whole new range of fundamental communication aspects that are not present in the classical point-to-point problem, such as *interference*, *user cooperation*, and *feedback*. The central goal of *network information theory* is to provide a thorough understanding of these basic mechanisms, by characterizing the fundamental limits of communication systems with multiple users. In this section, we discuss some results of network information theory that are useful for understanding information-theoretic security in subsequent chapters.

2.3.1 Distributed source coding

Consider a DMS $(\mathcal{UV}, p_{UV})$ that consists of two components U and V with joint distribution p_{UV}. As shown in Figure 2.4, the two components are to be processed by a joint encoder and transmitted to a common destination over two noiseless channels. The joint distribution p_{UV} can be arbitrary and the symbols produced by U and V at any given point in time are statistically dependent; therefore, we refer to U and V as *correlated sources*. Since the channels to the destination do not introduce any errors, we may ask the following question: at what rates R_1 and R_2 can we transmit information generated by U and V with an arbitrarily small probability of error? Since there is a common encoder and a common decoder, this problem reduces to the classical point-to-point problem and the solution follows naturally from Shannon's source coding theorem: the messages can be reconstructed with an arbitrarily small probability of error at the receiver if and only if

$$R_1 + R_2 > \mathbb{H}(UV);$$

that is, the sum rate must be greater than the joint entropy of U and V.

As illustrated in Figure 2.5, the problem becomes more challenging if instead of a joint encoder we consider two *separate* encoders. Here, each encoder observes only the realizations of the one source it is assigned to and does not know the output symbols of the other source. In this case, it is not immediately clear which encoding rates guarantee reconstruction with an arbitrarily small probability of error at the receiver. If we encode U at rate $R_1 > \mathbb{H}(U)$ and V at rate $R_2 > \mathbb{H}(V)$, then the source coding theorem guarantees once again that an arbitrarily small probability of error is possible. But, in this case, the sum rate satisfies $R_1 + R_2 > \mathbb{H}(U) + \mathbb{H}(V)$, which, in general, is greater than the joint entropy $\mathbb{H}(UV)$.

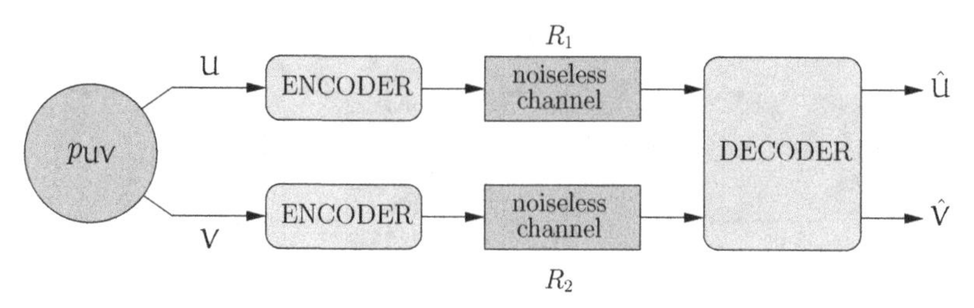

Figure 2.5 Separate encoding of correlated sources (the *Slepian–Wolf problem*).

Surprisingly, it turns out that the sum rate required by two separate encoders is the same as that required by a joint encoder, that is $R_1 + R_2 > \mathbb{H}(UV)$ is sufficient to reconstruct U and V with an arbitrarily small probability of error. In other words, there is no penalty in overall compression rate due to the fact that the encoders can observe only the realizations of the one source they have been assigned to. However, it is important to point out that the decoder does require a minimum amount of rate from each encoder; specifically, the average remaining uncertainty about the messages of one source given the messages of the other source, $\mathbb{H}(U|V)$ and $\mathbb{H}(V|U)$. Formally, a code for the distributed source coding problem is defined as follows.

Definition 2.21. *A $(2^{kR_1}, 2^{kR_2}, k)$ source code $\mathcal{C}_k$ for the DMS $(\mathcal{UV}, p_{UV})$ consists of*

- *two message sets $\mathcal{M}_1 = [\![1, 2^{kR_1}]\!]$ and $\mathcal{M}_2 = [\![1, 2^{kR_2}]\!]$;*
- *an encoding function $e_1 : \mathcal{U}^k \to \mathcal{M}_1$, which maps a sequence of k source symbols u^k to a message m_1;*
- *an encoding function $e_2 : \mathcal{V}^k \to \mathcal{M}_2$, which maps a sequence of k source symbols v^k to a message m_2;*
- *a decoding function $d : \mathcal{M}_1 \times \mathcal{M}_2 \to (\mathcal{U}^k \times \mathcal{V}^k) \cup \{?\}$, which maps a message pair (m_1, m_2) to a pair of source sequences $(\hat{u}^k, \hat{v}^k) \in \mathcal{U}^k \times \mathcal{V}^k$ or an error message* **?**.

The performance of a code $\mathcal{C}_k$ is measured in terms of the average probability of error

$$\mathbf{P}_e(\mathcal{C}_k) \triangleq \mathbb{P}\left[(\hat{U}^k, \hat{V}^k) \neq (U^k, V^k) \mid \mathcal{C}_k\right].$$

Definition 2.22. *A rate pair (R_1, R_2) is achievable if there exists a sequence of $(2^{kR_1}, 2^{kR_2}, k)$ codes $\{\mathcal{C}_k\}_{k\geqslant 1}$ such that*

$$\lim_{k\to\infty} \mathbf{P}_e(\mathcal{C}_k) = 0.$$

The achievable rate region is defined as

$$\mathcal{R}^{\text{SW}} \triangleq \text{cl}(\{(R_1, R_2) : (R_1, R_2) \text{ is achievable}\}).$$

The achievable rate region with separate encoding was first characterized by Slepian and Wolf; hence, the region is often called the Slepian–Wolf region and codes for the distributed source coding problem are often referred to as Slepian–Wolf codes.

Theorem 2.10 (Slepian–Wolf theorem). *The achievable rate region with separate encoding for a source $(\mathcal{UV}, p_{\mathsf{UV}})$ is*

$$\mathcal{R}^{\mathrm{SW}} \triangleq \left\{(R_1, R_2)\colon \begin{array}{c} R_1 \geqslant \mathbb{H}(\mathsf{U}|\mathsf{V}) \\ R_2 \geqslant \mathbb{H}(\mathsf{V}|\mathsf{U}) \\ R_1 + R_2 \geqslant \mathbb{H}(\mathsf{UV}) \end{array}\right\}.$$

Proof. We begin with the achievability part of the proof, which is based on joint typicality and random binning. Let $\epsilon > 0$ and $k \in \mathbb{N}^*$. Let $R_1 > 0$ and $R_2 > 0$ be rates to be specified later. We construct a $(2^{kR_1}, 2^{kR_2}, k)$ code $\mathcal{C}_k$ as follows.

- *Binning.* For each sequence $u^k \in \mathcal{T}_\epsilon^k(\mathsf{U})$, draw an index uniformly at random in the set $[\![1, 2^{kR_1}]\!]$. For each sequence $v^k \in \mathcal{T}_\epsilon^k(\mathsf{V})$, draw an index uniformly at random in the set $[\![1, 2^{kR_2}]\!]$. The index assignments define the encoding functions

$$e_1 : \mathcal{U}^k \to [\![1, 2^{kR_1}]\!] \quad \text{and} \quad e_2 : \mathcal{V}^k \to [\![1, 2^{kR_2}]\!],$$

  which are revealed to all parties.
- *Encoder 1.* Given the observation u^k, if $u^k \in \mathcal{T}_\epsilon^k(\mathsf{U})$, output $m_1 = e_1(u^k)$; otherwise output $m_1 = 1$.
- *Encoder 2.* Given the observation v^k, if $v^k \in \mathcal{T}_\epsilon^k(\mathsf{V})$, output $m_2 = e_2(v^k)$; otherwise output $m_2 = 1$.
- *Decoder.* Given messages m_1 and m_2, output $\hat{u}^k$ and $\hat{v}^k$ if they are the unique sequences such that $(\hat{u}^k, \hat{v}^k) \in \mathcal{T}_\epsilon^k(\mathsf{UV})$ and $e_1(\hat{u}^k) = m_1$, $e_2(\hat{v}^k) = m_2$; otherwise, output ?.

The random variables that represent the randomly generated functions e_1 and e_2 are denoted by E_1 and E_2, and the random variable that represents the randomly generated code $\mathcal{C}_k$ is denoted by C_k. We proceed to bound $\mathbb{E}[\mathbf{P}_e(\mathsf{C}_k)]$, which can be expressed in terms of the following events:

$$\begin{aligned}
\mathcal{E}_0 &= \{(\mathsf{U}^k, \mathsf{V}^k) \notin \mathcal{T}_\epsilon^k(\mathsf{UV})\},\\
\mathcal{E}_1 &= \{\exists \hat{u}^k \neq \mathsf{U}^k : \mathsf{E}_1(\hat{u}^k) = \mathsf{E}_1(\mathsf{U}^k) \text{ and } (\hat{u}^k, \mathsf{V}^k) \in \mathcal{T}_\epsilon^k(\mathsf{UV})\},\\
\mathcal{E}_2 &= \{\exists \hat{v}^k \neq \mathsf{V}^k : \mathsf{E}_2(\hat{v}^k) = \mathsf{E}_2(\mathsf{V}^k) \text{ and } (\hat{u}^k, \mathsf{V}^k) \in \mathcal{T}_\epsilon^k(\mathsf{UV})\},\\
\mathcal{E}_{12} &= \{\exists \hat{v}^k \neq \mathsf{V}^k, \hat{u}^k \neq \mathsf{U}^k, : \mathsf{E}_1(\hat{u}^k) = \mathsf{E}_1(\mathsf{U}^k),\ \mathsf{E}_2(\hat{v}^k) = \mathsf{E}_2(\mathsf{V}^k)\\
&\qquad \text{and } (\hat{u}^k, \hat{v}^k) \in \mathcal{T}_\epsilon^k(\mathsf{UV})\},
\end{aligned}$$

since $\mathbb{E}[\mathbf{P}_e(\mathsf{C}_k)] = \mathbb{P}[\mathcal{E}_0 \cup \mathcal{E}_1 \cup \mathcal{E}_2 \cup \mathcal{E}_{12}]$. By the union bound,

$$\mathbb{E}[\mathbf{P}_e(\mathsf{C}_k)] \leqslant \mathbb{P}[\mathcal{E}_0] + \mathbb{P}[\mathcal{E}_1] + \mathbb{P}[\mathcal{E}_2] + \mathbb{P}[\mathcal{E}_{12}]. \tag{2.7}$$

By the AEP,

$$\mathbb{P}[\mathcal{E}_0] \leqslant \delta_\epsilon(k). \tag{2.8}$$

Using Theorem 2.2,

$$\begin{aligned}
\mathbb{P}[\mathcal{E}_1] &= \sum_{u^k,v^k} p_{\mathsf{U}^k\mathsf{V}^k}(u^k,v^k)\,\mathbb{P}\left[\exists \hat{u}^k \neq u^k : \mathsf{E}_1(\hat{u}^k) = \mathsf{E}_1(u^k) \text{ and } (\hat{u}^k, v^k) \in \mathcal{T}_\epsilon^k(\mathsf{UV})\right] \\
&\leqslant \sum_{u^k,v^k} p_{\mathsf{U}^k\mathsf{V}^k}(u^k,v^k) \sum_{\substack{\hat{u}^k \in \mathcal{T}_\epsilon^k(\mathsf{UV}|v^k) \\ \hat{u}^k \neq u^k}} \mathbb{P}\left[\mathsf{E}_1(\hat{u}^k) = \mathsf{E}_1(u^k)\right] \\
&= \sum_{u^k,v^k} p_{\mathsf{U}^k\mathsf{V}^k}(u^k,v^k) \sum_{\substack{\hat{u}^k \in \mathcal{T}_\epsilon^k(\mathsf{UV}|v^k) \\ \hat{u}^k \neq u^k}} \frac{1}{\lceil 2^{kR_1} \rceil} \\
&\leqslant \sum_{u^k,v^k} p_{\mathsf{U}^k\mathsf{V}^k}(u^k,v^k) \frac{1}{\lceil 2^{kR_1} \rceil} \left|\mathcal{T}_\epsilon^k(\mathsf{UV}|v^k)\right| \\
&\leqslant \sum_{u^k,v^k} p_{\mathsf{U}^k\mathsf{V}^k}(u^k,v^k) \frac{1}{\lceil 2^{kR_1} \rceil} 2^{k(\mathbb{H}(\mathsf{U}|\mathsf{V})+\delta(\epsilon))} \\
&\leqslant 2^{k(\mathbb{H}(\mathsf{U}|\mathsf{V})+\delta(\epsilon)-R_1)}.
\end{aligned} \tag{2.9}$$

Similarly, we obtain the following bounds for $\mathbb{P}[\mathcal{E}_2]$ and $\mathbb{P}[\mathcal{E}_{12}]$:

$$\mathbb{P}[\mathcal{E}_2] \leqslant 2^{k(\mathbb{H}(\mathsf{V}|\mathsf{U})+\delta(\epsilon)-R_2)}, \tag{2.10}$$

$$\mathbb{P}[\mathcal{E}_{12}] \leqslant 2^{k(\mathbb{H}(\mathsf{UV})+\delta(\epsilon)-(R_1+R_2))}. \tag{2.11}$$

Hence, if we choose the rates R_1 and R_2 such that

$$\begin{aligned}
R_1 &> \mathbb{H}(\mathsf{U}|\mathsf{V}) + \delta(\epsilon), \\
R_2 &> \mathbb{H}(\mathsf{V}|\mathsf{U}) + \delta(\epsilon), \\
R_1 + R_2 &> \mathbb{H}(\mathsf{UV}) + \delta(\epsilon),
\end{aligned}$$

and substitute (2.8)–(2.11) into (2.7), we obtain $\mathbb{E}[\mathbf{P}_e(\mathsf{C}_k)] \leqslant \delta_\epsilon(k)$. By applying the selection lemma to the random variable C_k and the function $\mathbf{P}_e$, we conclude that there exists a specific code $\mathcal{C}_k$ such that $\mathbf{P}_e(\mathcal{C}_k) \leqslant \delta_\epsilon(k)$. Since ϵ can be chosen arbitrarily small, we conclude that

$$\left\{ (R_1, R_2) : \begin{array}{l} R_1 \geqslant \mathbb{H}(\mathsf{U}|\mathsf{V}) \\ R_2 \geqslant \mathbb{H}(\mathsf{V}|\mathsf{U}) \\ R_1 + R_2 \geqslant \mathbb{H}(\mathsf{UV}) \end{array} \right\} \subseteq \mathcal{R}^{\text{SW}}.$$

The converse part of the proof follows from the converse of the source coding theorem and is omitted. □

Figure 2.6 illustrates the typical shape of the Slepian–Wolf region $\mathcal{R}^{\text{SW}}$. A special case of the Slepian–Wolf problem is when one of the components of the DMS $(\mathcal{UV}, p_{\mathsf{UV}})$, say V, is directly available at the decoder as side information and only U should be compressed. This problem is known as source coding with side information. The characterization of the minimum compression rate required to reconstruct U reliably at the decoder follows from Theorem 2.10.

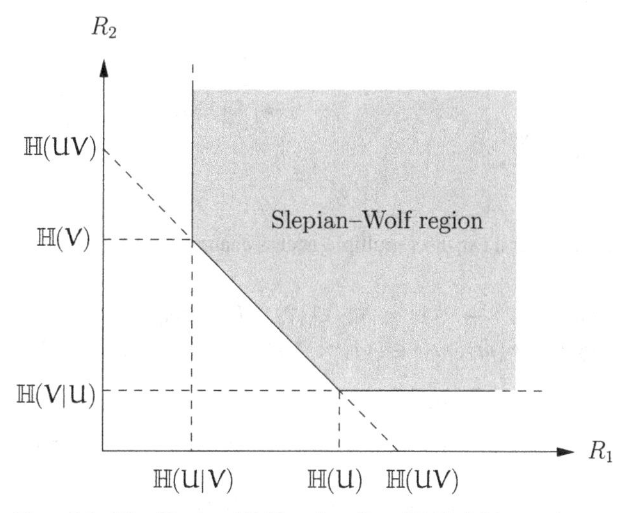

Figure 2.6 The Slepian–Wolf region for a DMS $(\mathcal{UV}, p_{\mathsf{UV}})$.

Corollary 2.4 (Source coding with side information). *Consider a DMS $(\mathcal{UV}, p_{\mathsf{UV}})$ and assume that $(\mathcal{U}, p_{\mathsf{U}})$ should be compressed knowing that $(\mathcal{V}, p_{\mathsf{V}})$ is available as side information at the decoder. Then,*

$$\inf\{R : R \text{ is an achievable compression rate}\} = \mathbb{H}(\mathsf{U}|\mathsf{V}).$$

Corollary 2.4 plays a fundamental role for secret-key agreement in Chapters 4 and 6.

2.3.2 The multiple-access channel

In the previous problem, we assumed that the information generated by multiple sources is transmitted over noiseless channels. If these data are to be communicated over a common noisy channel to a single destination, we call this type of channel a *multiple-access channel* (MAC). As illustrated in Figure 2.7, a discrete memoryless multiple access channel $(\mathcal{X}_1, \mathcal{X}_2, p_{\mathsf{Y}|\mathsf{X}_1\mathsf{X}_2}, \mathcal{Y})$ consists of two finite input alphabets $\mathcal{X}_1$ and $\mathcal{X}_2$, one finite output alphabet $\mathcal{Y}$, and transition probabilities $p_{\mathsf{Y}|\mathsf{X}_1\mathsf{X}_2}$ such that

$$\forall n \geqslant 1 \quad \forall (x_1^n, x_2^n, y^n) \in \mathcal{X}_1^n \times \mathcal{X}_2^n \times \mathcal{Y}^n$$
$$p_{\mathsf{Y}^n|\mathsf{X}_1^n\mathsf{X}_2^n}\left(y^n|x_1^n x_2^n\right) = \prod_{i=1}^{n} p_{\mathsf{Y}|\mathsf{X}_1\mathsf{X}_2}(y_i|x_{1,i}, x_{2,i}).$$

Definition 2.23. *A $(2^{nR_1}, 2^{nR_2}, n)$ code $\mathcal{C}_n$ for the MAC consists of*

- *two message sets $\mathcal{M}_1 = [\![1, 2^{nR_1}]\!]$ and $\mathcal{M}_2 = [\![1, 2^{nR_2}]\!]$;*
- *two encoding functions, $f_1 : \mathcal{M}_1 \to \mathcal{X}_1^n$ and $f_2 : \mathcal{M}_2 \to \mathcal{X}_2^n$, which map a message m_1 or m_2 to a codeword x_1^n or x_2^n;*

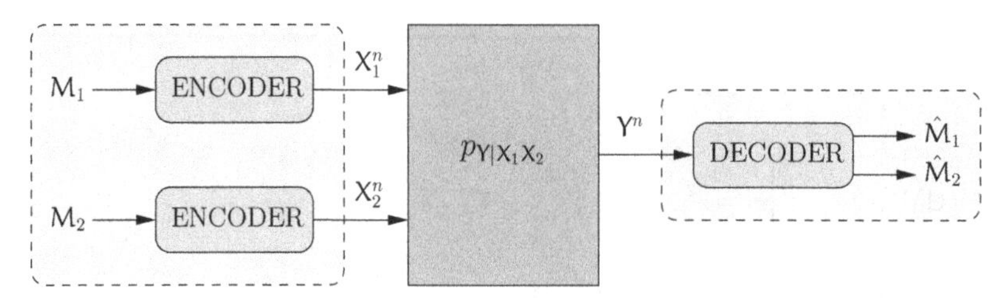

Figure 2.7 Communication over a two-user multiple-access channel.

- *a decoding function* $g : \mathcal{Y}^n \to \mathcal{M}_1 \times \mathcal{M}_2 \cup \{?\}$*, which maps each channel observation* y^n *to a message pair* $(\hat{m}_1, \hat{m}_2) \in \mathcal{M}_1 \times \mathcal{M}_2$ *or an error message* **?**.

The messages M_1 and M_2 are assumed uniformly distributed in their respective sets, and the performance of a code $\mathcal{C}_n$ is measured in terms of the average probability of error

$$\mathbf{P}_e(\mathcal{C}_n) \triangleq \mathbb{P}\left[(\hat{\mathsf{M}}_1, \hat{\mathsf{M}}_2) \neq (\mathsf{M}_1, \mathsf{M}_2)) \middle| \mathcal{C}_n\right].$$

Definition 2.24. *A rate pair* (R_1, R_2) *is achievable for the MAC if there exists a sequence of* $(2^{nR_1}, 2^{nR_2}, n)$ *codes* $\{\mathcal{C}_n\}_{n\geqslant 1}$ *such that*

$$\lim_{n\to\infty} \mathbf{P}_e(\mathcal{C}_n) = 0.$$

The capacity region of a MAC is defined as

$$\mathcal{C}^{\mathsf{MAC}} \triangleq \mathrm{cl}(\{(R_1, R_2) : (R_1, R_2) \text{ is achievable}\}).$$

The characterization of the capacity region requires the notion of a *convex hull*, which we define below.

Definition 2.25. *The convex hull of a set* $\mathcal{S} \subseteq \mathbb{R}^n$ *is the set*

$$\mathrm{co}(\mathcal{S}) \triangleq \left\{ \sum_{i=1}^{k} \lambda_i x_i^n : k \geqslant 1,\ \{\lambda_i\}_k \in [0,1]^k,\ \sum_{i=1}^{k} \lambda_i = 1,\ \{x_i^n\}_k \in \mathcal{S}^k \right\}.$$

Theorem 2.11 (Ahlswede and Liao). *Consider a MAC* $\left(\mathcal{X}_1, \mathcal{X}_2, p_{\mathsf{Y}|\mathsf{X}_1\mathsf{X}_2}, \mathcal{Y}\right)$. *For any independent distributions* p_{X_1} *on* $\mathcal{X}_1$ *and* p_{X_2} *on* $\mathcal{X}_2$*, define the set* $\mathcal{R}(p_{\mathsf{X}_1}p_{\mathsf{X}_2})$ *as*

$$\mathcal{R}(p_{\mathsf{X}_1}p_{\mathsf{X}_2}) \triangleq \left\{ (R_1, R_2) \colon \begin{array}{l} 0 \leqslant R_1 \leqslant \mathbb{I}(\mathsf{X}_1;\mathsf{Y}|\mathsf{X}_2) \\ 0 \leqslant R_2 \leqslant \mathbb{I}(\mathsf{X}_2;\mathsf{Y}|\mathsf{X}_1) \\ 0 \leqslant R_1 + R_2 \leqslant \mathbb{I}(\mathsf{X}_1\mathsf{X}_2;\mathsf{Y}) \end{array} \right\},$$

where the joint distribution of X_1, X_2, *and* Y *factorizes as* $p_{\mathsf{X}_1}p_{\mathsf{X}_2}p_{\mathsf{Y}|\mathsf{X}_1\mathsf{X}_2}$. *Then, the capacity region of a MAC is*

$$\mathcal{C}^{\mathsf{MAC}} \triangleq \mathrm{co}\left(\bigcup_{p_{\mathsf{X}_1}p_{\mathsf{X}_2}} \mathcal{R}(p_{\mathsf{X}_1}p_{\mathsf{X}_2}) \right).$$

Proof. We provide only the achievability part of the proof, which is similar to the proof of Shannon's channel coding theorem and is based on joint typicality and random coding. Fix two independent probability distributions, p_{X_1} on $\mathcal{X}_1$ and p_{X_2} on $\mathcal{X}_2$. Let $0 < \epsilon < \mu_{X_1X_2Y}$, where

$$\mu_{X_1X_2Y} \triangleq \min_{(x_1,x_2,y)\in\mathcal{X}_1\times\mathcal{X}_2\times\mathcal{Y}} p_{X_1}(x_1)p_{X_2}(x_2)p_{Y|X_1X_2}(y|x_1,x_2),$$

and let $n \in \mathbb{N}^*$. Let $R_1 > 0$ and $R_2 > 0$ be rates to be specified later. We construct a $(2^{nR_1}, 2^{nR_2}, n)$ code $\mathcal{C}_n$ as follows.

- *Codebook construction.* Construct a codebook for user 1 with $\lceil 2^{nR_1}\rceil$ codewords, labeled $x_1^n(m_1)$ with $m_1 \in [\![1, 2^{nR_1}]\!]$, by generating the symbols $x_{1,i}(m_1)$ for $i \in [\![1, n]\!]$ and $m_1 \in [\![1, 2^{nR_1}]\!]$ independently according to p_{X_1}. Similarly, construct a codebook for user 2 with $\lceil 2^{nR_2}\rceil$ codewords, labeled $x_2^n(m_2)$ with $m_2 \in [\![1, 2^{nR_2}]\!]$, by generating the symbols $x_{2,i}(m_2)$ for $i \in [\![1, n]\!]$ and $m_2 \in [\![1, 2^{nR_2}]\!]$ independently according to p_{X_2}. The codebooks are revealed to all encoders and decoders.
- *Encoder 1.* Given m_1, transmit $x_1^n(m_1)$.
- *Encoder 2.* Given m_2, transmit $x_2^n(m_2)$.
- *Decoder.* Given y^n, output $(\hat{m}_1, \hat{m}_2)$ if it is the unique message pair such that $(x_1^n(\hat{m}_1), x_2^n(\hat{m}_2), y^n) \in \mathcal{T}_\epsilon^n(X_1X_2Y)$; otherwise, output an error **?**.

The random variable that represents the randomly generated code $\mathcal{C}_n$ is denoted by C_n and we proceed to bound $\mathbb{E}[\mathbf{P}_e(C_n)]$. By virtue of the symmetry of the random code construction

$$\begin{aligned}\mathbb{E}[\mathbf{P}_e(C_n)] &= \mathbb{E}_{C_n}\left[\mathbb{P}\left[(\hat{M}_1, \hat{M}_2) \neq (M_1, M_2)\middle|C_n\right]\right]\\ &= \mathbb{E}_{C_n}\left[\mathbb{P}\left[(\hat{M}_1, \hat{M}_2) \neq (M_1, M_2)\middle|M_1 = 1, M_2 = 1, C_n\right]\right].\end{aligned}$$

Therefore, the probability of error can be expressed in terms of the error events

$$\mathcal{E}_{ij} \triangleq \{(X_1^n(i), X_2^n(j), Y^n) \in \mathcal{T}_\epsilon^n(X_1X_2Y)\} \quad \text{for } i \in [\![1, 2^{nR_1}]\!] \text{ and } j \in [\![1, 2^{nR_2}]\!]$$

as

$$\mathbb{E}[\mathbf{P}_e(C_n)] = \mathbb{P}\left[\mathcal{E}_{11}^c \cup \bigcup_{i\neq 1}\mathcal{E}_{i1} \cup \bigcup_{j\neq 1}\mathcal{E}_{1j} \cup \bigcup_{(i,j)\neq(1,1)}\mathcal{E}_{ij}\right].$$

By the union bound,

$$\mathbb{E}[\mathbf{P}_e(C_n)] \leqslant \mathbb{P}\left[\mathcal{E}_{11}^c\right] + \sum_{i\neq 1}\mathbb{P}[\mathcal{E}_{i1}] + \sum_{j\neq 1}\mathbb{P}\left[\mathcal{E}_{1j}\right] + \sum_{(i,j)\neq(1,1)}\mathbb{P}\left[\mathcal{E}_{ij}\right]. \tag{2.12}$$

By the joint AEP,

$$\mathbb{P}\left[\mathcal{E}_{11}^c\right] \leqslant \delta_\epsilon(n). \tag{2.13}$$

For $i \neq 1$, $X_1^n(i)$ is conditionally independent of Y^n given $X_2^n(1)$; therefore, by Corollary 2.3,

$$\mathbb{P}[\mathcal{E}_{i1}] \leqslant 2^{-n(\mathbb{I}(X_1;Y|X_2)-\delta(\epsilon))} \quad \text{for } i \neq 1. \tag{2.14}$$

Similarly, we can show

$$\mathbb{P}\left[\mathcal{E}_{1j}\right] \leqslant 2^{-n(\mathbb{I}(X_2;Y|X_1)-\delta(\epsilon))} \quad \text{for } j \neq 1, \tag{2.15}$$

$$\mathbb{P}\left[\mathcal{E}_{ij}\right] \leqslant 2^{-n(\mathbb{I}(X_1X_2;Y)-\delta(\epsilon))} \quad \text{for } i \neq 1 \text{ and } j \neq 1. \tag{2.16}$$

On substituting (2.13)–(2.16) into (2.12), we obtain

$$\begin{aligned}\mathbb{E}[\mathbf{P}_e(C_n)] \leqslant \delta_\epsilon(n) &+ \lceil 2^{nR_1}\rceil 2^{-n(\mathbb{I}(X_1;Y|X_2)-\delta(\epsilon))} + \lceil 2^{nR_2}\rceil 2^{-n(\mathbb{I}(X_2;Y|X_1)-\delta(\epsilon))}\\ &+ \lceil 2^{nR_1}\rceil\lceil 2^{nR_2}\rceil 2^{-n(\mathbb{I}(X_1X_2;Y)-\delta(\epsilon))}.\end{aligned}$$

Hence, if we choose R_1 and R_2 to satisfy

$$\begin{aligned}R_1 &< \mathbb{I}(X_1;Y|X_2) - \delta(\epsilon),\\ R_2 &< \mathbb{I}(X_2;Y|X_1) - \delta(\epsilon),\\ R_1 + R_2 &< \mathbb{I}(X_1X_2;Y) - \delta(\epsilon),\end{aligned}$$

we obtain $\mathbb{E}[\mathbf{P}_e(C_n)] \leqslant \delta_\epsilon(n)$. By applying the selection lemma to the random variable C_n and the function $\mathbf{P}_e$, we conclude that there exists a $(2^{nR_1}, 2^{nR_2}, n)$ code $\mathcal{C}_n$ such that $\mathbf{P}_e(\mathcal{C}_n) \leqslant \delta_\epsilon(n)$. Since ϵ can be chosen arbitrarily small and since the distributions p_{X_1} and p_{X_2} are arbitrary, we conclude that

$$\bigcup_{p_{X_1}p_{X_2}} \left\{ (R_1, R_2)\colon \begin{array}{c} 0 \leqslant R_1 \leqslant \mathbb{I}(X_1;Y|X_2) \\ 0 \leqslant R_2 \leqslant \mathbb{I}(X_2;Y|X_1) \\ 0 \leqslant R_1 + R_2 \leqslant \mathbb{I}(X_1X_2;Y) \end{array} \right\} \subseteq \mathcal{C}^{\text{MAC}}$$

is achievable. Finally, it can be shown that time-sharing between different codes achieves the entire convex hull [3, Section 15.3]. We refer the reader to [3, 6] for the converse part of the proof. □

The typical shape of the region $\mathcal{R}(p_{X_1}p_{X_2})$ is illustrated in Figure 2.8. The boundaries of the capacity region can be explained in a very intuitive way. When encoder 1 views the signals sent by encoder 2 as noise, its maximum achievable rate is on the order of $\mathbb{I}(X_1;Y)$, which is a direct consequence of the channel coding theorem. Then, the decoder can estimate the codeword x_1^n and subtract it from the channel output sequence y_1^n, thus allowing encoder 2 to achieve a maximum rate on the order of $\mathbb{I}(X_2;Y|X_1)$. This procedure is sometimes called *successive cancellation* and leads to the upper corner point of the region. The lower corner point corresponds to the symmetric case, in which encoder 2 views the signals sent by encoder 1 as noise.

2.3.3 The broadcast channel

While a multiple-access channel considers multiple sources and one destination, the *broadcast channel* (BC for short) considers a single information source that transmits to multiple users. Applications of the BC include the downlink channel of a satellite or of a base station in a mobile communication network, and the wiretap channel which is studied in detail in Chapter 3 and Chapter 5. As illustrated in Figure 2.9, a discrete

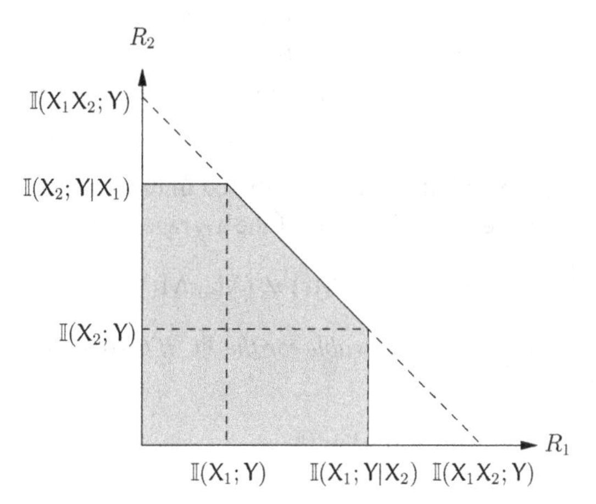

Figure 2.8 Typical shape of the rate region $\mathcal{R}(p_{X_1}p_{X_2})$ of the multiple-access channel for fixed input distributions p_{X_1} and p_{X_2}.

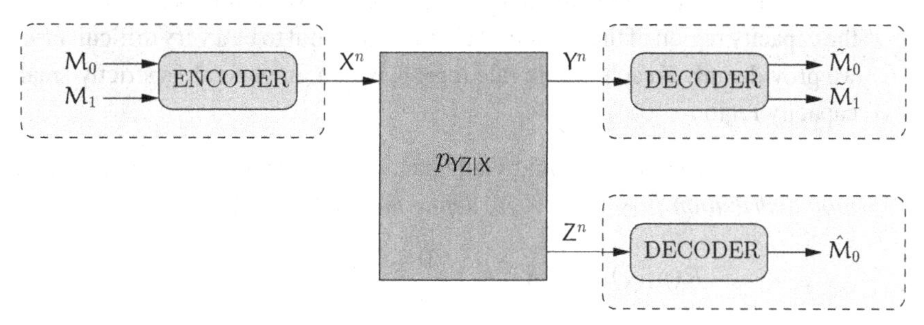

Figure 2.9 Communication over a two-user broadcast channel.

memoryless two-user broadcast channel $(\mathcal{X}, p_{YZ|X}, \mathcal{Y}, \mathcal{Z})$ consists of a finite input alphabet $\mathcal{X}$, two finite output alphabets $\mathcal{Y}$ and $\mathcal{Z}$, and transition probabilities $p_{YZ|X}$ such that

$$\forall n \geqslant 1 \quad \forall (x^n, y^n, z^n) \in \mathcal{X}^n \times \mathcal{Y}^n \times \mathcal{Z}^n$$

$$p_{Y^nZ^n|X^n}(y^n, z^n|x^n) = \prod_{i=1}^{n} p_{YZ|X}(y_i, z_i|x_i).$$

We assume that the transmitter wants to send a *common message* M_0 to both receivers and a *private message* M_1 to the receiver observing Y^n. The receiver observing Z^n is called a "weak" user, while the receiver observing Y^n is called the "strong" user.

Definition 2.26. *A* $\left(2^{nR_0}, 2^{nR_1}, n\right)$ *code* $\mathcal{C}_n$ *for the BC consists of*

- *two message sets* $\mathcal{M}_0 = \llbracket 1, 2^{nR_0} \rrbracket$ *and* $\mathcal{M}_1 = \llbracket 1, 2^{nR_1} \rrbracket$;
- *an encoding function* $f : \mathcal{M}_0 \times \mathcal{M}_1 \to \mathcal{X}^n$, *which maps a message pair* (m_0, m_1) *to a codeword* x^n;

- *a decoding function* $g : \mathcal{Y}^n \to (\mathcal{M}_0 \times \mathcal{M}_1) \cup \{?\}$, *which maps each channel observation* y^n *to a message pair* $(\tilde{m}_0, \tilde{m}_1) \in \mathcal{M}_0 \times \mathcal{M}_1$ *or an error message* ?;
- *a decoding function* $h : \mathcal{Z}^n \to \mathcal{M}_0 \cup \{?\}$, *which maps each channel observation* z^n *to a message* $\hat{m}_0 \in \mathcal{M}_0$ *or an error message* ?.

Messages M_0 and M_1 are assumed uniformly distributed in their respective sets and the performance of a code $\mathcal{C}_n$ is measured in terms of the average probability of error

$$\mathbf{P}_e(\mathcal{C}_n) \triangleq \mathbb{P}\left[\hat{\mathsf{M}}_0 \neq \mathsf{M}_0 \text{ or } (\tilde{\mathsf{M}}_0, \tilde{\mathsf{M}}_1) \neq (\mathsf{M}_0, \mathsf{M}_1) \mid \mathcal{C}_n\right].$$

Definition 2.27. *A rate pair* (R_0, R_1) *is achievable for the BC if there exists a sequence of* $\left(2^{nR_0}, 2^{nR_1}, n\right)$ *codes* $\{\mathcal{C}_n\}_{n\geqslant 1}$ *such that*

$$\lim_{n\to\infty} \mathbf{P}_e(\mathcal{C}_n) = 0.$$

The capacity region of a BC is defined as

$$\mathcal{C}^{\text{BC}} \triangleq \text{cl}(\{(R_0, R_1) : (R_0, R_1) \text{ is achievable}\}).$$

As in many other fundamental problems of network information theory, determining the capacity region of the broadcast channel turns out to be a very difficult task. Therefore, we provide only an achievable rate region, which, in general, is strictly smaller than the capacity region.

Theorem 2.12 (Bergsman and Gallager). *Consider a BC* $(\mathcal{X}, p_{\mathsf{YZ}|\mathsf{X}}, \mathcal{Y}, \mathcal{Z})$. *For any joint distribution* p_{UX} *on* $\mathcal{U} \times \mathcal{X}$, *define the set* $\mathcal{R}(p_{\mathsf{UX}})$ *as*

$$\mathcal{R}(p_{\mathsf{UX}}) \triangleq \left\{(R_0, R_1)\colon \begin{array}{l} 0 \leqslant R_0 \leqslant \min(\mathbb{I}(\mathsf{U};\mathsf{Y}), \mathbb{I}(\mathsf{U};\mathsf{Z})) \\ 0 \leqslant R_1 \leqslant \mathbb{I}(\mathsf{X};\mathsf{Y}|\mathsf{U}) \end{array}\right\},$$

where the joint distribution of U, X, Y, *and* Z *factorizes as* $p_{\mathsf{UX}}p_{\mathsf{YZ}|\mathsf{X}}$. *Then,*

$$\mathcal{R}^{\text{BC}} \triangleq \text{co}\left(\bigcup_{p_{\mathsf{UX}}} \mathcal{R}(p_{\mathsf{UX}})\right) \subseteq \mathcal{C}^{\text{BC}}.$$

In addition, the cardinality of the auxiliary random variable U *can be limited to* $|\mathcal{U}| \leqslant \min(|\mathcal{X}|, |\mathcal{Y}|, |\mathcal{Z}|)$.

Proof. The proof that $\mathcal{R}^{\text{BC}} \subseteq \mathcal{C}^{\text{BC}}$ is based on joint typicality, random coding, and a code construction called *superposition coding*. As illustrated in Figure 2.10, the idea of superposition coding is to create a codebook with $\lceil 2^{nR_0} \rceil$ codewords for the weakest user and to superpose a codebook with $\lceil 2^{nR_1} \rceil$ codewords for the strongest user to *every* codeword. The codewords u^n are often called "cloud centers," while the codewords x^n are called "satellite codewords." Formally, fix a joint probability distribution p_{UX} on $\mathcal{U} \times \mathcal{X}$. Let $0 < \epsilon < \mu_{\mathsf{XYU}}$, where $\mu_{\mathsf{XYU}} \triangleq \min p_{\mathsf{XU}}(x, u)p_{\mathsf{Y}|\mathsf{XU}}(y|x, u)$ and let $n \in \mathbb{N}^*$. Let $R_0 > 0$ and $R_1 > 0$ be rates to be specified later. We construct a $\left(2^{nR_0}, 2^{nR_1}, n\right)$ code $\mathcal{C}_n$ as follows.

- *Codebook construction.* Construct a codebook with $\lceil 2^{nR_0} \rceil$ codewords, labeled $u^n(m_0)$ with $m_0 \in \left[\!\left[1, 2^{nR_0}\right]\!\right]$, by generating the symbols $u_i(m_0)$ for $i \in [\![1, n]\!]$ and

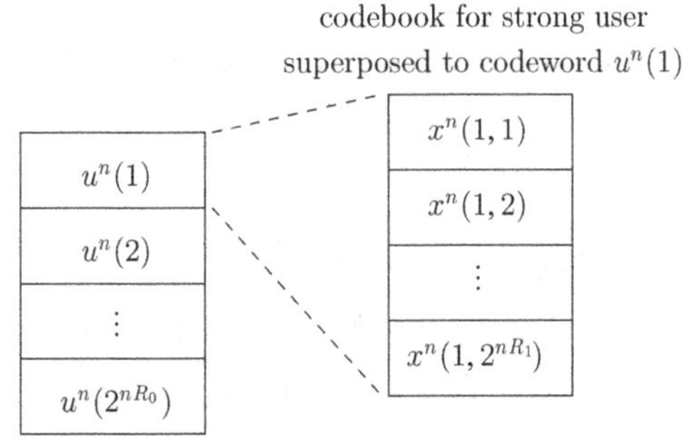

Figure 2.10 Superposition coding. A codebook for the strong user is superposed to every codeword of the codebook for the weak user.

$m_0 \in [\![1, 2^{nR_0}]\!]$ independently according to p_{U}. For each $u^n(m_0)$ with $m_0 \in [\![1, 2^{nR_0}]\!]$, generate another codebook with $\lceil 2^{nR_1} \rceil$ codewords, labeled $x^n(m_0, m_1)$ with $m_1 \in [\![1, 2^{nR_1}]\!]$, by generating the symbols $x_i(m_0, m_1)$ for $i \in [\![1, n]\!]$ and $m_1 \in [\![1, 2^{nR_1}]\!]$ independently according to $p_{\mathsf{X}|\mathsf{U}=u_i(m_0)}$. The codebooks are revealed to the encoder and both decoders.

- *Encoder.* Given (m_0, m_1), transmit $x^n(m_0, m_1)$.
- *Decoder for weak user.* Given z^n, output $\hat{m}_0$ if it is the unique message such that $(u^n(\hat{m}_0), z^n) \in \mathcal{T}_\epsilon^n(\mathsf{UZ})$; otherwise, output an error **?**.
- *Decoder for strong user.* Given y^n, output $(\tilde{m}_0, \tilde{m}_1)$ if it is the unique message pair such that $(u^n(\tilde{m}_0), y^n) \in \mathcal{T}_\epsilon^n(\mathsf{UY})$ and $(u^n(\tilde{m}_0), x^n(\tilde{m}_0, \tilde{m}_1), y^n) \in \mathcal{T}_\epsilon^n(\mathsf{UXY})$; otherwise, output an error **?**.

The random variable that denotes the randomly generated codebook $\mathcal{C}_n$ is denoted by C_n, and we proceed to bound $\mathbb{E}[\mathbf{P}_e(\mathsf{C}_n)]$. From the symmetry of the random code construction, notice that

$$\mathbb{E}[\mathbf{P}_e(\mathsf{C}_n)] = \mathbb{E}_{\mathsf{C}_n}\left[\mathbb{P}\left[\hat{\mathsf{M}}_0 \neq \mathsf{M}_0 \text{ or } (\tilde{\mathsf{M}}_0, \tilde{\mathsf{M}}_1) \neq (\mathsf{M}_0, \mathsf{M}_1) \mid \mathsf{C}_n\right]\right]$$
$$= \mathbb{E}_{\mathsf{C}_n}\left[\mathbb{P}\left[\hat{\mathsf{M}}_0 \neq \mathsf{M}_0 \text{ or } (\tilde{\mathsf{M}}_0, \tilde{\mathsf{M}}_1) \neq (\mathsf{M}_0, \mathsf{M}_1) \mid \mathsf{M}_0 = 1, \mathsf{M}_1 = 1, \mathsf{C}_n\right]\right].$$

Therefore, $\mathbb{E}[\mathbf{P}_e(\mathsf{C}_n)]$ can be expressed in terms of the events

$$\mathcal{E}_i = \left\{(\mathsf{U}^n(i), \mathsf{Y}^n) \in \mathcal{T}_\epsilon^n(\mathsf{UY})\right\} \quad \text{for } i \in [\![1, 2^{nR_0}]\!],$$
$$\mathcal{F}_i = \left\{(\mathsf{U}^n(i), \mathsf{Z}^n) \in \mathcal{T}_\epsilon^n(\mathsf{UZ})\right\} \quad \text{for } i \in [\![1, 2^{nR_0}]\!],$$
$$\mathcal{G}_{ij} = \left\{(\mathsf{U}^n(i), \mathsf{X}^n(j), \mathsf{Y}^n) \in \mathcal{T}_\epsilon^n(\mathsf{UXY})\right\} \quad \text{for } i \in [\![1, 2^{nR_0}]\!] \text{ and } j \in [\![1, 2^{nR_1}]\!]$$

as

$$\mathbb{E}[\mathbf{P}_e(\mathsf{C}_n)] = \mathbb{P}\left[\mathcal{E}_1^c \cup \bigcup_{i\neq 1} \mathcal{E}_i \cup \mathcal{F}_1^c \cup \bigcup_{i\neq 1} \mathcal{F}_i \cup \bigcup_{j\neq 1} \mathcal{G}_{1j}\right]$$

and, by the union bound,

$$\mathbb{E}[\mathbf{P}_e(C_n)] \leqslant \mathbb{P}\left[\mathcal{E}_1^c\right] + \sum_{i\neq 1}\mathbb{P}[\mathcal{E}_i] + \mathbb{P}\left[\mathcal{F}_1^c\right] + \sum_{i\neq 1}\mathbb{P}[\mathcal{F}_i] + \sum_{j\neq 1}\mathbb{P}\left[\mathcal{G}_{1j}\right]. \tag{2.17}$$

By the joint AEP,

$$\mathbb{P}\left[\mathcal{E}_1^c\right] \leqslant \delta_\epsilon(n) \quad \text{and} \quad \mathbb{P}\left[\mathcal{F}_1^c\right] \leqslant \delta_\epsilon(n). \tag{2.18}$$

For $i \neq 1$, notice that $\mathsf{U}^n(i)$ is independent of Y^n and Z^n; therefore, by Corollary 2.2,

$$\mathbb{P}[\mathcal{E}_i] \leqslant 2^{-n(\mathbb{I}(\mathsf{U};\mathsf{Y})-\delta(\epsilon))} \quad \text{and} \quad \mathbb{P}[\mathcal{F}_i] \leqslant 2^{-n(\mathbb{I}(\mathsf{U};\mathsf{Z})-\delta(\epsilon))} \quad \text{for } i \neq 1. \tag{2.19}$$

For $j \neq 1$, $\mathsf{X}^n(1, j)$ is conditionally independent of Z^n given $\mathsf{U}^n(1)$; therefore, by Corollary 2.3,

$$\mathbb{P}\left[\mathcal{G}_{1j}\right] \leqslant 2^{-n(\mathbb{I}(\mathsf{X};\mathsf{Y}|\mathsf{U})-\delta(\epsilon))} \quad \text{for } j \neq 1. \tag{2.20}$$

On substituting (2.18), (2.19), and (2.20) into (2.17), we obtain

$$\begin{aligned}\mathbb{E}[\mathbf{P}_e(C_n)] \leqslant \delta_\epsilon(n) &+ \lceil 2^{nR_0}\rceil 2^{-n(\mathbb{I}(\mathsf{U};\mathsf{Y})-\delta(\epsilon))} + \lceil 2^{nR_0}\rceil 2^{-n(\mathbb{I}(\mathsf{U};\mathsf{Z})-\delta(\epsilon))}\\ &+ \lceil 2^{nR_1}\rceil 2^{-n(\mathbb{I}(\mathsf{X};\mathsf{Y}|\mathsf{U})-\delta(\epsilon))}.\end{aligned}$$

Hence, if we choose the rates R_0 and R_1 to satisfy

$$\begin{aligned}R_0 &< \min(\mathbb{I}(\mathsf{U};\mathsf{Y}), \mathbb{I}(\mathsf{U};\mathsf{Z})) - \delta(\epsilon),\\ R_1 &< \mathbb{I}(\mathsf{X};\mathsf{Y}|\mathsf{U}) - \delta(\epsilon),\end{aligned}$$

we obtain $\mathbb{E}[C_n] \leqslant \delta_\epsilon(n)$. By applying the selection lemma to the random variable C_n and the function $\mathbf{P}_e$, we conclude that there exists a $\left(2^{nR_0}, 2^{nR_1}, n\right)$ code $\mathcal{C}_n$ such that $\mathbf{P}_e(\mathcal{C}_n) \leqslant \delta_\epsilon(n)$. Since ϵ can be chosen arbitrarily small and since the distribution p_{UX} is arbitrary, we conclude that

$$\left\{(R_0, R_1)\colon \begin{array}{l} 0 \leqslant R_0 \leqslant \min(\mathbb{I}(\mathsf{U};\mathsf{Y}), \mathbb{I}(\mathsf{U};\mathsf{Z}))\\ 0 \leqslant R_1 \leqslant \mathbb{I}(\mathsf{X};\mathsf{Y}|\mathsf{U})\end{array}\right\} \subseteq \mathcal{C}^{\text{BC}}.$$

Since p_{UX} is arbitrary and since it is possible to perform time-sharing, the theorem follows. The bound for the cardinality of the random variable U follows from Caratheodory's theorem, and we refer the reader to [3] for details. □

2.4 Bibliographical notes

The definitions of typical sequences and their properties are those described in the textbooks [3, 4, 6]. The notion of d-separation is a known result in statistical inference, and we have used the definition provided by Kramer [8].

There are several ways of proving the channel coding theorem, see for instance [2, 3, 4, 5], and our presentation is based on the approach in [3, 4]. The Slepian–Wolf theorem was established by Slepian and Wolf in [9]. Additional examples of results proved with random binning with side information can be found in [10] and [11].

The capacity of the two-user multiple-access channel with independent messages was obtained independently by Ahlswede [12] and Liao [13]. The broadcast channel model was proposed by Cover [14]. Bergmans [15] and Gallager [16] proved that the achievable region obtained in this chapter is the capacity region of a subclass of broadcast channels called *physically degraded* broadcast channels. Surveys of known results about network information theory can be found in Cover's survey paper [17] or Kramer's monograph [6].

Part II

Information-theoretic security

3 Secrecy capacity

In this chapter, we develop the notion of *secrecy capacity*, which plays a central role in physical-layer security. The secrecy capacity characterizes the fundamental limit of secure communications over noisy channels, and it is essentially the counterpart to the usual point-to-point channel capacity when communications are subject not only to reliability constraints but also to an information-theoretic secrecy requirement. It is inherently associated with a channel model called the *wiretap channel*, which is a broadcast channel in which one of the receivers is treated as an adversary. This adversarial receiver, which we call the *eavesdropper* to emphasize its passiveness, should remain ignorant of the messages transmitted over the channel. The mathematical tools, and especially the random-coding argument, presented in this chapter are the basis for most of the theoretical research in physical-layer security, and we use them extensively in subsequent chapters.

We start with a review of Shannon's model of secure communications (Section 3.1), and then we informally discuss the problem of secure communications over noisy channels (Section 3.2). The intuition we develop from loose arguments is useful to grasp the concepts underlying the proofs of the secrecy capacity and motivates a discussion of the choice of an information-theoretic secrecy metric (Section 3.3). We then study in detail the fundamental limits of secure communication over degraded wiretap channels (Section 3.4) and broadcast channels with confidential messages (Section 3.5). We also discuss the multiplexing of secure and non-secure messages as well as the role of feedback for securing communications (Section 3.6). Finally, we conclude the chapter with a summary of the lessons learned from the analysis of fundamental limits and a review of the explicit and implicit assumptions used in the models (Section 3.7). The Gaussian wiretap channel and its extensions to multiple-input multiple-output channels and wireless channels are considered separately in Chapter 5.

3.1 Shannon's cipher system

Shannon proposed the idea of measuring quantitatively the secrecy level of encryption systems on the basis of his mathematical theory of communication. Shannon's model of secure communications, which is often called *Shannon's cipher system*, is illustrated in Figure 3.1; it considers a situation in which a transmitter communicates with a legitimate receiver over a noiseless channel, while an eavesdropper overhears all signals sent over

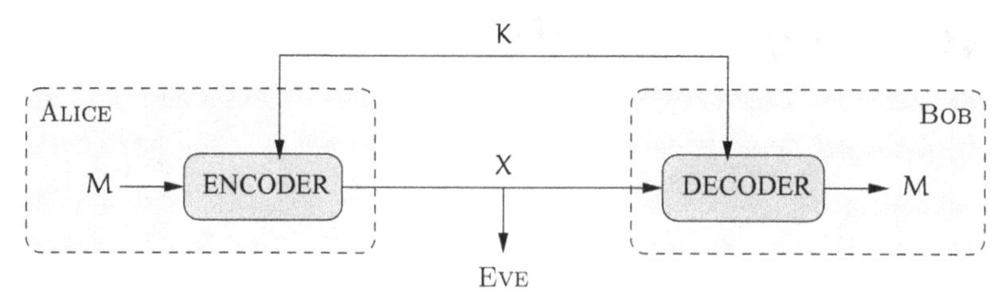

Figure 3.1 Shannon's cipher system.

the channel. To prevent the eavesdropper from retrieving information, the transmitter encodes his messages into codewords by means of a secret key, which is known to the legitimate receiver but unknown to the eavesdropper.[1] Messages, codewords, and keys are represented by the random variables $\mathsf{M} \in \mathcal{M}$, $\mathsf{X} \in \mathcal{X}$, and $\mathsf{K} \in \mathcal{K}$, respectively, and we assume that K is independent of M. The encoding function is denoted by $e : \mathcal{M} \times \mathcal{K} \to \mathcal{X}$, the decoding function is denoted by $d : \mathcal{X} \times \mathcal{K} \to \mathcal{M}$, and we refer to the pair (e, d) as a *coding scheme.* The legitimate receiver is assumed to retrieve messages without error, that is

$$\mathsf{M} = d(\mathsf{X}, \mathsf{K}) \quad \text{if} \quad \mathsf{X} = e(\mathsf{M}, \mathsf{K}).$$

Although the eavesdropper has no knowledge about the secret key, he is assumed to know the encoding function e and the decoding function d.

To measure secrecy with respect to Eve in terms of an information-theoretic quantity, it is natural to consider the conditional entropy $\mathbb{H}(\mathsf{M}|\mathsf{X})$, which we call the eavesdropper's *equivocation*. Intuitively, the equivocation represents Eve's uncertainty about the messages after intercepting the codewords. A coding scheme is said to achieve *perfect secrecy* if

$$\mathbb{H}(\mathsf{M}|\mathsf{X}) = \mathbb{H}(\mathsf{M}) \qquad \text{or, equivalently,} \qquad \mathbb{I}(\mathsf{M};\mathsf{X}) = 0.$$

We call the quantity $\mathbb{I}(\mathsf{M};\mathsf{X})$ the *leakage* of information to the eavesdropper. In other words, perfect secrecy is achieved if codewords X are statistically independent of messages M. This definition of security differs from the traditional assessment based on computational complexity not only because it provides a *quantitative* metric to measure secrecy but also because it disregards the computational power of the eavesdropper. Perfect secrecy guarantees that the eavesdropper's optimal attack is to guess the message M at random and that there exists no algorithm that could extract any information about M from X.

Proposition 3.1. *If a coding scheme for Shannon's cipher system achieves perfect secrecy, then*

$$\mathbb{H}(\mathsf{K}) \geqslant \mathbb{H}(\mathsf{M}).$$

[1] In cryptography, it is customary to call a message a *plaintext*, and a codeword a *ciphertext* or a *cryptogram*. We adopt instead the nomenclature prevalent in information theory and coding theory.

Proof. Consider a coding scheme that achieves perfect secrecy. By assumption, $\mathbb{H}(\mathsf{M}|\mathsf{X}) = \mathbb{H}(\mathsf{M})$; in addition, since messages M are decoded without errors upon observing X and K, Fano's inequality also ensures that $\mathbb{H}(\mathsf{M}|\mathsf{XK}) = 0$. Consequently,

$$\begin{aligned}\mathbb{H}(\mathsf{K}) &\overset{(a)}{\geqslant} \mathbb{H}(\mathsf{K}) - \mathbb{H}(\mathsf{K}|\mathsf{XM}) \\ &\overset{(b)}{\geqslant} \mathbb{H}(\mathsf{K}|\mathsf{X}) - \mathbb{H}(\mathsf{K}|\mathsf{XM}) \\ &= \mathbb{I}(\mathsf{K};\mathsf{M}|\mathsf{X}) \\ &= \mathbb{H}(\mathsf{M}|\mathsf{X}) - \mathbb{H}(\mathsf{M}|\mathsf{KX}) \\ &= \mathbb{H}(\mathsf{M}|\mathsf{X}) \\ &= \mathbb{H}(\mathsf{M}).\end{aligned}$$

Inequality (a) follows from $\mathbb{H}(\mathsf{K}|\mathsf{XM}) \geqslant 0$ and inequality (b) follows from $\mathbb{H}(\mathsf{K}) \geqslant \mathbb{H}(\mathsf{K}|\mathsf{X})$ because conditioning does not increase entropy. □

In other words, Proposition 3.1 states that it is necessary to use at least one secret-key bit for each message bit to achieve perfect secrecy. If the number of possible messages, keys, and codewords is the same, it is possible to obtain a more precise result and to establish necessary and sufficient conditions for communication in perfect secrecy.

Theorem 3.1. *If $|\mathcal{M}| = |\mathcal{X}| = |\mathcal{K}|$, a coding scheme for Shannon's cipher system achieves perfect secrecy if and only if*

- *for each pair $(m, x) \in \mathcal{M} \times \mathcal{X}$, there exists a unique key $k \in \mathcal{K}$ such that $x = e(m, k)$;*
- *the key K is uniformly distributed in $\mathcal{K}$.*

Proof. First, we establish that the conditions of Theorem 3.1 are necessary. Consider a coding scheme that achieves perfect secrecy with $|\mathcal{M}| = |\mathcal{X}| = |\mathcal{K}|$. Note that $p_{\mathsf{X}}(x) > 0$ for all $x \in \mathcal{X}$, otherwise some codewords would never be used and could be removed from $\mathcal{X}$, which would violate the assumption $|\mathcal{M}| = |\mathcal{X}|$. Since M and X are independent, this implies $p_{\mathsf{X}|\mathsf{M}}(x|m) = p_{\mathsf{X}}(x) > 0$ for all pairs $(m, x) \in \mathcal{M} \times \mathcal{X}$. In other words, for all messages $m \in \mathcal{M}$, the encoder can output all possible codewords in $\mathcal{X}$; therefore,

$$\forall m \in \mathcal{M} \quad \mathcal{X} = \{e(m, k) : k \in \mathcal{K}\}.$$

Because we have assumed $|\mathcal{X}| = |\mathcal{K}|$, for all pairs $(m, x) \in \mathcal{M} \times \mathcal{X}$ there must be a unique key $k \in \mathcal{K}$ such that $x = e(m, k)$. Now, fix an arbitrary codeword $x^* \in \mathcal{X}$. For every message $m \in \mathcal{M}$, let k_m be the unique key such that $x^* = e(m, k_m)$. Then $p_{\mathsf{K}}(k_m) = p_{\mathsf{X}|\mathsf{M}}(x^*|m)$ and $\mathcal{K} = \{k_m : m \in \mathcal{M}\}$. Using Bayes' rule, we obtain

$$p_{\mathsf{K}}(k_m) = p_{\mathsf{X}|\mathsf{M}}(x^*|m) = \frac{p_{\mathsf{M}|\mathsf{X}}(m|x^*)\, p_{\mathsf{X}}(x^*)}{p_{\mathsf{M}}(m)} = p_{\mathsf{X}}(x^*),$$

where the last equality follows from $p_{\mathsf{M}|\mathsf{X}}(m|x^*) = p_{\mathsf{M}}(m)$ by virtue of the independence of M and X. Therefore, $p_{\mathsf{K}}(k_m)$ takes on the same value for all $m \in \mathcal{M}$, which implies that K is uniformly distributed in $\mathcal{K}$.

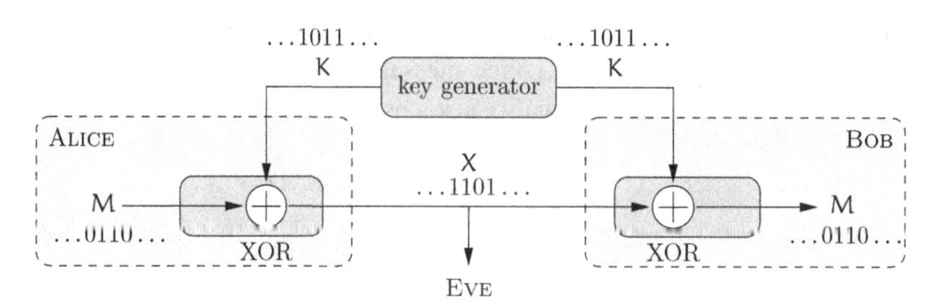

Figure 3.2 Vernam's cipher (one-time pad) illustrated for $\mathcal{M} = \{0, 1\}$.

We now show that the conditions of Theorem 3.1 are also sufficient. Since $|\mathcal{M}| = |\mathcal{X}| = |\mathcal{K}|$, we can assume without loss of generality that $\mathcal{M} = \mathcal{X} = \mathcal{K} = [\![0, |\mathcal{M}| - 1]\!]$. Consider now the coding scheme illustrated in Figure 3.2, called a *Vernam cipher* or *one-time pad*. To send a message $m \in \mathcal{M}$, Alice transmits $x = m \oplus k$, where k is the realization of a key K, which is independent of the message and with uniform distribution on $\mathcal{M}$, and $\oplus$ is the modulo-$|\mathcal{M}|$ addition. Since k is known to Bob, he can decode the message m from the codeword x without error by computing

$$x \oplus k = m \oplus k \ominus k = m,$$

where $\ominus$ is the modulo-$|\mathcal{M}|$ subtraction. In addition, this encoding procedure guarantees that, for all $x \in \mathcal{X}$,

$$p_{\mathsf{X}}(x) = \sum_{k \in \mathcal{M}} p_{\mathsf{X}|\mathsf{K}}(x|k) p_{\mathsf{K}}(k) = \sum_{k \in \mathcal{M}} p_{\mathsf{M}}(x \oplus k) \frac{1}{|\mathcal{M}|} = \frac{1}{|\mathcal{M}|},$$

and, consequently,

$$\begin{aligned} \mathbb{I}(\mathsf{M};\mathsf{X}) &= \mathbb{H}(\mathsf{X}) - \mathbb{H}(\mathsf{X}|\mathsf{M}) \\ &\overset{(a)}{=} \mathbb{H}(\mathsf{X}) - \mathbb{H}(\mathsf{K}|\mathsf{M}) \\ &\overset{(b)}{=} \mathbb{H}(\mathsf{X}) - \mathbb{H}(\mathsf{K}) \\ &= \log|\mathcal{M}| - \log|\mathcal{M}| \\ &= 0, \end{aligned}$$

where (a) follows from $\mathbb{H}(\mathsf{X}|\mathsf{M}) = \mathbb{H}(\mathsf{K}|\mathsf{M})$ because there is a one-to-one mapping between X and K given M and (b) follows from $\mathbb{H}(\mathsf{K}|\mathsf{M}) = \mathbb{H}(\mathsf{K})$ because M and K are independent. Notice that this result holds for any probability distribution of the message p_{M} for which $\forall m \in \mathcal{M} p_{\mathsf{M}}(m) > 0$. □

The fact that a one-time pad guarantees perfect secrecy is a result usually referred to as the "crypto lemma," which holds under very general conditions; in particular, the finite alphabet $\mathcal{M}$ can be replaced by a compact abelian group $\mathcal{G}$.[2]

[2] An abelian group $\mathcal{G}$ is a commutative group that need not be finite. The assumption that $\mathcal{G}$ is compact guarantees that its Haar measure is finite so that it is possible to define a uniform probability distribution over $\mathcal{G}$.

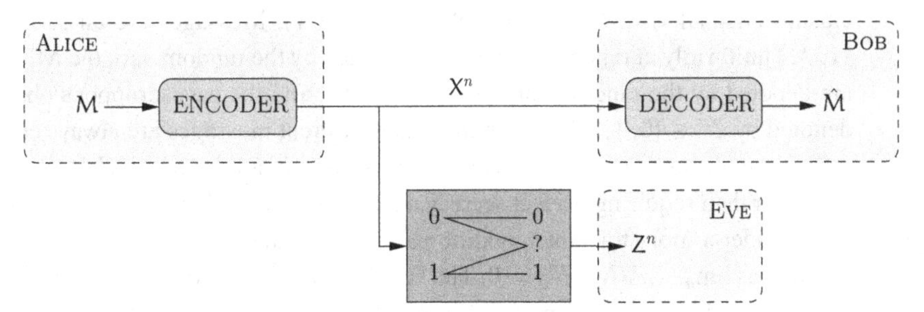

Figure 3.3 Communication over a binary erasure wiretap channel.

Lemma 3.1 (Crypto lemma, Forney). *Let $(\mathcal{G}, +)$ be a compact abelian group with binary operation $+$ and let $X = M + K$, where M and K are random variables over $\mathcal{G}$ and K is independent of M and uniform over $\mathcal{G}$. Then X is independent of M and uniform over $\mathcal{G}$.*

Although Theorem 3.1 shows the existence of coding schemes that achieve perfect secrecy, it provides an unsatisfactory result. In fact, since a one-time pad requires a new key bit for each message bit, it essentially replaces the problem of secure communication by that of secret-key distribution. Nevertheless, we show in the next sections that this disappointing result stems from the absence of noise at the physical layer in the model; in particular, Shannon's cipher system does not take into account the noise affecting the eavesdropper's observation of the codewords.

Remark 3.1. *Requiring perfect secrecy is much more stringent than preventing the eavesdropper from decoding correctly. To see this, assume for simplicity that messages are taken from the set $[\![1, M]\!]$ and that each of them is equally likely, in which case the eavesdropper minimizes his probability of decoding error $\mathbf{P}_e$ by performing maximum-likelihood decoding. Since the a-priori distribution of the message M is uniform over $[\![1, M]\!]$, the condition $\mathbb{H}(M|X) = \mathbb{H}(M)$ ensures that $p_{M|X}(m|x) = 1/M$ for all messages $m \in \mathcal{M}$ and codewords $x \in \mathcal{M}$ or, equivalently, that the probability of error under maximum-likelihood decoding is $\mathbf{P}_e = (M-1)/M$. In contrast, evaluating secrecy in terms of the non-decodability of the messages would merely guarantee that the probability of error under maximum-likelihood decoding is bounded away from zero, that is $\mathbf{P}_e > \epsilon$ for some fixed $\epsilon > 0$.*

3.2 Secure communication over a noisy channel

Before we study secrecy capacity in detail, it is instructive to consider the effect of noise with the simple model illustrated in Figure 3.3, which is called a *binary erasure wiretap channel*. This channel is a special case of more general models that are studied in Section 3.4 and Section 3.5. Here, a transmitter communicates messages to a legitimate receiver by sending binary codewords of length n over a noiseless channel, while an eavesdropper observes a corrupted version of these codewords at the output of a binary erasure channel with erasure probability $\epsilon \in (0, 1)$. Messages are taken from the set $[\![1, M]\!]$ uniformly at random, and are represented by the random variable M. Codewords are denoted by the random variable $X^n \in \{0, 1\}^n$ and the eavesdropper's observation is denoted by $Z^n \in \{0, 1, ?\}^n$. We assume that different messages are always encoded into different codewords, so that the reliable transmission rate is $(1/n)\mathbb{H}(M) = (1/n)\log M$.

Rather than requiring perfect secrecy and exact statistical independence of M and X^n, we consider a more tractable condition and we say that a coding scheme is secure if it guarantees $\lim_{n\to\infty} \mathbb{I}(M; Z^n) = 0$. The key difficulty is now that of how to determine the type of encoder that could enforce this condition. To obtain some insight, we consider a specific coding scheme for the model in Figure 3.3.

Example 3.1. Assume that messages are taken uniformly at random from the set $[\![1, 2]\!]$ so that $\mathbb{H}(M) = 1$, and let n be arbitrary. Let $\mathcal{C}_1$ be the set of binary sequences of length n with odd parity and let $\mathcal{C}_2$ be the set of binary sequences of length n with even parity. To send a message $m \in \{1, 2\}$, the emitter transmits a sequence x^n chosen uniformly at random in $\mathcal{C}_m$. The rate of the coding scheme is simply $1/n$. Now, assume that the eavesdropper observes a sequence Z^n with k erasures. If $k > 0$, the parity of the erased bits is just as likely to be even as it is to be odd. If $k = 0$, the eavesdropper knows perfectly which codeword was sent and thus knows its parity. To analyze the eavesdropper's equivocation formally, we introduce the random variable $E \in \{0, 1\}$ such that

$$E = \begin{cases} 0 & \text{if } Z^n \text{ contains no erasure;} \\ 1 & \text{otherwise.} \end{cases}$$

We can then lower bound the equivocation as

$$\begin{aligned} \mathbb{H}(M|Z^n) &\geqslant \mathbb{H}(M|Z^n E) \\ &\overset{(a)}{=} \mathbb{H}(M|Z^n E = 1)(1 - (1 - \epsilon)^n) \\ &= \mathbb{H}(M)(1 - (1 - \epsilon)^n) \\ &\overset{(b)}{=} \mathbb{H}(M) - (1 - \epsilon)^n. \end{aligned}$$

Equality (a) follows from the fact that $\mathbb{H}(M|Z^n E = 0) = 0$ and equality (b) follows from $\mathbb{H}(M) = 1$. Hence, we obtain

$$\mathbb{I}(M; Z^n) = \mathbb{H}(M) - \mathbb{H}(M|Z^n) \leqslant (1 - \epsilon)^n,$$

which vanishes exponentially fast with n; therefore, the coding scheme is secure.

In practice, the coding scheme of Example 3.1 is not really useful because the code rate vanishes with n as well, albeit more slowly than does $\mathbb{I}(M; Z^n)$; nevertheless, the example suggests that assigning multiple codewords to every message and selecting them randomly is useful to *confuse* the eavesdropper and to guarantee secrecy.

3.3 Perfect, weak, and strong secrecy

As mentioned in the previous section, the notion of perfect secrecy is too stringent and is not easily amenable to further analysis. It is convenient to replace the requirement of exact statistical independence between messages M and the eavesdropper's observations Z^n by *asymptotic* statistical independence as the codeword length n goes to infinity. In principle, this asymptotic independence can be measured in terms of any distance d defined on the set of joint probability distributions on $\mathcal{M} \times \mathcal{Z}^n$ as

$$\lim_{n\to\infty} d(p_{\mathsf{M}\mathsf{Z}^n}, p_{\mathsf{M}} p_{\mathsf{Z}^n}) = 0.$$

For instance, in the previous section we implicitly used the Kullback–Leibler divergence[3] and we required

$$\lim_{n\to\infty} \mathbb{D}(p_{\mathsf{M}\mathsf{Z}^n} \| p_{\mathsf{M}} p_{\mathsf{Z}^n}) = \lim_{n\to\infty} \mathbb{I}(\mathsf{M}; \mathsf{Z}^n) = 0.$$

This condition, which we call the *strong secrecy condition*, requires the amount of information leaked to the eavesdropper to vanish. For technical purposes, it is also convenient to consider the condition

$$\lim_{n\to\infty} \frac{1}{n} \mathbb{I}(\mathsf{M}; \mathsf{Z}^n) = 0,$$

which requires only the *rate* of information leaked to the eavesdropper to vanish. This condition is weaker than strong secrecy since it is satisfied as long as $\mathbb{I}(\mathsf{M}; \mathsf{Z}^n)$ grows at most sub-linearly with n. We call it the *weak secrecy condition*.

From an information-theoretic perspective, the specific measure of asymptotic statistical independence may seem irrelevant, and we may be tempted to choose a metric solely on the basis of its mathematical tractability; unfortunately, the weak secrecy condition and the strong secrecy condition are not equivalent and, more importantly, it is possible to construct examples of coding schemes with evident security flaws that satisfy the weak secrecy condition.

Example 3.2. Let $n \geqslant 1$ and $t \triangleq \lfloor \sqrt{n} \rfloor$. Suppose that Alice encodes message bits $\mathsf{M}^n \in \{0, 1\}^n$ into a codeword $\mathsf{X}^n \in \{0, 1\}^n$ with $n - t$ secret-key bits $\mathsf{K}^{n-t} \in \{0, 1\}^{n-t}$ as

$$\mathsf{X}_i = \begin{cases} \mathsf{M}_i \oplus \mathsf{K}_i & \text{for } i \in [\![1, n-t]\!], \\ \mathsf{M}_i & \text{for } i \in [\![n-t+1, n]\!]. \end{cases}$$

The key bits K_i for $i \in [\![1, n-t]\!]$ are assumed i.i.d. according to $\mathcal{B}(\frac{1}{2})$ and known to Bob. In other words, Alice performs a one-time pad of the first $n - t$ bits of M with the $n - t$ key bits and she appends the remaining t bits unprotected. Eve is assumed to intercept the codeword X^n directly.

[3] Strictly speaking, the Kullback–Leibler divergence is not a distance because it is not symmetric; nevertheless, $\mathbb{D}(p_{\mathsf{M}\mathsf{Z}^n} \| p_{\mathsf{M}} p_{\mathsf{Z}^n}) = 0$ if and only if M is independent of Z^n, and we ignore this subtlety.

Using the crypto lemma, we obtain

$$\forall n \geqslant 1 \qquad \mathbb{H}(\mathsf{M}|\mathsf{X}^n) = n - t = \mathbb{H}(\mathsf{M}) - t.$$

Therefore, $\mathbb{I}(\mathsf{M};\mathsf{X}^n) = t = \lfloor\sqrt{n}\rfloor$ and this scheme does not satisfy the strong secrecy criterion; even worse, the information leaked to the eavesdropper grows unbounded with n. However, notice that

$$\lim_{n\to\infty}\frac{1}{n}\mathbb{I}(\mathsf{M};\mathsf{X}^n) = \lim_{n\to\infty}\frac{\lfloor\sqrt{n}\rfloor}{n} = 0.$$

Hence, this scheme satisfies the weak secrecy criterion.

Example 3.3. Suppose that Alice encodes messages $\mathsf{M} = (\mathsf{M}_1 \dots \mathsf{M}_n)$ uniformly distributed on $\{0,1\}^n$ into codewords $\mathsf{X}^n \in \{0,1\}^n$ with secret keys $\mathsf{K}^n \in \{0,1\}^n$ as

$$\mathsf{X}_i = \mathsf{M}_i \oplus \mathsf{K}_i \quad \text{for } i \in [\![1,n]\!].$$

The secret key K^n, which we assume is known to Bob, is such that the all-zero n-bit sequence $\bar{0}^n$ has probability $1/n$ and all non-zero sequences are equally likely. Formally, the probability distribution of the secret key is

$$p_{\mathsf{K}^n}(k^n) = \begin{cases} \dfrac{1}{n} & \text{if } k^n = \bar{0}^n, \\ \dfrac{1-1/n}{2^n-1} & \text{if } k^n \neq \bar{0}^n. \end{cases}$$

Since K^n is not uniformly distributed, this encryption scheme no longer guarantees perfect secrecy. As in the previous example, we assume that Eve directly intercepts X^n.

We first prove that this scheme satisfies the weak secrecy criterion. We introduce a random variable J such that

$$\mathsf{J} \triangleq \begin{cases} 0 & \text{if } \mathsf{K}^n = \bar{0}^n, \\ 1 & \text{otherwise.} \end{cases}$$

Since conditioning does not increase entropy, we can write

$$\begin{aligned} \mathbb{H}(\mathsf{M}|\mathsf{X}^n) &\geqslant \mathbb{H}(\mathsf{M}|\mathsf{X}^n\mathsf{J}) \\ &= \mathbb{H}(\mathsf{M}|\mathsf{X}^n, \mathsf{J}=0)p_{\mathsf{J}}(0) + \mathbb{H}(\mathsf{M}|\mathsf{X}^n, \mathsf{J}=1)p_{\mathsf{J}}(1). \end{aligned} \tag{3.1}$$

By definition, $\mathsf{K}^n = \bar{0}^n$ if $\mathsf{J} = 0$; hence, $\mathbb{H}(\mathsf{M}|\mathsf{X}^n, \mathsf{J}=0) = 0$ and we can restrict our attention to the term

$$\mathbb{H}(\mathsf{M}|\mathsf{X}^n, \mathsf{J}=1)p_{\mathsf{J}}(1) = -\sum_{m,x^n} p(m, x^n, j=1)\log p(m|x^n, j=1).$$

For any $m \in \{0,1\}^n$ and $x^n \in \{0,1\}^n$, the joint probability $p(m, x^n, j=1)$ can be written as

$$p(x^n, m, j=1) = p(m|x^n, j=1)p(x^n|j=1)p(j=1)$$

with

$$p(m|x^n, j = 1) = \begin{cases} 0 & \text{if } m = x^n, \\ 1/(2^n - 1) & \text{otherwise,} \end{cases}$$

$$p(x^n|j = 1) = \frac{1}{2^n},$$

$$p(j = 1) = 1 - \frac{1}{n}.$$

On substituting these values into (3.1), we obtain

$$\begin{aligned}
\mathbb{H}(\mathsf{M}|\mathsf{X}^n) &\geqslant -\sum_{x^n}\sum_{m\neq x^n} \frac{1}{2^n - 1}\frac{1}{2^n}\left(1 - \frac{1}{n}\right)\log\left(\frac{1}{2^n - 1}\right) \\
&= -\left(1 - \frac{1}{n}\right)\log\left(\frac{1}{2^n - 1}\right) \\
&= \log(2^n - 1) - \frac{\log(2^n - 1)}{n} \\
&\geqslant \log(2^n - 1) - 1.
\end{aligned}$$

Since $\mathbb{H}(\mathsf{M}) = n$, we obtain

$$\begin{aligned}
\lim_{n\to\infty} \frac{1}{n}\mathbb{I}(\mathsf{M};\mathsf{X}^n) &= 1 - \lim_{n\to\infty} \frac{1}{n}\mathbb{H}(\mathsf{M}|\mathsf{X}^n) \\
&\leqslant 1 - \lim_{n\to\infty} \frac{\log(2^n - 1) - 1}{n} \\
&= 0.
\end{aligned}$$

Hence, this scheme satisfies the weak secrecy criterion. However,

$$\begin{aligned}
\mathbb{H}(\mathsf{M}|\mathsf{X}^n) &= \mathbb{H}(\mathsf{X}^n \oplus \mathsf{K}|\mathsf{X}^n) \\
&= \mathbb{H}(\mathsf{K}|\mathsf{X}^n) \\
&\leqslant \mathbb{H}(\mathsf{K}) \\
&= -\frac{1}{n}\log\left(\frac{1}{n}\right) - (2^n - 1)\cdot\frac{1 - 1/n}{2^n - 1}\log\left(\frac{1 - 1/n}{2^n - 1}\right) \\
&= \mathbb{H}_b(1/n) + (1 - 1/n)\log(2^n - 1) \\
&< \mathbb{H}_b(1/n) + n - 1 \\
&< n - 0.5 \quad \text{for } n \text{ large enough.}
\end{aligned}$$

Therefore, $\lim_{n\to\infty} \mathbb{I}(\mathsf{M};\mathsf{X}^n) > 0.5$, and this scheme does not satisfy the strong secrecy criterion.

One could argue that Example 3.2 and Example 3.3 have been constructed *ad hoc* to exhibit flaws. In Example 3.2, the eavesdropper always obtains a fraction of the

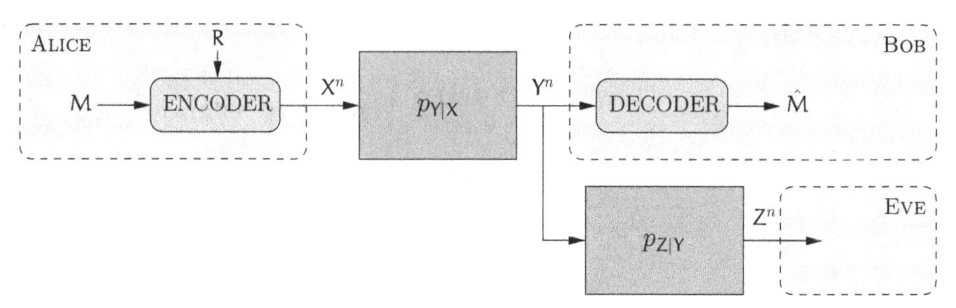

Figure 3.4 Communication over a DWTC. M represents the message to be transmitted securely, whereas R represents randomness used to randomize the encoder.

message bits without errors; in Example 3.3, the distribution of the key is skewed in such a way that the all-zero key is with overwhelming probability more likely than any other. Therefore, these examples do not imply that all weakly secure schemes are useless, but merely suggest that not all measures of asymptotic statistical independence are meaningful from a cryptographic perspective. In particular, this is a good indication that the weak secrecy criterion is likely not appropriate and, consequently, we should strive to prove all results with a strong secrecy criterion.

3.4 Wyner's wiretap channel

Secrecy capacity was originally introduced by Wyner for a channel model called a *degraded wiretap channel* (DWTC for short). Although this model is a special case of the broadcast channel with confidential messages studied in Section 3.5, it allows us to introduce many of the mathematical tools of information-theoretic security without the additional complexity of fully general models. As illustrated in Figure 3.4, a DWTC models a situation in which a sender (Alice) tries to communicate with a legitimate receiver (Bob) over a noisy channel, while an eavesdropper (Eve) observes a *degraded* version of the signal obtained by the legitimate receiver.

Formally, a discrete memoryless DWTC $\left(\mathcal{X}, p_{Z|Y}\, p_{Y|X}, \mathcal{Y}, \mathcal{Z}\right)$ consists of a finite input alphabet $\mathcal{X}$, two finite output alphabets $\mathcal{Y}$ and $\mathcal{Z}$, and transition probabilities $p_{Y|X}$ and $p_{Z|Y}$ such that

$$\forall n \geqslant 1 \quad \forall (x^n, y^n, z^n) \in \mathcal{X}^n \times \mathcal{Y}^n \times \mathcal{Z}^n$$

$$p_{Y^nZ^n|X^n}(y^n, z^n|x^n) = \prod_{i=1}^{n} p_{Y|X}(y_i|x_i)p_{Z|Y}(z_i|y_i). \quad (3.2)$$

The DMC $\left(\mathcal{X}, p_{Y|X}, \mathcal{Y}\right)$ characterized by the marginal transition probabilities $p_{Y|X}$ is referred to as the *main channel*, while the DMC $\left(\mathcal{X}, p_{Z|X}, \mathcal{Z}\right)$ characterized by the marginal transition probabilities $p_{Z|X}$ is referred to as the *eavesdropper's channel*. The eavesdropper is sometimes called the *wiretapper*, and, accordingly, its channel is called the *wiretapper's channel*, but we avoid this terminology because it makes limited sense for the wireless channels discussed in Chapter 5. Throughout this book, we always assume that the transmitter, the receiver, and the eavesdropper know the channel statistics ahead of time.

As hinted in Section 3.2, randomness in the encoding process is what enables secure communications. It is convenient to represent this randomness by the realization of a DMS $(\mathcal{R}, p_{\mathrm{R}})$, which is independent of the channel and of the messages to be transmitted. Because the source is available to Alice but not to Bob and Eve, we call it a source of *local randomness*.

Definition 3.1. *A $\left(2^{nR}, n\right)$ code $\mathcal{C}_n$ for a DWTC consists of*

- *a message set $\mathcal{M} = \llbracket 1, 2^{nR} \rrbracket$;*
- *a source of local randomness at the encoder $(\mathcal{R}, p_{\mathrm{R}})$;*
- *an encoding function $f : \mathcal{M} \times \mathcal{R} \to \mathcal{X}^n$, which maps a message m and a realization of the local randomness r to a codeword x^n;*
- *a decoding function $g : \mathcal{Y}^n \to \mathcal{M} \cup \{?\}$, which maps each channel observation y^n to a message $\hat{m} \in \mathcal{M}$ or an error message ?.*

Note that the DMS $(\mathcal{R}, p_{\mathrm{R}})$ is included in the definition because we implicitly assume that it can be optimized as part of the code design. The $(2^{nR}, n)$ code $\mathcal{C}_n$ is assumed known by Alice, Bob, and Eve, and this knowledge includes the statistics of the DMS $(\mathcal{R}, p_{\mathrm{R}})$; however, the realizations of the DMS used for encoding are accessible only to Alice. We also assume that the message M is uniformly distributed in $\mathcal{M}$, so that the code rate is $(1/n)\mathbb{H}(\mathrm{M}) = R + \delta(n)$. The reliability performance of $\mathcal{C}_n$ is measured in terms of its average probability of error

$$\mathbf{P}_{\mathrm{e}}(\mathcal{C}_n) \triangleq \mathbb{P}\left[\hat{\mathrm{M}} \neq \mathrm{M} \mid \mathcal{C}_n\right],$$

while its secrecy performance is measured in terms of the equivocation

$$\mathbf{E}(\mathcal{C}_n) \triangleq \mathbb{H}(\mathrm{M}|\mathrm{Z}^n \mathcal{C}_n).$$

We emphasize that equivocation is conditioned on the code $\mathcal{C}_n$ because the eavesdropper knows the code ahead of time. Equivalently, the secrecy performance of the code $\mathcal{C}_n$ can be measured in terms of the leakage

$$\mathbf{L}(\mathcal{C}_n) \triangleq \mathbb{I}(\mathrm{M}; \mathrm{Z}^n|\mathcal{C}_n),$$

which measures the information leaked to the eavesdropper instead of the uncertainty of the eavesdropper.

Remark 3.2. *In the literature, it is common to introduce the local randomness implicitly by considering a* stochastic *encoder $f : \mathcal{M} \to \mathcal{X}^n$, which maps a message $m \in \mathcal{M}$ to a codeword $x^n \in \mathcal{X}^n$ according to transition probabilities $p_{\mathrm{X}^n|\mathrm{M}}$.*

Remark 3.3. *Stochastic encoding is crucial to enable secure communications but there is no point in considering a stochastic decoder. To see this, consider a stochastic decoder that maps each channel observation $y^n \in \mathcal{Y}^n$ to a symbol $v \in \mathcal{V}$ according* *to transition probabilities $p_{V|Y^n}$, where $\mathcal{V}$ is an arbitrary alphabet. From the data-processing inequality, we have $\mathbb{I}(M; V) \leqslant \mathbb{I}(M; Y^n)$; therefore, according to the channel coding theorem, stochastic decoding can only reduce the rate of reliable transmission over the main channel, while having no effect on the eavesdropper's equivocation.*

Definition 3.2. *A weak rate–equivocation pair (R, R_e) is achievable for the DWTC if there exists a sequence of $(2^{nR}, n)$ codes $\{\mathcal{C}_n\}_{n\geqslant 1}$ such that*

$$\lim_{n\to\infty} \mathbf{P}_e(\mathcal{C}_n) = 0 \qquad \textit{(reliability condition)}, \tag{3.3}$$

$$\varliminf_{n\to\infty} \frac{1}{n}\mathbf{E}(\mathcal{C}_n) \geqslant R_e \qquad \textit{(weak secrecy condition)}. \tag{3.4}$$

The weak rate–equivocation region of a DWTC is

$$\mathcal{R}^{\text{DWTC}} \triangleq \text{cl}(\{(R, R_e) : (R, R_e) \text{ is achievable}\}),$$

and the weak secrecy capacity of a DWTC is

$$C_s^{\text{DWTC}} \triangleq \sup_R \{R : (R, R) \in \mathcal{R}^{\text{DWTC}}\}.$$

Remark 3.4. *According to our definition, if a rate–equivocation pair (R, R_e) is achievable, then any pair (R, R_e') with $R_e' \leqslant R_e$ is achievable as well. In particular, note that $(R, 0)$ is always achievable.*

The rate–equivocation region $\mathcal{R}^{\text{DWTC}}$ encompasses rate–equivocation pairs for which R_e is not equal to R; it characterizes the equivocation rate that can be guaranteed for an arbitrary rate R. If a pair (R, R_e) with $R_e = R$ is achievable, we say that R is a *full secrecy* rate. In this case, notice that a sequence of $(2^{nR}, n)$ codes $\{\mathcal{C}_n\}_{n\geqslant 1}$ that achieves a full secrecy rate satisfies

$$\lim_{n\to\infty} \frac{1}{n}\mathbf{L}(\mathcal{C}_n) = 0.$$

Full secrecy is of practical importance because the messages transmitted are then entirely hidden from the eavesdropper. In the literature, full secrecy is sometimes called "perfect secrecy." In this book, the term "perfect secrecy" is restricted to Shannon's definition of information-theoretic security, which requires *exact* statistical independence.

The secrecy condition (3.4) is weak because it is based on the equivocation rate $(1/n)\mathbf{E}(\mathcal{C}_n)$. As discussed in Section 3.3, it would be preferable to use a stronger condition and to rely on the following definition.

Definition 3.3. *A strong rate–equivocation pair (R, R_e) is achievable for the DWTC if there exists a sequence of $(2^{nR}, n)$ codes $\{\mathcal{C}_n\}_{n\geqslant 1}$ such that*

$$\lim_{n\to\infty} \mathbf{P}_e(\mathcal{C}_n) = 0 \qquad \textit{(reliability condition)}, \tag{3.5}$$

$$\varliminf_{n\to\infty} (\mathbf{E}(\mathcal{C}_n) - nR_e) \geqslant 0 \qquad \textit{(strong secrecy condition)}. \tag{3.6}$$

The strong rate–equivocation region of a DWTC is

$$\overline{\mathcal{R}}^{\text{DWTC}} \triangleq \text{cl}(\{(R, R_e) : (R, R_e) \textit{ is achievable}\}),$$

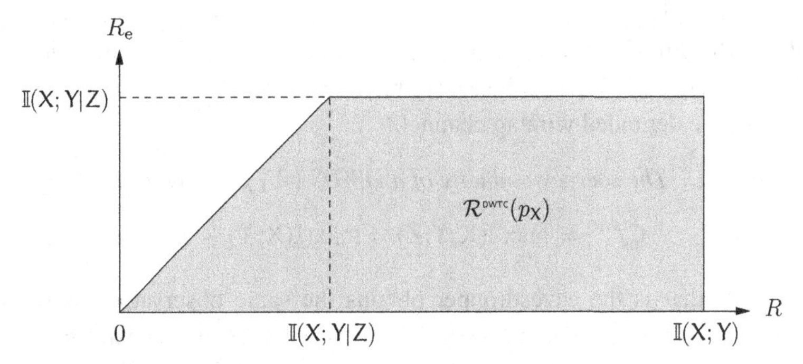

Figure 3.5 Typical shape of the rate–equivocation region $\mathcal{R}^{\text{DWTC}}(p_X)$.

and the strong secrecy capacity of a DWTC is

$$\overline{C}_s^{\text{DWTC}} \triangleq \sup_R \left\{ R : (R, R) \in \overline{\mathcal{R}}^{\text{DWTC}} \right\}.$$

Unfortunately, directly dealing with the stronger condition (3.6) is more arduous than dealing with the weak secrecy condition (3.4). Moreover, we will show in Section 4.5 that $\mathcal{R}^{\text{DWTC}} = \overline{\mathcal{R}}^{\text{DWTC}}$ and $C_s^{\text{DWTC}} = \overline{C}_s^{\text{DWTC}}$; therefore, we will content ourselves with (3.4) for now, but the reader should keep in mind that this is mainly for mathematical tractability.

It is not a priori obvious whether the reliability condition (3.3) and the secrecy condition (3.4) can be satisfied simultaneously. On the one hand, reliability calls for the introduction of redundancy to mitigate the effect of channel noise; on the other hand, creating too much redundancy is likely to jeopardize secrecy. Perhaps surprisingly, the balance between reliability and secrecy can be precisely controlled with appropriate coding schemes and the rate–equivocation region can be characterized exactly.

Theorem 3.2 (Wyner). *Consider a DWTC $\left(\mathcal{X}, p_{Z|Y}\, p_{Y|X}, \mathcal{Y}, \mathcal{Z}\right)$. For any distribution p_X on $\mathcal{X}$, define the set $\mathcal{R}^{\text{DWTC}}(p_X)$ as*

$$\mathcal{R}^{\text{DWTC}}(p_X) \triangleq \left\{ (R, R_e) \colon \begin{array}{l} 0 \leqslant R_e \leqslant R \leqslant \mathbb{I}(X;Y) \\ 0 \leqslant R_e \leqslant \mathbb{I}(X;Y|Z) \end{array} \right\},$$

where the joint distribution of X, Y, *and* Z *factorizes as* $p_X p_{Y|X} p_{Z|Y}$. *Then, the rate–equivocation region of the DWTC is the convex region*

$$\mathcal{R}^{\text{DWTC}} = \bigcup_{p_X} \mathcal{R}^{\text{DWTC}}(p_X). \tag{3.7}$$

The typical shape of $\mathcal{R}^{\text{DWTC}}(p_X)$ is illustrated in Figure 3.5. At transmission rates below $\mathbb{I}(X;Y|Z)$, it is always possible to find codes achieving full secrecy rates. It is also possible to transmit at rates above $\mathbb{I}(X;Y|Z)$, but the equivocation rate saturates at $R_e = \mathbb{I}(X;Y|Z)$, and there is no secrecy guaranteed for the remaining fraction of the rate.

Remark 3.5. *In Wyner's original work, the equivocation rate is defined per source symbol as $\Delta \triangleq (1/k)\mathbb{H}(M|Z^n C_n)$ with $k = \log\lceil 2^{nR} \rceil$. Since $\Delta = R_e/R$, the rate–equivocation region (R, Δ) can be obtained from the rate–equivocation region (R, R_e), but, in general, the region (R, Δ) is not convex.*

Theorem 3.2 is proved in Section 3.4.1 and Section 3.4.2. Before getting into the details of the proof, it is instructive to consider some of its implications. First, by specializing Theorem 3.2 to full secrecy rates for which $R_e = R$, we obtain the secrecy capacity of the degraded wiretap channel.

Corollary 3.1. *The secrecy capacity of a DWTC* $\left(\mathcal{X}, p_{Z|Y}\, p_{Y|X}, \mathcal{Y}, \mathcal{Z}\right)$ *is*

$$C_s^{\text{DWTC}} = \max_{p_X} \mathbb{I}(X;Y|Z) = \max_{p_X}(\mathbb{I}(X;Y) - \mathbb{I}(X;Z)). \tag{3.8}$$

If $Y = Z$, that is the eavesdropper obtains the same observation as the legitimate receiver, then $\mathbb{I}(X;Y|Z) = 0$ and thus $C_s^{\text{DWTC}} = 0$. This result is consistent with the analysis of Shannon's cipher system in Section 3.1 and the idea that information-theoretic security cannot be achieved over noiseless channels without secret keys.

Remark 3.6. *Theorem 3.2 and Corollary 3.1 also hold for a vector channel on replacing random variables by random vectors where appropriate.*

Corollary 3.1 is quite appealing because the secrecy capacity is expressed as the difference between an information rate conveyed to the legitimate receiver and an information rate leaked to the eavesdropper. To obtain an even simpler and more intuitive characterization, it is also useful to relate the secrecy capacity to the main channel capacity $C_m \triangleq \max_{p_X} \mathbb{I}(X;Y)$ and to the eavesdropper's channel capacity $C_e \triangleq \max_{p_X} \mathbb{I}(X;Z)$. For a generic DWTC $\left(\mathcal{X}, p_{Z|Y}\, p_{Y|X}, \mathcal{Y}, \mathcal{Z}\right)$, we have

$$\begin{aligned} C_s^{\text{DWTC}} &= \max_{p_X}(\mathbb{I}(X;Y) - \mathbb{I}(X;Z)) \\ &\geqslant \max_{p_X} \mathbb{I}(X;Y) - \max_{p_X} \mathbb{I}(X;Z) \\ &= C_m - C_e; \end{aligned}$$

that is, the secrecy capacity is *at least* as large as the difference between the main channel capacity and the eavesdropper's channel capacity. The inequality can be strict, as illustrated by the following example.

Example 3.4. Consider the DWTC illustrated in Figure 3.6, in which the main channel is a "Z-channel" with parameter p, while the eavesdropper's channel is a binary symmetric channel with cross-over probability p. One can check that

$$\begin{aligned} C_m &= \max_{q\in[0,1]} (\mathbb{H}_b(q(1-p)) - q\mathbb{H}_b(p)), \\ C_e &= 1 - \mathbb{H}_b(p), \\ C_s^{\text{DWTC}} &= \max_{q\in[0,1]} (\mathbb{H}_b(q(1-p)) + (1-q)\mathbb{H}_b(p) - \mathbb{H}_b(p+q-2pq)). \end{aligned}$$

For $p = 0.1$, we obtain numerically $C_m - C_e \approx 0.232$ bits, whereas $C_s^{\text{DWTC}} \approx 0.246$ bits.

Nevertheless, there are DWTCs for which the lower bound $C_m - C_e$ turns out to be exactly the secrecy capacity.

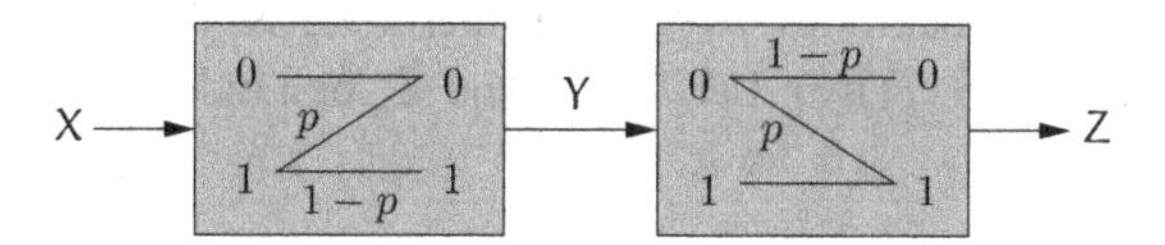

Figure 3.6 Example of a DWTC with non-symmetric channels.

Definition 3.4 (Weakly symmetric channels). *A DMC $(\mathcal{X}, p_{Y|X}, \mathcal{Y})$ is weakly symmetric if the rows of the channel transition-probability matrix are permutations of each other and the column sums $\sum_{x\in\mathcal{X}} p_{Y|X}(y|x)$ are independent of y.*

An important characteristic of weakly symmetric channels is captured by the following lemma.

Lemma 3.2. *The capacity-achieving input distribution of a weakly symmetric channel $(\mathcal{X}, p_{Y|X}, \mathcal{Y})$ is the uniform distribution over $\mathcal{X}$.*

Proof. For an input distribution p_X, the mutual information $\mathbb{I}(X;Y)$ is

$$\mathbb{I}(X;Y) = \mathbb{H}(Y) - \mathbb{H}(Y|X) = \mathbb{H}(Y) - \sum_{x\in\mathcal{X}} \mathbb{H}(Y|X=x)p_X(x).$$

Notice that $\mathbb{H}(Y|X=x)$ is a constant, say H, that is independent of x because the rows of the channel transition-probability matrix are permutations of each other. Thus,

$$\mathbb{I}(X;Y) = \mathbb{H}(Y) - H \leqslant \log|\mathcal{Y}| - H,$$

with equality if Y is uniform. We show that choosing $p_X(x) = 1/|\mathcal{X}|$ for all $x \in \mathcal{X}$ induces a uniform distribution for Y. In fact,

$$p_Y(y) = \sum_{x\in\mathcal{X}} p_{Y|X}(y|x)p_X(x) = \frac{1}{|\mathcal{X}|}\sum_{x\in\mathcal{X}} p_{Y|X}(y|x).$$

Since $\sum_x p_{Y|X}(y|x)$ is independent of y by assumption, $p_Y(y)$ is a constant. By the law of total probability, it must hold that $p_Y(y) = 1/|\mathcal{Y}|$ for all $y \in \mathcal{Y}$. □

Proposition 3.2 (Leung-Yan-Cheong). *If the main channel and the eavesdropper's channel of a DWTC $(\mathcal{X}, p_{Z|Y}\, p_{Y|X}, \mathcal{Y}, \mathcal{Z})$ are both weakly symmetric, then*

$$C_s^{\text{DWTC}} = C_m - C_e, \tag{3.9}$$

where C_m is the capacity of the main channel and C_e is that of the eavesdropper's channel.

The proof of Proposition 3.2 hinges on a general concavity property of the conditional mutual information $\mathbb{I}(X;Y|Z)$, which we establish in the following lemma.

Lemma 3.3. *Let $X \in \mathcal{X}$, $Y \in \mathcal{Y}$, and $Z \in \mathcal{Z}$ be three random variables with joint probability distribution p_{XYZ}. Then, $\mathbb{I}(X;Y|Z)$ is a concave function of p_X for fixed $p_{YZ|X}$.*

Proof. For fixed transition probabilities $p_{YZ|X}$, we interpret $\mathbb{I}(X;Y|Z)$ as a function of p_X and we write $\mathbb{I}(X;Y|Z) \triangleq f(p_X)$. Let X_1, Y_1, and Z_1 be random variables such that

$$\forall (x,y,z) \in \mathcal{X} \times \mathcal{Y} \times \mathcal{Z} \quad p_{X_1Y_1Z_1}(x,y,z) = p_{YZ|X}(y,z|x)p_{X_1}(x).$$

Similarly, let X_2, Y_2, and Z_2 be random variables such that

$$\forall (x,y,z) \in \mathcal{X} \times \mathcal{Y} \times \mathcal{Z} \quad p_{X_2Y_2Z_2}(x,y,z) = p_{YZ|X}(y,z|x)p_{X_2}(x).$$

We introduce the random variable $Q \in \{1,2\}$ which is independent of all others such that

$$Q \triangleq \begin{cases} 1 & \text{with probability } \lambda, \\ 2 & \text{with probability } 1-\lambda, \end{cases}$$

and we define the random variables

$$X \triangleq X_Q, \quad Y \triangleq Y_Q, \quad \text{and} \quad Z \triangleq Z_Q.$$

Note that $Q \rightarrow X \rightarrow YZ$ forms a Markov chain and that, for all $x \in \mathcal{X}$, $p_X(x) = \lambda p_{X_1}(x) + (1-\lambda)p_{X_2}(x)$. Then,

$$\begin{aligned} \mathbb{I}(X;Y|Z) &= \mathbb{H}(Y|Z) - \mathbb{H}(Y|XZ) \\ &\geqslant \mathbb{H}(Y|ZQ) - \mathbb{H}(Y|XZQ), \end{aligned}$$

where the inequality follows from $\mathbb{H}(Y|Z) \geqslant \mathbb{H}(Y|ZQ)$, since conditioning does not increase entropy, and $\mathbb{H}(Y|ZX) = \mathbb{H}(Y|ZXQ)$, since Q is independent of Y given X. Therefore,

$$\mathbb{I}(X;Y|Z) \geqslant \mathbb{I}(X;Y|ZQ) = \lambda\mathbb{I}(X_1;Y_1|Z_1) + (1-\lambda)\mathbb{I}(X_2;Y_2|Z_2),$$

or, equivalently,

$$f\left(\lambda p_{X_1} + (1-\lambda)p_{X_2}\right) \geqslant \lambda f(p_{X_1}) + (1-\lambda) f(p_{X_2}),$$

which is the desired result. □

Note that Lemma 3.3 holds for *any* transition probabilities $p_{YZ|X}$, not just those corresponding to a degraded channel.

Proof of Proposition 3.2. The DMCs $\left(\mathcal{X}, p_{Y|X}, \mathcal{Y}\right)$ and $\left(\mathcal{X}, p_{Z|X}, \mathcal{Z}\right)$ are weakly symmetric; therefore, by Lemma 3.2, $\mathbb{I}(X;Y)$ and $\mathbb{I}(X;Z)$ are both maximized if X is uniformly distributed over $\mathcal{X}$. For a degraded channel, $\mathbb{I}(X;Y) - \mathbb{I}(X;Z) = \mathbb{I}(X;Y|Z)$, which is a concave function of p_X by Lemma 3.3. Therefore, $\mathbb{I}(X;Y|Z)$ is also maximized if X is uniformly distributed and

$$C_s^{\text{DWTC}} = \max_{p_X} \mathbb{I}(X;Y|Z) = \max_{p_X} \mathbb{I}(X;Y) - \max_{p_X} \mathbb{I}(X;Z) = C_m - C_e. \quad \square$$

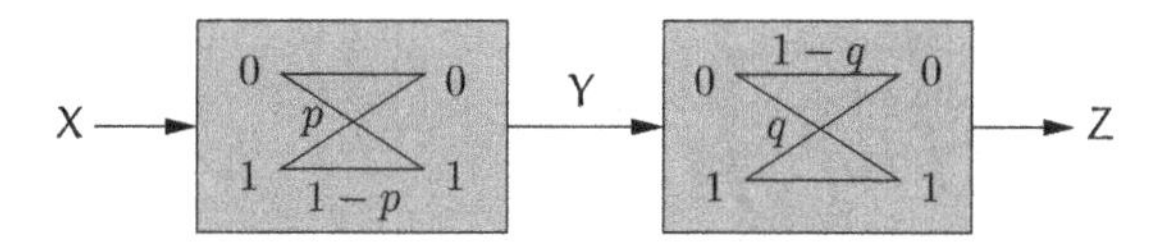

Figure 3.7 Example of a DWTC with weakly symmetric channels.

Remark 3.7. *From the proof of Proposition 3.2, we see that a sufficient condition to obtain $C_s^{\text{DWTC}} = C_m - C_e$ is that $\mathbb{I}(X;Y)$ and $\mathbb{I}(X;Z)$ are maximized for the same input distribution p_X. Nevertheless, checking that the channels $(\mathcal{X}, p_{Y|X}, \mathcal{Y})$ and $(\mathcal{X}, p_{Z|X}, \mathcal{Z})$ are weakly symmetric is, in general, a much simpler task.*

Proposition 3.2 is useful because many channels of interest (binary symmetric channels) are indeed weakly symmetric and their secrecy capacity then follows easily.

Example 3.5. Consider the DWTC illustrated in Figure 3.7, which is obtained by cascading two binary symmetric channels BSC(p) and BSC(q). The main channel is symmetric by construction, and the eavesdropper's channel is a BSC($p + q - 2pq$), which is also symmetric. Therefore, by Proposition 3.2,

$$\begin{aligned} C_s^{\text{DWTC}} &= C_m - C_e \\ &= 1 - \mathbb{H}_b(p) - (1 - \mathbb{H}_b(p + q - 2pq)) \\ &= \mathbb{H}_b(p + q - 2pq) - \mathbb{H}_b(p). \end{aligned}$$

3.4.1 Achievability proof for the degraded wiretap channel

In this section, we prove that the rate pairs in $\mathcal{R}^{\text{DWTC}}$ given by Theorem 3.2 are achievable. As is usual in information theory, we use a random-coding argument, and we show the existence of codes for the DWTC without constructing them explicitly; nevertheless, before we can start the proof, it is still necessary to identify a generic code structure that can guarantee secrecy and reliability simultaneously. In the next paragraphs, we do so by developing two desirable properties that wiretap codes should satisfy to guarantee full secrecy.

The discussion and example in Section 3.2 suggest that several codewords should represent the same message and that the choice of which codeword to transmit should be random, to "confuse" the eavesdropper. This statement can be made somewhat more precise; we can argue that, in general, a wiretap code *must* possess this property. To see this, assume that we use a wiretap code $\mathcal{C}_n$ that guarantees communication with full secrecy. It is reasonable to assume that messages are determined uniquely by codewords, that is $\mathbb{H}(M|X^n\mathcal{C}_n) = 0$. In addition, assume that the encoding function is a one-to-one

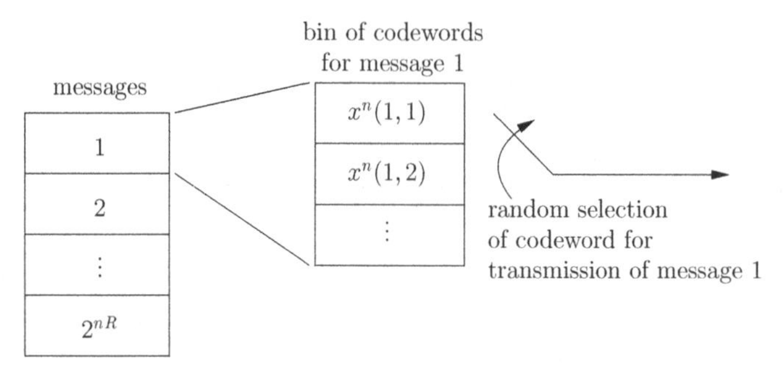

Figure 3.8 Binning structure and encoding process for a wiretap code.

mapping, that is $\mathbb{H}(X^n|MC_n) = 0$. We can write the leakage $(1/n)\mathbf{L}(C_n)$ as

$$\begin{aligned}\frac{1}{n}\mathbf{L}(C_n) &= \frac{1}{n}\mathbb{I}(M;Z^n|C_n)\\ &= \frac{1}{n}\mathbb{I}(MX^n;Z^n|C_n) - \frac{1}{n}\mathbb{I}(X^n;Z^n|MC_n)\\ &= \frac{1}{n}\mathbb{I}(X^n;Z^n|C_n) + \frac{1}{n}\mathbb{I}(M;Z^n|X^nC_n) - \frac{1}{n}\mathbb{I}(X^n;Z^n|MC_n).\end{aligned}$$

Since we have assumed that $\mathbb{H}(M|X^nC_n) = 0$ and $\mathbb{H}(X^n|MC_n) = 0$, we have also $\mathbb{I}(M;Z^n|X^nC_n) = 0$ and $\mathbb{I}(X^n;Z^n|MC_n) = 0$; therefore,

$$\frac{1}{n}\mathbb{I}(X^n;Z^n|C_n) = \frac{1}{n}\mathbf{L}(C_n),$$

which means that the information leaked to the eavesdropper about codewords is equal to the information leaked to the eavesdropper about messages. If C_n allows communication in full secrecy, then for some small $\epsilon > 0$ we have $(1/n)\mathbf{L}(C_n) \leqslant \epsilon$ and, consequently, $(1/n)\mathbb{I}(X^n;Z^n|C_n) \leqslant \epsilon$ as well. Notice that the relation between X^n and Z^n is determined in part by the channel, over which the transmitter does not have full control. For a DMC, guaranteeing that $(1/n)\mathbb{I}(X^n;Z^n|C_n) \leqslant \epsilon$ is in general[4] possible only if $(1/n)\mathbb{H}(X^n|C_n) \leqslant \delta(\epsilon)$; that is, the transmission rate must be negligible. Therefore, to transmit at a non-negligible rate, we need $(1/n)\mathbb{H}(X^n|MC_n)$ to be non-zero. In other words, the encoder should select a codeword at random among a set of codewords representing the same message. As illustrated in Figure 3.8, we can think of such a set as a sub-codebook or as a "bin" of codewords within the codebook; hence, we will say that a wiretap code should possess a *binning structure*.

Remark 3.8. *Despite the similarity between Figure 3.8 and Figure 2.10, the binning structure of a wiretap code is different from the superposition coding structure introduced for the broadcast channel in Section 2.3.3. A wiretap code consists of a* single *codebook partitioned into bins, whereas a superposition codebook for the broadcast channel consists of* several *codebooks that are superposed.*

[4] It is possible to prove that, with high probability, the good random codes identified by random-coding arguments leak an information rate that grows linearly with n over a DMC.

The second desirable property of wiretap codes concerns the local randomness R used in the encoding process. Since codewords are a function of messages and local randomness, note that $\mathbb{I}(\mathsf{X}^n; \mathsf{Z}^n|\mathcal{C}_n) = \mathbb{I}(\mathsf{MR}; \mathsf{Z}^n|\mathcal{C}_n)$. In addition, since the local randomness R is independent of M, we have $\mathbb{H}(\mathsf{R}|\mathsf{M}\mathcal{C}_n) = \mathbb{H}(\mathsf{R}|\mathcal{C}_n)$. Therefore,

$$\begin{aligned}
\frac{1}{n}\mathbf{L}(\mathcal{C}_n) &= \frac{1}{n}\mathbb{I}(\mathsf{M}; \mathsf{Z}^n|\mathcal{C}_n) \\
&= \frac{1}{n}\mathbb{I}(\mathsf{MR}; \mathsf{Z}^n|\mathcal{C}_n) - \frac{1}{n}\mathbb{I}(\mathsf{R}; \mathsf{Z}^n|\mathsf{M}\mathcal{C}_n) \\
&= \frac{1}{n}\mathbb{I}(\mathsf{X}^n; \mathsf{Z}^n|\mathcal{C}_n) - \frac{1}{n}\mathbb{H}(\mathsf{R}|\mathsf{M}\mathcal{C}_n) + \frac{1}{n}\mathbb{H}(\mathsf{R}|\mathsf{Z}^n\mathsf{M}\mathcal{C}_n) \\
&= \frac{1}{n}\mathbb{I}(\mathsf{X}^n; \mathsf{Z}^n|\mathcal{C}_n) - \frac{1}{n}\mathbb{H}(\mathsf{R}|\mathcal{C}_n) + \frac{1}{n}\mathbb{H}(\mathsf{R}|\mathsf{Z}^n\mathsf{M}\mathcal{C}_n). \qquad (3.10)
\end{aligned}$$

If the code $\mathcal{C}_n$ allows communication in full secrecy, then it must be that $(1/n)\mathbf{L}(\mathcal{C}_n) \leqslant \epsilon$ for some $\epsilon > 0$. Note that (3.10) suggests that this is indeed possible, because the confusion introduced by the source of local randomness is represented by the term $(1/n)\mathbb{H}(\mathsf{R}|\mathcal{C}_n)$, which compensates in part for the information rate leaked to the eavesdropper $(1/n)\mathbb{I}(\mathsf{X}^n; \mathsf{Z}^n|\mathcal{C}_n)$. However, to ensure that the confusion *cancels out* the information rate leaked, it seems desirable to design a code such that $(1/n)\mathbb{H}(\mathsf{R}|\mathsf{Z}^n\mathsf{M}\mathcal{C}_n)$ is small.

Remark 3.9. *Now that generic properties of wiretap codes have been identified, it would be tempting to start a random-coding argument and to analyze both the error probability and the equivocation rate for a random-code ensemble. However, there is a subtle but critical detail to which we must pay attention. Let C_n be the random variable that denotes the choice of a code $\mathcal{C}_n$ in the code ensemble. The probability of error averaged over the ensemble is then*

$$\mathbb{P}\left[\mathsf{M} \neq \hat{\mathsf{M}}\right] = \mathbb{E}_{\mathsf{C}_n}\left[\mathbb{P}\left[\mathsf{M} \neq \hat{\mathsf{M}}|\mathsf{C}_n\right]\right] = \sum_{\mathcal{C}_n} p_{\mathsf{C}_n}(\mathcal{C}_n)\mathbf{P}_{\mathrm{e}}(\mathcal{C}_n).$$

In other words, the probability of error averaged over the ensemble is equal *to the average of the probability of error of individual codes. Consequently, if the probability of error averaged over the ensemble is smaller than some $\epsilon > 0$, there must exist at least one* specific *code $\mathcal{C}_n$ with $\mathbf{P}_{\mathrm{e}}(\mathcal{C}_n) \leqslant \epsilon$. In contrast, for the equivocation of the code ensemble $\mathbb{H}(\mathsf{M}|\mathsf{Z}^n)$, we have*

$$\frac{1}{n}\mathbb{H}(\mathsf{M}|\mathsf{Z}^n) \geqslant \frac{1}{n}\mathbb{H}(\mathsf{M}|\mathsf{Z}^n\mathsf{C}_n) = \sum_{\mathcal{C}_n} p_{\mathsf{C}_n}(\mathcal{C}_n)\frac{1}{n}\mathbf{E}(\mathcal{C}_n).$$

The equivocation of the code ensemble is greater *than the average of equivocations of individual codes. Therefore, even if $(1/n)\mathbb{H}(\mathsf{M}|\mathsf{Z}^n)$ is greater than some value R_{e}, this does not ensure the existence of a specific code $\mathcal{C}_n$ such that $(1/n)\mathbf{E}(\mathcal{C}_n) \geqslant R_{\mathrm{e}}$.*

Consequently, our proof must somehow analyze $\mathbb{H}(\mathsf{M}|\mathsf{Z}^n\mathsf{C}_n)$ or $\mathbb{H}(\mathsf{M}|\mathsf{Z}^n\mathcal{C}_n)$ directly. Wyner's original approach was to study the equivocation $\mathbb{H}(\mathsf{M}|\mathsf{Z}^n\mathcal{C}_n)$ of a *well-constructed code $\mathcal{C}_n$; in this book we choose to study $\mathbb{H}(\mathsf{M}|\mathsf{Z}^n\mathcal{C}_n)$ directly with a random-coding argument.*

For ease of reading, we carry out the proof in three distinct steps.

1. For a fixed distribution p_X on $\mathcal{X}$ and $R < \mathbb{I}(\mathsf{X};\mathsf{Y}|\mathsf{Z})$, we use a random-coding argument and show the existence of a sequence of $(2^{nR}, n)$ codes $\{\mathcal{C}_n\}_{n\geqslant 1}$ that possess the binning structure illustrated in Figure 3.8 and are such that

$$\lim_{n\to\infty} \mathbf{P}_\mathrm{e}(\mathcal{C}_n) = 0, \quad \lim_{n\to\infty} \frac{1}{n}\mathbb{H}(\mathsf{R}|\mathsf{Z}^n\mathsf{M}\mathcal{C}_n) = 0, \quad \text{and} \quad \lim_{n\to\infty} \frac{1}{n}\mathbf{L}(\mathcal{C}_n) \leqslant \epsilon$$

   for some arbitrary $\epsilon > 0$. This shows the existence of wiretap codes with "close" to full secrecy and

$$\mathcal{R}'(p_\mathsf{X}) \triangleq \left\{(R, R_\mathrm{e}): \begin{array}{c} 0 \leqslant R \leqslant \mathbb{I}(\mathsf{X};\mathsf{Y}|\mathsf{Z}) \\ 0 \leqslant R_\mathrm{e} \leqslant R \end{array}\right\} \subseteq \mathcal{R}^{\mathrm{DWTC}}.$$

   The region $\mathcal{R}'(p_\mathsf{X})$ contains the full secrecy rate $R < \mathbb{I}(\mathsf{X};\mathsf{Y}|\mathsf{Z})$, but is, in general, strictly smaller than $\mathcal{R}^{\mathrm{DWTC}}(p_\mathsf{X})$ defined in Theorem 3.2.
2. We show that $\mathcal{R}^{\mathrm{DWTC}}(p_\mathsf{X}) \subseteq \mathcal{R}^{\mathrm{DWTC}}$ with a minor modification of the codes $\{\mathcal{C}_n\}_{n\geqslant 1}$ analyzed in Step 1.
3. We show that $\mathcal{R}^{\mathrm{DWTC}}$ is convex.

Step 1. Random-coding argument

We prove the existence of a sequence of $(2^{nR}, n)$ codes $\{\mathcal{C}_n\}_{n\geqslant 1}$ for the DWTC with a binning structure as in Figure 3.8 such that

$$\lim_{n\to\infty} \mathbf{P}_\mathrm{e}(\mathcal{C}_n) = 0, \quad \lim_{n\to\infty} \frac{1}{n}\mathbb{H}(\mathsf{R}|\mathsf{Z}^n\mathsf{M}\mathcal{C}_n) = 0, \tag{3.11}$$

$$\lim_{n\to\infty} \frac{1}{n}\mathbf{L}(\mathcal{C}_n) \leqslant \delta(\epsilon). \tag{3.12}$$

The existence of these codes is established by choosing a specific source of local randomness and by combining the two constraints in (3.11) into a single reliability constraint for the *enhanced* DWTC illustrated in Figure 3.9. This channel enhances the original DWTC by

- introducing a *virtual receiver*, hereafter named Charlie, who observes the same channel output Z^n as Eve in the original DWTC, but who also has access to M through an error-free side channel;
- using a message M_d with uniform distribution over $[\![1, 2^{nR_\mathrm{d}}]\!]$ in place of the source of local randomness $(\mathcal{R}, p_\mathsf{R})$, and by requiring M_d to be reliably decoded by both Bob and Charlie.

Formally, a code for the enhanced channel is defined as follows.

Definition 3.5. *A $(2^{nR}, 2^{nR_\mathrm{d}}, n)$ code $\mathcal{C}_n$ for the enhanced DWTC consists of*

- *two message sets, $\mathcal{M} = [\![1, 2^{nR}]\!]$ and $\mathcal{M}_\mathrm{d} = [\![1, 2^{nR_\mathrm{d}}]\!]$;*

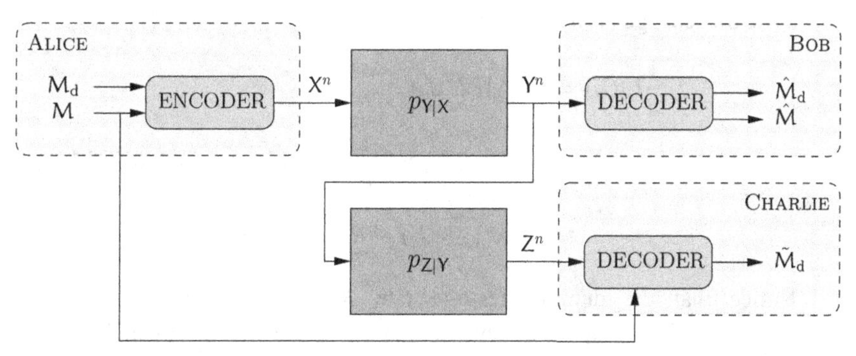

Figure 3.9 Enhanced DWTC.

- *an encoding function* $f : \mathcal{M} \times \mathcal{M}_d \to \mathcal{X}^n$, *which maps each message pair* (m, m_d) *to a codeword* x^n*;*
- *a decoding function* $g : \mathcal{Y}^n \to (\mathcal{M} \times \mathcal{M}_d) \cup \{?\}$, *which maps each channel observation* y^n *to a message pair* $(\hat{m}, \hat{m}_d) \in \mathcal{M} \times \mathcal{M}_d$ *or an error message* **?***;*
- *a decoding function* $h : \mathcal{Z}^n \times \mathcal{M} \to \mathcal{M}_d \cup \{?\}$, *which maps each channel observation* z^n *and its corresponding message* m *to a message* $\tilde{m}_d \in \mathcal{M}_d$ *or an error message* **?**.

We assume that M and M_d are uniformly distributed in their respective sets. The reliability performance of a $(2^{nR}, 2^{nR_d}, n)$ code $\mathcal{C}_n$ is measured in terms of its average probability of error

$$\mathbf{P}_e(\mathcal{C}_n) \triangleq \mathbb{P}\left[(\hat{M}, \hat{M}_d) \neq (M, M_d) \text{ or } \tilde{M}_d \neq M_d \middle| \mathcal{C}_n\right].$$

Because the message M_d is a *dummy message* that corresponds to a specific choice for the source of local randomness $(\mathcal{R}, p_R)$ in the original DWTC, a $\left(2^{nR}, 2^{nR_d}, n\right)$ code $\mathcal{C}_n$ for the enhanced channel is also a $\left(2^{nR}, n\right)$ code $\mathcal{C}_n$ for the original DWTC. By construction, the probability of error for the DWTC does not exceed the probability of error for the enhanced DWTC, since

$$\mathbb{P}\left[\hat{M} \neq M \middle| \mathcal{C}_n\right] \leqslant \mathbb{P}\left[(\hat{M}, \hat{M}_d) \neq (M, M_d) \text{ or } \tilde{M}_d \neq M_d \middle| \mathcal{C}_n\right] = \mathbf{P}_e(\mathcal{C}_n).$$

In addition, from Fano's inequality, we have

$$\frac{1}{n}\mathbb{H}(M_d|Z^n M \mathcal{C}_n) \leqslant \delta(\mathbf{P}_e(\mathcal{C}_n)).$$

Therefore, if $\lim_{n\to\infty} \mathbf{P}_e(\mathcal{C}_n) = 0$, the constraints of (3.11) are automatically satisfied with M_d in place of R.

The leakage guaranteed by the codes $\mathcal{C}_n$ is formally calculated in the next paragraphs; nevertheless, it is useful to understand intuitively why the structure of $\mathcal{C}_n$ makes this calculation possible. By using the condition $(1/n)\mathbb{H}(M_d|Z^n M \mathcal{C}_n) \approx 0$ in (3.10), we write the leakage $(1/n)\mathbf{L}(\mathcal{C}_n)$ as

$$\begin{aligned}\frac{1}{n}\mathbf{L}(\mathcal{C}_n) &= \frac{1}{n}\mathbb{I}(\mathsf{M};\mathsf{Z}^n\mathcal{C}_n)\\ &= \frac{1}{n}\mathbb{I}(\mathsf{X}^n;\mathsf{Z}^n\mathcal{C}_n) - \frac{1}{n}\mathbb{H}(\mathsf{M}_\mathrm{d}|\mathcal{C}_n) + \frac{1}{n}\mathbb{H}(\mathsf{M}_\mathrm{d}|\mathsf{Z}^n\mathsf{M}\mathcal{C}_n)\\ &\approx \frac{1}{n}\mathbb{I}(\mathsf{X}^n;\mathsf{Z}^n|\mathcal{C}_n) - R_\mathrm{d}.\end{aligned}$$

Notice that the dummy-message rate R_d counterbalances the information rate $(1/n)\mathbb{I}(\mathsf{X}^n;\mathsf{Z}^n|\mathcal{C}_n)$ about codewords leaked to the eavesdropper. In what follows, we design $\mathcal{C}_n$ carefully so that the dummy-message rate almost *cancels out* the information leaked to the eavesdropper.

Remark 3.10. *Consider a DWTC such that for any distribution p_X on $\mathcal{X}$*

$$\mathbb{I}(\mathsf{X};\mathsf{Y}) - \mathbb{I}(\mathsf{X};\mathsf{Z}) = 0 \quad or \quad \mathbb{I}(\mathsf{X};\mathsf{Z}) = 0.$$

If $\mathbb{I}(\mathsf{X};\mathsf{Y}) - \mathbb{I}(\mathsf{X};\mathsf{Z}) = 0$, the equivocation bound in (3.7) *reduces to $R_\mathrm{e} = 0$, which is always achieved, as has already been discussed in Remark 3.4. If $\mathbb{I}(\mathsf{X};\mathsf{Z}) = 0$, the eavesdropper's observation is independent of the channel input, which automatically ensures full secrecy $R_\mathrm{e} = R$. In both cases, the achievability proof reduces to that of the channel coding theorem for the DMC $\left(\mathcal{X}, p_{\mathsf{Y}|\mathsf{X}}, \mathcal{Y}\right)$.*

We now go back to the construction of codes for the enhanced DWTC. We begin by choosing a distribution p_X on $\mathcal{X}$ and, following the discussion in Remark 3.10, we assume without losing generality that p_X is such that

$$\mathbb{I}(\mathsf{X};\mathsf{Y}) - \mathbb{I}(\mathsf{X};\mathsf{Z}) > 0 \quad \text{and} \quad \mathbb{I}(\mathsf{X};\mathsf{Z}) > 0.$$

Let $0 < \epsilon < \mu_{\mathsf{XYZ}}$, where

$$\mu_{\mathsf{XYZ}} \triangleq \min_{(x,y,z)\in\mathcal{X}\times\mathcal{Y}\times\mathcal{Z}} p_{\mathsf{XYZ}}(x,y,z),$$

and let $n \in \mathbb{N}^*$. Let $R > 0$ and $R_\mathrm{d} > 0$ be rates to be specified later. We construct a $(2^{nR}, 2^{nR_\mathrm{d}}, n)$ code for the enhanced DWTC as follows.

- *Codebook construction.* Construct a codebook $\mathcal{C}_n$ with $\lceil 2^{nR}\rceil\lceil 2^{nR_\mathrm{d}}\rceil$ codewords labeled $x^n(m, m_\mathrm{d})$ with $m \in \left[\!\left[1, 2^{nR}\right]\!\right]$ and $m_\mathrm{d} \in \left[\!\left[1, 2^{nR_\mathrm{d}}\right]\!\right]$, by generating the symbols $x_i(m, m_\mathrm{d})$ for $i \in [\![1, n]\!]$, $m \in \left[\!\left[1, 2^{nR}\right]\!\right]$, and $m_\mathrm{d} \in \left[\!\left[1, 2^{nR_\mathrm{d}}\right]\!\right]$ independently according to p_X. In terms of the binning structure of Figure 3.8, m_d indexes the codewords within the bin corresponding to message m. The codebook is revealed to Alice, Bob, and Charlie.
- *Alice's encoder f.* Given (m, m_d), transmit $x^n(m, m_\mathrm{d})$.
- *Bob's decoder g.* Given y^n, output $(\hat{m}, \hat{m}_\mathrm{d})$ if it is the unique message pair such that $(x^n(\hat{m}, \hat{m}_\mathrm{d}), y^n) \in \mathcal{T}_\epsilon^n(\mathsf{XY})$; otherwise, output an error **?**.
- *Charlie's decoder h.* Given z^n and m, output $\tilde{m}_\mathrm{d}$ if it is the unique message such that $(x^n(m, \tilde{m}_\mathrm{d}), z^n) \in \mathcal{T}_\epsilon^n(\mathsf{XZ})$; otherwise, output an error **?**.

The random variable that represents the randomly generated codebook $\mathcal{C}_n$ is denoted by C_n. First, we develop an upper bound for $\mathbb{E}[\mathbf{P}_e(C_n)]$ as in the proof of the channel coding theorem. Note that, from the symmetry of the random-coding construction, we have

$$\mathbb{E}[\mathbf{P}_e(C_n)] = \mathbb{E}_{C_n}\left[\mathbb{P}\left[\hat{M} \neq M | C_n, M = 1\right]\right].$$

Thus, without loss of generality, we can assume that $M = 1$ and $M_d = 1$, and we can express $\mathbb{E}[\mathbf{P}_e(C_n)]$ in terms of the events

$$\mathcal{E}_{ij} = \left\{(X^n(i, j), Y^n) \in \mathcal{T}_\epsilon^n(XY)\right\} \text{ for } (i, j) \in \llbracket 1, 2^R \rrbracket \times \llbracket 1, 2^{nR_d} \rrbracket,$$
$$\mathcal{F}_i = \left\{(X^n(1, i), Z^n) \in \mathcal{T}_\epsilon^n(XZ)\right\} \text{ for } i \in \llbracket 1, 2^{nR_d} \rrbracket$$

as

$$\mathbb{E}[\mathbf{P}_e(C_n)] = \mathbb{P}\left[\mathcal{E}_{11}^c \cup \bigcup_{(i,j)\neq(1,1)} \mathcal{E}_{ij} \cup \mathcal{F}_1^c \cup \bigcup_{i\neq 1} \mathcal{F}_i\right].$$

By the union bound,

$$\mathbb{E}[\mathbf{P}_e(C_n)] \leqslant \mathbb{P}\left[\mathcal{E}_{11}^c\right] + \sum_{(i,j)\neq(1,1)} \mathbb{P}\left[\mathcal{E}_{ij}\right] + \mathbb{P}\left[\mathcal{F}_1^c\right] + \sum_{i\neq 1} \mathbb{P}[\mathcal{F}_i]. \tag{3.13}$$

By the AEP, we know that

$$\mathbb{P}\left[\mathcal{E}_{11}^c\right] \leqslant \delta_\epsilon(n) \quad \text{and} \quad \mathbb{P}\left[\mathcal{F}_1^c\right] \leqslant \delta_\epsilon(n). \tag{3.14}$$

For $(i, j) \neq (1, 1)$, $X^n(i, j)$ is independent of Y^n; hence, Corollary 2.2 applies and

$$\mathbb{P}\left[\mathcal{E}_{ij}\right] \leqslant 2^{-n(\mathbb{I}(X;Y)-\delta(\epsilon))} \quad \text{for } (i, j) \neq (1, 1). \tag{3.15}$$

Similarly, for $i \neq 1$, $X^n(1, i)$ is independent of Z^n and, by Corollary 2.2,

$$\mathbb{P}[\mathcal{F}_i] \leqslant 2^{-n(\mathbb{I}(X;Z)-\delta(\epsilon))} \quad \text{for } i \neq 1. \tag{3.16}$$

On substituting (3.14), (3.15), and (3.16) into (3.13), we obtain

$$\mathbb{E}[\mathbf{P}_e(C_n)] \leqslant \delta_\epsilon(n) + \lceil 2^{nR} \rceil \lceil 2^{nR_d} \rceil 2^{-n(\mathbb{I}(X;Y)-\delta(\epsilon))} + \lceil 2^{nR_d} \rceil 2^{-n(\mathbb{I}(X;Z)-\delta(\epsilon))}. \tag{3.17}$$

Hence, if we choose the rates R and R_d to satisfy

$$R + R_d < \mathbb{I}(X; Y) - \delta(\epsilon) \quad \text{and} \quad R_d < \mathbb{I}(X; Z) - \delta(\epsilon), \tag{3.18}$$

then (3.17) implies that

$$\mathbb{E}[\mathbf{P}_e(C_n)] \leqslant \delta_\epsilon(n). \tag{3.19}$$

Next, we compute an upper bound for $(1/n)\mathbb{E}[\mathbf{L}(C_n)]$. Following the same steps as in (3.10), we obtain

$$\begin{aligned}\frac{1}{n}\mathbb{E}[\mathbf{L}(C_n)] &= \frac{1}{n}\mathbb{I}(M; Z^n | C_n) \\ &= \frac{1}{n}\mathbb{I}(X^n; Z^n | C_n) - \frac{1}{n}\mathbb{H}(M_d | C_n) + \frac{1}{n}\mathbb{H}(M_d | Z^n M C_n).\end{aligned} \tag{3.20}$$

We proceed to lower bound each of the three terms on the right-hand side of (3.20) separately. First, notice that, by construction, all randomly generated codes contain the same number of codewords; in addition, all these codewords are used with equal probability. Therefore,

$$\begin{aligned}\frac{1}{n}\mathbb{H}(\mathsf{M_d}|\mathsf{C}_n) &= \sum_{\mathcal{C}_n} p_{\mathsf{C}_n}(\mathcal{C}_n)\,\frac{1}{n}\mathbb{H}(\mathsf{M_d}|\mathcal{C}_n)\\ &= \frac{1}{n}\log(\lceil 2^{nR_d}\rceil)\\ &\geqslant R_d.\end{aligned} \tag{3.21}$$

Next, by Fano's inequality,

$$\begin{aligned}\frac{1}{n}\mathbb{H}(\mathsf{X}^n|\mathsf{M}\mathsf{Z}^n\mathsf{C}_n) &= \sum_{\mathcal{C}_n} p_{\mathsf{C}_n}(\mathcal{C}_n)\frac{1}{n}\mathbb{H}(\mathsf{X}^n|\mathsf{M}\mathsf{Z}^n\mathcal{C}_n)\\ &\leqslant \sum_{\mathcal{C}_n} p_{\mathsf{C}_n}(\mathcal{C}_n)\left(\frac{1}{n}+\mathbf{P}_e(\mathcal{C}_n)\frac{1}{n}\log\lceil 2^{nR_d}\rceil\right)\\ &= \delta(n)+\mathbb{E}[\mathbf{P}_e(\mathsf{C}_n)](R_d+\delta(n))\\ &= \delta_\epsilon(n),\end{aligned} \tag{3.22}$$

where the last inequality follows from (3.19). Finally, note that $\mathsf{C}_n \rightarrow \mathsf{X}^n \rightarrow \mathsf{Z}^n$ forms a Markov chain. Therefore,

$$\frac{1}{n}\mathbb{I}(\mathsf{X}^n;\mathsf{Z}^n|\mathsf{C}_n) \leqslant \frac{1}{n}\mathbb{I}(\mathsf{X}^n;\mathsf{Z}^n) = \mathbb{I}(\mathsf{X};\mathsf{Z}), \tag{3.23}$$

since $(\mathsf{X}^n, \mathsf{Z}^n)$ is i.i.d. according to p_{XZ}. On substituting (3.21), (3.22), and (3.23) into (3.20), we obtain

$$\mathbb{E}\left[\frac{1}{n}\mathbf{L}(\mathsf{C}_n)\right] \leqslant \mathbb{I}(\mathsf{X};\mathsf{Z}) - R_d + \delta_\epsilon(n).$$

In particular, for the specific choice

$$R < \mathbb{I}(\mathsf{X};\mathsf{Y}) - \mathbb{I}(\mathsf{X};\mathsf{Z}) \quad \text{and} \quad R_d = \mathbb{I}(\mathsf{X};\mathsf{Z}) - \delta(\epsilon), \tag{3.24}$$

which is compatible with the conditions in (3.18), we obtain

$$\mathbb{E}\left[\frac{1}{n}\mathbf{L}(\mathsf{C}_n)\right] \leqslant \delta(\epsilon) + \delta_\epsilon(n).$$

Finally, by applying the selection lemma to the random variable C_n and the functions $\mathbf{P}_e$ and $\mathbf{L}$, we conclude that there exists a specific code $\mathcal{C}_n$ such that $\mathbf{P}_e(\mathcal{C}_n) \leqslant \delta_\epsilon(n)$ and $(1/n)\mathbf{L}(\mathcal{C}_n) \leqslant \delta(\epsilon) + \delta_\epsilon(n)$; consequently, there exists a sequence of $(2^{nR}, n)$ codes $\{\mathcal{C}_n\}_{n\geqslant 1}$ such that

$$\lim_{n\to\infty} \mathbf{P}_e(\mathcal{C}_n) = 0 \quad \text{and} \quad \lim_{n\to\infty} \frac{1}{n}\mathbf{L}(\mathcal{C}_n) \leqslant \delta(\epsilon),$$

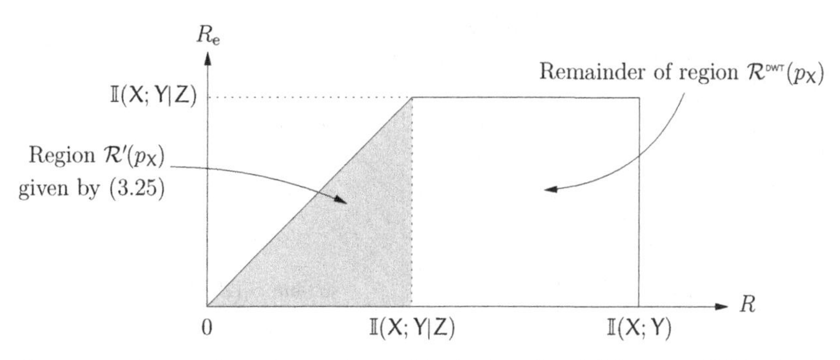

Figure 3.10 Rate–equivocation regions $\mathcal{R}'(p_X)$ given by (3.25) and $\mathcal{R}^{DWTC}(p_X)$ given in Theorem 3.2.

and the rate–equivocation pair $(R, R - \delta(\epsilon))$ is achievable. Since ϵ can be chosen arbitrarily small, since R satisfies the conditions in (3.24), and since $\mathbb{I}(X;Y) - \mathbb{I}(X;Z) = \mathbb{I}(X;Y|Z)$ for a DWTC, we conclude that

$$\mathcal{R}'(p_X) \triangleq \left\{(R, R_e)\colon \begin{array}{l} 0 \leqslant R \leqslant \mathbb{I}(X;Y|Z) \\ 0 \leqslant R_e \leqslant R \end{array}\right\} \subseteq \mathcal{R}^{DWTC}. \tag{3.25}$$

Remark 3.11. *The fact that the channel is degraded has not been used to obtain* (3.25)*; therefore,*

$$\left\{(R, R_e)\colon \begin{array}{l} 0 \leqslant R \leqslant \mathbb{I}(X;Y) - \mathbb{I}(X;Z) \\ 0 \leqslant R_e \leqslant R \end{array}\right\} \subseteq \mathcal{R}^{DWTC}$$

for any DMC $(\mathcal{X}, p_{YZ|X}, \mathcal{Y}, \mathcal{Z})$*, not just for those of the form* $\left(\mathcal{X}, p_{Z|Y}\, p_{Y|X}, \mathcal{Y}, \mathcal{Z}\right)$.

Step 2. Achieving the entire region $\mathcal{R}^{DWTC}$

As illustrated in Figure 3.10, the region $\mathcal{R}'(p_X)$ given in (3.25) is only a *subset* of the region $\mathcal{R}^{DWTC}(p_X)$. The key idea to achieve the full region is to modify the $(2^{nR}, 2^{nR_d}, n)$ codes identified in Step 1 and to exploit part of the dummy message M_d to transmit additional information. However, we have to be careful because the analysis of the probability of error and leakage for the $(2^{nR}, 2^{nR_d}, n)$ codes assumed that M and M_d were uniformly distributed; hence, we must check that our modifications do not affect the results. In the following paragraphs, we prove that this is indeed the case, but in the remainder of the book, we overlook this subtlety.

Consider a $(2^{nR}, 2^{nR_d}, n)$ code $\mathcal{C}_n$ identified in Step 1 with $R_d = \mathbb{I}(X;Z) - \delta(\epsilon)$ and $R < \mathbb{I}(X;Y) - \mathbb{I}(X;Z)$ and such that $\mathbf{P}_e(\mathcal{C}_n) \leqslant \delta_\epsilon(n)$ and $(1/n)\mathbf{L}(\mathcal{C}_n) \leqslant \delta(\epsilon) + \delta_\epsilon(n)$ provided that M and M_d are uniformly distributed. For $R' \leqslant R_d$, note that $\lceil 2^{nR'} \rceil$ might not divide $\lceil 2^{nR_d} \rceil$, and, by Euclidean division,

$$\lceil 2^{nR_d} \rceil = q \lceil 2^{nR'} \rceil + r,$$

for some integer $q > 0$ and some integer $0 \leqslant r < \lceil 2^{nR'} \rceil$. As illustrated in Figure 3.11, for each $m \in [\![1, 2^{nR}]\!]$, we distribute the codewords $x^n(m, m_d)$ with $m_d \in [\![1, 2^{nR_d}]\!]$, into sub-bins $\mathcal{B}_m(i)$ with $i \in [\![1, 2^{nR'}]\!]$, such that r of the sub-bins have size $q + 1$

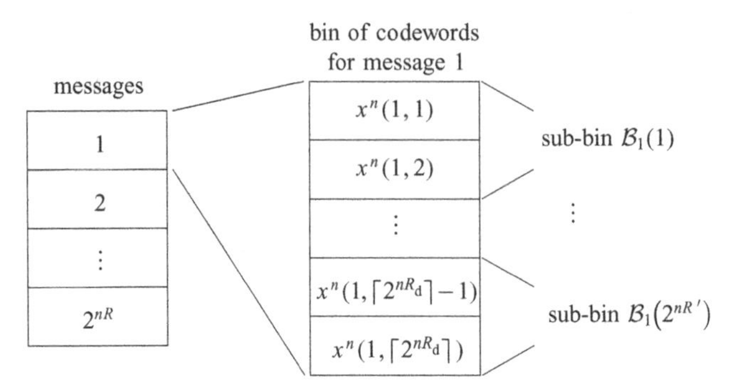

Figure 3.11 Sub-binning of codewords in $\mathcal{C}_n$.

while the remaining $\lceil 2^{nR'}\rceil - r$ have size q. We relabel the codewords $x^n(m, m', k)$ with $m \in [\![1, 2^{nR}]\!]$, $m' \in [\![1, 2^{nR'}]\!]$, and $k \in [\![1, q+1]\!]$ or $k \in [\![1, q]\!]$. The sub-binning is revealed to all parties, and we consider the following encoding/decoding procedure.

- *Encoder.* Given m and m', transmit a codeword $x^n(m, m', k) \in \mathcal{C}_n$ chosen uniformly at random in $\mathcal{B}_m(m')$.
- *Decoder.* Given y^n, use the decoding procedure of $\mathcal{C}_n$.

The sub-binning together with the encoding defines a $(2^{nR''}, n)$ code $\tilde{\mathcal{C}}_n$ for the DWTC, with $R'' = R + R' + \delta(n)$. We assume that M and M′ are uniformly distributed but, because the sub-bins have different sizes, the distribution of the codewords is now slightly non-uniform. In fact, with our encoding scheme, some codewords x^n are selected with probability $p_{X^n|\mathcal{C}_n}(x^n) = 1/(\lceil 2^{nR}\rceil\lceil 2^{nR'}\rceil q)$ while others are selected with probability $p_{X^n|\mathcal{C}_n}(x^n) = 1/[\lceil 2^{nR}\rceil\lceil 2^{nR'}\rceil(q+1)]$. Nevertheless, the reader can check that

$$\sum_{x^n \in \mathcal{C}_n} \left| p_{X^n|\mathcal{C}_n}(x^n) - \frac{1}{\lceil 2^{nR}\rceil\lceil 2^{nR_d}\rceil} \right| \leqslant \delta(n);$$

that is, the variational distance between the distribution of codewords $p_{X^n|\mathcal{C}_n}$ and the uniform distribution over $\mathcal{C}_n$ vanishes for n large enough. Consequently, the probability of decoding $\mathbf{P}_e(\tilde{\mathcal{C}}_n)$ satisfies

$$\mathbf{P}_e(\tilde{\mathcal{C}}_n) \leqslant \mathbf{P}_e(\mathcal{C}_n)(1 + \delta(n)) \leqslant \delta_\epsilon(n).$$

In addition, the equivocation $(1/n)\mathbf{E}(\tilde{\mathcal{C}}_n)$ satisfies

$$\begin{aligned} \frac{1}{n}\mathbf{E}(\tilde{\mathcal{C}}_n) &= \frac{1}{n}\mathbb{H}(\mathsf{M}\mathsf{M}'|\mathsf{Z}^n\tilde{\mathcal{C}}_n) \\ &\geqslant \frac{1}{n}\mathbb{H}(\mathsf{M}|\mathsf{Z}^n\tilde{\mathcal{C}}_n) \\ &= \frac{1}{n}\mathbb{H}(\mathsf{M}|\tilde{\mathcal{C}}_n) - \frac{1}{n}\mathbb{I}(\mathsf{M};\mathsf{Z}^n|\tilde{\mathcal{C}}_n). \end{aligned}$$

Since M is chosen uniformly at random, $\mathbb{H}(\mathsf{M}|\tilde{\mathcal{C}}_n) = \mathbb{H}(\mathsf{M}|\mathcal{C}_n)$. Also, $\mathbb{I}(\mathsf{M};\mathsf{Z}^n|\tilde{\mathcal{C}}_n)$ is a continuous function of p_{X^n}; therefore,

$$\mathbb{I}(\mathsf{M};\mathsf{Z}^n|\tilde{\mathcal{C}}_n) \leqslant \mathbb{I}(\mathsf{M};\mathsf{Z}^n|\mathcal{C}_n) + \delta(n) = \mathbf{L}(\mathcal{C}_n) + \delta(n).$$

Hence,

$$\frac{1}{n}\mathbf{E}(\tilde{\mathcal{C}}_n) \geqslant \frac{1}{n}\mathbb{H}(\mathsf{M}|\mathcal{C}_n) - \frac{1}{n}\mathbf{L}(\mathcal{C}_n) - \delta(n)$$
$$\geqslant R - \delta(\epsilon) - \delta_\epsilon(n).$$

Therefore, the rate–equivocation pair $(R + R', R - \delta(\epsilon))$ is achievable. Since R' can be chosen as large as $\mathbb{I}(\mathsf{X};\mathsf{Z}) - \delta(\epsilon)$ and since ϵ can be chosen arbitrarily small, we conclude that

$$\mathcal{R}^{\text{DWTC}}(p_{\mathsf{X}}) = \left\{(R, R_{\text{e}})\colon \begin{array}{l} 0 \leqslant R_{\text{e}} \leqslant R \leqslant \mathbb{I}(\mathsf{X};\mathsf{Y}) \\ 0 \leqslant R_{\text{e}} \leqslant \mathbb{I}(\mathsf{X};\mathsf{Y}|\mathsf{Z}) \end{array}\right\} \subseteq \mathcal{R}^{\text{DWTC}}. \tag{3.26}$$

Step 3. Convexity of the rate–equivocation region

We show that $\mathcal{R}^{\text{DWTC}}$ is convex by proving that, for any distributions p_{X_1} and p_{X_2} on $\mathcal{X}$, the convex hull of $\mathcal{R}^{\text{DWTC}}(p_{\mathsf{X}_1}) \cup \mathcal{R}^{\text{DWTC}}(p_{\mathsf{X}_2})$ is in $\mathcal{R}^{\text{DWTC}}$.

Let $(R_1, R_{\text{e}1}) \in \mathcal{R}^{\text{DWTC}}(p_{\mathsf{X}_1})$ be a rate–equivocation pair satisfying the inequalities in (3.26) for some random variables X_1, Y_1, and Z_1 whose joint distribution is such that

$$\forall(x, y, z) \in \mathcal{X} \times \mathcal{Y} \times \mathcal{Z} \quad p_{\mathsf{X}_1\mathsf{Y}_1\mathsf{Z}_1}(x, y, z) = p_{\mathsf{X}_1}(x)p_{\mathsf{Y}|\mathsf{X}}(y|x)p_{\mathsf{Z}|\mathsf{Y}}(z|y).$$

Similarly, let $(R_2, R_{\text{e}2}) \in \mathcal{R}^{\text{DWTC}}(p_{\mathsf{X}_2})$ be a rate–equivocation pair satisfying the inequalities in (3.26) for some random variables X_2, Y_2, and Z_2 whose joint distribution is such that

$$\forall(x, y, z) \in \mathcal{X} \times \mathcal{Y} \times \mathcal{Z} \quad p_{\mathsf{X}_2\mathsf{Y}_2\mathsf{Z}_2}(x, y, z) = p_{\mathsf{X}_2}(x)p_{\mathsf{Y}|\mathsf{X}}(y|x)p_{\mathsf{Z}|\mathsf{Y}}(z|y).$$

Our objective is to show that, for any $\lambda \in [0, 1]$, there exists a distribution p_{X_λ} on $\mathcal{X}$ such that

$$(\lambda R_1 + (1-\lambda)R_2, \lambda R_{\text{e}1} + (1-\lambda)R_{\text{e}2}) \in \mathcal{R}^{\text{DWTC}}(p_{\mathsf{X}_\lambda}).$$

We define a random variable $\mathsf{Q} \in \{1, 2\}$ that is independent of all others such that

$$\mathsf{Q} \triangleq \begin{cases} 1 & \text{with probability } \lambda, \\ 2 & \text{with probability } 1-\lambda. \end{cases}$$

By construction, $\mathsf{Q} \rightarrow \mathsf{X}_\mathsf{Q} \rightarrow \mathsf{Y}_\mathsf{Q} \rightarrow \mathsf{Z}_\mathsf{Q}$ forms a Markov chain and the joint distribution of X_Q, Y_Q, and Z_Q satisfies

$$\forall(x, y, z) \in \mathcal{X} \times \mathcal{Y} \times \mathcal{Z} \quad p_{\mathsf{X}_\mathsf{Q}\mathsf{Y}_\mathsf{Q}\mathsf{Z}_\mathsf{Q}}(x, y, z) = p_{\mathsf{X}_\mathsf{Q}}(x)p_{\mathsf{Y}|\mathsf{X}}(y|x)p_{\mathsf{Z}|\mathsf{Y}}(z|y).$$

We set $\mathsf{X}_\lambda \triangleq \mathsf{X}_\mathsf{Q}$, $\mathsf{Y}_\lambda \triangleq \mathsf{Y}_\mathsf{Q}$, and $\mathsf{Z}_\lambda \triangleq \mathsf{Z}_\mathsf{Q}$. Then,

$$\begin{aligned}\mathbb{I}(\mathsf{X}_\lambda, \mathsf{Y}_\lambda) &= \mathbb{I}(\mathsf{X}_\mathsf{Q};\mathsf{Y}_\mathsf{Q}) \\ &\geqslant \mathbb{I}(\mathsf{X}_\mathsf{Q};\mathsf{Y}_\mathsf{Q}|\mathsf{Q}) \\ &= \lambda\,\mathbb{I}(\mathsf{X}_1;\mathsf{Y}_1) + (1-\lambda)\mathbb{I}(\mathsf{X}_2;\mathsf{Y}_2) \\ &\geqslant \lambda R_1 + (1-\lambda)R_2,\end{aligned}$$

and, similarly,

$$\begin{aligned}\mathbb{I}(\mathsf{X}_\lambda;\mathsf{Y}_\lambda|\mathsf{Z}_\lambda) &= \mathbb{I}(\mathsf{X}_\mathsf{Q};\mathsf{Y}_\mathsf{Q}|\mathsf{Z}_\mathsf{Q})\\ &\geqslant \mathbb{I}(\mathsf{X}_\mathsf{Q};\mathsf{Y}_\mathsf{Q}|\mathsf{Z}_\mathsf{Q}\mathsf{Q})\\ &= \lambda\,\mathbb{I}(\mathsf{X}_1;\mathsf{Y}_1|\mathsf{Z}_1) + (1-\lambda)\mathbb{I}(\mathsf{X}_2;\mathsf{Y}_2|\mathsf{Z}_2)\\ &\geqslant \lambda R_{\mathrm{e}1} + (1-\lambda)R_{\mathrm{e}2}.\end{aligned}$$

Hence, for any $\lambda \in [0, 1]$, there exists X_λ such that

$$(\lambda R_1 + (1-\lambda)R_2, \lambda R_{\mathrm{e}1} + (1-\lambda)R_{\mathrm{e}2}) \in \mathcal{R}^{\mathrm{DWTC}}(p_{\mathsf{X}_\lambda}) \subseteq \mathcal{R}^{\mathrm{DWTC}}.$$

Therefore, the convex hull of $\mathcal{R}^{\mathrm{DWTC}}(p_{\mathsf{X}_1}) \cup \mathcal{R}^{\mathrm{DWTC}}(p_{\mathsf{X}_2})$ is included in $\mathcal{R}^{\mathrm{DWTC}}$ and $\mathcal{R}^{\mathrm{DWTC}}$ is convex.

3.4.2 Converse proof for the degraded wiretap channel

Let $(R, R_\mathrm{e}) \in \mathcal{R}^{\mathrm{DWTC}}$ be an achievable rate–equivocation pair and let $\epsilon > 0$. For n sufficiently large, there exists a $(2^{nR}, n)$ code $\mathcal{C}_n$ such that

$$\frac{1}{n}\mathbb{H}(\mathsf{M}|\mathcal{C}_n) \geqslant R, \quad \frac{1}{n}\mathbf{E}(\mathcal{C}_n) \geqslant R_\mathrm{e} - \delta(\epsilon), \quad \mathbf{P}_\mathrm{e}(\mathcal{C}_n) \leqslant \delta(\epsilon).$$

In the remainder of this section, we drop the conditioning on $\mathcal{C}_n$ to simplify the notation. By Fano's inequality, we have

$$\frac{1}{n}\mathbb{H}(\mathsf{M}|\mathsf{Y}^n\mathsf{Z}^n) \leqslant \frac{1}{n}\mathbb{H}(\mathsf{M}|\mathsf{Y}^n) \leqslant \delta(\mathbf{P}_\mathrm{e}(\mathcal{C}_n)) = \delta(\epsilon).$$

Therefore,

$$\begin{aligned}R &\leqslant \frac{1}{n}\mathbb{H}(\mathsf{M})\\ &= \frac{1}{n}\mathbb{I}(\mathsf{M};\mathsf{Y}^n) + \frac{1}{n}\mathbb{H}(\mathsf{M}|\mathsf{Y}^n)\\ &\overset{(a)}{\leqslant} \frac{1}{n}\mathbb{I}(\mathsf{X}^n;\mathsf{Y}^n) + \delta(\epsilon)\\ &= \frac{1}{n}\mathbb{H}(\mathsf{Y}^n) - \frac{1}{n}\mathbb{H}(\mathsf{Y}^n|\mathsf{X}^n) + \delta(\epsilon)\\ &\overset{(b)}{=} \frac{1}{n}\sum_{i=1}^n\left(\mathbb{H}(\mathsf{Y}_i|\mathsf{Y}^{i-1}) - \frac{1}{n}\mathbb{H}(\mathsf{Y}_i|\mathsf{X}_i\mathsf{Y}^{i-1})\right) + \delta(\epsilon)\\ &= \frac{1}{n}\sum_{i=1}^n\mathbb{I}(\mathsf{X}_i;\mathsf{Y}_i|\mathsf{Y}^{i-1}) + \delta(\epsilon),\end{aligned} \tag{3.27}$$

where (a) follows from the data-processing inequality applied to the Markov chain $\mathsf{M} \rightarrow \mathsf{X}^n \rightarrow \mathsf{Y}^n$ and the bound $(1/n)\mathbb{H}(\mathsf{M}|\mathsf{Y}^n) \leqslant \delta(\epsilon)$, and (b) follows because the channel is memoryless. Similarly, we bound the equivocation rate R_e as

$$\begin{aligned}
R_e &\leqslant \frac{1}{n}\mathbb{H}(M|Z^n) + \delta(\epsilon)\\
&= \frac{1}{n}\mathbb{I}(M;Y^n|Z^n) + \frac{1}{n}\mathbb{H}(M|Y^nZ^n) + \delta(\epsilon)\\
&\overset{(a)}{\leqslant} \frac{1}{n}\mathbb{I}(M;Y^n|Z^n) + \delta(\epsilon)\\
&\overset{(b)}{\leqslant} \frac{1}{n}\mathbb{I}(X^n;Y^n|Z^n) + \delta(\epsilon)\\
&= \frac{1}{n}\mathbb{H}(Y^n|Z^n) - \frac{1}{n}\mathbb{H}(Y^n|X^nZ^n) + \delta(\epsilon)\\
&= \frac{1}{n}\sum_{i=1}^{n}\left(\mathbb{H}\left(Y_i|Y^{i-1}Z^n\right) - \mathbb{H}\left(Y_i|Y^{i-1}X^nZ^n\right)\right) + \delta(\epsilon)\\
&\overset{(c)}{\leqslant} \frac{1}{n}\sum_{i=1}^{n}\left(\mathbb{H}\left(Y_i|Y^{i-1}Z_i\right) - \mathbb{H}\left(Y_i|Y^{i-1}X_iZ_i\right)\right) + \delta(\epsilon)\\
&= \frac{1}{n}\sum_{i=1}^{n}\mathbb{I}\left(X_i;Y_i|Z_iY^{i-1}\right) + \delta(\epsilon),
\end{aligned} \tag{3.28}$$

where (a) follows from the bound $(1/n)\mathbb{H}(M|Y^nZ^n) \leqslant \delta(\epsilon)$, (b) follows from the data-processing inequality applied to the Markov chain $M \rightarrow X^n \rightarrow Y^n \rightarrow Z^n$, and (c) follows from $\mathbb{H}\left(Y_i|Y^{i-1}X^nZ^n\right) = \mathbb{H}\left(Y_i|Y^{i-1}X_iZ_i\right)$ because the channel is memoryless and $\mathbb{H}\left(Y_i|Y^{i-1}Z^n\right) \leqslant \mathbb{H}\left(Y_i|Y^{i-1}Z_i\right)$ since conditioning does not increase entropy.

We now introduce a random variable Q, which is independent of all other random variables and uniformly distributed in $[\![1,n]\!]$, so that we can rewrite (3.27) and (3.28) as

$$\begin{aligned}
R &\leqslant \sum_{i=1}^{n}\frac{1}{n}\mathbb{I}\left(X_i;Y_i|Y^{i-1}\right) + \delta(\epsilon) = \mathbb{I}\left(X_Q;Y_Q|Y^{Q-1}Q\right) + \delta(\epsilon),\\
R_e &\leqslant \sum_{i=1}^{n}\frac{1}{n}\mathbb{I}\left(X_i;Y_i|Z_iY^{i-1}\right) + \delta(\epsilon) = \mathbb{I}\left(X_Q;Y_Q|Z_QY^{Q-1}Q\right) + \delta(\epsilon).
\end{aligned} \tag{3.29}$$

Finally, we define the random variables

$$X \triangleq X_Q, \quad Y \triangleq Y_Q, \quad Z \triangleq Z_Q, \quad \text{and} \quad U \triangleq Y^{Q-1}Q. \tag{3.30}$$

Note that $U \rightarrow X \rightarrow Y \rightarrow Z$ forms a Markov chain and that the transition probabilities $p_{Z|Y}$ and $p_{Y|X}$ are the same as those of the original DWTC. On substituting (3.30) into (3.29), we obtain the conditions

$$\begin{aligned}
&0 \leqslant R_e \leqslant R \leqslant \mathbb{I}(X;Y|U) + \delta(\epsilon),\\
&0 \leqslant R_e \leqslant \mathbb{I}(X;Y|ZU) + \delta(\epsilon).
\end{aligned}$$

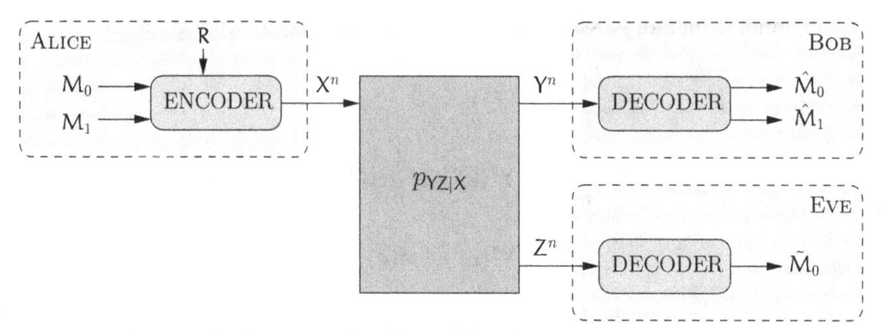

Figure 3.12 Communication over a broadcast channel with confidential messages. M_0 represents a common message for both Bob and Eve. M_1 represents an individual message for Bob, which should be kept secret from Eve. R represents local randomness used in Alice's encoder.

Since $U \to X \to Y \to Z$ forms a Markov chain, $\mathbb{I}(X;Y|U) \leqslant \mathbb{I}(X;Y)$ and $\mathbb{I}(X;Y|ZU) \leqslant \mathbb{I}(X;Y|Z)$; therefore,

$$0 \leqslant R_e \leqslant R \leqslant \mathbb{I}(X;Y) + \delta(\epsilon),$$
$$0 \leqslant R_e \leqslant \mathbb{I}(X;Y|Z) + \delta(\epsilon).$$

Since ϵ can be chosen arbitrarily small, we conclude that

$$\mathcal{R}^{\text{DWTC}} \subseteq \bigcup_{p_X} \left\{ (R, R_e) \colon \begin{array}{l} 0 \leqslant R_e \leqslant R \leqslant \mathbb{I}(X;Y) \\ 0 \leqslant R_e \leqslant \mathbb{I}(X;Y|Z) \end{array} \right\} = \bigcup_{p_X} \mathcal{R}^{\text{DWTC}}(p_X).$$

3.5 Broadcast channel with confidential messages

The DWTC model is not entirely satisfactory because it explicitly puts the eavesdropper at a disadvantage. Although the achievability proof of Section 3.4.1 does not exploit the degraded nature of the channel, the converse proof does and it is not obvious whether the specific stochastic encoding used in Section 3.4.1 is still optimal for non-degraded channels. In addition, it is useful to characterize the trade-off between reliability and security more precisely; in particular, we would like to investigate whether one could transmit reliable messages to the eavesdropper and conceal other messages simultaneously.

These issues are resolved by analyzing a more general model than the DWTC. As illustrated in Figure 3.12, we consider a broadcast channel with two receivers for which a sender attempts to send two messages simultaneously: a *common message*, which is intended for both receivers, and an *individual secret message*, which is intended for only one receiver, treating the other receiver as an eavesdropper. This channel model was termed the *broadcast channel with confidential messages* (BCC for short) by Csiszár and Körner. In the absence of a common message, the channel is called a *wiretap channel* (WTC for short).

Formally, a *discrete memoryless* BCC $(\mathcal{X}, p_{YZ|X}, \mathcal{Y}, \mathcal{Z})$ consists of a finite input alphabet $\mathcal{X}$, two finite output alphabets $\mathcal{Y}$ and $\mathcal{Z}$, and transition probabilities $p_{YZ|X}$ such that

$$\forall n \geqslant 1 \quad \forall (x^n, y^n, z^n) \in \mathcal{X}^n \times \mathcal{Y}^n \times \mathcal{Z}^n$$

$$p_{\mathsf{Y}^n\mathsf{Z}^n|\mathsf{X}^n}(y^n, z^n|x^n) = \prod_{i=1}^{n} p_{\mathsf{YZ}|\mathsf{X}}(y_i, z_i|x_i).$$

The marginal probabilities $p_{\mathsf{Y}|\mathsf{X}}$ and $p_{\mathsf{Z}|\mathsf{X}}$ define two DMCs. By convention, the DMC $(\mathcal{X}, p_{\mathsf{Y}|\mathsf{X}}, \mathcal{Y})$ is the main channel and the DMC $(\mathcal{X}, p_{\mathsf{Z}|\mathsf{X}}, \mathcal{Z})$ is the eavesdropper's channel.

Definition 3.6. *A $(2^{nR_0}, 2^{nR_1}, n)$ code $\mathcal{C}_n$ for the BCC consists of*

- *a common message set $\mathcal{M}_0 = [\![1, 2^{nR_0}]\!]$ and an individual message set $\mathcal{M}_1 = [\![1, 2^{nR_1}]\!]$;*
- *a source of local randomness $(\mathcal{R}, p_{\mathsf{R}})$;*
- *an encoding function $f : \mathcal{M}_0 \times \mathcal{M}_1 \times \mathcal{R} \to \mathcal{X}^n$, which maps a message pair (m_0, m_1) and a realization of local randomness r to a codeword x^n;*
- *a decoding function $g : \mathcal{Y}^n \to (\mathcal{M}_0 \times \mathcal{M}_1) \cup \{?\}$, which maps each channel observation y^n to a message pair $(\hat{m}_0, \hat{m}_1) \in \mathcal{M}_0 \times \mathcal{M}_1$ or an error message $\{?\}$;*
- *a decoding function $h : \mathcal{Z}^n \to \mathcal{M}_0 \cup \{?\}$, which maps each channel observation z^n to a message $\tilde{m}_0 \in \mathcal{M}_0$ or an error message $\{?\}$.*

The $(2^{nR_0}, 2^{nR_1}, n)$ code $\mathcal{C}_n$ is known to Alice, Bob, and Eve, and we assume that messages M_0 and M_1 are chosen uniformly at random. The reliability performance of the code $\mathcal{C}_n$ is measured in terms of its average probability of error

$$\mathbf{P}_{\mathrm{e}}(\mathcal{C}_n) \triangleq \mathbb{P}\left[(\hat{\mathsf{M}}_0, \hat{\mathsf{M}}_1) \neq (\mathsf{M}_0, \mathsf{M}_1) \text{ or } \tilde{\mathsf{M}}_0 \neq \mathsf{M}_0 \middle| \mathcal{C}_n\right],$$

while its secrecy performance is measured in terms of the equivocation

$$\mathbf{E}(\mathcal{C}_n) \triangleq \mathbb{H}(\mathsf{M}_1|\mathsf{Z}^n\mathcal{C}_n).$$

Definition 3.7. *A rate tuple $(R_0, R_1, R_{\mathrm{e}})$ is achievable for the BCC if there exists a sequence of $(2^{nR_0}, 2^{nR_1}, n)$ codes $\{\mathcal{C}_n\}_{n\geqslant 1}$ such that*

$$\lim_{n\to\infty} \mathbf{P}_{\mathrm{e}}(\mathcal{C}_n) = 0 \qquad \textit{(reliability condition)}, \tag{3.31}$$

$$\varliminf_{n\to\infty} \frac{1}{n}\mathbf{E}(\mathcal{C}_n) \geqslant R_{\mathrm{e}} \qquad \textit{(weak secrecy condition)}. \tag{3.32}$$

The rate–equivocation region of the BCC is

$$\mathcal{R}^{\mathrm{BCC}} \triangleq \mathrm{cl}(\{(R_0, R_1, R_{\mathrm{e}}) : (R_0, R_1, R_{\mathrm{e}}) \text{ is achievable}\}).$$

The secrecy-capacity region is

$$\mathcal{C}^{\mathrm{BCC}} \triangleq \mathrm{cl}\big(\{(R_0, R_1) : (R_0, R_1, R_1) \in \mathcal{R}^{\mathrm{BCC}}\}\big),$$

the rate–equivocation region of the wiretap channel is

$$\mathcal{R}^{\text{WT}} \triangleq \text{cl}\big(\{(R_1, R_e) : (0, R_1, R_e) \in \mathcal{R}^{\text{BCC}}\}\big),$$

and the secrecy capacity is

$$C_s^{\text{WT}} \triangleq \sup_R \{R : (0, R, R) \in \mathcal{R}^{\text{BCC}}\}.$$

The regions $\mathcal{C}^{\text{BCC}}$ and $\mathcal{R}^{\text{WT}}$ are specializations of the region $\mathcal{R}^{\text{BCC}}$ that highlight different characteristics of a BCC. The region $\mathcal{C}^{\text{BCC}}$ captures the fundamental trade-off between reliable communication with both Bob and Eve and communication in full secrecy with Bob, while the region $\mathcal{R}^{\text{WT}}$ is just the generalization of the rate–equivocation region $\mathcal{R}^{\text{DWTC}}$ defined in Section 3.4 for DWTCs. By replacing the weak secrecy condition (3.32) by the stronger requirement $\underline{\lim}_{n\to\infty}(\mathbf{E}(\mathcal{C}_n) - nR_e) \geqslant 0$, we obtain the strong rate–equivocation region $\overline{\mathcal{R}}^{\text{BCC}}$, the strong secrecy-capacity region $\overline{\mathcal{C}}^{\text{BCC}}$, and strong secrecy capacity $\overline{C}_s^{\text{WT}}$.

Theorem 3.3 (Csiszár and Körner). *Consider a BCC $(\mathcal{X}, p_{\text{YZ}|\text{X}}, \mathcal{Y}, \mathcal{Z})$. For any joint distribution p_{UVX} on $\mathcal{U} \times \mathcal{V} \times \mathcal{X}$ that factorizes as $p_{\text{U}}p_{\text{V}|\text{U}}p_{\text{X}|\text{V}}$, define the set $\mathcal{R}^{\text{BCC}}(p_{\text{U}}p_{\text{V}|\text{U}}p_{\text{X}|\text{V}})$ as*

$$\mathcal{R}^{\text{BCC}}(p_{\text{U}}p_{\text{V}|\text{U}}p_{\text{X}|\text{V}}) \triangleq \left\{ (R_0, R_1, R_e) : \begin{array}{l} 0 \leqslant R_e \leqslant R_1 \\ 0 \leqslant R_e \leqslant \mathbb{I}(\text{V};\text{Y}|\text{U}) - \mathbb{I}(\text{V};\text{Z}|\text{U}) \\ 0 \leqslant R_0 \leqslant \min\left(\mathbb{I}(\text{U};\text{Y}), \mathbb{I}(\text{U};\text{Z})\right) \\ 0 \leqslant R_1 + R_0 \leqslant \mathbb{I}(\text{V};\text{Y}|\text{U}) + \min(\mathbb{I}(\text{U};\text{Y}), \mathbb{I}(\text{U};\text{Z})) \end{array} \right\},$$

where the joint distribution of U, V, X, Y, *and* Z *factorizes as $p_{\text{U}}p_{\text{V}|\text{U}}p_{\text{X}|\text{V}}p_{\text{YZ}|\text{X}}$. Then, the rate–equivocation region of the BCC is the convex region*

$$\mathcal{R}^{\text{BCC}} = \bigcup_{p_{\text{U}}p_{\text{V}|\text{U}}p_{\text{X}|\text{V}}} \mathcal{R}^{\text{BCC}}(p_{\text{U}}p_{\text{V}|\text{U}}p_{\text{X}|\text{V}}).$$

In addition, the cardinality of the sets $\mathcal{U}$ and $\mathcal{V}$ can be limited to

$$|\mathcal{U}| \leqslant |\mathcal{X}| + 3 \quad \textit{and} \quad |\mathcal{V}| \leqslant |\mathcal{X}|^2 + 4\,|\mathcal{X}| + 3.$$

The typical shape of $\mathcal{R}^{\text{BCC}}(p_{\text{U}}p_{\text{V}|\text{U}}p_{\text{X}|\text{V}})$ is illustrated in Figure 3.13. Note that the upper bound for the equivocation R_e is similar to that obtained in Theorem 3.2 for the DWTC and involves the difference between two information rates. However, the expression includes the *auxiliary random variables* U and V. Exactly how and why U and V appear in the expressions will become clear when we discuss the details of the proof in Section 3.5.2 and Section 3.5.3; at this point, suffice it to say that U, which appears as a conditioning random variable, represents the common message decodable by both the legitimate receiver and the eavesdropper while V accounts for additional randomization in the encoder.

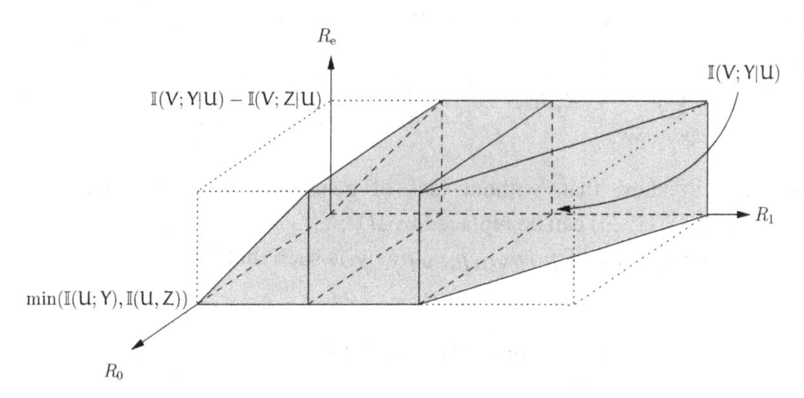

Figure 3.13 Typical shape of $\mathcal{R}^{\text{BCC}}(p_U p_{V|U} p_{X|V})$.

Theorem 3.3 leads to the following characterizations of $\mathcal{C}^{\text{BCC}}$, $\mathcal{R}^{\text{WT}}$, and $\mathcal{C}^{\text{WT}}$.

Corollary 3.2. *Consider a BCC $(\mathcal{X}, p_{YZ|X}, \mathcal{Y}, \mathcal{Z})$. For any joint distribution p_{UVX} on $\mathcal{U} \times \mathcal{V} \times \mathcal{X}$ that factorizes as $p_U p_{V|U} p_{X|V}$, define the set $\mathcal{C}^{\text{BCC}}(p_U p_{V|U} p_{X|V})$ as*

$$\mathcal{C}^{\text{BCC}}(p_U p_{V|U} p_{X|V}) \triangleq \left\{(R_0, R_1): \begin{array}{l} 0 \leqslant R_1 \leqslant \mathbb{I}(V;Y|U) - \mathbb{I}(V;Z|U) \\ 0 \leqslant R_0 \leqslant \min(\mathbb{I}(U;Y), \mathbb{I}(U;Z)) \end{array}\right\},$$

where the joint distribution of U, V, X, Y, *and* Z *factorizes as $p_U p_{V|U} p_{X|V} p_{YZ|X}$. Then, the secrecy-capacity region of the BCC is the convex set*

$$\mathcal{C}^{\text{BCC}} = \bigcup_{p_U p_{V|U} p_{X|V}} \mathcal{C}^{\text{BCC}}(p_U p_{V|U} p_{X|V}).$$

Corollary 3.3. *Consider a WTC $(\mathcal{X}, p_{YZ|X}, \mathcal{Y}, \mathcal{Z})$. For any joint distribution p_{UVX} on $\mathcal{U} \times \mathcal{V} \times \mathcal{X}$ that factorizes as $p_U p_{V|U} p_{X|V}$, define the set $\mathcal{R}^{\text{WT}}(p_U p_{V|U} p_{X|V})$ as*

$$\mathcal{R}^{\text{WT}}(p_U p_{V|U} p_{X|V}) \triangleq \left\{(R, R_e): \begin{array}{l} 0 \leqslant R_e \leqslant R \leqslant \mathbb{I}(V;Y) \\ 0 \leqslant R_e \leqslant \mathbb{I}(V;Y|U) - \mathbb{I}(V;Z|U) \end{array}\right\}.$$

Then, the rate–equivocation region of the WTC is the convex set

$$\mathcal{R}^{\text{WT}} = \bigcup_{p_U p_{V|U} p_{X|V}} \mathcal{R}^{\text{WT}}(p_U p_{V|U} p_{X|V}).$$

In addition, the distributions $p_U p_{V|U} p_{X|V}$ can be limited to those such that $\mathbb{I}(U;Y) \leqslant \mathbb{I}(U;Z)$.

Remark 3.12. *The additional condition $\mathbb{I}(U;Y) \leqslant \mathbb{I}(U;Z)$ in Corollary 3.3 may seem unnecessary since $\mathcal{R}^{\text{WT}}$ is already completely characterized by taking the union over all distributions factorizing as $p_U p_{V|U} p_{X|V}$; however, if we were to evaluate $\mathcal{R}^{\text{WT}}$ numerically, we could speed up computations tremendously by focusing on the subset of distributions satisfying the condition $\mathbb{I}(U;Y) \leqslant \mathbb{I}(U;Z)$.*

Proof of Corollary 3.3. Substituting $R_0 = 0$ into Theorem 3.3 shows that

$$\mathcal{R}^{\mathrm{WT}} = \bigcup_{p_{\mathsf{U}} p_{\mathsf{V}|\mathsf{U}} p_{\mathsf{X}|\mathsf{V}}} \left\{ (R, R_e)\colon \begin{array}{l} 0 \leqslant R_e \leqslant R_1 \leqslant \mathbb{I}(\mathsf{V};\mathsf{Y}|\mathsf{U}) + \min(\mathbb{I}(\mathsf{U};\mathsf{Y}), \mathbb{I}(\mathsf{U};\mathsf{Z})) \\ 0 \leqslant R_e \leqslant \mathbb{I}(\mathsf{V};\mathsf{Y}|\mathsf{U}) - \mathbb{I}(\mathsf{V};\mathsf{Z}|\mathsf{U}) \end{array} \right\}.$$

We need to show that, without loss of generality, the upper bound $\mathbb{I}(\mathsf{V};\mathsf{Y}|\mathsf{U}) + \min(\mathbb{I}(\mathsf{U};\mathsf{Y}), \mathbb{I}(\mathsf{U};\mathsf{Z}))$ can be replaced by $\mathbb{I}(\mathsf{V};\mathsf{Y})$.

This is always true if $p_{\mathsf{U}} p_{\mathsf{V}|\mathsf{U}} p_{\mathsf{X}|\mathsf{V}} p_{\mathsf{YZ}|\mathsf{X}}$ is such that $\mathbb{I}(\mathsf{U};\mathsf{Y}) \leqslant \mathbb{I}(\mathsf{U};\mathsf{Z})$ because, in this case,

$$\begin{aligned} \mathbb{I}(\mathsf{V};\mathsf{Y}|\mathsf{U}) + \min(\mathbb{I}(\mathsf{U};\mathsf{Y}), \mathbb{I}(\mathsf{U};\mathsf{Z})) &= \mathbb{I}(\mathsf{UV};\mathsf{Y}) \\ &= \mathbb{I}(\mathsf{V};\mathsf{Y}) + \mathbb{I}(\mathsf{U};\mathsf{Y}|\mathsf{V}) \\ &= \mathbb{I}(\mathsf{V};\mathsf{Y}), \end{aligned}$$

where the last equality follows from $\mathbb{I}(\mathsf{U};\mathsf{Y}|\mathsf{V}) = 0$ since $\mathsf{U} \rightarrow \mathsf{V} \rightarrow \mathsf{Y}$ forms a Markov chain. In particular, note that the distributions $p_{\mathsf{U}} p_{\mathsf{V}|\mathsf{U}} p_{\mathsf{X}|\mathsf{V}} p_{\mathsf{YZ}|\mathsf{X}}$ for which U is a constant satisfy $\mathbb{I}(\mathsf{U};\mathsf{Y}) = \mathbb{I}(\mathsf{U};\mathsf{Z}) = 0$ and thus $\mathbb{I}(\mathsf{U};\mathsf{Y}) \leqslant \mathbb{I}(\mathsf{U};\mathsf{Z})$.

Consider now a distribution $p_{\mathsf{U}} p_{\mathsf{V}|\mathsf{U}} p_{\mathsf{X}|\mathsf{V}} p_{\mathsf{YZ}|\mathsf{X}}$ such that $\mathbb{I}(\mathsf{U};\mathsf{Y}) > \mathbb{I}(\mathsf{U};\mathsf{Z})$. Then,

$$\begin{aligned} \mathbb{I}(\mathsf{V};\mathsf{Y}|\mathsf{U}) + \min(\mathbb{I}(\mathsf{U};\mathsf{Y}), \mathbb{I}(\mathsf{U};\mathsf{Z})) &= \mathbb{I}(\mathsf{V};\mathsf{Y}|\mathsf{U}) + \mathbb{I}(\mathsf{U};\mathsf{Z}) \\ &< \mathbb{I}(\mathsf{V};\mathsf{Y}|\mathsf{U}) + \mathbb{I}(\mathsf{U};\mathsf{Y}) \\ &= \mathbb{I}(\mathsf{V};\mathsf{Y}), \end{aligned}$$

and, similarly,

$$\begin{aligned} \mathbb{I}(\mathsf{V};\mathsf{Y}|\mathsf{U}) - \mathbb{I}(\mathsf{V};\mathsf{Z}|\mathsf{U}) &= \sum_{u \in \mathcal{U}} p_{\mathsf{U}}(u)(\mathbb{I}(\mathsf{V};\mathsf{Y}|\mathsf{U}=u) - \mathbb{I}(\mathsf{V};\mathsf{Z}|\mathsf{U}=u)) \\ &\leqslant \max_{u \in \mathcal{U}}(\mathbb{I}(\mathsf{V};\mathsf{Y}|\mathsf{U}=u) - \mathbb{I}(\mathsf{V};\mathsf{Z}|\mathsf{U}=u)). \end{aligned}$$

Therefore, the rates (R_1, R_e) obtained with a distribution $p_{\mathsf{U}} p_{\mathsf{V}|\mathsf{U}} p_{\mathsf{X}|\mathsf{V}} p_{\mathsf{YZ}|\mathsf{X}}$ such that $\mathbb{I}(\mathsf{U};\mathsf{Y}) > \mathbb{I}(\mathsf{U};\mathsf{Z})$ are upper bounded by the rates obtained with a distribution $p_{\mathsf{U}} p_{\mathsf{V}|\mathsf{U}} p_{\mathsf{X}|\mathsf{V}} p_{\mathsf{YZ}|\mathsf{X}}$ in which U is a constant. Therefore, without loss of generality, we can obtain the entire region $\mathcal{R}^{\mathrm{WT}}$ by restricting the union to the distributions $p_{\mathsf{U}} p_{\mathsf{V}|\mathsf{U}} p_{\mathsf{X}|\mathsf{V}} p_{\mathsf{YZ}|\mathsf{X}}$ that satisfy $\mathbb{I}(\mathsf{U};\mathsf{Y}) \leqslant \mathbb{I}(\mathsf{U};\mathsf{Z})$ and we can replace the upper bound $\mathbb{I}(\mathsf{V};\mathsf{Y}|\mathsf{U}) + \min(\mathbb{I}(\mathsf{U};\mathsf{Y}), \mathbb{I}(\mathsf{U};\mathsf{Z}))$ by $\mathbb{I}(\mathsf{V};\mathsf{Y})$. □

Corollary 3.4. *The secrecy capacity of a WTC $(\mathcal{X}, p_{\mathsf{YZ}|\mathsf{X}}, \mathcal{Y}, \mathcal{Z})$ is*

$$C_s^{\mathrm{WT}} = \max_{p_{\mathsf{VX}}}(\mathbb{I}(\mathsf{V};\mathsf{Y}) - \mathbb{I}(\mathsf{V};\mathsf{Z})).$$

Proof. By Corollary 3.3,

$$C_s^{\mathrm{WT}} = \max_{p_{\mathsf{U}} p_{\mathsf{V}|\mathsf{U}} p_{\mathsf{X}|\mathsf{V}}} (\mathbb{I}(\mathsf{V};\mathsf{Y}|\mathsf{U}) - \mathbb{I}(\mathsf{V};\mathsf{Z}|\mathsf{U})),$$

which can be expanded as

$$\begin{aligned}
C_s^{\mathrm{WT}} &= \max_{p_{\mathsf{U}} p_{\mathsf{V}|\mathsf{U}} p_{\mathsf{X}|\mathsf{V}}} \left(\mathbb{I}(\mathsf{V};\mathsf{Y}|\mathsf{U}) - \mathbb{I}(\mathsf{V};\mathsf{Z}|\mathsf{U})\right) \\
&= \max_{p_{\mathsf{U}} p_{\mathsf{V}|\mathsf{U}} p_{\mathsf{X}|\mathsf{V}}} \left(\sum_{u\in\mathcal{U}} p_{\mathsf{U}}(u)\left(\mathbb{I}(\mathsf{V};\mathsf{Y}|\mathsf{U}=u) - \mathbb{I}(\mathsf{V};\mathsf{Z}|\mathsf{U}=u)\right)\right) \\
&= \max_{p_{\mathsf{VX}}} (\mathbb{I}(\mathsf{V};\mathsf{Y}) - \mathbb{I}(\mathsf{V};\mathsf{Z})).
\end{aligned}$$

□

3.5.1 Channel comparison

Although Corollary 3.4 provides an exact characterization of the secrecy capacity, the auxiliary random variable V makes the evaluation of C_s^{WT} arduous and prevents us from developing much intuition about the possibility of secure communication. Nevertheless, it is possible to establish the following general result.

Lemma 3.4 (Liang *et al.*). *The secrecy capacity of a WTC* $(\mathcal{X}, p_{\mathsf{YZ}|\mathsf{X}}, \mathcal{Y}, \mathcal{Z})$ *depends on the transition probabilities* $p_{\mathsf{YZ}|\mathsf{X}}$ *only through the marginal transition probabilities* $p_{\mathsf{Y}|\mathsf{X}}$ *and* $p_{\mathsf{Z}|\mathsf{X}}$.

Proof. Consider a code $\mathcal{C}_n$ designed for a WTC $(\mathcal{X}, p_{\mathsf{YZ}|\mathsf{X}}, \mathcal{Y}, \mathcal{Z})$. By definition, the average error probability $\mathbf{P}_e(\mathcal{C}_n) = \mathbb{P}\left[\mathsf{M} \neq \hat{\mathsf{M}} \mid \mathcal{C}_n\right]$ is determined by the distribution $p_{\mathsf{MX}^n\mathsf{Y}^n}$ and hence depends on the transition probabilities $p_{\mathsf{Y}|\mathsf{X}}$ but not on the transition probabilities $p_{\mathsf{Z}|\mathsf{X}}$. Similarly, by definition, the equivocation $\mathbf{E}(\mathcal{C}_n) = \mathbb{H}(\mathsf{M}|\mathsf{Z}^n\mathcal{C}_n)$ is determined by the distribution $p_{\mathsf{MX}^n\mathsf{Z}^n}$ and hence depends on the transition probabilities $p_{\mathsf{Z}|\mathsf{X}}$ but not on the transition probabilities $p_{\mathsf{Y}|\mathsf{X}}$. Consequently, whether a rate is achievable or not depends only on the marginal transition probabilities $p_{\mathsf{Y}|\mathsf{X}}$ and $p_{\mathsf{Z}|\mathsf{X}}$. □

Intuitively, Lemma 3.4 states that we can understand whether secure communication is possible or not if we can somehow compare the main channel $(\mathcal{X}, p_{\mathsf{Y}|\mathsf{X}}, \mathcal{Y})$ with the eavesdropper's channel $(\mathcal{X}, p_{\mathsf{Z}|\mathsf{X}}, \mathcal{Z})$.

We have already studied a specific relation between the main channel and the eavesdropper's channel when we analyzed the DWTC in Section 3.4. In fact, the transition probabilities $p_{\mathsf{YZ}|\mathsf{X}}$ factorize as $p_{\mathsf{Z}|\mathsf{Y}}p_{\mathsf{Y}|\mathsf{X}}$ for a DWTC. We formalize this relation between the eavesdropper's channel and the main channel by introducing the notion of *physically degraded* channels.

Definition 3.8 (Physically degraded channel). *We say that* $(\mathcal{X}, p_{\mathsf{Z}|\mathsf{X}}, \mathcal{Z})$ *is physically degraded with respect to* $(\mathcal{X}, p_{\mathsf{Y}|\mathsf{X}}, \mathcal{Y})$ *if*

$$\forall (x,y,z) \in \mathcal{X}\times\mathcal{Y}\times\mathcal{Z} \quad p_{\mathsf{YZ}|\mathsf{X}}(y,z|x) = p_{\mathsf{Z}|\mathsf{Y}}(z|y)p_{\mathsf{Y}|\mathsf{X}}(y|x)$$

for some transition probabilities $p_{\mathsf{Z}|\mathsf{Y}}$. *In other words,* $(\mathcal{X}, p_{\mathsf{Z}|\mathsf{X}}, \mathcal{Z})$ *is physically degraded with respect to* $(\mathcal{X}, p_{\mathsf{Y}|\mathsf{X}}, \mathcal{Y})$ *if* $\mathsf{X}\rightarrow\mathsf{Y}\rightarrow\mathsf{Z}$ *forms a Markov chain.*

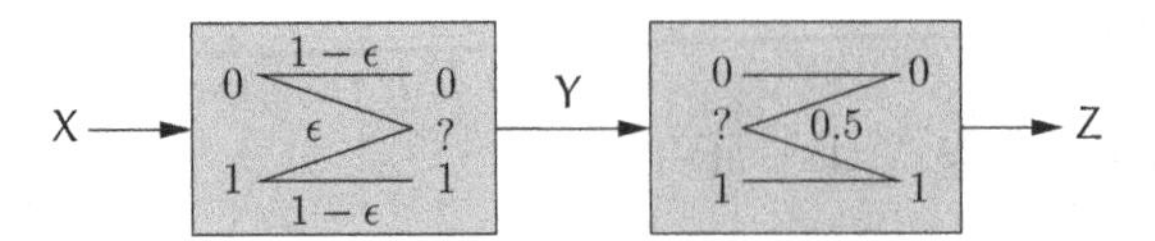

Figure 3.14 Example of a physically degraded channel.

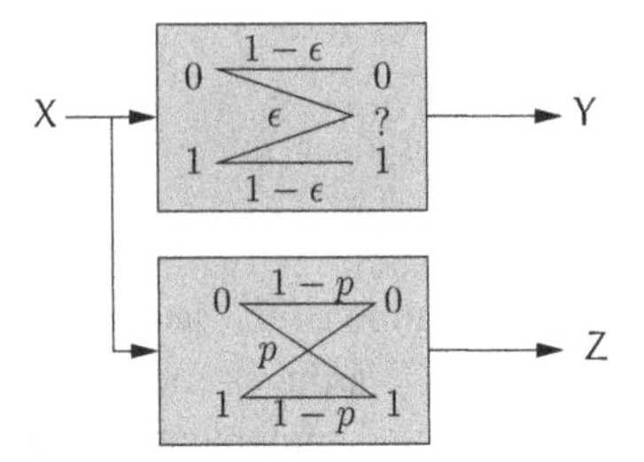

Figure 3.15 Example of a non-physically degraded channel.

Therefore, a DWTC is simply a WTC in which the eavesdropper's channel is physically degraded with respect to the main channel (see (3.2)).

Example 3.6. Consider the concatenated channels illustrated in Figure 3.14, which are such that $(\mathcal{X}, p_{Y|X}, \mathcal{Y})$ is a binary erasure channel BEC(ϵ) and $(\mathcal{X}, p_{Z|X}, \mathcal{Z})$ is a binary symmetric channel BSC($\epsilon/2$). By construction, $(\mathcal{X}, p_{Z|X}, \mathcal{Z})$ is physically degraded with respect to $(\mathcal{X}, p_{Y|X}, \mathcal{Y})$.

Since physical degradedness is a stringent constraint, it is useful to consider weaker relations between the channels $(\mathcal{X}, p_{Y|X}, \mathcal{Y})$ and $(\mathcal{X}, p_{Z|X}, \mathcal{Z})$ obtained from the marginals of a broadcast channel $(\mathcal{X}, p_{YZ|X}, \mathcal{Y}, \mathcal{Z})$.

Definition 3.9 (Stochastically degraded channel). *We say that $(\mathcal{X}, p_{Z|X}, \mathcal{Z})$ is stochastically degraded with respect to $(\mathcal{X}, p_{Y|X}, \mathcal{Y})$ if there exists a channel $(\mathcal{Y}, p_{Z|Y}, \mathcal{Z})$ such that*

$$\forall (x, z) \in \mathcal{X} \times \mathcal{Z} \quad p_{Z|X}(z|x) = \sum_{y \in \mathcal{Y}} p_{Z|Y}(z|y) p_{Y|X}(y|x).$$

In other words, $(\mathcal{X}, p_{Z|X}, \mathcal{Z})$ has the same marginal as a channel that is physically degraded with respect to $(\mathcal{X}, p_{Y|X}, \mathcal{Y})$.

Example 3.7. Consider the broadcast channel illustrated in Figure 3.15 in which $(\mathcal{X}, p_{Y|X}, \mathcal{Y})$ is a binary erasure channel BEC(ϵ) and $(\mathcal{X}, p_{Z|X}, \mathcal{Z})$ is a binary symmetric channel BSC(p) with $p \in [0, \frac{1}{2}]$. If $0 \leqslant \epsilon \leqslant 2p$, then $(\mathcal{X}, p_{Z|X}, \mathcal{Z})$ is stochastically degraded with respect to $(\mathcal{X}, p_{Y|X}, \mathcal{Y})$. This fact is the consequence of a more general result that we derive in Proposition 6.4.

Note that there is no real difference between stochastically degraded channels and physically degraded channels because, by Lemma 3.4, the secrecy capacity depends only on the marginal transition probabilities of the WTC. For stochastically degraded channels, it is possible to relax the assumptions made about the eavesdropper's channel as follows.

Definition 3.10 (Class of stochastically degraded channels). *A class of channels is said to be stochastically degraded with respect to a worst channel* $(\mathcal{X}, p_{\mathsf{Y}_0|\mathsf{X}}, \mathcal{Y}_0)$ *if and only if every channel* $(\mathcal{X}, p_{\mathsf{Y}|\mathsf{X}}, \mathcal{Y})$ *in the class is stochastically degraded with respect to the worst channel.*

Proposition 3.3 (Robustness of worst-case design). *Given a class of stochastically degraded eavesdropper's channels, a wiretap code ensuring equivocation* R_{e} *for the worst channel guarantees at least the same equivocation for any eavesdropper's channel in the class.*

Proof. The result is a direct consequence of the data-processing inequality. Let M be the message sent, let Y_0^n denote the output of the worst channel, and let Y^n denote the output of the eavesdropper's channel. Note that Y^n is statistically indistinguishable from the output of a physically degraded channel for which $\mathsf{M} \rightarrow \mathsf{Y}_0^n \rightarrow \mathsf{Y}^n$; therefore, the data-processing inequality ensures that

$$\frac{1}{n}\mathbb{H}(\mathsf{M}|\mathsf{Y}^n) \geqslant \frac{1}{n}\mathbb{H}\left(\mathsf{M}|\mathsf{Y}_0^n\right) \geqslant R_{\mathrm{e}}. \qquad \square$$

Despite is simplicity, Proposition 3.3 has useful applications, as illustrated by the following examples.

Example 3.8. Consider a binary erasure wiretap channel BEC(ϵ) as in Figure 3.3, for which only a lower bound ϵ^* of the eavesdropper's erasure probability is known. It is easy to verify that the set of erasure channels with erasure probability $\epsilon \geqslant \epsilon^*$ is a class of stochastically degraded channels, for which the worst channel is the one with erasure probability ϵ^*. Proposition 3.3 ensures that a wiretap code designed for the worst channel guarantees secrecy no matter what the actual erasure probability is.

Example 3.9. Another application of Proposition 3.3 is the situation in which we do not know how to design a code for a specific channel. For instance, we will see in Chapter 6 that designing wiretap codes for channels other than erasure channels is challenging. If we can show that the channel is stochastically degraded with respect to another "simpler" channel C^*, we can design a wiretap code for C^*. For instance, consider a wiretap channel with a noiseless main channel and a binary symmetric channel BSC(p) ($p < \frac{1}{2}$) as the eavesdropper's channel. The binary symmetric channel is degraded with respect to a binary erasure channel BEC($2p$), hence any wiretap code designed to operate for the latter channel also provides secrecy for the former.

Definition 3.11 (Noisier channel). *$(\mathcal{X}, p_{Z|X}, \mathcal{Z})$ is noisier than $(\mathcal{X}, p_{Y|X}, \mathcal{Y})$ if for every random variable* V *such that* $V \rightarrow X \rightarrow YZ$ *we have* $\mathbb{I}(V;Y) \geqslant \mathbb{I}(V;Z)$.

Recall from Corollary 3.4 that $C_s^{WT} = \max_{p_{VX}}(\mathbb{I}(V;Y) - \mathbb{I}(V;Z))$; therefore, $C_s = 0$ if and only if $\mathbb{I}(V;Y) \leqslant \mathbb{I}(V;Z)$ for all Markov chains $V \rightarrow X \rightarrow YZ$, which is exactly the definition of the eavesdropper's channel being noisier than the main channel. This result is summarized in the following proposition.

Proposition 3.4. *The secrecy capacity of a WTC $(\mathcal{X}, p_{YZ|X}, \mathcal{Y}, \mathcal{Z})$ is zero if and only if the main channel $(\mathcal{X}, p_{Y|X}, \mathcal{Y})$ is noisier than the eavesdropper's channel $(\mathcal{X}, p_{Z|X}, \mathcal{Z})$.*

Because the definition of "being noisier" involves an auxiliary random variable V, it may be much harder to verify that a channel is noisier than to verify that it is physically or stochastically degraded. Fortunately, the property "being noisier" admits the following characterization, which is sometimes simpler to check.

Proposition 3.5 (Van Dijk). *$(\mathcal{X}, p_{Z|X}, \mathcal{Z})$ is noisier than $(\mathcal{X}, p_{Y|X}, \mathcal{Y})$ if and only if $\mathbb{I}(X;Y) - \mathbb{I}(X;Z)$ is a concave function of the input probability distribution p_X.*

Proof. Suppose $V \rightarrow X \rightarrow Y$ forms a Markov chain. Using the conditional independence of V and Y given X we have

$$\begin{aligned}\mathbb{I}(V;Y) &= \mathbb{I}(VX;Y) - \mathbb{I}(X;Y|V)\\ &= \mathbb{I}(X;Y) + \mathbb{I}(V;Y|X) - \mathbb{I}(X;Y|V)\\ &= \mathbb{I}(X;Y) - \mathbb{I}(X;Y|V).\end{aligned}$$

The same equality also holds if we replace Y by Z; therefore,

$$\begin{aligned}\mathbb{I}(V;Z) \leqslant \mathbb{I}(V;Y) &\Leftrightarrow \mathbb{I}(X;Z) - \mathbb{I}(X;Z|V) \leqslant \mathbb{I}(X;Y) - \mathbb{I}(X;Y|V)\\ &\Leftrightarrow \mathbb{I}(X;Y|V) - \mathbb{I}(X;Z|V) \leqslant \mathbb{I}(X;Y) - \mathbb{I}(X;Z).\end{aligned} \tag{3.33}$$

For any $v \in \mathcal{V}$, we define the random variable X_v, whose distribution satisfies

$$\forall x \in \mathcal{X} \quad p_{X_v}(x) \triangleq p_{X|V}(x|v).$$

Since $V \rightarrow X \rightarrow YZ$ forms a Markov chain, note that

$$\mathbb{I}(X;Y|V) = \sum_{v\in\mathcal{V}} p_V(v)\mathbb{I}(X;Y|V=v) = \sum_{v\in\mathcal{V}} p_V(v)\mathbb{I}(X_v;Y),$$

$$\mathbb{I}(X;Z|V) = \sum_{v\in\mathcal{V}} p_V(v)\mathbb{I}(X;Z|V=v) = \sum_{v\in\mathcal{V}} p_V(v)\mathbb{I}(X_v;Z);$$

therefore, we can rewrite (3.33) as

$$\mathbb{I}(V;Z) \leqslant \mathbb{I}(V;Y) \Leftrightarrow \sum_{v\in\mathcal{V}} p_V(v)\left(\mathbb{I}(X_v;Y) - \mathbb{I}(X_v;Z)\right) \leqslant \mathbb{I}(X;Y) - \mathbb{I}(X;Z).$$

Noting that

$$\forall x \in \mathcal{X} \quad p_X(x) = \sum_{v\in\mathcal{V}} p_{X|V}(x|v)p_V(v) = \sum_{v\in\mathcal{V}} p_{X_v}(x)p_V(v),$$

and treating $\mathbb{I}(X;Y) - \mathbb{I}(X;Z)$ as a function of the input distribution p_X, the condition $\sum_{v\in\mathcal{V}} p_V(v)(\mathbb{I}(X_v;Y) - \mathbb{I}(X_v;Z)) \leqslant \mathbb{I}(X;Y) - \mathbb{I}(X;Z)$ means that $\mathbb{I}(X;Y) - \mathbb{I}(X;Z)$ is a concave function of the input distribution p_X. □

Example 3.10. Let $(\mathcal{X}, p_{Y|X}, \mathcal{Y})$ be a BEC(ϵ) and let $(\mathcal{X}, p_{Z|X}, \mathcal{Z})$ be a BSC(p) with $p \in [0, \frac{1}{2}]$ as in Figure 3.15. We show that $(\mathcal{X}, p_{Z|X}, \mathcal{Z})$ is noisier than $(\mathcal{X}, p_{Y|X}, \mathcal{Y})$ if and only if

$$0 \leqslant \epsilon \leqslant 4p(1-p).$$

Proof. The result holds trivially if $\epsilon = 1$ or $p = \frac{1}{2}$. Hence, we assume $\epsilon > 0$ and $p < \frac{1}{2}$. Let $X \sim \mathcal{B}(q)$ for some $q \in [0, 1]$. Then, $\mathbb{I}(X;Y) - \mathbb{I}(X;Z)$ is a differentiable function of q given by

$$f : q \mapsto (1-\epsilon)\mathbb{H}_b(q) - \mathbb{H}_b(p + q(1-2p)) + \mathbb{H}_b(p).$$

By Proposition 3.5, it suffices to determine conditions for f to be concave. After some algebra, one obtains

$$\frac{d^2 f}{dq^2}(q) \leqslant 0 \Leftrightarrow -\epsilon q^2 + \epsilon q - \frac{(1-\epsilon)p(1-p)}{(1-2p)^2} \leqslant 0.$$

The quadratic polynomial in q

$$-\epsilon q^2 + \epsilon q - \frac{(1-\epsilon)(1-p)}{(1-2p)^2}$$

is negative if and only if its discriminant Δ is negative; one can check that

$$\Delta \leqslant 0 \Leftrightarrow \epsilon \leqslant 4p(1-p).$$ □

Notice that, for $2p < \epsilon \leqslant 4p(1-p)$, $(\mathcal{X}, p_{Z|X}, \mathcal{Z})$ is noisier than but not stochastically degraded with respect to $(\mathcal{X}, p_{Y|X}, \mathcal{Y})$.

Definition 3.12 (Less capable channel). *$(\mathcal{X}, p_{Z|X}, \mathcal{Z})$ is less capable than $(\mathcal{X}, p_{Y|X}, \mathcal{Y})$ if for every input X we have $\mathbb{I}(X;Y) \geqslant \mathbb{I}(X;Z)$.*

Example 3.11. Let $(\mathcal{X}, p_{Y|X}, \mathcal{Y})$ be a BEC(ϵ) and let $(\mathcal{X}, p_{Z|X}, \mathcal{Z})$ be a BSC(p) with $p \in [0, \frac{1}{2}]$ as in Figure 3.15. We show that $(\mathcal{X}, p_{Z|X}, \mathcal{Z})$ is less capable than $(\mathcal{X}, p_{Y|X}, \mathcal{Y})$ if

$$0 \leqslant \epsilon \leqslant \mathbb{H}_b(p).$$

Proof. The result holds trivially if $p = \frac{1}{2}$, hence without loss of generality we assume $p < \frac{1}{2}$. Assume $\epsilon \leqslant \mathbb{H}_b(p)$, let $X \sim \mathcal{B}(q)$ for some $q \in [0, 1]$, and let f be defined as in Example 3.10. Since $f(q) = f(1-q)$, the function f is symmetric around $q = \frac{1}{2}$. In addition, because $f(0) = 0$ and $f(\frac{1}{2}) = \mathbb{H}_b(p) - \epsilon \geqslant 0$, we prove that $f(q) \geqslant 0$ for $q \in [0, 1]$ by showing that $df/dq(0) \geqslant 0$ and that df/dq changes sign at most once in the interval $[0, \frac{1}{2}]$. After some algebra, one obtains

$$\begin{aligned}\frac{\mathrm{d}f}{\mathrm{d}q}(q) \geqslant 0 &\Leftrightarrow (1-\epsilon)\log\left(\frac{1-q}{q}\right) - (1-2p)\log\left(\frac{1-p-q(1-2p)}{p+q(1-2p)}\right) \geqslant 0\\ &\Leftrightarrow a(p+q(1-2p))(1-q) - ((1-p)-q(1-2p))q \geqslant 0\\ &\Leftrightarrow -(a-1)(1-2p)q^2 + q(a(1-3p)-(1-p)) + ap \geqslant 0,\end{aligned}$$

with

$$a \triangleq \exp\left(\frac{1-\epsilon}{1-2p}\right) \geqslant 1.$$

The quadratic polynomial in q

$$P(q) \triangleq -(a-1)(1-2p)q^2 + q(a(1-3p)-(1-p)) + ap$$

is such that $P(0) = ap \geqslant 0$; therefore, $\mathrm{d}f/\mathrm{d}q(0) \geqslant 0$. In addition, $P(q)$ has at most one root in the interval $[0, \frac{1}{2}]$; therefore, $\mathrm{d}f/\mathrm{d}q$ changes sign at most once, which establishes the result. □

Notice that, for $4p(1-p) < \epsilon \leqslant \mathbb{H}_b(p)$, $(\mathcal{X}, p_{Z|X}, \mathcal{Z})$ is less capable than but not noisier than $(\mathcal{X}, p_{Y|X}, \mathcal{Y})$.

The following proposition shows how Definitions 3.8–3.12 relate to one another.

Proposition 3.6. *Let $(\mathcal{X}, p_{Y|X}, \mathcal{Y})$ and $(\mathcal{X}, p_{Z|X}, \mathcal{Z})$ be two DMCs and consider the following statements:*

(1) $(\mathcal{X}, p_{Z|X}, \mathcal{Z})$ *is physically degraded w.r.t.* $(\mathcal{X}, p_{Y|X}, \mathcal{Y})$
(2) $(\mathcal{X}, p_{Z|X}, \mathcal{Z})$ *is stochastically degraded w.r.t.* $(\mathcal{X}, p_{Y|X}, \mathcal{Y})$
(3) $(\mathcal{X}, p_{Z|X}, \mathcal{Z})$ *is noisier than* $(\mathcal{X}, p_{Y|X}, \mathcal{Y})$
(4) $(\mathcal{X}, p_{Z|X}, \mathcal{Z})$ *is less capable than* $(\mathcal{X}, p_{Y|X}, \mathcal{Y})$.

Then,

$$(1) \Rightarrow (2) \Rightarrow (3) \Rightarrow (4).$$

Examples 3.6–3.11 show that the implications of Proposition 3.6 are strict.

Corollary 3.5. *The secrecy capacity of a WTC $(\mathcal{X}, p_{YZ|X}, \mathcal{Y}, \mathcal{Z})$ in which the eavesdropper's channel is less capable than the main channel is*

$$C_s^{\mathrm{WT}} = \max_{p_X}(\mathbb{I}(X;Y) - \mathbb{I}(X;Z)).$$

Proof. The achievability of rates below C_s^{WT} follows directly from Corollary 3.3 for general non-degraded channels on choosing $V = X$. To obtain the converse, note that

$$\begin{aligned}\mathbb{I}(V;Y) &= \mathbb{I}(VX;Y) - \mathbb{I}(X;Y|V)\\ &= \mathbb{I}(X;Y) + \mathbb{I}(V;Y|X) - \mathbb{I}(X;Y|V)\\ &= \mathbb{I}(X;Y) - \mathbb{I}(X;Y|V)\end{aligned}$$

since $\mathsf{V} \to \mathsf{X} \to \mathsf{Y}$ forms a Markov chain. Similarly,

$$\mathbb{I}(\mathsf{V};\mathsf{Z}) = \mathbb{I}(\mathsf{X};\mathsf{Z}) - \mathbb{I}(\mathsf{X};\mathsf{Z}|\mathsf{V}).$$

Therefore,

$$\mathbb{I}(\mathsf{V};\mathsf{Y}) - \mathbb{I}(\mathsf{V};\mathsf{Z}) = \mathbb{I}(\mathsf{X};\mathsf{Y}) - \mathbb{I}(\mathsf{X};\mathsf{Z}) + \mathbb{I}(\mathsf{X};\mathsf{Z}|\mathsf{V}) - \mathbb{I}(\mathsf{X};\mathsf{Y}|\mathsf{V}).$$

Now, the difference $\mathbb{I}(\mathsf{X};\mathsf{Z}|\mathsf{V}) - \mathbb{I}(\mathsf{X};\mathsf{Y}|\mathsf{V})$ can be upper bounded as

$$\begin{aligned}
\mathbb{I}(\mathsf{X};\mathsf{Z}|\mathsf{V}) - \mathbb{I}(\mathsf{X};\mathsf{Y}|\mathsf{V}) &\leqslant \max_{p_{\mathsf{V}\mathsf{X}}} \left(\mathbb{I}(\mathsf{X};\mathsf{Z}|\mathsf{V}) - \mathbb{I}(\mathsf{X};\mathsf{Y}|\mathsf{V})\right) \\
&= \max_{p_{\mathsf{V}\mathsf{X}}} \left(\sum_{v \in \mathcal{V}} p_{\mathsf{V}}(v) \left(\mathbb{I}(\mathsf{X};\mathsf{Z}|\mathsf{V}=v) - \mathbb{I}(\mathsf{X};\mathsf{Y}|\mathsf{V}=v)\right)\right) \\
&= \max_{p_{\mathsf{X}}} \left(\mathbb{I}(\mathsf{X};\mathsf{Z}) - \mathbb{I}(\mathsf{X};\mathsf{Y})\right) \\
&\leqslant 0,
\end{aligned}$$

where the last inequality follows from the assumption that the eavesdropper's channel is less capable than the main channel. Consequently,

$$\mathbb{I}(\mathsf{V};\mathsf{Y}) - \mathbb{I}(\mathsf{V};\mathsf{Z}) \leqslant \mathbb{I}(\mathsf{X};\mathsf{Y}) - \mathbb{I}(\mathsf{X};\mathsf{Z}),$$

which proves that the choice $\mathsf{V} = \mathsf{X}$ in Corollary 3.3 is optimal. □

From a practical standpoint, the fact that the choice $\mathsf{V} = \mathsf{X}$ is optimum means that achieving the secrecy capacity does not require additional randomization at the encoder, which is convenient because we do not have an explicit characterization of this randomization. Because of Proposition 3.6, the expression for the secrecy capacity given in Corollary 3.5 also holds if the eavesdropper's channel is noisier than, stochastically degraded with respect to, or physically degraded with respect to the main channel. In retrospect, it might be surprising that this expression remains the same on weakening the advantage of the main channel over the eavesdropper's channel, but this suggests that additional randomization in the encoder is unnecessary for a large class of channels.

If the eavesdropper's channel is noisier than the main channel and both channels are weakly symmetric, then the secrecy capacity is the difference between the main channel capacity and the eavesdropper's channel capacity. This result generalizes Proposition 3.2, which was established for DWTCs.

Proposition 3.7 (Van Dijk). *For a WTC $(\mathcal{X}, p_{\mathsf{YZ}|\mathsf{X}}, \mathcal{Y}, \mathcal{Z})$, if the eavesdropper's channel is noisier than the main channel and both channels are weakly symmetric, then*

$$C_{\mathrm{s}}^{\mathrm{WT}} = C_{\mathrm{m}} - C_{\mathrm{e}},$$

where C_{m} is the channel capacity of the main channel and C_{e} that of the eavesdropper's channel.

Proof. From Corollary 3.5, $C_{\mathrm{s}}^{\mathrm{WT}} = \max_{p_{\mathsf{X}}} (\mathbb{I}(\mathsf{X};\mathsf{Y}) - \mathbb{I}(\mathsf{X};\mathsf{Z}))$ and, from Proposition 3.5, $\mathbb{I}(\mathsf{X};\mathsf{Y}) - \mathbb{I}(\mathsf{X};\mathsf{Z})$ is a concave function of p_{X}; therefore, we can reiterate the proof of Proposition 3.2. We note that the result also holds if the channels are not weakly symmetric, provided that $\mathbb{I}(X;Y)$ and $\mathbb{I}(X;Z)$ are maximized by the same input distribution p_X. □

As illustrated by Example 3.12 below, Proposition 3.7 does *not* hold if we replace "noisier" by "less capable."

Example 3.12. Let $(\mathcal{X}, p_{Y|X}, \mathcal{Y})$ be a BEC(ϵ) and let $(\mathcal{X}, p_{Z|X}, \mathcal{Z})$ be a BSC(p) as in Figure 3.15. For $\epsilon = \mathbb{H}_b(p)$, we know from Example 3.11 that $(\mathcal{X}, p_{Z|X}, \mathcal{Z})$ is less capable than but not noisier than $(\mathcal{X}, p_{Y|X}, \mathcal{Y})$. Notice that $C_m = 1 - \epsilon = 1 - \mathbb{H}_b(p) = C_e$; therefore, $C_m - C_e = 0$ bits. On the other hand, one can check numerically that $C_s^{WT} \approx 0.026$ bits.

To conclude this section, we illustrate the usefulness of Corollary 3.5 by computing the secrecy capacities of several wiretap channels.

Example 3.13. Consider a broadcast channel in which the main channel is a BSC(p) and the eavesdropper's channel is a BSC(r) with $r > p$. Then, the eavesdropper's channel is stochastically degraded with respect to the main channel and the secrecy capacity is $C_s^{WT} = \mathbb{H}_b(r) - \mathbb{H}_b(p)$.

Example 3.14. Consider a broadcast channel in which the main channel is a BEC(ϵ_1) and the eavesdropper's channel is a BEC(ϵ_2) with $\epsilon_2 > \epsilon_1$. Although the two channels could be *correlated* erasure channels, the secrecy capacity is $C_s^{WT} = \epsilon_2 - \epsilon_1$ because the BEC(ϵ_2) is stochastically degraded with respect to the BEC(ϵ_1).

3.5.2 Achievability proof for the broadcast channel with confidential messages

The idea of the proof is similar to Section 3.4.1. The presence of a common message and absence of physical degradedness require the introduction of two auxiliary random variables and make the proof slightly more technical, but the intuition developed earlier still applies. We carry out the proof in four steps.

1. For a fixed distribution p_{UX} on $\mathcal{U} \times \mathcal{X}$, some arbitrary $\epsilon > 0$, and rates (R_0, R_1) such that

$$R_0 < \min(\mathbb{I}(U;Y), \mathbb{I}(U;Z)), \quad \text{and} \quad R_1 < \mathbb{I}(X;Y|U) - \mathbb{I}(X;Z|U),$$

we use a random-coding argument to show the existence of a sequence of $(2^{nR_0}, 2^{nR_1}, n)$ codes $\{\mathcal{C}_n\}_{n\geqslant 1}$ such that

$$\lim_{n\to\infty} \mathbf{P}_e(\mathcal{C}_n) = 0, \quad \lim_{n\to\infty} \mathbb{H}(R|Z^n M_1 M_0 \mathcal{C}_n) = 0, \quad \text{and} \quad \lim_{n\to\infty} \frac{1}{n}\mathbf{L}(\mathcal{C}_n) \leqslant \delta(\epsilon).$$

The proof combines the superposition coding technique introduced in Section 2.3.3 for broadcast channels with the binning structure of wiretap codes identified in Section 3.4.1. This shows the existence of wiretap codes with rate close to full secrecy and guarantees that $\mathcal{R}_1(p_{\mathsf{UX}}) \subseteq \mathcal{R}^{\text{BCC}}$, where

$$\mathcal{R}_1(p_{\mathsf{UX}}) = \left\{(R_0, R_1, R_e) \colon \begin{array}{l} 0 \leqslant R_0 \leqslant \min(\mathbb{I}(\mathsf{U};\mathsf{Y}), \mathbb{I}(\mathsf{U};\mathsf{Z})) \\ 0 \leqslant R_e \leqslant R_1 \leqslant \mathbb{I}(\mathsf{X};\mathsf{Y}|\mathsf{U}) - \mathbb{I}(\mathsf{X};\mathsf{Z}|\mathsf{U}) \end{array}\right\}.$$

2. With a minor modification of the codes identified in Step 1, which can be thought of as an outer code construction, we show that $\mathcal{R}_2(p_{\mathsf{UX}}) \subseteq \mathcal{R}^{\text{BCC}}$, where

$$\mathcal{R}_2(p_{\mathsf{UX}}) = \left\{(R_0, R_1, R_e) \colon \begin{array}{l} 0 \leqslant R_e \leqslant R_1 \\ R_e \leqslant \mathbb{I}(\mathsf{X};\mathsf{Y}|\mathsf{U}) - \mathbb{I}(\mathsf{X};\mathsf{Z}|\mathsf{U}) \\ R_1 + R_0 \leqslant \mathbb{I}(\mathsf{X};\mathsf{Y}|\mathsf{U}) + \min(\mathbb{I}(\mathsf{U};\mathsf{Y}), \mathbb{I}(\mathsf{U};\mathsf{Z})) \\ 0 \leqslant R_0 \leqslant \min(\mathbb{I}(\mathsf{U};\mathsf{Y}), \mathbb{I}(\mathsf{U};\mathsf{Z})) \end{array}\right\}.$$

3. We show that the region $\bigcup_{p_{\mathsf{UX}}} \mathcal{R}_2(p_{\mathsf{UX}})$ is convex.
4. We introduce a "prefix channel" $(\mathcal{V}, p_{\mathsf{X}|\mathsf{V}}, \mathcal{X})$ before the BCC $(\mathcal{X}, p_{\mathsf{YZ}|\mathsf{X}}, \mathcal{Y}, \mathcal{Z})$ to create a BCC $(\mathcal{V}, p_{\mathsf{YZ}|\mathsf{V}}, \mathcal{Y}, \mathcal{Z})$. This prefix channel introduces the auxiliary random variable V and accounts for more sophisticated encoders than those used in Step 1. This shows the achievability of the entire rate–equivocation region $\mathcal{R}^{\text{BCC}}$.

Step 1. Random coding argument

We prove the existence of a sequence of $(2^{nR_0}, 2^{nR_1}, n)$ codes $\{\mathcal{C}_n\}_{n\geqslant 1}$ such that

$$\lim_{n\to\infty} \mathbf{P}_e(\mathcal{C}_n) = 0, \qquad \lim_{n\to\infty} \mathbb{H}(\mathsf{R}|\mathsf{Z}^n\mathsf{M}_0\mathsf{M}_1\mathcal{C}_n) = 0, \tag{3.34}$$

$$\lim_{n\to\infty} \frac{1}{n}\mathbf{L}(\mathcal{C}_n) \leqslant \delta(\epsilon). \tag{3.35}$$

The proof combines the technique of superposition coding introduced in Section 2.3.3 with the technique of wiretap coding discussed in Section 3.4.1. As in Section 3.4.1, we start by combining the two constraints in (3.34) into a single reliability constraint by

- introducing a virtual receiver, hereafter named Charlie, who observes the same channel output Z^n as Eve in the original BCC, but who also has access to M_1 and M_0 through an error-free side channel;
- using a message M_d in place of the source of local randomness $(\mathcal{R}, p_{\mathsf{R}})$ and requiring M_d to be decoded by both Bob and Charlie.

A code for this "enhanced" BCC is then defined as follows.

Definition 3.13. *A $(2^{nR_0}, 2^{nR_1}, 2^{nR_d}, n)$ code $\mathcal{C}_n$ for the enhanced channel consists of the following:*

- *three message sets, $\mathcal{M}_0 = [\![1, 2^{nR_0}]\!]$, $\mathcal{M}_1 = [\![1, 2^{nR_1}]\!]$, and $\mathcal{M}_d = [\![1, 2^{nR_d}]\!]$;*
- *an encoding function $f : \mathcal{M}_0 \times \mathcal{M}_1 \times \mathcal{M}_d \to \mathcal{X}^n$, which maps each message triple (m_0, m_1, m_d) to a codeword x^n;*

- *a decoding function* $g : \mathcal{Y}^n \to (\mathcal{M}_0 \times \mathcal{M}_1 \times \mathcal{M}_d) \cup \{?\}$, *which maps each channel observation* y^n *to a message triple* $(\hat{m}_0, \hat{m}_1, \hat{m}_d) \in \mathcal{M}_0 \times \mathcal{M}_1 \times \mathcal{M}_d$ *or an error message* ?;
- *a decoding function* $h : \mathcal{Z}^n \to \mathcal{M}_0 \cup \{?\}$, *which maps each channel observation* z^n *to a message* $\tilde{m}_0 \in \mathcal{M}_0$ *or an error message* ?;
- *a decoding function* $k : \mathcal{Z}^n \times \mathcal{M}_0 \times \mathcal{M}_1 \to \mathcal{M}_d \cup \{?\}$, *which maps each message pair* (m_0, m_1) *and the corresponding channel observation* z^n *to a message* $\tilde{m}_d \in \mathcal{M}_d$ *or an error message* ?.

We assume that messages M_0, M_1, and M_d are uniformly distributed. The reliability performance of a $\left(2^{nR_0}, 2^{nR_1}, 2^{nR_d}, n\right)$ code $\mathcal{C}_n$ is measured in terms of its average probability of error

$$\mathbf{P}_e(\mathcal{C}_n) \triangleq \mathbb{P}\left[(\hat{\mathsf{M}}_0, \hat{\mathsf{M}}_1, \hat{\mathsf{M}}_d) \neq (\mathsf{M}_0, \mathsf{M}_1, \mathsf{M}_d) \text{ or } \tilde{\mathsf{M}}_0 \neq \mathsf{M}_0 \text{ or } \tilde{\mathsf{M}}_d \neq \mathsf{M}_d \middle| \mathcal{C}_n\right],$$

while its secrecy performance is still measured in terms of the leakage

$$\mathbf{L}(\mathcal{C}_n) \triangleq \mathbb{I}(\mathsf{M}_1; \mathsf{Z}^n | \mathcal{C}_n).$$

Because M_d is a dummy message that corresponds to a specific choice for the source of local randomness $(\mathcal{R}, p_\mathsf{R})$, a $\left(2^{nR_0}, 2^{nR_1}, 2^{nR_d}, n\right)$ code $\mathcal{C}_n$ for the enhanced BCC is also a $\left(2^{nR_0}, 2^{nR_1}, n\right)$ code for the original BCC. By construction, the probability of error over the original BCC is at most $\mathbf{P}_e(\mathcal{C}_n)$, since

$$\mathbb{P}\left[(\hat{\mathsf{M}}_0, \hat{\mathsf{M}}_1) \neq (\mathsf{M}_0, \mathsf{M}_1) \text{ or } \tilde{\mathsf{M}}_0 \neq \mathsf{M}_0 \middle| \mathcal{C}_n\right] \leqslant \mathbf{P}_e(\mathcal{C}_n).$$

In addition, using Fano's inequality, we have

$$\frac{1}{n}\mathbb{H}(\mathsf{M}_d | \mathsf{Z}^n \mathsf{M}_0 \mathsf{M}_1 \mathcal{C}_n) \leqslant \delta(\mathbf{P}_e(\mathcal{C}_n)).$$

Therefore, if $\lim_{n\to\infty} \mathbf{P}_e(\mathcal{C}_n) = 0$, the constraints (3.34) are automatically satisfied.

We begin by choosing a joint distribution p_{UX} on $\mathcal{U} \times \mathcal{X}$ and we assume, without loss of generality, that

$$\mathbb{I}(\mathsf{X}; \mathsf{Y}|\mathsf{U}) - \mathbb{I}(\mathsf{X}; \mathsf{Z}|\mathsf{U}) > 0 \qquad \text{and} \qquad \mathbb{I}(\mathsf{X}; \mathsf{Z}|\mathsf{U}) > 0,$$

otherwise the result follows from the channel coding theorem as discussed in Remark 3.10. Let $0 < \epsilon < \mu_{\mathsf{UXYZ}}$, where

$$\mu_{\mathsf{UXYZ}} \triangleq \min_{(u,x,y,z)\in\mathcal{U}\times\mathcal{X}\times\mathcal{Y}\times\mathcal{Z}} p_{\mathsf{UXYZ}}(u, x, y, z),$$

and let $n \in \mathbb{N}^*$. Let $R_0 > 0$, $R_1 > 0$, and $R_d > 0$ be rates to be specified later. We construct a $\left(2^{nR_0}, 2^{nR_1}, 2^{nR_d}, n\right)$ code for the enhanced BCC by combining superposition coding and binning as follows.

- *Codebook construction.* Construct codewords $u^n(m_0)$ for $m_0 \in \left[\!\left[1, 2^{nR_0}\right]\!\right]$, by generating symbols $u_i(m_0)$ with $i \in [\![1, n]\!]$ and $m_0 \in \left[\!\left[1, 2^{nR_0}\right]\!\right]$ independently according to p_U. Then, for every $u^n(m_0)$, generate codewords $x^n(m_0, m_1, m_d)$ for $m_1 \in \left[\!\left[1, 2^{nR_1}\right]\!\right]$ *and* $m_d \in \left[\!\left[1, 2^{nR_d}\right]\!\right]$ by generating symbols $x_i(m_0, m_1, m_d)$ with $i \in [\![1, n]\!]$, $m_1 \in \left[\!\left[1, 2^{nR_1}\right]\!\right]$, and $m_d \in \left[\!\left[1, 2^{nR_d}\right]\!\right]$ independently at random according to $p_{\mathsf{X}|\mathsf{U}=u_i(m_0)}$.

- *Alice's encoder f.* Given (m_0, m_1, m_d), transmit $x^n(m_0, m_1, m_d)$.
- *Bob's decoder g.* Given y^n, output $(\hat{m}_0, \hat{m}_1, \hat{m}_d)$ if it is the unique triple such that $(u^n(\hat{m}_0), x^n(\hat{m}_0, \hat{m}_1, \hat{m}_d), y^n) \in \mathcal{T}_\epsilon^n(\mathsf{UXY})$. Otherwise, output an error ?.
- *Eve's decoder h.* Given z^n, output $\tilde{m}_0$ if it is the unique message such that $(u^n(\tilde{m}_0), z^n) \in \mathcal{T}_\epsilon^n(\mathsf{UZ})$. Otherwise, output an error ?.
- *Charlie's decoder k.* Given z^n, m_0, and m_1, output $\tilde{m}_d$ if it is the unique message such that $(u^n(m_0), x^n(m_0, m_1, \tilde{m}_d), z^n) \in \mathcal{T}_\epsilon^n(\mathsf{UXZ})$. Otherwise, output an error ?.

The random variable that represents the randomly generated codebook $\mathcal{C}_n$ is denoted by C_n. By combining the analysis of superposition coding in Section 2.3.3 with the analysis of wiretap coding in Section 3.4.1, we can prove that, if

$$\begin{aligned} R_0 &< \min(\mathbb{I}(\mathsf{U};\mathsf{Y}), \mathbb{I}(\mathsf{U};\mathsf{Z})) - \delta(\epsilon), \\ R_1 + R_d &< \mathbb{I}(\mathsf{X};\mathsf{Y}|\mathsf{U}) - \delta(\epsilon), \\ R_d &< \mathbb{I}(\mathsf{X};\mathsf{Z}|\mathsf{U}) - \delta(\epsilon), \end{aligned} \tag{3.36}$$

then

$$\mathbb{E}[\mathbf{P}_e(\mathsf{C}_n)] \leqslant \delta_\epsilon(n). \tag{3.37}$$

Next, we compute an upper bound for $\mathbb{E}[(1/n)\mathbf{L}(\mathsf{C}_n)]$. Note that

$$\begin{aligned} \mathbb{E}\left[\frac{1}{n}\mathbf{L}(\mathsf{C}_n)\right] &= \frac{1}{n}\mathbb{I}(\mathsf{M}_1;\mathsf{Z}^n|\mathsf{C}_n) \\ &\leqslant \frac{1}{n}\mathbb{I}(\mathsf{M}_1;\mathsf{Z}^n\mathsf{M}_0|\mathsf{C}_n) \\ &= \frac{1}{n}\mathbb{I}(\mathsf{M}_1\mathsf{X}^n;\mathsf{Z}^n\mathsf{M}_0|\mathsf{C}_n) - \frac{1}{n}\mathbb{I}(\mathsf{X}^n;\mathsf{Z}^n\mathsf{M}_0|\mathsf{M}_1\mathsf{C}_n) \\ &= \frac{1}{n}\mathbb{I}(\mathsf{X}^n;\mathsf{Z}^n\mathsf{M}_0|\mathsf{C}_n) + \frac{1}{n}\mathbb{I}(\mathsf{M}_1;\mathsf{Z}^n\mathsf{M}_0|\mathsf{X}^n\mathsf{C}_n) - \frac{1}{n}\mathbb{I}(\mathsf{X}^n;\mathsf{Z}^n\mathsf{M}_0|\mathsf{M}_1\mathsf{C}_n) \\ &\overset{(a)}{=} \frac{1}{n}\mathbb{I}(\mathsf{X}^n;\mathsf{M}_0|\mathsf{C}_n) + \frac{1}{n}\mathbb{I}(\mathsf{X}^n;\mathsf{Z}^n|\mathsf{M}_0\mathsf{C}_n) - \frac{1}{n}\mathbb{I}(\mathsf{X}^n;\mathsf{Z}^n\mathsf{M}_0|\mathsf{M}_1\mathsf{C}_n) \\ &\overset{(b)}{=} \frac{1}{n}\mathbb{H}(\mathsf{M}_0|\mathsf{C}_n) + \frac{1}{n}\mathbb{I}(\mathsf{X}^n;\mathsf{Z}^n|\mathsf{M}_0\mathsf{C}_n) - \frac{1}{n}\mathbb{H}(\mathsf{X}^n|\mathsf{M}_1\mathsf{C}_n) \\ &\quad + \frac{1}{n}\mathbb{H}(\mathsf{X}^n|\mathsf{Z}^n\mathsf{M}_0\mathsf{M}_1\mathsf{C}_n) \\ &\overset{(c)}{=} \frac{1}{n}\mathbb{I}(\mathsf{X}^n;\mathsf{Z}^n|\mathsf{M}_0\mathsf{C}_n) - \frac{1}{n}\mathbb{H}(\mathsf{M}_d|\mathsf{C}_n) + \frac{1}{n}\mathbb{H}(\mathsf{X}^n|\mathsf{Z}^n\mathsf{M}_0\mathsf{M}_1\mathsf{C}_n), \end{aligned} \tag{3.38}$$

where (a) follows from $\mathbb{I}(\mathsf{M}_1;\mathsf{Z}^n\mathsf{M}_0|\mathsf{X}^n\mathsf{C}_n) = 0$, (b) follows from $\mathbb{I}(\mathsf{X}^n;\mathsf{M}_0|\mathsf{C}_n) = \mathbb{H}(\mathsf{M}_0|\mathsf{C}_n)$, and (c) follows from $\mathbb{H}(\mathsf{X}^n|\mathsf{M}_1\mathsf{C}_n) = \mathbb{H}(\mathsf{M}_0|\mathsf{C}_n) + \mathbb{H}(\mathsf{M}_d|\mathsf{C}_n)$. We now bound each of the terms on the right-hand side of (3.38) individually. First, notice that the code construction ensures that

$$\frac{1}{n}\mathbb{H}(\mathsf{M}_d|\mathsf{C}_n) = \sum_{\mathcal{C}_n} p_{\mathsf{C}_n}(\mathcal{C}_n)\frac{1}{n}\mathbb{H}(\mathsf{M}_d|\mathcal{C}_n) \geqslant R_d. \tag{3.39}$$

Next, using Fano's inequality,

$$\begin{aligned}
\frac{1}{n}\mathbb{H}(\mathsf{X}^n|\mathsf{Z}^n\mathsf{M}_0\mathsf{M}_1\mathsf{C}_n) &= \sum_{\mathcal{C}_n} p_{\mathsf{C}_n}(\mathcal{C}_n)\,\frac{1}{n}\mathbb{H}(\mathsf{X}^n|\mathsf{Z}^n\mathsf{M}_0\mathsf{M}_1\mathcal{C}_n)\\
&\leqslant \sum_{\mathcal{C}_n} p_{\mathsf{C}_n}(\mathcal{C}_n)\left(\frac{1}{n}+\mathbf{P}_{\mathrm{e}}(\mathcal{C}_n)\frac{1}{n}\log\lceil 2^{nR_d}\rceil\right)\\
&= \delta(n)+\mathbb{E}[\mathbf{P}_{\mathrm{e}}(\mathsf{C}_n)](R_d+\delta(n))\\
&\leqslant \delta_\epsilon(n),
\end{aligned} \tag{3.40}$$

where the last inequality follows from (3.37). Finally, note that, given a code C_n, there is a one-to-one mapping between the message M_0 and the codeword U^n. In addition, $\mathsf{C}_n\mathsf{U}^n \rightarrow \mathsf{X}^n \rightarrow \mathsf{Z}^n$ forms a Markov chain; therefore,

$$\begin{aligned}
\frac{1}{n}\mathbb{I}(\mathsf{Z}^n;\mathsf{X}^n|\mathsf{M}_0\mathsf{C}_n) &= \frac{1}{n}\mathbb{I}(\mathsf{Z}^n;\mathsf{X}^n|\mathsf{U}^n\mathsf{C}_n)\\
&= \frac{1}{n}\mathbb{H}(\mathsf{Z}^n|\mathsf{U}^n\mathsf{C}_n)-\frac{1}{n}\mathbb{H}(\mathsf{Z}^n|\mathsf{X}^n\mathsf{U}^n)\\
&\leqslant \frac{1}{n}\mathbb{H}(\mathsf{Z}^n|\mathsf{U}^n)-\frac{1}{n}\mathbb{H}(\mathsf{Z}^n|\mathsf{X}^n\mathsf{U}^n)\\
&= \frac{1}{n}\mathbb{I}(\mathsf{X}^n;\mathsf{Z}^n|\mathsf{U}^n)\\
&= \mathbb{I}(\mathsf{X};\mathsf{Z}|\mathsf{U}),
\end{aligned} \tag{3.41}$$

where we have used the fact that $(\mathsf{U}^n, \mathsf{X}^n, \mathsf{Z}^n)$ is i.i.d. according to p_{UXZ}. On substituting (3.39), (3.40), and (3.41) into (3.38), we obtain

$$\mathbb{E}\left[\frac{1}{n}\mathbf{L}(\mathsf{C}_n)\right] \leqslant \mathbb{I}(\mathsf{X};\mathsf{Z}|\mathsf{U}) - R_{\mathrm{d}} + \delta_\epsilon(n).$$

In particular, if we choose R_1 and R_{d} such that

$$R_1 < \mathbb{I}(\mathsf{X};\mathsf{Y}|\mathsf{U}) - \mathbb{I}(\mathsf{X};\mathsf{Z}|\mathsf{U}) \quad \text{and} \quad R_{\mathrm{d}} = \mathbb{I}(\mathsf{X};\mathsf{Z}|\mathsf{U}) - \delta(\epsilon), \tag{3.42}$$

which is compatible with the constraints in (3.36), then R_{d} almost cancels out the information rate leaked to the eavesdropper and

$$\mathbb{E}\left[\frac{1}{n}\mathbf{L}(\mathsf{C}_n)\right] \leqslant \delta(\epsilon) + \delta_\epsilon(n). \tag{3.43}$$

From (3.37) and (3.43) and by applying the selection lemma to the random variable C_n and the functions $\mathbf{P}_{\mathrm{e}}$ and $\mathbf{L}$, we conclude that there exists a specific code $\mathcal{C}_n$, such that

$$\mathbf{P}_{\mathrm{e}}(\mathcal{C}_n) \leqslant \delta_\epsilon(n) \quad \text{and} \quad \frac{1}{n}\mathbf{L}(\mathcal{C}_n) \leqslant \delta(\epsilon) + \delta_\epsilon(n).$$

Consequently, there exists a sequence of $(2^{nR_0}, 2^{nR_1}, n)$ codes $\{\mathcal{C}_n\}_{n\geqslant 1}$ such that

$$\lim_{n\to\infty} \mathbf{P}_{\mathrm{e}}(\mathcal{C}_n) = 0 \quad \text{and} \quad \lim_{n\to\infty} \frac{1}{n}\mathbf{L}(\mathcal{C}_n) \leqslant \delta(\epsilon),$$

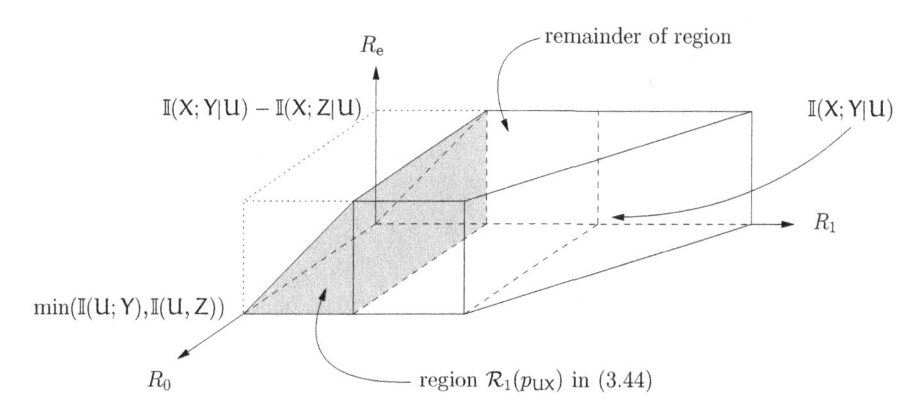

Figure 3.16 Region $\mathcal{R}_1(p_{UX})$.

which proves that $(R_0, R_1, R_1 - \delta(\epsilon))$ is achievable. Since R_0 and R_1 must satisfy the constraints imposed by (3.36) and (3.42), and since ϵ can be chosen arbitrarily small, we conclude that $\mathcal{R}_1\,(p_{UX}) \subseteq \mathcal{R}^{BCC}$, where

$$\mathcal{R}_1(p_{UX}) \triangleq \left\{(R_0, R_1, R_e)\colon \begin{array}{c} 0 \leqslant R_0 \leqslant \min(\mathbb{I}(U;Y), \mathbb{I}(U;Z)) \\ 0 \leqslant R_e \leqslant R_1 \leqslant \mathbb{I}(X;Y|U) - \mathbb{I}(X;Z|U) \end{array}\right\}. \tag{3.44}$$

By construction, the joint distribution of the random variables U, X, Y, and Z in $\mathcal{R}_1\,(p_{UX})$ factorizes as $p_{UX}p_{YZ|X}$.

Step 2. Outer code construction

As illustrated in Figure 3.16, the rate region $\mathcal{R}_1(p_{UX})$ in (3.44) represents only a subset of $\mathcal{R}^{BCC}(p_{UVX})$; nevertheless, the entire region can be obtained with minor modifications of the $(2^{nR_0}, 2^{nR_1}, 2^{nR_d})$ codes $\mathcal{C}_n$ identified in Step 1. The key idea is to exploit part of the dummy message M_d and part of the common message M_0 as individual messages. This can be performed by introducing sub-bins for M_0 and M_d as done in Section 3.4.1, and we provide a sketch of the proof only.

- By using a fraction of the dummy message rate $R_d = \mathbb{I}(X;Z|U) - \delta(\epsilon)$, we can increase the individual-message rate without changing the equivocation and without changing the common-message rate. Hence, the region $\mathcal{R}_1'(p_{UX})$ defined as

$$\mathcal{R}_1'(p_{UX}) \triangleq \left\{(R_0, R_1, R_e)\colon \begin{array}{c} 0 \leqslant R_e \leqslant R_1 \\ 0 \leqslant R_e \leqslant \mathbb{I}(X;Y|U) - \mathbb{I}(X;Z|U) \\ 0 \leqslant R_0 \leqslant \min(\mathbb{I}(U;Y), \mathbb{I}(U;Z)) \\ 0 \leqslant R_1 \leqslant \mathbb{I}(X;Y|U) \end{array}\right\}$$

  satisfies $\mathcal{R}_1'(p_{UX}) \subseteq \mathcal{R}^{BCC}$.
- By sacrificing a fraction of the common-message rate R_0, we can further increase the individual-message rate without changing the equivocation; however, because of the constraints (3.36), the trade-off between the individual-message rate and the common-message rate is limited by

$$R_0 + R_1 \leqslant \min(\mathbb{I}(U;Y); \mathbb{I}(U;Y)) + \mathbb{I}(X;Y|U).$$

Hence, the region $\mathcal{R}_2(p_{\mathsf{UX}})$ defined as

$$\mathcal{R}_2(p_{\mathsf{UX}}) \triangleq \left\{ (R_0, R_1, R_e) \colon \begin{array}{l} 0 \leqslant R_e \leqslant R_1 \\ 0 \leqslant R_e \leqslant \mathbb{I}(\mathsf{X};\mathsf{Y}|\mathsf{U}) - \mathbb{I}(\mathsf{X};\mathsf{Z}|\mathsf{U}) \\ 0 \leqslant R_0 + R_1 \leqslant \mathbb{I}(\mathsf{X};\mathsf{Y}|\mathsf{U}) + \min(\mathbb{I}(\mathsf{U};\mathsf{Y}), \mathbb{I}(\mathsf{U};\mathsf{Z})) \end{array} \right\} \tag{3.45}$$

satisfies $\mathcal{R}_2(p_{\mathsf{UX}}) \subseteq \mathcal{R}^{\text{BCC}}$.

Finally, since the distribution p_{UX} is arbitrary, we obtain

$$\bigcup_{p_{\mathsf{UX}}} \mathcal{R}_2(p_{\mathsf{UX}}) \subseteq \mathcal{R}^{\text{BCC}}.$$

Step 3. Convexity of $\bigcup_{p_{\mathsf{UX}}} \mathcal{R}_2(p_{\mathsf{UX}})$

We show that $\bigcup_{p_{\mathsf{UX}}} \mathcal{R}_2(p_{\mathsf{UX}})$ is convex by proving that, for any distributions $p_{\mathsf{U}_1\mathsf{X}_1}$ and $p_{\mathsf{U}_2\mathsf{X}_2}$ on $\mathcal{U} \times \mathcal{X}$, the convex hull of $\mathcal{R}_2(p_{\mathsf{U}_1\mathsf{X}_1}) \cup \mathcal{R}_2(p_{\mathsf{U}_2\mathsf{X}_2})$ is included in $\bigcup_{p_{\mathsf{UX}}} \mathcal{R}_2(p_{\mathsf{UX}})$.

Let $(R_{0,1}, R_{1,1}, R_{e,1}) \in \mathcal{R}_2(p_{\mathsf{U}_1\mathsf{X}_1})$ be a rate triple satisfying the inequalities in (3.45) for some random variables U_1, X_1, Y_1, and Z_1 whose joint distribution satisfies

$$\forall (u,x,y,z) \in \mathcal{U} \times \mathcal{X} \times \mathcal{Y} \times \mathcal{Z} \quad p_{\mathsf{U}_1\mathsf{X}_1\mathsf{Y}_1\mathsf{Z}_1}(u,x,y,z) = p_{\mathsf{U}_1}(u) p_{\mathsf{X}_1|\mathsf{U}_1}(x|u) p_{\mathsf{YZ}|\mathsf{X}}(y,z|x).$$

Let $(R_{0,2}, R_{1,2}, R_{e,2}) \in \mathcal{R}_2(p_{\mathsf{U}_2\mathsf{X}_2})$ be another rate triple satisfying the inequalities in (3.45) for some random variables U_2, X_2, Y_2, and Z_2 whose joint distribution satisfies

$$\forall (u,x,y,z) \in \mathcal{U} \times \mathcal{X} \times \mathcal{Y} \times \mathcal{Z} \quad p_{\mathsf{U}_2\mathsf{X}_2\mathsf{Y}_2\mathsf{Z}_2}(u,x,y,z) = p_{\mathsf{U}_2}(u) p_{\mathsf{X}_2|\mathsf{U}_2}(x|u) p_{\mathsf{YZ}|\mathsf{X}}(y,z|x).$$

Our objective is to show that, for any $\lambda \in [0,1]$, there exist random variables U_λ and X_λ such that

$$(\lambda R_{0,1} + (1-\lambda) R_{0,2}, \lambda R_{1,1} + (1-\lambda) R_{1,2}, \lambda R_{e,1} + (1-\lambda) R_{e,2}) \in \mathcal{R}_2(p_{\mathsf{U}_\lambda \mathsf{X}_\lambda}).$$

We introduce a random variable $\mathsf{Q} \in \{1,2\}$ that is independent of all others such that

$$\mathsf{Q} \triangleq \begin{cases} 1 & \text{with probability } \lambda, \\ 2 & \text{with probability } 1-\lambda. \end{cases}$$

By construction, $\mathsf{Q} \rightarrow \mathsf{U}_\mathsf{Q} \rightarrow \mathsf{X}_\mathsf{Q} \rightarrow \mathsf{Y}_\mathsf{Q}\mathsf{Z}_\mathsf{Q}$ forms a Markov chain, and the joint distribution of U_Q, X_Q, Y_Q, and Z_Q satisfies

$$\forall (u,x,y,z) \in \mathcal{U} \times \mathcal{X} \times \mathcal{Y} \times \mathcal{Z} \quad p_{\mathsf{U}_\mathsf{Q}\mathsf{X}_\mathsf{Q}\mathsf{Y}_\mathsf{Q}\mathsf{Z}_\mathsf{Q}}(u,x,y,z) = p_{\mathsf{U}_\mathsf{Q}}(u) p_{\mathsf{X}_\mathsf{Q}|\mathsf{U}_\mathsf{Q}}(x|u) p_{\mathsf{YZ}|\mathsf{X}}(y,z|x).$$

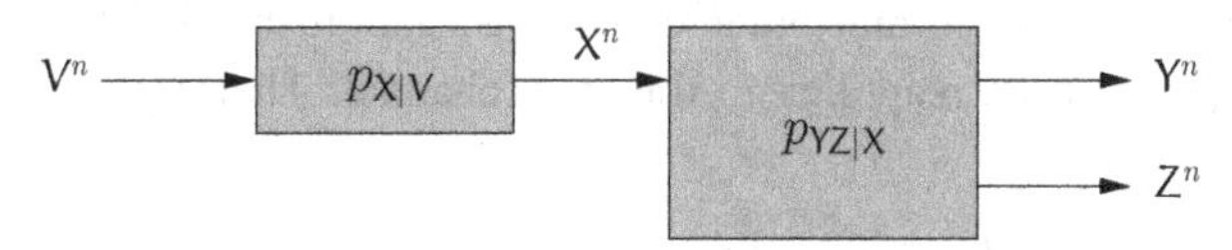

Figure 3.17 Addition of prefix channel to BCC.

Since $Q \to U_Q \to X_Q \to Y_Q Z_Q$ forms a Markov chain, notice that

$$\begin{aligned}
\mathbb{I}(X_Q;Y_Q|U_Q) &= \mathbb{I}(X_Q;Y_Q|U_Q Q),\\
\mathbb{I}(X_Q;Z_Q|U_Q) &= \mathbb{I}(X_Q;Z_Q|U_Q Q),\\
\mathbb{I}(U_Q;Y_Q) &\geqslant \mathbb{I}(U_Q;Y_Q|Q),\\
\mathbb{I}(U_Q;Z_Q) &\geqslant \mathbb{I}(U_Q;Z_Q|Q).
\end{aligned}$$

We set $U_\lambda \triangleq U_Q$, $X_\lambda \triangleq X_Q$, $Y_\lambda \triangleq Y_Q$, and $Z_\lambda \triangleq Z_Q$. Then,

$$\begin{aligned}
&\lambda(R_{0,1} + R_{1,1}) + (1-\lambda)(R_{0,2} + R_{1,2})\\
&\quad\leqslant \lambda(\mathbb{I}(X_1;Y_1|U_1) + \min(\mathbb{I}(U_1;Y_1), \mathbb{I}(U_1;Z_1)))\\
&\qquad + (1-\lambda)(\mathbb{I}(X_2;Y_2|U_2) + \min(\mathbb{I}(U_2;Y_2), \mathbb{I}(U_2;Z_2)))\\
&\quad= \lambda\mathbb{I}(X_1;Y_1|U_1) + (1-\lambda)\mathbb{I}(X_2;Y_2|U_2)\\
&\qquad + \lambda\min(\mathbb{I}(U_1;Y_1), \mathbb{I}(U_1;Z_1)) + (1-\lambda)\min(\mathbb{I}(U_2;Y_2), \mathbb{I}(U_2;Z_2))\\
&\quad\leqslant \mathbb{I}(X_Q;Y_Q|U_Q) + \min(\mathbb{I}(U_Q;Y_Q|Q), \mathbb{I}(U_Q;Z_Q|Q))\\
&\quad\leqslant \mathbb{I}(X_Q;Y_Q|U_Q) + \min(\mathbb{I}(U_Q;Y_Q), \mathbb{I}(U_Q;Z_Q))\\
&\quad= \mathbb{I}(X_\lambda;Y_\lambda|U_\lambda) + \min(\mathbb{I}(U_\lambda;Y_\lambda), \mathbb{I}(U_\lambda;Z_\lambda)).
\end{aligned}$$

Similarly, we can show,

$$\begin{aligned}
\lambda R_{e,1} + (1-\lambda)R_{e,2} &\leqslant \mathbb{I}(X_Q;Y_Q|U_Q) - \mathbb{I}(X_Q;Z_Q|U_Q)\\
&= \mathbb{I}(X_\lambda;Y_\lambda|U_\lambda) - \mathbb{I}(X_\lambda;Z_\lambda|U_\lambda).
\end{aligned}$$

Hence, for any $\lambda \in [0, 1]$, there exist U_λ and X_λ such that

$$(\lambda R_{0,1} + (1-\lambda)R_{0,2}, \lambda R_{1,1} + (1-\lambda)R_{1,2}, \lambda R_{e,1} + (1-\lambda)R_{e,2}) \in \mathcal{R}_2(p_{U_\lambda X_\lambda}) \subseteq \mathcal{R}'''.$$

Therefore, the convex hull of $\mathcal{R}_2(p_{U_1X_1}) \cup \mathcal{R}_2(p_{U_2X_2})$ is in $\bigcup_{p_{UX}} \mathcal{R}_2(p_{UX})$ and $\bigcup_{p_{UX}} \mathcal{R}_2(p_{UX})$ is convex.

Step 4. Addition of prefix channel

Consider an arbitrary DMC $(\mathcal{V}, p_{X|V}, \mathcal{X})$ that we append before the BCC $(\mathcal{X}, p_{YZ|X}, \mathcal{Y}, \mathcal{Z})$. This can be done using the source of local randomness. As illustrated in Figure 3.17, the concatenation defines a new BCC $(\mathcal{V}, p_{YZ|V}, \mathcal{Y}, \mathcal{Z})$ such that

$$\forall (v, y, z) \in \mathcal{V} \times \mathcal{Y} \times \mathcal{Z} \quad p_{YZ|V}(y, z|v) = \sum_{x\in\mathcal{X}} p_{YZ|X}(y, z|x)p_{X|V}(x|v).$$

The random-coding argument and outer code construction that led to the characterization of $\mathcal{R}_2(p_{\mathsf{UX}})$ can be reapplied to this new channel. Therefore, we conclude that

$$\bigcup_{p_{\mathsf{UVX}}} \left\{ (R_0, R_1, R_e): \begin{array}{l} 0 \leqslant R_e \leqslant R_1 \\ 0 \leqslant R_e \leqslant \mathbb{I}(\mathsf{V};\mathsf{Y}|\mathsf{U}) - \mathbb{I}(\mathsf{V};\mathsf{Z}|\mathsf{U}) \\ 0 \leqslant R_0 + R_1 \leqslant \mathbb{I}(\mathsf{V};\mathsf{Y}|\mathsf{U}) + \min(\mathbb{I}(\mathsf{U};\mathsf{Y}), \mathbb{I}(\mathsf{U};\mathsf{Z})) \end{array} \right\} \subseteq \mathcal{R}^{\text{BCC}}.$$

By construction, $\mathsf{U} \rightarrow \mathsf{V} \rightarrow \mathsf{X} \rightarrow \mathsf{YZ}$ forms a Markov chain. It remains to prove that we can restrict the cardinality of U and V to $|\mathcal{U}| \leqslant |\mathcal{X}| + 3$ and $|\mathcal{V}| \leqslant |\mathcal{X}|^2 + |\mathcal{X}| + 1$. This follows from Caratheodory's theorem, and we refer interested readers to [18, Appendix] for details.

At this point, the introduction of a prefix channel is somewhat artificial, but the converse part of the proof given in the next section shows that this additional randomization of the stochastic encoder is *required* in order to match the rate–equivocation region obtained in the converse proof. From a practical perspective, the prefix channel suggests that additional randomization may be required in the encoder; this might not be surprising, because the equivocation calculation in Step 2 relies on a *specific* stochastic encoding scheme that might not be optimal. Fortunately, Corollary 3.5 shows that this additional randomization is not necessary if the eavesdropper's channel is less capable than the main channel.

3.5.3 Converse proof for the broadcast channel with confidential messages

Consider an achievable rate triple (R_0, R_1, R_e) and let $\epsilon > 0$. For n sufficiently large, there exists a $(2^{nR_0}, 2^{nR_1}, n)$ code $\mathcal{C}_n$ such that

$$\frac{1}{n}\mathbb{H}(\mathsf{M}_0|\mathcal{C}_n) \geqslant R_0, \qquad \frac{1}{n}\mathbb{H}(\mathsf{M}_1|\mathcal{C}_n) \geqslant R_1,$$

$$\frac{1}{n}\mathbf{E}(\mathcal{C}_n) \geqslant R_e - \delta(\epsilon), \qquad \mathbf{P}_e(\mathcal{C}_n) \leqslant \delta(\epsilon).$$

In the remainder of this section, we drop the conditioning on $\mathcal{C}_n$ to simplify the notation. Using Fano's inequality, we obtain

$$\frac{1}{n}\mathbb{H}(\mathsf{M}_0\mathsf{M}_1|\mathsf{Y}^n) \leqslant \delta(\epsilon) \quad \text{and} \quad \frac{1}{n}\mathbb{H}(\mathsf{M}_0|\mathsf{Z}^n) \leqslant \delta(\epsilon).$$

Therefore,

$$\begin{aligned} R_e &\leqslant \frac{1}{n}\mathbf{E}(\mathcal{C}_n) + \delta(\epsilon) \\ &= \frac{1}{n}\mathbb{H}(\mathsf{M}_1|\mathsf{Z}^n) + \delta(\epsilon) \\ &= \frac{1}{n}\mathbb{H}(\mathsf{M}_1|\mathsf{Z}^n\mathsf{M}_0) + \frac{1}{n}\mathbb{I}(\mathsf{M}_0;\mathsf{M}_1|\mathsf{Z}^n) + \delta(\epsilon) \\ &\leqslant \frac{1}{n}\mathbb{H}(\mathsf{M}_1|\mathsf{M}_0) - \frac{1}{n}\mathbb{I}(\mathsf{M}_1;\mathsf{Z}^n|\mathsf{M}_0) + \frac{1}{n}\mathbb{H}(\mathsf{M}_0|\mathsf{Z}^n) + \delta(\epsilon) \end{aligned}$$

$$
\begin{aligned}
&\leqslant \frac{1}{n}\mathbb{H}(M_1|M_0) - \frac{1}{n}\mathbb{I}(M_1; Z^n|M_0) + \delta(\epsilon) \\
&= \frac{1}{n}\mathbb{I}(M_1; Y^n|M_0) - \frac{1}{n}\mathbb{I}(M_1; Z^n|M_0) + \frac{1}{n}\mathbb{H}(M_1|Y^n M_0) + \delta(\epsilon) \\
&\leqslant \frac{1}{n}\mathbb{I}(M_1; Y^n|M_0) - \frac{1}{n}\mathbb{I}(M_1; Z^n|M_0) + \delta(\epsilon).
\end{aligned} \tag{3.46}
$$

Single-letterizing (3.46) is more arduous than in Section 3.4.2 because the channel is not degraded. The solution to circumvent this difficulty is a standard technique from multi-user information theory, which consists of symmetrizing the expression by introducing the vectors

$$
Y^{i-1} \triangleq (Y_1, \ldots, Y_{i-1}) \qquad \text{and} \qquad \tilde{Z}^{i+1} \triangleq (Z_{i+1}, \ldots, Z_n) \quad \text{for } i \in [\![1, n]\!],
$$

with the convention that $Y^0 = 0$ and $\tilde{Z}^{n+1} = 0$. We introduce Y^{i-1} and $\tilde{Z}^{i+1}$ in $\mathbb{I}(M_1; Y^n|M_0)$ as follows:

$$
\begin{aligned}
\mathbb{I}(M_1; Y^n|M_0) &= \sum_{i=1}^{n} \mathbb{I}(M_1; Y_i|M_0 Y^{i-1}) \\
&= \sum_{i=1}^{n} \left(\mathbb{I}(M_1 \tilde{Z}^{i+1}; Y_i|M_0 Y^{i-1}) - \mathbb{I}(\tilde{Z}^{i+1}; Y_i|M_0 M_1 Y^{i-1})\right) \\
&= \sum_{i=1}^{n} \left(\mathbb{I}(M_1; Y_i|M_0 Y^{i-1}\tilde{Z}^{i+1}) + \mathbb{I}(\tilde{Z}^{i+1}; Y_i|M_0 Y^{i-1}) \right.\\
&\qquad \left. - \mathbb{I}(\tilde{Z}^{i+1}; Y_i|M_0 M_1 Y^{i-1})\right).
\end{aligned} \tag{3.47}
$$

Similarly, we introduce Y^{i-1} and $\tilde{Z}^{i+1}$ in $\mathbb{I}(M_1; Z^n|M_0)$ as follows:

$$
\begin{aligned}
\mathbb{I}(M_1; Z^n|M_0) &= \sum_{i=1}^{n} \left(\mathbb{I}(M_1; Z_i|M_0 \tilde{Z}^{i+1})\right) \\
&= \sum_{i=1}^{n} \left(\mathbb{I}(M_1 Y^{i-1}; Z_i|M_0 \tilde{Z}^{i+1}) - \mathbb{I}(Y^{i-1}; Z_i|M_0 M_1 \tilde{Z}^{i+1})\right) \\
&= \sum_{i=1}^{n} \left(\mathbb{I}(M_1; Z_i|M_0 Y^{i-1}\tilde{Z}^{i+1}) + \mathbb{I}(Y^{i-1}; Z_i|M_0 \tilde{Z}^{i+1}) \right.\\
&\qquad \left. - \mathbb{I}(Y^{i-1}; Z_i|M_0 M_1 \tilde{Z}^{i+1})\right).
\end{aligned} \tag{3.48}
$$

The key observation to simplify these expressions is the following lemma

Lemma 3.5.

$$
\sum_{i=1}^{n} \mathbb{I}(\tilde{Z}^{i+1}; Y_i|M_0 Y^{i-1}) = \sum_{j=1}^{n} \mathbb{I}(Z_j; Y^{j-1}|M_0 \tilde{Z}^{j+1}),
$$

$$
\sum_{i=1}^{n} \mathbb{I}(\tilde{Z}^{i+1}; Y_i|M_0 M_1 Y^{i-1}) = \sum_{j=1}^{n} \mathbb{I}(Z_j; Y^{j-1}|M_0 M_1 \tilde{Z}^{j+1}).
$$

Proof. This result follows from the chain rule of mutual information using an appropriate change of indices:

$$\begin{aligned}\sum_{i=1}^{n} \mathbb{I}(\tilde{Z}^{i+1}; Y_i | M_0 Y^{i-1}) &= \sum_{i=1}^{n} \sum_{j=i+1}^{n} \mathbb{I}(Z_j; Y_i | M_0 Y^{i-1} \tilde{Z}^{j+1}) \\ &= \sum_{j=1}^{n} \sum_{i=1}^{j-1} \mathbb{I}(Z_j; Y_i | M_0 Y^{i-1} \tilde{Z}^{j+1}) \\ &= \sum_{j=1}^{n} \mathbb{I}(Z_j; Y^{j-1} | M_0 \tilde{Z}^{j+1}). \end{aligned} \tag{3.49}$$

Similarly, one can show that

$$\sum_{i=1}^{n} \mathbb{I}(\tilde{Z}^{i+1}; Y_i | M_0 M_1 Y^{i-1}) = \sum_{j=1}^{n} \mathbb{I}(Z_j; Y^{j-1} | M_0 M_1 \tilde{Z}^{j+1}). \tag{3.50}$$

□

Hence, on substituting (3.47) and (3.48) into (3.46), we obtain, with the help of Lemma 3.5,

$$R_e \leqslant \frac{1}{n} \sum_{i=1}^{n} \left(\mathbb{I}(M_1; Y_i | M_0 Y^{i-1} \tilde{Z}^{i+1}) - \mathbb{I}(M_1; Z_i | M_0 Y^{i-1} \tilde{Z}^{i+1}) \right) + \delta(\epsilon). \tag{3.51}$$

The common message rate R_0 can be bounded in a similar manner as

$$\begin{aligned} R_0 \leqslant \frac{1}{n}\mathbb{H}(M_0) &= \frac{1}{n}\mathbb{I}(M_0; Y^n) + \frac{1}{n}\mathbb{H}(M_0|Y^n) \\ &\leqslant \frac{1}{n}\mathbb{I}(M_0; Y^n) + \delta(\epsilon) \\ &= \frac{1}{n}\sum_{i=1}^{n} \mathbb{I}(M_0; Y_i | Y^{i-1}) + \delta(\epsilon) \\ &= \frac{1}{n}\sum_{i=1}^{n} \mathbb{I}(M_0 \tilde{Z}^{i+1}; Y_i | Y^{i-1}) - \frac{1}{n}\sum_{i=1}^{n} \mathbb{I}(\tilde{Z}^{i+1}; Y_i | Y^{i-1} M_0) + \delta(\epsilon). \end{aligned} \tag{3.52}$$

On substituting the simple bounds

$$\mathbb{I}(\tilde{Z}^{i+1}; Y_i | Y^{i-1} M_0) \geqslant 0 \quad \text{and} \quad \mathbb{I}(M_0 \tilde{Z}^{i+1}; Y_i | Y^{i-1}) \leqslant \mathbb{I}(M_0 \tilde{Z}^{i+1} Y^{i-1}; Y_i)$$

into (3.52), we obtain

$$R_0 \leqslant \frac{1}{n}\sum_{i=1}^{n} \mathbb{I}(M_0 \tilde{Z}^{i+1} Y^{i-1}; Y_i) + \delta(\epsilon). \tag{3.53}$$

By repeating the steps above with the observation Z^n instead of Y^n we obtain a second bound for R_0:

$$\begin{aligned} R_0 &\leqslant \frac{1}{n}\mathbb{H}(\mathrm{M}_0) \\ &\leqslant \frac{1}{n}\sum_{i=1}^{n}\mathbb{I}\left(\mathrm{M}_0\tilde{\mathrm{Z}}^{i+1}\mathrm{Y}^{i-1};\mathrm{Z}_i\right) - \frac{1}{n}\sum_{i=1}^{n}\mathbb{I}\left(\mathrm{Y}^{i-1};\mathrm{Z}_i|\tilde{\mathrm{Z}}^{i+1}\mathrm{M}_0\right) + \delta(\epsilon) \qquad (3.54) \\ &\leqslant \frac{1}{n}\sum_{i=1}^{n}\mathbb{I}\left(\mathrm{M}_0\tilde{\mathrm{Z}}^{i+1}\mathrm{Y}^{i-1};\mathrm{Z}_i\right) + \delta(\epsilon). \qquad (3.55) \end{aligned}$$

Finally, we bound the sum-rate $R_0 + R_1$ as follows:

$$\begin{aligned} R_1 + R_0 \leqslant \frac{1}{n}\mathbb{H}(\mathrm{M}_0\mathrm{M}_1) &= \frac{1}{n}\mathbb{H}(\mathrm{M}_1|\mathrm{M}_0) + \mathbb{H}(\mathrm{M}_0) \\ &= \frac{1}{n}\mathbb{I}(\mathrm{M}_1;\mathrm{Y}^n|\mathrm{M}_0) + \frac{1}{n}\mathbb{H}(\mathrm{M}_1|\mathrm{Y}^n\mathrm{M}_0) + \mathbb{H}(\mathrm{M}_0) \\ &\leqslant \frac{1}{n}\mathbb{I}(\mathrm{M}_1;\mathrm{Y}^n|\mathrm{M}_0) + \frac{1}{n}\mathbb{H}(\mathrm{M}_0) + \delta(\epsilon). \end{aligned}$$

From (3.52), we know that

$$\frac{1}{n}\mathbb{H}(\mathrm{M}_0) \leqslant \frac{1}{n}\sum_{i=1}^{n}\mathbb{I}\left(\mathrm{M}_0\tilde{\mathrm{Z}}^{i+1};\mathrm{Y}_i|\mathrm{Y}^{i-1}\right) - \frac{1}{n}\sum_{i=1}^{n}\mathbb{I}\left(\tilde{\mathrm{Z}}^{i+1};\mathrm{Y}_i|\mathrm{Y}^{i-1}\mathrm{M}_0\right) + \delta(\epsilon),$$

and from (3.47) we have

$$\frac{1}{n}\mathbb{I}(\mathrm{M}_1;\mathrm{Y}^n|\mathrm{M}_0) \leqslant \frac{1}{n}\sum_{i=1}^{n}\mathbb{I}\left(\mathrm{M}_1;\mathrm{Y}_i|\mathrm{M}_0\mathrm{Y}^{i-1}\tilde{\mathrm{Z}}^{i+1}\right) + \frac{1}{n}\sum_{i=1}^{n}\mathbb{I}\left(\tilde{\mathrm{Z}}^{i+1};\mathrm{Y}_i|\mathrm{M}_0\mathrm{Y}^{i-1}\right).$$

On combining the two inequalities, we obtain

$$R_1 + R_0 \leqslant \frac{1}{n}\sum_{i=1}^{n}\mathbb{I}\left(\mathrm{M}_0\tilde{\mathrm{Z}}^{i+1};\mathrm{Y}_i|\mathrm{Y}^{i-1}\right) + \frac{1}{n}\sum_{i=1}^{n}\mathbb{I}\left(\mathrm{M}_1;\mathrm{Y}_i|\mathrm{M}_0\mathrm{Y}^{i-1}\tilde{\mathrm{Z}}^{i+1}\right) + \delta(\epsilon). \qquad (3.56)$$

Similarly, using (3.54) in place of (3.52) and using Lemma 3.5, we obtain a second bound

$$R_1 + R_0 \leqslant \frac{1}{n}\sum_{i=1}^{n}\mathbb{I}\left(\mathrm{M}_0\mathrm{Y}^{i-1};\mathrm{Z}_i|\tilde{\mathrm{Z}}^{i+1}\right) + \frac{1}{n}\sum_{i=1}^{n}\mathbb{I}\left(\mathrm{M}_1;\mathrm{Y}_i|\mathrm{M}_0\mathrm{Y}^{i-1}\tilde{\mathrm{Z}}^{i+1}\right) + \delta(\epsilon). \qquad (3.57)$$

Let us now introduce the random variables

$$\mathrm{U}_i \triangleq \mathrm{Y}^{i-1}\tilde{\mathrm{Z}}^{i+1}\mathrm{M}_0 \quad \text{and} \quad \mathrm{V}_i \triangleq \mathrm{U}_i\mathrm{M}_1.$$

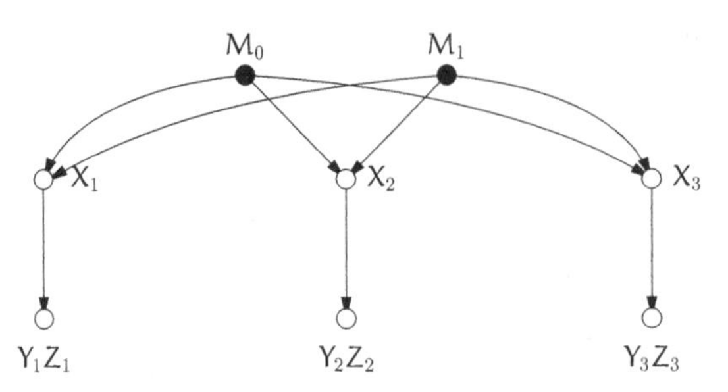

Figure 3.18 Functional dependence graph of random variables involved in the coding scheme for $n = 3$.

Using the functional dependence graph illustrated in Figure 3.18, one can verify that the joint distribution of U_i, V_i, X_i, Y_i, and Z_i is such that

$$\forall (u, v, x, y, z) \in \mathcal{U} \times \mathcal{V} \times \mathcal{X} \times \mathcal{Y} \times \mathcal{Z}$$
$$p_{\mathsf{U}_i\mathsf{V}_i\mathsf{X}_i\mathsf{Y}_i\mathsf{Z}_i}(u, v, x, y, z) = p_{\mathsf{U}_i}(u) p_{\mathsf{V}_i|\mathsf{U}_i}(v|u) p_{\mathsf{X}_i|\mathsf{U}_i}(x|u) p_{\mathsf{YZ}|\mathsf{X}}(yz|x),$$

where $p_{\mathsf{YZ}|\mathsf{X}}$ are the transition probabilities of the BCC $(\mathcal{X}, p_{\mathsf{YZ}|\mathsf{X}}, \mathcal{Y}, \mathcal{Z})$. We now introduce a random variable Q that is uniformly distributed over $[\![1, n]\!]$ and independent of $\mathsf{M}_0\mathsf{M}_1\mathsf{X}^n\mathsf{Z}^n\mathsf{Y}^n$, and we define

$$\mathsf{U} \triangleq \mathsf{U}_\mathsf{Q}\mathsf{Q}, \quad \mathsf{V} \triangleq \mathsf{U}\mathsf{M}_1, \quad \mathsf{X} \triangleq \mathsf{X}_\mathsf{Q}, \quad \mathsf{Y} \triangleq \mathsf{Y}_\mathsf{Q}, \quad \text{and} \quad \mathsf{Z} \triangleq \mathsf{Z}_\mathsf{Q}. \tag{3.58}$$

Note that $\mathsf{Q} \to \mathsf{U} \to \mathsf{X} \to \mathsf{YZ}$ forms a Markov chain. On substituting U, V, X, Y, and Z defined by (3.58) into (3.53) and (3.55), we obtain

$$\begin{aligned}
R_0 &\leqslant \min\left(\frac{1}{n}\sum_{i=1}^{n}\mathbb{I}(\mathsf{M}_0\tilde{\mathsf{Z}}^{i+1}\mathsf{Y}^{i-1};\mathsf{Z}_i), \frac{1}{n}\sum_{i=1}^{n}\mathbb{I}(\mathsf{M}_0\tilde{\mathsf{Z}}^{i+1}\mathsf{Y}^{i-1};\mathsf{Y}_i)\right) + \delta(\epsilon)\\
&= \min\left(\frac{1}{n}\sum_{i=1}^{n}\mathbb{I}(\mathsf{U}_i;\mathsf{Z}_i), \frac{1}{n}\sum_{i=1}^{n}\mathbb{I}(\mathsf{U}_i;\mathsf{Y}_i)\right) + \delta(\epsilon)\\
&= \min(\mathbb{I}(\mathsf{U};\mathsf{Z}|\mathsf{Q}), \mathbb{I}(\mathsf{U};\mathsf{Y}|\mathsf{Q})) + \delta(\epsilon)\\
&\leqslant \min(\mathbb{I}(\mathsf{U};\mathsf{Z}), \mathbb{I}(\mathsf{U};\mathsf{Y})) + \delta(\epsilon),
\end{aligned} \tag{3.59}$$

where the last inequality follows because $\mathsf{Q} \to \mathsf{U} \to \mathsf{YZ}$ forms a Markov chain. Similarly, on substituting (3.58) into (3.51),

$$\begin{aligned}
R_e &\leqslant \frac{1}{n}\sum_{i=1}^{n}\left(\mathbb{I}(\mathsf{M}_1;\mathsf{Y}_i|\mathsf{M}_0\mathsf{Y}^{i-1}\tilde{\mathsf{Z}}^{i+1}) - \mathbb{I}(\mathsf{M}_1;\mathsf{Z}_i|\mathsf{M}_0\mathsf{Y}^{i-1}\tilde{\mathsf{Z}}^{i+1})\right) + \delta(\epsilon)\\
&= \frac{1}{n}\sum_{i=1}^{n}(\mathbb{I}(\mathsf{V}_i;\mathsf{Y}_i|\mathsf{U}_i) - \mathbb{I}(\mathsf{V}_i;\mathsf{Z}_i|\mathsf{U}_i)) + \delta(\epsilon)\\
&= \mathbb{I}(\mathsf{V};\mathsf{Y}|\mathsf{U}) - \mathbb{I}(\mathsf{V};\mathsf{Z}|\mathsf{U}) + \delta(\epsilon).
\end{aligned} \tag{3.60}$$

Finally, on substituting (3.58) into (3.56) and (3.57),

$$R_0 + R_1 \leqslant \mathbb{I}(\mathsf{V};\mathsf{Y}|\mathsf{U}) + \min(\mathbb{I}(\mathsf{U};\mathsf{Z}), \mathbb{I}(\mathsf{U};\mathsf{Y})) + \delta(\epsilon).$$

Since ϵ can be chosen arbitrarily small, we finally obtain the converse result

$$\mathcal{R}^{\text{BCC}} \subseteq \bigcup_{p_{\mathsf{UVX}}} \left\{ (R_0, R_1, R_e): \begin{array}{l} 0 \leqslant R_e \leqslant R_1 \\ 0 \leqslant R_e \leqslant R_1 \leqslant \mathbb{I}(\mathsf{V};\mathsf{Y}|\mathsf{U}) - \mathbb{I}(\mathsf{V};\mathsf{Z}|\mathsf{U}) \\ 0 \leqslant R_0 + R_1 \leqslant \mathbb{I}(\mathsf{V};\mathsf{Y}|\mathsf{U}) + \min(\mathbb{I}(\mathsf{U};\mathsf{Y}), \mathbb{I}(\mathsf{U};\mathsf{Z})) \end{array} \right\}.$$

3.6 Multiplexing and feedback

3.6.1 Multiplexing secure and non-secure messages

The expression of the secrecy-capacity region $\mathcal{C}^{\text{BCC}}$ in Corollary 3.2 tells us that, for a fixed distribution p_{UVX} on $\mathcal{U} \times \mathcal{V} \times \mathcal{X}$ that factorizes as $p_{\mathsf{U}}p_{\mathsf{V}|\mathsf{U}}p_{\mathsf{X}|\mathsf{V}}$, we can transmit a common message to Eve and Bob at a rate arbitrarily close to $\min(\mathbb{I}(\mathsf{U};\mathsf{Z}), \mathbb{I}(\mathsf{U};\mathsf{Y}))$ while simultaneously transmitting an individual secret message to Bob at a rate arbitrarily close to $\mathbb{I}(\mathsf{V};\mathsf{Y}|\mathsf{U}) - \mathbb{I}(\mathsf{V};\mathsf{Z}|\mathsf{U})$. However, this result provides only a partial view of what can be transmitted over the channel, because we know from the proof of Theorem 3.3 that Alice can use a dummy message M_d as her source of local randomness; this message is decodable by Bob and is transmitted at a rate arbitrarily close to $\mathbb{I}(\mathsf{V};\mathsf{Z}|\mathsf{U})$. Although message M_d is not decodable by Eve, it is not secure either. Hence, we call M_d a *public message* to distinguish it from the common message and the individual secret message. Therefore, for a BCC $(\mathcal{X}, p_{\mathsf{YZ}|\mathsf{X}}, \mathcal{Y}, \mathcal{Z})$, there exists a code that conveys three messages reliably:

- a *common message* for Bob and Eve at a rate close to $\min(\mathbb{I}(\mathsf{U};\mathsf{Z}), \mathbb{I}(\mathsf{U};\mathsf{Y}))$;
- a *secret message* for Bob at a rate close to $\mathbb{I}(\mathsf{V};\mathsf{Y}|\mathsf{U}) - \mathbb{I}(\mathsf{V};\mathsf{Z}|\mathsf{U})$;
- a *public message* for Bob at a rate close to $\mathbb{I}(\mathsf{V};\mathsf{Z}|\mathsf{U})$.

The total transmission rate to Bob, R_{tot}, is then arbitrarily close to

$$\min(\mathbb{I}(\mathsf{U};\mathsf{Z}), \mathbb{I}(\mathsf{U};\mathsf{Y})) + \mathbb{I}(\mathsf{V};\mathsf{Y}|\mathsf{U}).$$

In particular, for the specific choice $\mathsf{U} = 0$, we obtain a total rate to Bob on the order of $\mathbb{I}(\mathsf{V};\mathsf{Y})$, of which a fraction $\mathbb{I}(\mathsf{V};\mathsf{Y}) - \mathbb{I}(\mathsf{V};\mathsf{Z})$ corresponds to a secure message. As illustrated in Figure 3.19, this allows us to interpret a wiretap code for the WTC as a means to create two parallel "pipes," a first pipe transmitting secure messages hidden from the eavesdropper and a second pipe transmitting public messages. From this perspective, note that transmitting secret messages incurs little rate penalty and comes almost "for free." For some specific WTCs, it is possible to show that secrecy comes exactly "for free."

Proposition 3.8. *Consider a WTC $(\mathcal{X}, p_{\mathsf{YZ}|\mathsf{X}}, \mathcal{Y}, \mathcal{Z})$ in which the eavesdroppers's channel $(\mathcal{X}, p_{\mathsf{Z}|\mathsf{X}}, \mathcal{Z})$ with capacity C_e is noisier than the main channel $(\mathcal{X}, p_{\mathsf{Y}|\mathsf{X}}, \mathcal{Y})$ with capacity C_m. Assume that both channels are weakly symmetric. Then, there exists a code that transmits simultaneously*

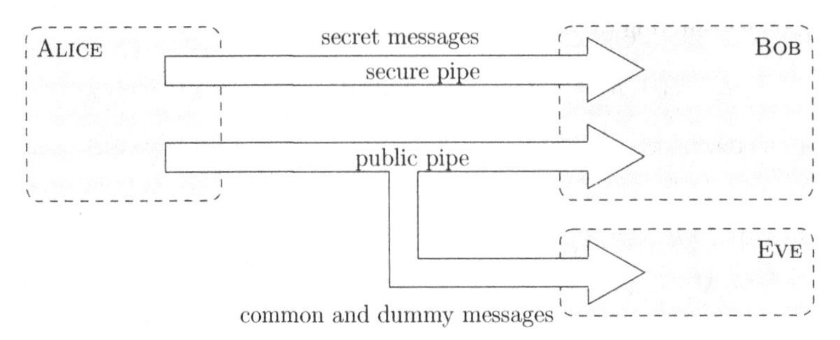

Figure 3.19 Communication over a broadcast channel with confidential messages viewed as parallel bit-pipes.

- *a secret message at a rate* R_s *arbitrarily close to* $C_s^{WT} = C_m - C_e$ *and*
- *a public message at a rate* R_p *arbitrarily close to* C_e.

In other words, it is possible to transmit a secret message at a rate arbitrarily close to the secrecy capacity C_s^{WT} *and still achieve a total reliable transmission rate arbitrarily close to the capacity of the main channel* C_m.

Proof. Let $\epsilon > 0$. For the specific choice $U = 0$ and $V = X$ in Section 3.5.2, we know that there exists a $(2^{nR_s}, n)$ wiretap code $\mathcal{C}_n$ with $R_s = \mathbb{I}(X;Y) - \mathbb{I}(X;Z) - \delta(\epsilon)$ that reliably transmits a dummy message at rate $R_d = \mathbb{I}(X;Z) - \delta(\epsilon)$. The total transmission rate is then

$$R_{tot} = R_s + R_d = \mathbb{I}(X;Y) - \delta(\epsilon).$$

In general, this does not imply that we can exhaust the capacity of the main channel and transmit at the secrecy capacity because the distribution maximizing $\mathbb{I}(X;Y)$ might not maximize $\mathbb{I}(X;Y) - \mathbb{I}(X;Z)$ simultaneously. However, if all channels are weakly symmetric and the eavesdropper's channel is noisier than the main channel, the maximizing distribution is the same, and it is possible to transmit *simultaneously* a secure message at rate

$$\max_{p_X}(\mathbb{I}(X;Y) - \mathbb{I}(X;Z)) - \delta(\epsilon) = C_m - C_e - \delta(\epsilon)$$

and a public message at rate

$$\max_{p_X} \mathbb{I}(X;Z) - \delta(\epsilon) = C_e - \delta(\epsilon),$$

such that the total rate is arbitrarily close to the capacity of the main channel. □

3.6.2 Feedback and secrecy

Although feedback does not increase the channel capacity of a DMC, the situation is quite different for the secrecy capacity. In many situations, it is possible to show that *feedback increases the secrecy capacity*; however, a more precise statement hinges on additional assumptions regarding the nature of the feedback link. The most general approach to analyze feedback would be to consider a two-way communication channel, in which both

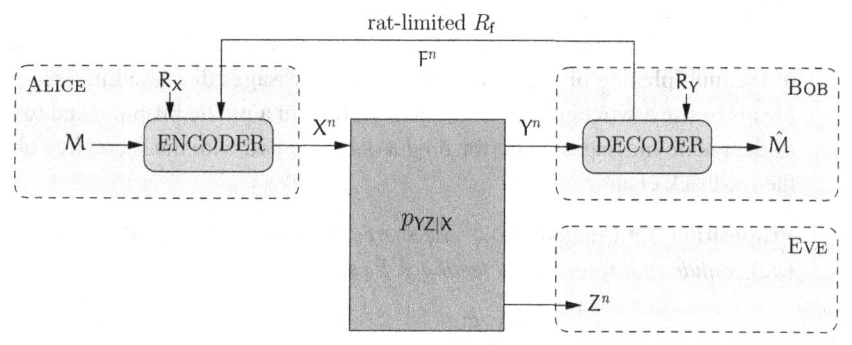

Figure 3.20 WTC with confidential rate-limited feedback.

the forward channel and the reverse channel are broadcast channels overheard by the eavesdropper. A specific instance of a two-way wiretap channel is analyzed in Chapter 8 in the context of multi-user wiretap channels, but a general solution remains elusive. Even if we simplify the model by considering extreme situations in which the feedback link is either *confidential* (unheard by the eavesdropper) or *public* (perfectly heard by the eavesdropper), the fundamental limit is unknown for arbitrary discrete memoryless channels.

In this section, we determine achievable full secrecy rates for a WTC with confidential but rate-limited feedback, which is illustrated in Figure 3.20. This model is a variation of the WTC, in which Bob has the opportunity to transmit confidential messages to Alice at a rate less than R_f. Although the model may seem overly optimistic, it provides valuable insight into how confidential feedback should be used for secrecy. The case of public feedback is studied in Chapter 4 in the context of secret-key agreement.

Definition 3.14. *A $\left(2^{nR}, n\right)$ code $\mathcal{C}_n$ for a WTC with confidential rate-limited feedback R_f consists of*

- *a message set $\mathcal{M} = [\![1, 2^{nR}]\!]$;*
- *a source of local randomness $\left(\mathcal{R}_{\mathcal{X}}, p_{R_X}\right)$ at the encoder;*
- *a source of local randomness $\left(\mathcal{R}_{\mathcal{Y}}, p_{R_Y}\right)$ at the receiver;*
- *a feedback alphabet $\mathcal{F} = [\![1, F]\!]$ such that $\log F \leqslant R_f$;*
- *a sequence of n encoding functions $e_i : \mathcal{M} \times \mathcal{F}^{i-1} \times \mathcal{R}_{\mathcal{X}} \to \mathcal{X}$ for $i \in [\![1, n]\!]$, which map a message m, the causally known feedback symbols f^{i-1}, and a realization r_x of the local randomness to a symbol $x_i \in \mathcal{X}$;*
- *a sequence of n feedback functions $g_i : \mathcal{Y}^{i-1} \times \mathcal{R}_{\mathcal{Y}} \to \mathcal{F}$ for $i \in [\![1, n]\!]$, which map past observations y^{i-1} and a realization r_y of the local randomness to a feedback symbol $f_i \in \mathcal{K}$;*
- *a decoding function $g : \mathcal{Y}^n \times \mathcal{R}_{\mathcal{Y}} \to \mathcal{M} \cup \{?\}$, which maps each channel observation y^n and realization of the local randomness r_y to a message $\hat{m} \in \mathcal{M}$ or an error message ?.*

We restrict ourselves to characterization of the secrecy capacity. Note that it is not a priori obvious what the optimal way of exploiting the feedback is. Nevertheless, we can obtain a lower bound for the secrecy capacity by studying a simple yet effective strategy based on the exchange of a secret key over the confidential feedback channel. In terms of the multiplexing of secure and non-secure messages discussed in Section 3.6.1, the idea is to use a wiretap code to create a secret and a public bit-pipe, and to protect part of the public messages by performing a one-time pad with the secret key obtained over the feedback channel.

Proposition 3.9 (Yamamoto). *The secrecy capacity C_s^{FB} of a WTC $(\mathcal{X}, p_{YZ|X}, \mathcal{Y}, \mathcal{Z})$ with confidential rate-limited feedback R_f satisfies*

$$C_s^{FB} \geqslant \max_{p_{VX}} \min(\mathbb{I}(V;Y), \mathbb{I}(V;Y) - \mathbb{I}(V;Z) + R_f).$$

If the eavesdropper's channel is noisier than the main channel and both channels are weakly symmetric, the bound can be replaced by

$$C_s^{FB} \geqslant \min(C_m, C_m - C_e + R_f),$$

where C_m is the main channel capacity and C_e the eavesdropper's channel capacity.

Proof. Let $\epsilon > 0$, $B \in \mathbb{N}^*$, and $n \in \mathbb{N}^*$. We consider a block-Markov coding scheme over B blocks of length n, such that the selection of the transmitted codeword in each block $b \in [\![2, B]\!]$ is a function of the current message m_b and of an independent secret key k_{b-1} exchanged during the previous block over the secure feedback channel. The secret key allows us to encrypt otherwise insecure parts of the codewords with a one-time pad, and thus leads to higher secure communication rates.

Formally, the scheme operates as follows. We assume there exists a distribution p_X over $\mathcal{X}$ such that $\mathbb{I}(X;Y) - \mathbb{I}(X;Z) > 0$. We then consider a $(2^{nR}, 2^{nR_d}, n)$ code $\mathcal{C}_n$ identified by the random coding argument in Section 3.4.1 with

$$R = \mathbb{I}(X;Y) - \mathbb{I}(X;Z) - \delta(\epsilon), \qquad R_d = \mathbb{I}(X;Z) - \delta(\epsilon)$$

and such that

$$\mathbf{P}_e(\mathcal{C}_n) \leqslant \delta_\epsilon(n) \quad \text{and} \quad \frac{1}{n}\mathbf{L}(\mathcal{C}_n) \leqslant \delta(\epsilon) + \delta_\epsilon(n).$$

We set $R_o = \min(R_f, R_d)$ and, for each $m \in [\![1, 2^{nR}]\!]$, we distribute the codewords $x^n(m, m_d)$ with $m_d \in [\![1, 2^{nR_d}]\!]$ in $\lceil 2^{R_o} \rceil$ bins $\mathcal{B}_m(i)$ with $i \in [\![1, 2^{nR_o}]\!]$ and we relabel the codewords

$$x^n(i, j, k) \quad \text{with} \quad (i, j, k) \in [\![1, 2^{nR}]\!] \times [\![1, 2^{nR_o}]\!] \times [\![1, 2^{nR_d - nR_o}]\!].$$

This binning procedure is the same as the one illustrated in Figure 3.11. The sub-binning is revealed to all parties and we consider the following encoding/decoding procedure over B blocks of length n.

- *Encoder for block 1.* Given m_1 and m'_1, transmit $x^n(m_1, m'_1, k) \in \mathcal{C}_n$, where k is chosen uniformly at random in $\mathcal{B}_{m_1}(m'_1)$. During the transmission, receive secret key k_1 uniformly distributed in $[\![1, 2^{nR_o}]\!]$ over the feedback channel.
- *Encoder for block $b \in [\![2, B]\!]$.* Given m_b and m'_b, transmit $x^n(m_b, m'_b \oplus k_{b-1}, k) \in \mathcal{C}_n$, where $\oplus$ denotes modulo-$\lceil 2^{nR_o} \rceil$ addition and k is chosen uniformly at random in

$\mathcal{B}_{m_b}(m'_b)$. During the transmission, receive secret key k_b uniformly distributed in $[\![1, 2^{nR_o}]\!]$ over the feedback channel.

- *Decoder for block* $b = 1$. Given y^n, use the decoding procedure for $\mathcal{C}_n$ described in Section 3.4.1.
- *Decoder for block* $b \in [\![2, B]\!]$. Given y^n, use the decoding procedure for $\mathcal{C}_n$ to retrieve m_b and $m'_b \oplus k_{b-1}$. Use k_{b-1} to retrieve m'_b.

The sub-binning together with the encoding and decoding procedures above defines a $(2^{nBR'}, nB)$ code $\tilde{\mathcal{C}}_{nB}$ of length nB for the WTC with secure feedback. We assume that messages M_b and M'_b for $b \in [\![1, B]\!]$ are uniformly distributed so that the rate of $\tilde{\mathcal{C}}_{nB}$ is

$$R' = R + R_o - \delta(n) = \min(\mathbb{I}(\mathsf{X};\mathsf{Y}), \mathbb{I}(\mathsf{X};\mathsf{Y}) - \mathbb{I}(\mathsf{X};\mathsf{Z}) + R_f) - \delta(n).$$

We ignore the fact that the distribution of the codewords of $\mathcal{C}_n$ may be slightly non-uniform because of the binning and we refer the reader to Section 3.4.1 for details on how to deal with this subtlety.

Because the secret keys k_b are perfectly known both to the encoder and to the decoder, the probability of error for $\tilde{\mathcal{C}}_{nB}$ is at most B times that of $\mathcal{C}_n$ and

$$\mathbf{P}_e(\tilde{\mathcal{C}}_{nB}) \leqslant B\mathbf{P}_e(\mathcal{C}_n) \leqslant B\delta_\epsilon(n).$$

For $b \in [\![1, b]\!]$, we let Z_b^n denote the eavesdropper's observation in block b. The information leaked $\mathbf{L}(\tilde{\mathcal{C}}_{nB})$ is then

$$\begin{aligned}\frac{1}{nB}\mathbf{L}(\tilde{\mathcal{C}}_{nB}) &= \frac{1}{nB}\mathbb{I}\left(\mathsf{M}_1 \dots \mathsf{M}_B \mathsf{M}'_1 \dots \mathsf{M}'_B; \mathsf{Z}_1^n \dots \mathsf{Z}_B^n | \tilde{\mathcal{C}}_{nB}\right) \\ &= \frac{1}{nB}\sum_{b=1}^{B}\mathbb{I}\left(\mathsf{M}_b \mathsf{M}'_b; \mathsf{Z}_b^n | \tilde{\mathcal{C}}_{nB}\right) \\ &= \frac{1}{B}\sum_{b=1}^{B}\left(\frac{1}{n}\mathbb{I}\left(\mathsf{M}_b; \mathsf{Z}_b^n | \tilde{\mathcal{C}}_{nB}\right) + \frac{1}{n}\mathbb{I}\left(\mathsf{M}'_b; \mathsf{Z}_b^n | \mathsf{M}_b \tilde{\mathcal{C}}_{nB}\right)\right),\end{aligned}$$

where the first equality follows because M_b and M'_b depend only on Z_b^n. For $b \geqslant 2$, message M'_b is protected with a one-time pad and the crypto lemma guarantees that $(1/n)\mathbb{I}\left(\mathsf{M}'_b; \mathsf{Z}_b^n | \mathsf{M}_b \tilde{\mathcal{C}}_{nB}\right) = 0$. For $b = 1$, we can use the upper bound

$$\frac{1}{n}\mathbb{I}\left(\mathsf{M}'_1 \mathsf{Z}_1^n | \mathsf{M}_1 \tilde{\mathcal{C}}_{nB}\right) \leqslant R_o + \delta(n).$$

In addition, the construction of $\tilde{\mathcal{C}}_{nB}$ guarantees that

$$\frac{1}{n}\mathbb{I}\left(\mathsf{M}_b; \mathsf{Z}_b^n | \tilde{\mathcal{C}}_{nB}\right) = \frac{1}{n}\mathbf{L}(\mathcal{C}_n) \leqslant \delta(\epsilon) + \delta_\epsilon(n) \quad \text{for } b \in [\![1, B]\!].$$

Therefore,

$$\begin{aligned}\frac{1}{nB}\mathbf{L}(\tilde{\mathcal{C}}_{nB}) &\leqslant \frac{1}{B}\left(R_o + \delta(n) + \sum_{b=2}^{B}(\delta_\epsilon(n) + \delta(\epsilon))\right) \\ &= \delta(\epsilon) + \delta(B) + \delta_\epsilon(n).\end{aligned}$$

Hence, the rate–equivocation pair $(R', R' - \delta(\epsilon) - \delta(B))$ is achievable. Since ϵ can be chosen arbitrarily small and B can be chosen arbitrarily large, we conclude that

$$C_s^{\text{FB}} \geqslant \max_{p_X} \min(\mathbb{I}(X;Y), \mathbb{I}(X;Y) - \mathbb{I}(X;Z) + R_f).$$

One can also check that $\min(\mathbb{I}(X;Y), R_f)$ is an achievable rate if $\mathbb{I}(X;Y) - \mathbb{I}(X;Z) = 0$ for all X. As in Section 3.5.2, we can introduce a prefix channel $(\mathcal{V}, p_{X|V}, \mathcal{X})$ before the WTC $(\mathcal{X}, p_{YZ|X}, \mathcal{Y}, \mathcal{Z})$ to obtain the desired lower bound.

If all channels are weakly symmetric and the eavesdropper's channel is noisier than the main channel, we can remove the prefix channel as in the proof of Corollary 3.5, and the symmetry guarantees that both $\mathbb{I}(X;Y)$ and $\mathbb{I}(X;Y) - \mathbb{I}(X;Z)$ are maximized with X uniformly distributed. □

If the eavesdropper's channel is physically degraded with respect to the main channel, the pragmatic feedback strategy used in Proposition 3.9 turns out to be optimal. The proof of optimality is established in Section 4.4 on the basis of results about the secret-key capacity.

Theorem 3.4 (Ardestanizadeh *et al.*). *The secrecy capacity of a DWTC* $(\mathcal{X}, p_{YZ|X}, \mathcal{Y}, \mathcal{Z})$ *with confidential rate-limited feedback* R_f *is*

$$C_s^{\text{DWTC}} = \max_{p_X} \min(\mathbb{I}(X;Y), \mathbb{I}(X;Y|Z) + R_f).$$

Note that, even if $Z = Y$ and the secrecy capacity without feedback is zero, Proposition 3.9 guarantees a non-zero secure communication rate. This is not surprising because the feedback link is secure and can always be used to transmit a secure key and perform a one-time pad encryption; however, we will see in Chapter 4 and Chapter 8 that even public feedback enables more secure communications.

3.7 Conclusions and lessons learned

We conclude this chapter by summarizing the lessons learned from the analysis of secure communication over noisy DMCs. Most importantly, we proved the existence and identified the structure of codes that guarantee reliability and security simultaneously over wiretap channels. The specificity of these codes is that they map a given message to a *bin* of codewords, of which one is selected randomly by the encoder. Intuitively, the role of the randomness in the encoder is to "confuse" the eavesdropper by compensating for the information leaked about the codeword during transmission.

Unfortunately, as exemplified by Corollary 3.5, secure communications at a non-zero rate seem to be possible only when the legitimate receiver has a "physical advantage" over the eavesdropper, which we formalized with the notion of "noisier" channels. Although there are practical applications for which this condition is met, in particular if the Gaussian wiretap channel studied in Chapter 5 is a realistic model (near-field communications, RFID transmissions; see Chapter 7), it is fair to acknowledge that requiring an explicit advantage over the eavesdropper is a weakness in the model.

From the results of this chapter alone, one would even be inclined to conclude that the information-theoretic approach considered in this chapter does not improve on standard cryptography, and might merely be an alternative solution for which different trade-offs are made. Cryptography imposes restrictions on the computational power of the eavesdropper to relax assumptions on the communication channel (it considers the worst situation with a noiseless channel to the eavesdropper), whereas information-theoretic security seems to require an advantage at the physical layer in order to avoid restrictive assumptions on the abilities of the eavesdropper. Nevertheless, the somewhat unsatisfactory results obtained thus far hinge on the simplicity of the communication schemes, which do not account for powerful codes based on interactive communications. We will see in subsequent chapters that information-theoretically secure communications are sometimes possible without an explicit advantage at the physical layer.

We close this section with a review of the explicit and implicit assumptions used in the previous sections. Understanding these assumptions and their relevance is crucial to assess the potential of physical-layer security from a cryptographic perspective.

- **Knowledge of channel statistics.** In general, the lower bound of the eavesdropper's equivocation guaranteed by Theorem 3.2 is valid only if the code is tailored to the channel, which requires knowledge of the channel statistics *both* for the main channel and for the eavesdropper's channel. The assumption that the main channel is perfectly known is usually reasonable, since Alice and Bob can always cooperate to characterize their channel; however, the assumption that the eavesdropper's channel statistics are also known is more questionable. Nevertheless, as discussed in Section 3.5.1, this stringent requirement can be somewhat alleviated for a class of stochastically degraded channels.
- **Authentication.** The wiretap channel model assumes implicitly that the main channel is authenticated and, therefore, wiretap codes do not provide any protection against man-in-the-middle attacks; however, this assumption is not too restrictive, since authentication mechanisms can be implemented in the upper layers of the protocol stack. We shall see in Chapter 7 that it is possible to ensure unconditionally secure authentication with a negligible cost in terms of the secure communication rate, provided that a short secret key is available initially.
- **Passive attacker.** The scope of the results developed in this chapter is strictly restricted to eavesdropping strategies. Additional techniques are required for situations in which the adversary tampers with the channels, for instance by jamming the channel.
- **Perfect random-number generation.** The proofs developed in this chapter rely on the availability of a perfect random-number generator at the transmitter. The reader can check that the equivocation *decreases* if the entropy of the random-number generator is not maximum.
- **Weak secrecy.** As discussed in Section 3.3, weak secrecy is likely not an appropriate cryptographic metric. Therefore, the relevance of the results derived in this chapter will be fully justified in Chapter 4 when we show that the results do not change if the weak secrecy criterion is replaced by the strong secrecy criterion.

3.8 Bibliographical notes

The one-time pad was developed by Vernam during World War I for the Signal Corps of the US Army [19]. Although no rigorous analysis of the cipher was performed, the need for a random key as long as the message to be encrypted had already been identified. Shannon's cipher system formalized the problem of secure communications over noiseless channels in an information-theoretic framework [1]. Shannon's work includes an analysis of the one-time pad, but the crypto lemma in its most general form is due to Forney [20]. The degraded wiretap channel model and the notion of secrecy capacity were first introduced by Wyner in his seminal paper [21]. The simple expression of the secrecy capacity for symmetric channels was established by Leung-Yan-Cheong shortly after [22], and a simplified proof of Wyner's result was proposed by Massey [23]. The extension of Wyner's results to broadcast channels with confidential messages is due to Csiszár and Körner [18]. The proofs presented in this chapter are based on typical-set decoding in the spirit of [21], and can be easily combined with standard random-coding techniques for multi-user channels [3, 6]; however, this approach yields only weak secrecy results, and additional steps are required in order to strengthen the secrecy criterion. These steps are based on results from secret-key agreement that are presented in Section 4.5. Alternative proofs based on more powerful mathematical tools, such as graph-coloring techniques [24] or information-spectrum methods [25, 26], can be used to derive the secrecy capacity with strong secrecy directly.

Although stochastic encoders and codes with a binning structure are required for secure communications over memoryless channels, this need not be the case for other models. For instance, Dunn, Bloch, and Laneman investigated the possibility of secure communications over parallel timing channels [27] and showed that deterministic codes can achieve non-zero secure rates.

The various notions of partial ordering of channels were introduced by Körner and Marton [28], and the characterization of the relation "being noisier" in terms of the concavity of $\mathbb{I}(X; Y) - \mathbb{I}(X; Z)$ was obtained by Van Dijk [29]. The examples illustrating the notions of "noisier" and "less capable" channels given in Section 3.5.1 are due to Nair [30]. The combination of wiretap coding and one-time pad was proposed by Yamamoto [31], and the wiretap channel with a secure feedback link was investigated by Leung-Yan-Cheong [32], Ahlswede and Cai [33], and Ardestanizadeh, Franceschetti, Javidi, and Kim [34].

All of the results presented in this chapter assume a uniformly distributed source of messages; nevertheless, Theorem 3.2 and Theorem 3.3 can be generalized to account for arbitrarily distributed sources [18, 21], in which case a separation result holds: no rate is lost by first compressing the source and then transmitting it over a broadcast channel with confidential messages.

Although we mentioned that equivocation and probability of error are fundamentally different quantities, the relation between them was investigated more precisely by Feder and Merhav [35]. In particular, they provide lower bounds for the conditional entropy as a function of the probability of block error, which confirm that evaluating security on the basis of equivocation is a stronger requirement than simply requiring a decoding error for the eavesdropper.

An important issue that we have not addressed in this chapter is the numerical computation of the secrecy capacity. The presence of auxiliary random variables in Corollary 3.4 makes this computation difficult, and no efficient generic algorithm is known. Nevertheless, headway has been made in certain cases. For cases in which the eavesdropper's channel is noisier than the main channel, Yasui, Suko, and Matsushima proposed a Blahut–Arimoto-like algorithm to compute the secrecy capacity [36], which was also shown to be useful when the eavesdropper's channel is less capable than the main channel by Gowtham and Thangaraj [37].

4 Secret-key capacity

In Chapter 3, we considered the transmission of information over a noisy broadcast channel subject to reliability and security constraints; we showed that appropriate coding schemes can exploit the presence of noise to confuse the eavesdropper and guarantee some amount of information-theoretic security. It is important to note that the wiretap channel model assumes that all communications occur over the channel, hence communications are inherently rate-limited and one-way. Consequently, the results obtained do not fully capture the role of noise for secrecy; in particular, for situations in which the secrecy capacity is zero, it is not entirely clear whether this stems from the lack of any "physical advantage" over the eavesdropper or the restrictions imposed on the communication schemes.

The objective of this chapter is to study more precisely the fundamental role of noise in information-theoretic security. Instead of studying how we can communicate messages securely over a noisy channel, we now analyze how much secrecy we can extract from the noise itself in the form of a secret key. Specifically, we assume that the legitimate parties and the eavesdropper observe the realizations of correlated random variables and that the legitimate parties attempt to agree on a secret key unknown to the eavesdropper. To isolate the role played by noise, we remove restrictions on communication schemes and we assume that the legitimate parties can distill their key by communicating over a *two-way*, *public*, *noiseless*, and *authenticated* channel at no cost. Contrary to the case in Chapter 3, in which the natural metric of interest was the number of message bits that one could transmit securely and reliably per channel use, here the relevant metric is the number of secret-key bits distilled per observation of the correlated random variables.

We start this chapter by introducing two standard models for secret-key agreement, called the source model and the channel model, and by defining key-distillation strategies (Section 4.1). We then discuss the fundamental limits of key generation for a source model (Section 4.2) and we study in detail a specific type of key-distillation strategy, which we call "sequential key-distillation strategies" (Section 4.3). For these strategies, we prove results under a strong secrecy condition and we show how to construct practical strategies. Finally, we study the fundamental limit of secret-key generation over a channel model (Section 4.4) and show that the fundamental limit remains unchanged under a strong secrecy condition (Section 4.5). In other words, we prove that strong secrecy comes "for free," that is, a secure rate achievable with weak secrecy is also achievable with strong secrecy. This result is crucial to justify a posteriori the cryptographic relevance of the results derived in Chapter 3 for a weak-secrecy condition.

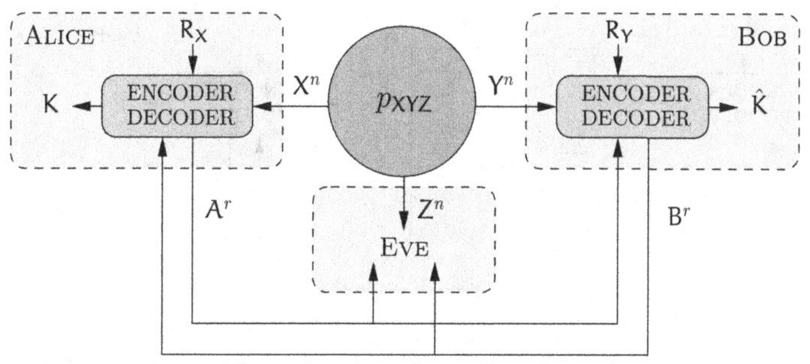

Figure 4.1 Source model for secret-key agreement.

4.1 Source and channel models for secret-key agreement

As illustrated in Figure 4.1, a *source model* for secret-key agreement represents a situation in which three parties, Alice, Bob, and Eve, observe the realizations of a DMS $(\mathcal{XYZ}, p_{XYZ})$ with three components. The DMS is assumed to be outside the control of all parties, but its statistics are known. By convention, component X is observed by Alice, component Y by Bob, and component Z by Eve. Alice and Bob's objective is to process their observations and agree on a key K about which Eve should have no information. To capture the essence of the problem, few restrictions are placed on the communication between Alice and Bob: they can exchange messages over a noiseless, two-way, and authenticated channel; however, to avoid trivializing the problem, the two-way channel is *public*, that is all messages are overheard by Eve and the existence of the public channel does not provide Alice and Bob with an explicit advantage over Eve. The only real simplifying assumption is the existence of an authentication mechanism that prevents Eve from tampering with communications over the public channel. Finally, we allow Alice and Bob to randomize the messages they transmit, which we model with sources of local randomness as done in Chapter 3. Alice has access to a DMS $(\mathcal{R}_\mathcal{X}, p_{R_X})$ and Bob has access to a DMS $(\mathcal{R}_\mathcal{Y}, p_{R_Y})$, which are mutually independent and independent of the DMS $(\mathcal{XYZ}, p_{XYZ})$. The rules by which Alice and Bob compute the messages they exchange over the public channel and agree on a key define a *key-distillation strategy*.[1]

Definition 4.1. *A $(2^{nR}, n)$ key-distillation strategy $\mathcal{S}_n$ for a source model with DMS $(\mathcal{XYZ}, p_{XYZ})$ consists of*

- *a key alphabet $\mathcal{K} = [\![1, 2^{nR}]\!]$;*
- *an alphabet $\mathcal{A}$ used by Alice to communicate over the public channel;*
- *an alphabet $\mathcal{B}$ used by Bob to communicate over the public channel;*
- *a source of local randomness for Alice $(\mathcal{R}_\mathcal{X}, p_{R_X})$;*
- *a source of local randomness for Bob $(\mathcal{R}_\mathcal{Y}, p_{R_Y})$;*

[1] We use the term "strategy" instead of "code" because a key does not carry information by itself.

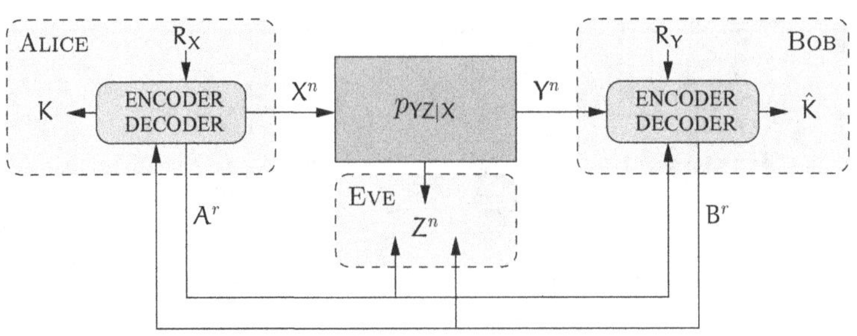

Figure 4.2 Channel model for secret-key agreement.

- *an integer $r \in \mathbb{N}^*$ that represents the number of rounds of communication;*
- *r encoding functions $f_i : \mathcal{X}^n \times \mathcal{B}^{i-1} \times \mathcal{R}_\mathcal{X} \to \mathcal{A}$ for $i \in [\![1, r]\!]$;*
- *r encoding functions $g_i : \mathcal{Y}^n \times \mathcal{A}^{i-1} \times \mathcal{R}_\mathcal{Y} \to \mathcal{B}$ for $i \in [\![1, r]\!]$;*
- *a key-distillation function $\kappa_\text{a} : \mathcal{X}^n \times \mathcal{B}^r \times \mathcal{R}_\mathcal{X} \to \mathcal{K}$;*
- *a key-distillation function $\kappa_\text{b} : \mathcal{Y}^n \times \mathcal{A}^r \times \mathcal{R}_\mathcal{Y} \to \mathcal{K}$;*

and operates as follows:

- *Alice observes n realizations of the source x^n while Bob observes y^n;*
- *Alice generates a realization r_x of her source of local randomness while Bob generates r_y from his;*
- *in round $i \in [\![1, r]\!]$, Alice transmits $a_i = f_i\left(x^n, b^{i-1}, r_x\right)$ while Bob transmits $b_i = g_i\left(y^n, a^{i-1}, r_y\right)$;*
- *after round r, Alice computes a key $k = \kappa_\text{a}(x^n, b^r, r_x)$ while Bob computes a key $\hat{k} = \kappa_\text{b}\left(y^n, a^r, r_y\right)$.*

By convention, we set $\mathcal{A}^0 \triangleq 0$ and $\mathcal{B}^0 \triangleq 0$. Note that the number of rounds r and the DMSs $\left(\mathcal{R}_\mathcal{X}, p_{\text{R}_\text{X}}\right)$ and $\left(\mathcal{R}_\mathcal{Y}, p_{\text{R}_\text{Y}}\right)$ can be optimized as part of the strategy design. The $(2^{nR}, n)$ key-distillation strategy $\mathcal{S}_n$ is assumed known ahead of time to Alice, Bob, and Eve.

The source model assumes the existence of an uncontrollable external source of randomness and abstracts the physical origin of the randomness completely. Although this is a strong assumption, there are several practical situations in which this would be a reasonable model. For instance, in wireless sensor networks, devices may monitor a physical phenomenon (change in temperature or pressure) whose statistics may be known but whose complexity is such that we can reasonably assume that it cannot be controlled. Nevertheless, it is legitimate to wonder what happens if the source of randomness is partially controlled by one of the parties. The analysis of situations in which the source is partially controlled by the eavesdropper is not fully understood and is not covered in this book. We refer the interested reader to the bibliographical notes at the end of the chapter for references to existing results.

It is somewhat less arduous to study the situation in which the source is partially controlled by one of the legitimate parties. In this case, the model is called a *channel model* for secret-key agreement and is as illustrated in Figure 4.2. Instead of observing the realizations of an external source, Bob and Eve now observe the outputs of a DMC $(\mathcal{X}, p_{YZ|X}, \mathcal{Y}, \mathcal{Z})$ whose input is controlled by Alice. Alice and Bob have again access to a public, noiseless, two-way, and authenticated channel over which they can exchange messages to agree on a secret key. We also assume that Alice and Bob have access to sources of local randomness $(\mathcal{R}_\mathcal{X}, p_{R_X})$ and $(\mathcal{R}_\mathcal{Y}, p_{R_Y})$ to randomize their communications. Key-distillation strategies for the channel model are more sophisticated than those for the source model, because Alice can use the feedback provided by Bob to adapt the symbols she sends in the channel. Despite the similarity between the channel model for secret-key agreement and the wiretap channel studied in Chapter 3, note that the problems are different: in a channel model for secret-key agreement, the broadcast channel is used not only to communicate messages but also to generate randomness.

Definition 4.2. *A $(2^{nR}, n)$ key-distillation strategy $\mathcal{S}_n$ for a channel model with DMC $(\mathcal{X}, p_{YZ|X}, \mathcal{Y}, \mathcal{Z})$ consists of*

- *a key alphabet $\mathcal{K} = [\![1, 2^{nR}]\!]$;*
- *an alphabet $\mathcal{A}$ used by Alice to communicate over the public channel;*
- *an alphabet $\mathcal{B}$ used by Bob to communicate over the public channel;*
- *a source of local randomness for Alice $(\mathcal{R}_\mathcal{X}, p_{R_X})$;*
- *a source of local randomness for Bob $(\mathcal{R}_\mathcal{Y}, p_{R_Y})$;*
- *an integer $r \in \mathbb{N}^*$ that represents the number of rounds of communication;*
- *a set of n distinct integers $\{i_j\}_n \subseteq [\![1, r]\!]$ that represents the rounds in which Alice transmits a symbol over the channel;*
- *$r - n$ encoding functions $f_i : \mathcal{B}^{i-1} \times \mathcal{R}_\mathcal{X} \to \mathcal{A}$ for $i \in [\![1, r]\!] \setminus \{i_j\}_n$;*
- *$r - n$ encoding functions g_i for $i \in [\![1, r]\!] \setminus \{i_j\}_n$ of the form $g_i : \mathcal{Y}^j \times \mathcal{A}^{i-1} \times \mathcal{R}_\mathcal{Y} \to \mathcal{B}$ if $i \in [\![i_j + 1, i_{j+1} - 1]\!]$;*
- *n functions $h_j : \mathcal{B}^{i_j-1} \times \mathcal{R}_\mathcal{X} \to \mathcal{X}$ for $j \in [\![1, n]\!]$ to generate channel inputs;*
- *a key-distillation function $\kappa_a : \mathcal{X}^n \times \mathcal{B}^r \times \mathcal{R}_\mathcal{X} \to \mathcal{K}$;*
- *a key-distillation function $\kappa_b : \mathcal{Y}^n \times \mathcal{A}^r \times \mathcal{R}_\mathcal{Y} \to \mathcal{K}$;*

and operates as follows:

- *Alice generates a realization r_x of her source of local randomness while Bob generates r_y from his;*
- *in round $i \in [\![1, i_1 - 1]\!]$, Alice transmits message $a_i = f_i(b^{i-1}, r_x)$ and Bob transmits message $b_i = g_i(a^{i-1}, r_y)$;*
- *in round i_j with $j \in [\![1, n]\!]$, Alice transmits symbol $x_j = h_j(b^{i_j-1}, r_x)$ over the channel, and Bob and Eve observe the symbols y_j and Z_j, respectively;*
- *in round $i \in [\![i_j + 1, i_{j+1} - 1]\!]$, Alice transmits message $a_i = f_i(x^j, b^{i-1}, r_x)$ and Bob transmits message $b_i = g_i(y^j, a^{i-1}, r_y)$;*
- *after the last round, Alice computes a key $k = \kappa_a(x^n, b^r, r_x)$ and Bob computes a key $\hat{k} = \kappa_b(y^n, a^r, r_y)$.*

By convention, we set $i_{n+1} \triangleq r + 1, i_0 = 0, \mathcal{A}^0 = 0$, and $\mathcal{B}^0 \triangleq 0$. Note that the number of rounds r, the indices $\{i_j\}_n$ of the rounds during which a symbol is transmitted over the channel, and the sources of local randomness $(\mathcal{R}_\mathcal{X}, p_{R_X})$ and $(\mathcal{R}_\mathcal{Y}, p_{R_Y})$ can be optimized as part of the strategy design. Again, the key-distillation strategy $\mathcal{S}_n$ is assumed known to Alice, Bob, and Eve ahead of time.

Remark 4.1. *A wiretap code $\mathcal{C}_n$ is a key-distillation strategy for a channel model since Alice can use $\mathcal{C}_n$ to transmit uniform secret keys to Bob directly without using the public channel. In general, key-distillation strategies that exploit the public channel are more powerful, but we will see an example of a channel model for which using a wiretap code is an optimal key-distillation strategy in Section 4.4.*

For both source and channel models, the performance of a $(2^{nR}, n)$ key-distillation strategy $\mathcal{S}_n$ is measured in terms of the average probability of error

$$\mathbf{P}_e(\mathcal{S}_n) \triangleq \mathbb{P}\left[\mathsf{K} \neq \hat{\mathsf{K}} \middle| \mathcal{S}_n\right],$$

in terms of the information leakage to the eavesdropper,

$$\mathbf{L}(\mathcal{S}_n) \triangleq \mathbb{I}(\mathsf{K}; \mathsf{Z}^n \mathsf{A}^r \mathsf{B}^r | \mathcal{S}_n),$$

and in terms of the uniformity of the keys,

$$\mathbf{U}(\mathcal{S}_n) \triangleq \log\lceil 2^{nR}\rceil - \mathbb{H}(\mathsf{K}|\mathcal{S}_n).$$

Note that, by definition, $\mathbf{U}(\mathcal{S}_n) \geqslant 0$ with equality if and only if the key is exactly uniformly distributed.

Remark 4.2. *It is possible to combine the information leaked to the eavesdropper and the uniformity of the key into a single quantity called the* security index *and defined as*

$$\log\lceil 2^{nR}\rceil - \mathbb{H}(\mathsf{K}|\mathsf{Z}^n \mathsf{A}^r \mathsf{B}^r \mathcal{S}_n).$$

The security index is equal to zero if and only if the key is uniformly distributed and unknown to the eavesdropper. However, we choose to study $\mathbf{U}(\mathcal{S}_n)$ and $\mathbf{L}(\mathcal{S}_n)$ independently to emphasize that these are different constraints.

Definition 4.3. *A weak secret-key rate R is achievable for a source or channel model if there exists a sequence of $(2^{nR}, n)$ key-distillation strategies $\{\mathcal{S}_n\}_{n\geqslant 1}$ such that*

$$\lim_{n\to\infty} \mathbf{P}_e(\mathcal{S}_n) = 0 \quad \textit{(reliability)}, \tag{4.1}$$

$$\lim_{n\to\infty} \frac{1}{n}\mathbf{L}(\mathcal{S}_n) = 0 \quad \textit{(weak secrecy)}, \tag{4.2}$$

$$\lim_{n\to\infty} \frac{1}{n}\mathbf{U}(\mathcal{S}_n) = 0 \quad \textit{(weak uniformity)}. \tag{4.3}$$

The corresponding keys are called weak secret keys. If the strategies $\{\mathcal{S}_n\}_{n\geqslant 1}$ exploit public messages sent either from Alice to Bob only or from Bob to Alice only, the secret-key rate R is said to be achievable with one-way *communication; otherwise, R is said to be achievable with* two-way *communication.*

Condition (4.1) means that Alice and Bob should agree on a common key with high probability. Condition (4.2) requires that Eve, who has access to information through her randomness Z^n and the messages exchanged over the public channel $\mathsf{A}^r\mathsf{B}^r$, obtains a negligible *rate* of information about the key; this condition is tantamount to the full secrecy rate condition studied in the context of wiretap channels. Finally, Condition (4.3) requires the key rate to be almost that of a uniform key, which is a necessary property of secret keys if they are are to be used to protect messages with a one-time pad as seen in Theorem 3.1. As already discussed in Section 3.3, Conditions (4.2) and (4.3) are called "weak" because they impose constraints on the rate of information leaked and on the rate at which the entropy of a key approaches that of a uniform distribution.

Remark 4.3. *In principle, we could attempt to characterize a rate–equivocation region for keys and consider partially secret keys for which* $(1/n)\mathbf{E}(\mathsf{K}|\mathsf{Z}^n\mathsf{A}^r\mathsf{B}^r\mathcal{S}_n) \geqslant R_e$ *with* $R_e \leqslant R$*; however, it is not clear what the cryptographic purpose of such keys would be. In addition, note that partially secret keys can be obtained by expanding secret keys with a known function or by adding known bits; therefore, without loss of generality, we restrict ourselves to the analysis of secret keys as in Definition 4.3.*

Before we investigate the fundamental limits of achievable weak secret-key rates, it is worth looking at how weak secret keys can be used. Consider a $(2^{nR}, n)$ key-distillation strategy $\mathcal{S}_n$ such that

$$\mathbf{P}_e(\mathcal{S}_n) \leqslant \epsilon, \quad \frac{1}{n}\mathbf{L}(\mathcal{S}_n) \leqslant \epsilon, \quad \text{and} \quad \frac{1}{n}\mathbf{U}(\mathcal{S}_n) \leqslant \epsilon$$

for some $\epsilon > 0$. Assume the resulting weak secret key K is used in the one-time-pad encryption of a message $\mathsf{M} \in \mathcal{K}$ that is independent of all variables involved in the key-generation process. Since K is not exactly uniform and not totally unknown to the adversary, we should not expect to obtain the perfect secrecy discussed in Chapter 3. In fact, note that

$$\begin{aligned}
\mathbb{I}(\mathsf{M}\oplus\mathsf{K}, \mathsf{A}^r, \mathsf{B}^r, \mathsf{Z}^n; \mathsf{M}|\mathcal{S}_n) &= \mathbb{I}(\mathsf{A}^r\mathsf{B}^r\mathsf{Z}^n; \mathsf{M}|\mathcal{S}_n) + \mathbb{I}(\mathsf{M}\oplus\mathsf{K}; \mathsf{M}|\mathsf{A}^r\mathsf{B}^r\mathsf{Z}^n\mathcal{S}_n)\\
&\overset{(a)}{=} \mathbb{I}(\mathsf{M}\oplus\mathsf{K}; \mathsf{M}|\mathsf{A}^r\mathsf{B}^r\mathsf{Z}^n\mathcal{S}_n)\\
&\overset{(b)}{=} \mathbb{H}(\mathsf{M}\oplus\mathsf{K}|\mathsf{A}^r\mathsf{B}^r\mathsf{Z}^n\mathcal{S}_n) - \mathbb{H}(\mathsf{K}|\mathsf{M}\mathsf{A}^r\mathsf{B}^r\mathsf{Z}^n\mathcal{S}_n)\\
&\leqslant \log\lceil 2^{nR}\rceil - \mathbb{H}(\mathsf{K}|\mathsf{M}\mathsf{A}^r\mathsf{B}^r\mathsf{Z}^n\mathcal{S}_n)\\
&\overset{(c)}{\leqslant} \mathbb{H}(\mathsf{K}|\mathcal{S}_n) + n\epsilon - \mathbb{H}(\mathsf{K}|\mathsf{A}^r\mathsf{B}^r\mathsf{Z}^n\mathcal{S}_n)\\
&= \mathbf{L}(\mathcal{S}_n) + n\epsilon\\
&\overset{(d)}{\leqslant} n\delta(\epsilon),
\end{aligned} \tag{4.4}$$

where (a) follows from $\mathbb{I}(\mathsf{A}^r\mathsf{B}^r\mathsf{Z}^n; \mathsf{M}|\mathcal{S}_n) = 0$ since M is independent of $\mathsf{A}^r\mathsf{B}^r\mathsf{Z}^n$, (b) follows from $\mathbb{H}(\mathsf{M}\oplus\mathsf{K}|\mathsf{M}\mathsf{A}^r\mathsf{B}^r\mathsf{Z}^n\mathcal{S}_n) = \mathbb{H}(\mathsf{K}|\mathsf{M}\mathsf{A}^r\mathsf{B}^r\mathsf{Z}^n\mathcal{S}_n)$, (c) follows from $(1/n)\mathbf{U}(\mathcal{S}_n) \leqslant \epsilon$, and (d) follows from $(1/n)\mathbf{L}(\mathcal{S}_n) \leqslant \epsilon$. Therefore, we can prove only that this encryption guarantees the weak secrecy condition

$$\frac{1}{n}\mathbb{I}(\mathsf{M}\oplus\mathsf{K}, \mathsf{A}^r, \mathsf{B}^r, \mathsf{Z}^n; \mathsf{M}|\mathcal{S}_n) \leqslant \delta(\epsilon).$$

This result is somewhat unsatisfactory, but it can be improved if we strengthen the notion of the achievable rate as follows.

Definition 4.4. *A strong secret-key rate R is achievable for a source or channel model if there exists a sequence of $(2^{nR}, n)$ key-distillation strategies $\{\mathcal{S}_n\}_{n\geqslant 1}$ such that*

$$\lim_{n\to\infty} \mathbf{P}_e(\mathcal{S}_n) = 0 \qquad \textit{(reliability)}, \tag{4.5}$$

$$\lim_{n\to\infty} \mathbf{L}(\mathcal{S}_n) = 0 \qquad \textit{(strong secrecy)}, \tag{4.6}$$

$$\lim_{n\to\infty} \mathbf{U}(\mathcal{S}_n) = 0 \qquad \textit{(strong uniformity)}. \tag{4.7}$$

The corresponding keys are called strong secret keys.

The secrecy condition and the uniformity condition for strong secret-key rates differ from their weak counterparts in that they do not involve any normalization by n; hence, achievable strong secret-key rates are also achievable weak secret-key rates. Consider now a $(2^{nR}, n)$ key-distillation strategy such that

$$\mathbf{P}_e(\mathcal{S}_n) \leqslant \epsilon, \quad \mathbf{L}(\mathcal{S}_n) \leqslant \epsilon, \quad \text{and} \quad \mathbf{U}(\mathcal{S}_n) \leqslant \epsilon,$$

and assume as done earlier that the resulting strong secret key K is used for the one-time-pad encryption of a message M that is independent of all random variables involved in the key-distillation process. We can reiterate the calculation in (4.4) and the reader can check that we can then guarantee the strong secrecy condition

$$\mathbb{I}(\mathsf{M} \oplus \mathsf{K}, \mathsf{A}^r, \mathsf{B}^r, \mathsf{Z}^n; \mathsf{M}|\mathcal{S}_n) \leqslant \delta(\epsilon).$$

Note that this still does not match the perfect secrecy condition, but does guarantee a reasonable secrecy level if ϵ is small enough.

4.2 Secret-key capacity of the source model

In this section, we study the *secret-key capacity*, which is defined as the supremum of secret-key rates achievable for a source model. Since a secret-key rate is defined as a number of secret-key bits per realization of a DMS, the secret-key capacity does not account for the amount of communication required to distill keys, which could be arbitrarily large.

Definition 4.5. *The weak secret-key capacity of a source model with DMS $(\mathcal{XYZ}, p_{\mathsf{XYZ}})$ is*

$$C_s^{\text{SM}} \triangleq \sup\{R : R \text{ is an achievable weak secret-key rate}\}.$$

Similarly, the strong secret-key capacity of a source model with DMS $(\mathcal{XYZ}, p_{\mathsf{XYZ}})$ is

$$\overline{C}_s^{\text{SM}} \triangleq \sup\{R : R \text{ is an achievable strong secret-key rate}\}.$$

When required, we write $C_s^{\text{SM}}(p_{\mathsf{XYZ}})$ in place of C_s^{SM} to explicitly specify the underlying distribution of the DMS under consideration. It follows directly from the definition of achievable secret-key rates that $\overline{C}_{\text{s}}^{\text{SM}} \leqslant C_{\text{s}}^{\text{SM}}$. We will show at the end of Section 4.3 that $\overline{C}_{\text{s}}^{\text{SM}} = C_{\text{s}}^{\text{SM}}$ but, until then, we restrict our attention to the weak secret-key capacity. Although this restriction will prove unnecessary, the study of weak secret-key capacity is less arduous than that of strong secret-key capacity and still provides useful insight into the design of key-distillation strategies. In addition, note that any upper bound we derive for C_{s}^{SM} is automatically an upper bound for $\overline{C}_{\text{s}}^{\text{SM}}$.

A closed-form expression for the secret-key capacity for a general source model remains elusive. Nevertheless, it is possible to obtain simple upper and lower bounds that are useful in many situations.

Theorem 4.1 (Maurer, Ahlswede, and Csiszár). *The weak secret-key capacity of a source model $(\mathcal{XYZ}, p_{\text{XYZ}})$ satisfies*

$$\mathbb{I}(\text{X};\text{Y}) - \min(\mathbb{I}(\text{X};\text{Z}), \mathbb{I}(\text{Y};\text{Z})) \leqslant C_{\text{s}}^{\text{SM}} \leqslant \min(\mathbb{I}(\text{X};\text{Y}), \mathbb{I}(\text{X}, \text{Y}|\text{Z})).$$

Moreover, the secret-key rate $\mathbb{I}(\text{X};\text{Y}) - \min(\mathbb{I}(\text{X};\text{Z}), \mathbb{I}(\text{Y};\text{Z}))$ is achievable with one-way communication.

Proof. We provide proofs of the result in Sections 4.2.1, 4.2.2, and 4.2.3. The proof in Section 4.2.1 leverages the results obtained for WTCs in Chapter 3, whereas the proof in Section 4.2.2 provides a more direct approach based on Slepian–Wolf codes. □

The lower bound $\mathbb{I}(\text{X};\text{Y}) - \min(\mathbb{I}(\text{X};\text{Z}), \mathbb{I}(\text{Y};\text{Z}))$ can be understood as the difference between the information rate between Alice and Bob and some information rate leaked to Eve. However, in contrast to the results obtained in Chapter 3, Alice and Bob can choose whether the information rate obtained by Eve is leaked from Alice ($\mathbb{I}(\text{X};\text{Y})$) or from Bob ($\mathbb{I}(\text{Y};\text{Z})$). We will see in the course of the proof that this result stems from the possibility of two-way communication over the public channel. Before we prove Theorem 4.1, it is also useful to note that, in general, the bounds in Theorem 4.1 are loose. In Section 4.3.1, we will see an example of a source model for which the lower bound is useless because $\mathbb{I}(\text{X};\text{Y}) - \mathbb{I}(\text{X};\text{Z}) < 0$ and $\mathbb{I}(\text{X};\text{Y}) - \mathbb{I}(\text{Y};\text{Z}) < 0$. In Section 4.2.4, we will provide an example of a source model for which $\mathbb{I}(\text{X};\text{Y}|\text{Z}) > 0$ and $\mathbb{I}(\text{X};\text{Y}) > 0$ while $C_{\text{s}}^{\text{SM}} = 0$. Nevertheless, there are several situations in which the bounds are tight.

Corollary 4.1. *Consider a source model with DMS $(\mathcal{XYZ}, p_{\text{XYZ}})$:*

- *if* Z *is independent of* (X, Y), *then* $C_{\text{s}}^{\text{SM}} = \mathbb{I}(\text{X};\text{Y})$*;*
- *if* $\text{X} \rightarrow \text{Y} \rightarrow \text{Z}$ *forms a Markov chain, then* $C_{\text{s}}^{\text{SM}} = \mathbb{I}(\text{X};\text{Y}) - \mathbb{I}(\text{X};\text{Z})$*;*
- *if* $\text{Y} \rightarrow \text{X} \rightarrow \text{Z}$ *forms a Markov chain, then* $C_{\text{s}}^{\text{SM}} = \mathbb{I}(\text{X};\text{Y}) - \mathbb{I}(\text{Y};\text{Z})$.

Proof. If Z is independent of X and Y, then both bounds in Theorem 4.1 are equal to $\mathbb{I}(\text{X};\text{Y})$. If $\text{X} \rightarrow \text{Y} \rightarrow \text{Z}$ forms a Markov chain then

$$\begin{aligned}
\mathbb{I}(\text{X};\text{Y}|\text{Z}) &= \mathbb{I}(\text{X};\text{YZ}) - \mathbb{I}(\text{X};\text{Z}) \\
&= \mathbb{I}(\text{X};\text{Y}) + \mathbb{I}(\text{X};\text{Z}|\text{Y}) - \mathbb{I}(\text{X};\text{Z}) \\
&= \mathbb{I}(\text{X};\text{Y}) - \mathbb{I}(\text{X};\text{Z}).
\end{aligned}$$

Finally, if $\mathsf{Y} \to \mathsf{X} \to \mathsf{Z}$, then Z and Y are conditionally independent given X, and

$$\begin{aligned}\mathbb{I}(\mathsf{X};\mathsf{Y}|\mathsf{Z}) &= \mathbb{I}(\mathsf{XZ};\mathsf{Y}) - \mathbb{I}(\mathsf{Y};\mathsf{Z})\\ &= \mathbb{I}(\mathsf{X};\mathsf{Y}) + \mathbb{I}(\mathsf{Z};\mathsf{Y}|\mathsf{X}) - \mathbb{I}(\mathsf{Y};\mathsf{Z})\\ &= \mathbb{I}(\mathsf{X};\mathsf{Y}) - \mathbb{I}(\mathsf{Y};\mathsf{Z}).\end{aligned}$$

□

Remark 4.4. *If we drop the term Z^n in the information leaked $\mathbf{L}(\mathcal{S}_n)$ and if we use $\mathbb{I}(\mathsf{K};\mathsf{A}^r\mathsf{B}^r|\mathcal{S}_n)$ as the measure of secrecy, then the secret-key capacity is called the* private-key capacity. *The private-key capacity measures the maximum key rate with respect to an eavesdropper who observes communications over the public channel only and who disregards the realizations of the source Z^n. The private-key capacity can also be viewed as a special case of the secret-key capacity for a source model in which Z is independent of X and Y.*

4.2.1 Secret-key distillation based on wiretap codes

In this section, we establish that the secret-key capacity is at least as large as $\mathbb{I}(\mathsf{X};\mathsf{Y}) - \min(\mathbb{I}(\mathsf{X};\mathsf{Z}), \mathbb{I}(\mathsf{Y};\mathsf{Z}))$ by leveraging the results of Chapter 3 about the secrecy capacity of WTCs. The basic idea of the proof is to analyze a specific key-distillation strategy that creates a *conceptual* WTC, for which we know the existence of wiretap codes and achievable secure communication rates.

We first show that $\mathbb{I}(\mathsf{X};\mathsf{Y}) - \mathbb{I}(\mathsf{X};\mathsf{Z})$ is an achievable secret-key rate. Assume that Alice wants to send a symbol $u \in \mathcal{X}$ that is independent of the DMS $(\mathcal{XYZ}, p_{\mathsf{XYZ}})$ over the public channel. Instead of transmitting u directly, she observes a realization $x \in \mathcal{X}$ of the source and transmits $u \oplus x$, where $\oplus$ denotes addition modulo-$|\mathcal{X}|$ over $\mathcal{X}$. At the same time, Bob observes y and Eve observes z. In effect, this operation creates a memoryless WTC with input U, for which Bob receives the *pair* of symbols $(\mathsf{Y}, \mathsf{U} \oplus \mathsf{X})$ and Eve receives the *pair* of symbols $(\mathsf{Z}, \mathsf{U} \oplus \mathsf{X})$. We know from Corollary 3.4 that the secrecy capacity of this WTC is at least

$$\mathbb{I}(\mathsf{U};\mathsf{Y},\mathsf{U}\oplus\mathsf{X}) - \mathbb{I}(\mathsf{U};\mathsf{Z},\mathsf{U}\oplus\mathsf{X}) = \mathbb{H}(\mathsf{U}|\mathsf{Z},\mathsf{U}\oplus\mathsf{X}) - \mathbb{H}(\mathsf{U}|\mathsf{Y},\mathsf{U}\oplus\mathsf{X}),$$

where the distribution p_{U} over $\mathcal{X}$ can be chosen arbitrarily. Here, we choose p_{U} to be the uniform distribution over $\mathcal{X}$. Then,

$$\begin{aligned}\mathbb{H}(\mathsf{U}|\mathsf{Z},\mathsf{U}\oplus\mathsf{X}) &= \mathbb{H}(\mathsf{U},\mathsf{U}\oplus\mathsf{X}|\mathsf{Z}) - \mathbb{H}(\mathsf{U}\oplus\mathsf{X}|\mathsf{Z})\\ &= \mathbb{H}(\mathsf{U}|\mathsf{Z}) + \mathbb{H}(\mathsf{U}\oplus\mathsf{X}|\mathsf{U},\mathsf{Z}) - \mathbb{H}(\mathsf{U}\oplus\mathsf{X}|\mathsf{Z}).\end{aligned}$$

Since U is independent of (X,Z), $\mathbb{H}(\mathsf{U}|\mathsf{Z}) = \mathbb{H}(\mathsf{U})$ and $\mathbb{H}(\mathsf{U}\oplus\mathsf{X}|\mathsf{U},\mathsf{Z}) = \mathbb{H}(\mathsf{X}|\mathsf{UZ}) = \mathbb{H}(\mathsf{X}|\mathsf{Z})$. Additionally, since U is uniformly distributed over $\mathcal{X}$, the crypto lemma applies and $\mathbb{H}(\mathsf{U}\oplus\mathsf{X}|\mathsf{Z}) = \mathbb{H}(\mathsf{U})$. Therefore,

$$\mathbb{H}(\mathsf{U}|\mathsf{Z},\mathsf{U}\oplus\mathsf{X}) = \mathbb{H}(\mathsf{X}|\mathsf{Z}).$$

Similarly, we can show that $\mathbb{H}(\mathsf{U}|\mathsf{Y}, \mathsf{U} \oplus \mathsf{X}) = \mathbb{H}(\mathsf{X}|\mathsf{Y})$; therefore,

$$\mathbb{I}(\mathsf{U}; \mathsf{Y}, \mathsf{U} \oplus \mathsf{X}) - \mathbb{I}(\mathsf{U}; \mathsf{Z}, \mathsf{U} \oplus \mathsf{X}) = \mathbb{H}(\mathsf{X}|\mathsf{Z}) - \mathbb{H}(\mathsf{X}|\mathsf{Y}).$$

Since the secret-key capacity is at least as large as the secrecy capacity of the conceptual WTC, we conclude that

$$C_{\mathrm{s}}^{\mathrm{SM}} \geqslant \mathbb{H}(\mathsf{X}|\mathsf{Z}) - \mathbb{H}(\mathsf{X}|\mathsf{Y}) = \mathbb{I}(\mathsf{X}; \mathsf{Y}) - \mathbb{I}(\mathsf{X}; \mathsf{Z}). \tag{4.8}$$

Because the public channel is two-way, we can reverse the roles of Alice and Bob and create another conceptual WTC from Bob to Alice. Following the same arguments as above, we obtain the second lower bound

$$C_{\mathrm{s}}^{\mathrm{SM}} \geqslant \mathbb{H}(\mathsf{Y}|\mathsf{Z}) - \mathbb{H}(\mathsf{Y}|\mathsf{X}) = \mathbb{I}(\mathsf{Y}; \mathsf{X}) - \mathbb{I}(\mathsf{Y}; \mathsf{Z}). \tag{4.9}$$

By combining (4.8) and (4.9) we obtain

$$\begin{aligned} C_{\mathrm{s}}^{\mathrm{SM}} &\geqslant \max(\mathbb{I}(\mathsf{X}; \mathsf{Y}) - \mathbb{I}(\mathsf{X}; \mathsf{Z}), \mathbb{I}(\mathsf{X}; \mathsf{Y}) - \mathbb{I}(\mathsf{Y}; \mathsf{Z})) \\ &= \mathbb{I}(\mathsf{X}; \mathsf{Y}) - \min(\mathbb{I}(\mathsf{X}; \mathsf{Z}), \mathbb{I}(\mathsf{Y}; \mathsf{Z})). \end{aligned}$$

4.2.2 Secret-key distillation based on Slepian–Wolf codes

In this section, we rederive the achievable rates given by Theorem 4.1 with a more constructive proof that is based on Slepian–Wolf codes. Although this alternative approach does not improve on the bounds already obtained, it is useful because it provides operational insight into the design of key-distillation strategies without relying on the existence of wiretap codes. The use of Slepian–Wolf codes makes the derivation slightly more involved than that in Section 4.2.1, but it bears some similarity to the achievability proof of Theorem 3.2, which is based on the notion of an enhanced channel.

In principle, the definition of key-distillation strategies allows many exchanges of messages over the public channel; however, to make the analysis tractable, we study simpler strategies. The first simplification consists of restricting our attention to strategies that exploit a *single* and *one-way* round of communication over the public channel and that do not rely on local randomness. If we assume that Alice is the one transmitting the public message, then such a strategy involves only

- a *single* encoding function $f : \mathcal{X}^n \to \mathcal{A}$ to compute the message a sent over the public channel;
- Alice's key-distillation function $\kappa_{\mathrm{a}} : \mathcal{X}^n \to \mathcal{K}$;
- Bob's key-distillation function $\kappa_{\mathrm{b}} : \mathcal{Y}^n \times \mathcal{A} \to \mathcal{K}$.

The second simplification consists of requiring Bob to decode Alice's observation x^n on the basis of his own observation y^n and the public message a instead of computing k directly. We will show that these simple strategies suffice to achieve the rates in Theorem 4.1 by means of a random-binning argument; however, before we do so, it is useful to develop an additional desirable property that these strategies should

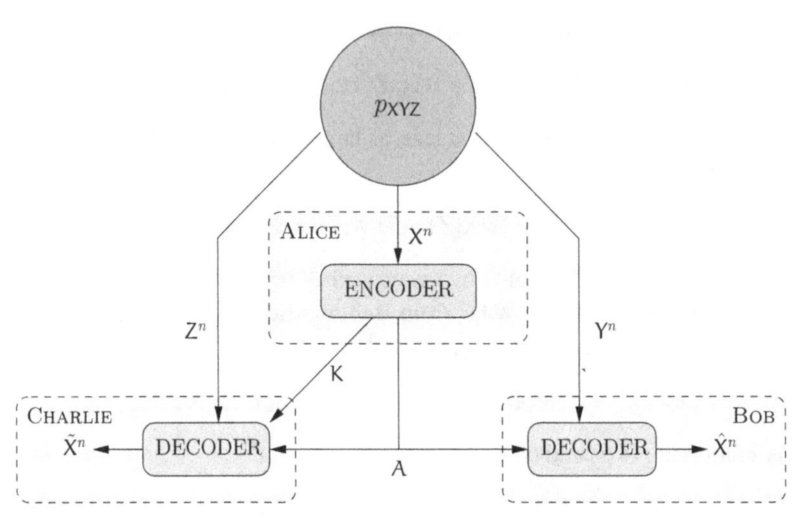

Figure 4.3 Enhanced source model for secret-key agreement.

possess. We expand the information rate $(1/n)\mathbf{L}(\mathcal{S}_n)$ leaked to the eavesdropper as follows:

$$\begin{aligned}
\frac{1}{n}\mathbf{L}(\mathcal{S}_n) &= \frac{1}{n}\mathbb{I}(\mathsf{K};\mathsf{Z}^n\mathsf{A}|\mathcal{S}_n) \\
&= \frac{1}{n}\mathbb{I}(\mathsf{K}\mathsf{X}^n;\mathsf{Z}^n\mathsf{A}|\mathcal{S}_n) - \frac{1}{n}\mathbb{I}(\mathsf{X}^n;\mathsf{Z}^n\mathsf{A}|\mathsf{K}\mathcal{S}_n) \\
&= \frac{1}{n}\mathbb{I}(\mathsf{X}^n;\mathsf{Z}^n\mathsf{A}|\mathcal{S}_n) + \frac{1}{n}\mathbb{I}(\mathsf{K};\mathsf{Z}^n\mathsf{A}|\mathsf{X}^n\mathcal{S}_n) - \frac{1}{n}\mathbb{I}(\mathsf{X}^n;\mathsf{Z}^n\mathsf{A}|\mathsf{K}\mathcal{S}_n) \\
&= \frac{1}{n}\mathbb{I}(\mathsf{X}^n;\mathsf{Z}^n\mathsf{A}|\mathcal{S}_n) - \frac{1}{n}\mathbb{I}(\mathsf{X}^n;\mathsf{Z}^n\mathsf{A}|\mathsf{K}\mathcal{S}_n),
\end{aligned}$$

where the last inequality follows from $\mathbb{I}(\mathsf{K};\mathsf{Z}^n\mathsf{A}|\mathsf{X}^n\mathcal{S}_n) = 0$ since K is a function of X^n. Note that, no matter how A is computed, the leakage is minimized if $(1/n)\mathbb{I}(\mathsf{X}^n;\mathsf{Z}^n\mathsf{A}|\mathsf{K}\mathcal{S}_n)$ is maximized, which happens if $(1/n)\mathbb{H}(\mathsf{X}^n|\mathsf{Z}^n\mathsf{A}\mathsf{K}\mathcal{S}_n)$ is small.

Therefore, we shall prove the existence of a sequence of $(2^{nR}, n)$ strategies $\{\mathcal{S}_n\}_{n\geqslant 1}$ with a single and one-way round of communication such that

$$\lim_{n\to\infty} \mathbf{P}_e(\mathcal{S}_n) = 0, \quad \lim_{n\to\infty} \frac{1}{n}\mathbb{H}(\mathsf{X}^n|\mathsf{Z}^n\mathsf{A}\mathsf{K}\mathcal{S}_n) = 0, \tag{4.10}$$

$$\lim_{n\to\infty} \frac{1}{n}\mathbf{U}(\mathcal{S}_n) = 0, \quad \lim_{n\to\infty} \frac{1}{n}\mathbf{L}(\mathcal{S}_n) = 0. \tag{4.11}$$

We now show that the two conditions in (4.10) can be combined into a single reliability constraint by considering the enhanced source model illustrated in Figure 4.3. This model enhances the original secret-key agreement problem by introducing a *virtual receiver*, hereafter named Charlie, who obtains the same observation Z^n as Eve, overhears the message A over the public channel, and has access to K through an error-free side channel.

Definition 4.6. *An $(2^{nR}, 2^{nR_p}, n)$ strategy $\mathcal{S}_n$ for the enhanced source model consists of*

- *two index sets $\mathcal{K} = [\![1, 2^{nR}]\!]$ and $\mathcal{A} = [\![1, 2^{nR_p}]\!]$;*
- *an encoding function $f : \mathcal{X}^n \to \mathcal{A}$;*
- *a key-distillation function $\kappa_a : \mathcal{X}^n \to \mathcal{K}$;*
- *a decoding function $g : \mathcal{Y}^n \times \mathcal{A} \to \mathcal{X}^n \cup \{?\}$ for Bob;*
- *a decoding function $h : \mathcal{Z}^n \times \mathcal{A} \times \mathcal{K} \to \mathcal{X}^n \cup \{?\}$ for Charlie.*

Bob's key-distillation function is implicitly defined as $\kappa_b \triangleq \kappa_a \circ g$ if $g(y^n, a) \neq ?$. The reliability performance of a $(2^{nR}, 2^{nR_p}, n)$ strategy $\mathcal{S}_n$ is measured in terms of its average probability of error

$$\mathbf{P}_e(\mathcal{S}_n) \triangleq \mathbb{P}\left[\hat{\mathsf{X}}^n \neq \mathsf{X}^n \text{ or } \tilde{\mathsf{X}}^n \neq \mathsf{X}^n \middle| \mathcal{S}_n\right],$$

its secrecy performance is measured in terms of the leakage

$$\mathbf{L}(\mathcal{S}_n) \triangleq \mathbb{I}(\mathsf{K}; \mathsf{Z}^n\mathsf{A}|\mathcal{S}_n),$$

and the uniformity of keys is measured in terms of

$$\mathbf{U}(\mathcal{S}_n) \triangleq \log\lceil 2^{nR}\rceil - \mathbb{H}(\mathsf{K}|\mathcal{S}_n).$$

Note that a $(2^{nR}, 2^{nR_p}, n)$ strategy $\mathcal{S}_n$ for the enhanced source model is a $(2^{nR}, n)$ secret-key distillation strategy for the original source model, which is subject to a more stringent reliability constraint and for which we are controlling explicitly the rate R_p of communication over the public channel. By construction, the probability of error for the original source model is at most $\mathbf{P}_e(\mathcal{S}_n)$ since

$$\mathbb{P}\left[\hat{\mathsf{X}}^n \neq \mathsf{X}^n \middle| \mathcal{S}_n\right] \leqslant \mathbb{P}\left[\hat{\mathsf{X}}^n \neq \mathsf{X}^n \text{ or } \tilde{\mathsf{X}}^n \neq \mathsf{X}^n \middle| \mathcal{S}_n\right] = \mathbf{P}_e(\mathcal{S}_n).$$

In addition, Fano's inequality guarantees that $(1/n)\mathbb{H}(\mathsf{X}^n|\mathsf{A}\mathsf{Z}^n\mathsf{K}\mathcal{S}_n) \leqslant \delta(\mathbf{P}_e(\mathcal{S}_n))$. Therefore, the two constraints in (4.10) are automatically satisfied if $\lim_{n\to\infty} \mathbf{P}_e(\mathcal{S}_n) = 0$ for the enhanced source model.

We are now ready to develop a random-binning argument and show the existence of a sequence of $(2^{nR}, 2^{nR_p}, n)$ strategies $\{\mathcal{S}_n\}_{n\geqslant 1}$ such that

$$\lim_{n\to\infty} \mathbf{P}_e(\mathcal{S}_n) = 0, \quad \lim_{n\to\infty} \frac{1}{n}\mathbf{L}(\mathcal{S}_n) = 0, \quad \text{and} \quad \lim_{n\to\infty} \frac{1}{n}\mathbf{U}(\mathcal{S}_n) = 0$$

for some appropriate choice of R and R_p.

Without loss of generality, we assume that $\mathbb{H}(\mathsf{X}) > \mathbb{H}(\mathsf{X}|\mathsf{Z}) - \mathbb{H}(\mathsf{X}|\mathsf{Y}) > 0$. Let $\epsilon > 0$ and $n \in \mathbb{N}^*$. Let $R > 0$ and $R_p > 0$ be rates to be specified later. We construct a $(2^{nR}, 2^{nR_p}, n)$ strategy as follows.

- *Codebook construction.* For each sequence $x^n \in \mathcal{T}_\epsilon^n(\mathsf{X})$, draw two indices uniformly at random in the sets $[\![1, 2^{nR_p}]\!]$ and $[\![1, 2^{nR}]\!]$; these index assignments define the functions $f : \mathcal{X}^n \to [\![1, 2^{nR_p}]\!]$ and $\kappa_a : \mathcal{X}^n \to [\![1, 2^{nR}]\!]$, which are revealed to all parties.
- *Alice's encoder.* Given an observation x^n, if $x^n \in \mathcal{T}_\epsilon^n(\mathsf{X})$, set $k = \kappa_a(x^n)$ and $a = f(x^n)$; otherwise, set $k = 1$ and $a = 1$.

- *Bob's decoder.* Given an observation y^n, output $\hat{x}^n$ if it is the unique sequence such that $(\hat{x}^n, y^n) \in \mathcal{T}_\epsilon^n(\mathsf{XY})$ and $f(\hat{x}^n) = a$; otherwise, output an error ?; if there is no error, distill a key $\hat{k} = \kappa_a(\hat{x}^n)$.
- *Charlie's decoder.* Given an observation z^n, output $\tilde{x}^n$ if it is the unique sequence such that $(\tilde{x}^n, z^n) \in \mathcal{T}_\epsilon^n(\mathsf{XZ})$, $f(\tilde{x}^n) = a$ and $\kappa_a(\tilde{x}^n) = k$; otherwise, output an error ?.

The random variable that represents the strategy defined by the randomly chosen index assignments is denoted by S_n. We proceed to bound the quantities $\mathbb{E}[\mathbf{P}_e(\mathsf{S}_n)]$, $\mathbb{E}[(1/n)\mathbf{L}(\mathsf{S}_n)]$, and $\mathbb{E}[(1/n)\mathbf{U}(\mathsf{S}_n)]$ separately.

The upper bound for $\mathbb{E}[\mathbf{P}_e(\mathsf{S}_n)]$ is obtained with the approach used in Section 2.3.1 for the Slepian–Wolf theorem. $\mathbb{E}[\mathbf{P}_e(\mathsf{S}_n)]$ can be expressed in terms of the events

$$\begin{aligned}
\mathcal{E}_0 &= \left\{\mathsf{X}^n \notin \mathcal{T}_\epsilon^n(\mathsf{X}) \text{ or } (\mathsf{X}^n, \mathsf{Y}^n) \notin \mathcal{T}_\epsilon^n(\mathsf{XY})\right\},\\
\mathcal{E}_1 &= \left\{\exists x^n \neq \mathsf{X}^n : f(x^n) = \mathsf{A} \text{ and } (x^n, \mathsf{Y}^n) \in \mathcal{T}_\epsilon^n(\mathsf{XY})\right\},\\
\mathcal{E}_2 &= \left\{\exists x^n \neq \mathsf{X}^n : \kappa_a(x^n) = \mathsf{K},\, f(x^n) = \mathsf{A} \text{ and } (x^n, \mathsf{Z}^n) \in \mathcal{T}_\epsilon^n(\mathsf{XZ})\right\}
\end{aligned}$$

since $\mathbb{E}[\mathbf{P}_e(\mathsf{S}_n)] = \mathbb{P}[\mathcal{E}_0 \cup \mathcal{E}_1 \cup \mathcal{E}_2]$. By the union bound, we obtain

$$\mathbb{E}[\mathbf{P}_e(\mathsf{S}_n)] \leqslant \mathbb{P}[\mathcal{E}_0] + \mathbb{P}[\mathcal{E}_1] + \mathbb{P}[\mathcal{E}_2],$$

and, following exactly the same approach as that used in Section 2.3.1 to prove the Slepian–Wolf theorem, we can show that, if

$$R_p > \mathbb{H}(\mathsf{X}|\mathsf{Y}) + \delta(\epsilon) \quad \text{and} \quad R + R_p > \mathbb{H}(\mathsf{X}|\mathsf{Z}) + \delta(\epsilon), \tag{4.12}$$

then $\mathbb{E}[\mathbf{P}_e(\mathsf{S}_n)] \leqslant \delta_\epsilon(n)$.

Next, we develop an upper bound for $\mathbb{E}[(1/n)\mathbf{L}(\mathsf{S}_n)]$. We expand $\mathbb{E}[(1/n)\mathbf{L}(\mathsf{S}_n)]$ as

$$\begin{aligned}
\mathbb{E}\left[\frac{1}{n}\mathbf{L}(\mathsf{S}_n)\right] &= \frac{1}{n}\mathbb{I}(\mathsf{K}; \mathsf{A}\mathsf{Z}^n|\mathsf{S}_n)\\
&= \frac{1}{n}\mathbb{H}(\mathsf{K}|\mathsf{S}_n) + \frac{1}{n}\mathbb{H}(\mathsf{A}\mathsf{Z}^n|\mathsf{S}_n) - \frac{1}{n}\mathbb{H}(\mathsf{A}\mathsf{K}\mathsf{Z}^n|\mathsf{S}_n)\\
&= \frac{1}{n}\mathbb{H}(\mathsf{K}|\mathsf{S}_n) + \frac{1}{n}\mathbb{H}(\mathsf{A}|\mathsf{S}_n) + \frac{1}{n}\mathbb{H}(\mathsf{Z}^n|\mathsf{A}\mathsf{S}_n) - \frac{1}{n}\mathbb{H}(\mathsf{A}\mathsf{K}\mathsf{Z}^n|\mathsf{S}_n)\\
&\leqslant \frac{1}{n}\mathbb{H}(\mathsf{K}|\mathsf{S}_n) + \frac{1}{n}\mathbb{H}(\mathsf{A}|\mathsf{S}_n) + \mathbb{H}(\mathsf{Z}) - \frac{1}{n}\mathbb{H}(\mathsf{A}\mathsf{K}\mathsf{Z}^n|\mathsf{S}_n),
\end{aligned} \tag{4.13}$$

where the last inequality follows from $\mathbb{H}(\mathsf{Z}^n|\mathsf{A}\mathsf{S}_n) \leqslant \mathbb{H}(\mathsf{Z}^n) = n\mathbb{H}(\mathsf{Z})$. We bound the remaining terms on the right-hand side separately. By construction of a $(2^{nR}, 2^{nR_p}, n)$ strategy,

$$\frac{1}{n}\mathbb{H}(\mathsf{K}|\mathsf{S}_n) \leqslant R + \delta(n), \quad \frac{1}{n}\mathbb{H}(\mathsf{A}|\mathsf{S}_n) \leqslant R_p + \delta(n). \tag{4.14}$$

In addition,

$$\begin{aligned}
\frac{1}{n}\mathbb{H}(\mathsf{A}\mathsf{K}\mathsf{Z}^n|\mathsf{S}_n) &= \frac{1}{n}\mathbb{H}(\mathsf{A}\mathsf{K}\mathsf{Z}^n\mathsf{X}^n|\mathsf{S}_n) - \frac{1}{n}\mathbb{H}(\mathsf{X}^n|\mathsf{A}\mathsf{K}\mathsf{Z}^n\mathsf{S}_n)\\
&= \frac{1}{n}\mathbb{H}(\mathsf{Z}^n\mathsf{X}^n|\mathsf{S}_n) + \frac{1}{n}\mathbb{H}(\mathsf{A}\mathsf{K}|\mathsf{X}^n\mathsf{Z}^n\mathsf{S}_n) - \frac{1}{n}\mathbb{H}(\mathsf{X}^n|\mathsf{A}\mathsf{K}\mathsf{Z}^n\mathsf{S}_n)\\
&= \frac{1}{n}\mathbb{H}(\mathsf{Z}^n\mathsf{X}^n|\mathsf{S}_n) - \frac{1}{n}\mathbb{H}(\mathsf{X}^n|\mathsf{A}\mathsf{K}\mathsf{Z}^n\mathsf{S}_n),
\end{aligned}$$

where the last equality follows from $\mathbb{H}(\mathsf{AK}|\mathsf{Z}^n\mathsf{X}^n\mathsf{S}_n) = 0$ because A and K are functions of X^n. By virtue of Fano's inequality,

$$\begin{aligned}
\frac{1}{n}\mathbb{H}(\mathsf{X}^n|\mathsf{AKZ}^n\mathsf{S}_n) &= \sum_{\mathcal{S}_n} p_{\mathsf{S}_n}(\mathcal{S}_n)\,\frac{1}{n}\mathbb{H}(\mathsf{X}^n|\mathsf{AKZ}^n\mathcal{S}_n)\\
&\leqslant \sum_{\mathcal{S}_n} p_{\mathsf{S}_n}(\mathcal{S}_n)\left(\frac{1}{n} + \mathbf{P}_{\mathrm{e}}(\mathcal{S}_n)\log|\mathcal{X}|\right)\\
&= \delta(n) + \mathbb{E}[\mathbf{P}_{\mathrm{e}}(\mathsf{S}_n)]\log|\mathcal{X}|\\
&= \delta_\epsilon(n).
\end{aligned}$$

Therefore,

$$\frac{1}{n}\mathbb{H}(\mathsf{AKZ}^n|\mathsf{S}_n) \geqslant \frac{1}{n}\mathbb{H}(\mathsf{Z}^n\mathsf{X}^n|\mathsf{S}_n) - \delta_\epsilon(n) = \mathbb{H}(\mathsf{ZX}) - \delta(\epsilon). \tag{4.15}$$

On substituting (4.14) and (4.15) into (4.13), we obtain

$$\mathbb{E}\left[\frac{1}{n}\mathbf{L}(\mathsf{S}_n)\right] \leqslant R + R_{\mathrm{p}} + \mathbb{H}(\mathsf{Z}) - \mathbb{H}(\mathsf{ZX}) + \delta_\epsilon(n),$$

for any R and R_{p} satisfying (4.12). In particular, the choice

$$R_{\mathrm{p}} = \mathbb{H}(\mathsf{X}|\mathsf{Y}) + \delta(\epsilon) \quad \text{and} \quad R = \mathbb{H}(\mathsf{X}|\mathsf{Z}) - \mathbb{H}(\mathsf{X}|\mathsf{Y}) + \delta(\epsilon) \tag{4.16}$$

is compatible with the constraints (4.12) and yields

$$\mathbb{E}\left[\frac{1}{n}\mathbf{L}(\mathsf{S}_n)\right] = \delta(\epsilon) + \delta_\epsilon(n).$$

Finally, we develop an upper bound for $\mathbb{E}[(1/n)\mathbf{U}(\mathsf{S}_n)]$ by establishing a lower bound for $(1/n)\mathbb{H}(\mathsf{K}|\mathsf{S}_n)$. Let us introduce the random variable Ξ, such that

$$\Xi \triangleq \begin{cases} 1 & \text{if } \mathsf{X}^n \in \mathcal{T}_\epsilon^n(\mathsf{X}),\\ 0 & \text{otherwise.}\end{cases}$$

By construction, Ξ is independent of S_n and, by the AEP, $\mathbb{P}[\Xi = 1] \geqslant 1 - \delta_\epsilon(n)$. Next, notice that,

$$\begin{aligned}
\frac{1}{n}\mathbb{H}(\mathsf{K}|\mathsf{S}_n) &\geqslant \frac{1}{n}\mathbb{H}(\mathsf{K}|\mathsf{S}_n\Xi)\\
&\geqslant \mathbb{P}[\Xi = 1]\frac{1}{n}\mathbb{H}(\mathsf{K}|\mathsf{S}_n, \Xi = 1)\\
&\geqslant (1 - \delta_\epsilon(n))\frac{1}{n}\mathbb{H}(\mathsf{K}|\mathsf{S}_n, \Xi = 1).
\end{aligned} \tag{4.17}$$

For a specific strategy $\mathcal{S}_n$, define $\mathsf{K}_{\mathcal{S}_n}$ as the random variable with distribution

$$p_{\mathsf{K}_{\mathcal{S}_n}} \triangleq p_{\mathsf{K}|\mathsf{S}_n=\mathcal{S}_n,\Xi=1}.$$

Then $\mathbb{H}(\mathsf{K}|\mathsf{S}_n = \mathcal{S}_n, \Xi = 1) = \mathbb{H}(\mathsf{K}_{\mathcal{S}_n})$, which can be written explicitly in terms of the probability $p_{\mathsf{K}_{\mathcal{S}_n}}$ as

$$\frac{1}{n}\mathbb{H}(\mathsf{K}_{\mathcal{S}_n}) = -\frac{1}{n}\sum_{k\in\mathcal{K}} p_{\mathsf{K}_{\mathcal{S}_n}}(k)\log p_{\mathsf{K}_{\mathcal{S}_n}}(k).$$

By virtue of the symmetry of the random-binning construction, the quantity $\mathbb{E}_{S_n}\left[-p_{K_{S_n}}(k)\log p_{K_{S_n}}(k)\right]$ is independent of k; therefore,

$$\mathbb{E}_{S_n}\left[\frac{1}{n}\mathbb{H}(K_{S_n})\right] = \frac{1}{n}\sum_{k\in\mathcal{K}}\mathbb{E}_S\left[-p_{K_{S_n}}(k)\log p_{K_{S_n}}(k)\right]$$
$$= \frac{\lceil 2^{nR}\rceil}{n}\mathbb{E}_S\left[-p_{K_{S_n}}(1)\log p_{K_{S_n}}(1)\right]. \tag{4.18}$$

Intuitively, because we use a random binning in which keys are assigned uniformly at random, we expect $p_{K_{\mathcal{S}_n}}(1)$ to be on the order of 2^{-nR} for most strategies $\mathcal{S}_n$. This idea is formalized in the following lemma, whose proof is relegated to the appendix to this chapter.

Lemma 4.1. *Let χ be a function of a key-distillation strategy $\mathcal{S}_n$ defined as*

$$\chi(\mathcal{S}_n) = \begin{cases} 1 & \text{if } \left|p_{K_{\mathcal{S}_n}}(1) - 2^{-nR}\right| \leqslant \epsilon 2^{-nR}, \\ 0 & \text{otherwise.}\end{cases}$$

If $R < \mathbb{H}(X) - \delta(\epsilon)$ then $\mathbb{P}_{S_n}[\chi(S_n) = 1] \geqslant 1 - \delta_\epsilon(n)$.

Using Lemma 4.1 with R as defined by (4.16), we then bound $\mathbb{E}_{S_n}\left[-p_{K_{S_n}}(1)\log p_{K_{S_n}}(1)\right]$ as

$$\mathbb{E}_{S_n}\left[-p_{K_{S_n}}(1)\log p_{K_{S_n}}(1)\right] \geqslant \mathbb{E}_{S_n}\left[-p_{K_{S_n}}(1)\log p_{K_{S_n}}(1)|\chi(S_n) = 1\right]\mathbb{P}_{S_n}[\chi(S_n) = 1]$$
$$\geqslant (1-\delta_\epsilon(n))(1-\epsilon)2^{-nR}\log\left(\frac{2^{nR}}{1+\epsilon}\right). \tag{4.19}$$

On combining (4.17), (4.18), and (4.19), we obtain

$$\frac{1}{n}\mathbb{H}(K|S_n) \geqslant R - \delta(\epsilon),$$

and, therefore,

$$\mathbb{E}\left[\frac{1}{n}\mathbf{U}(S_n)\right] \leqslant \delta_\epsilon(n) + \delta(\epsilon).$$

By applying the selection lemma to the random variable S_n and the functions $\mathbf{P}_e$, $\mathbf{L}$, and $\mathbf{U}$, we conclude that there exists a specific strategy $\mathcal{S}_n$ with rate R given by (4.16) and such that

$$\mathbf{P}_e(\mathcal{S}_n) \leqslant \delta_\epsilon(n), \quad \frac{1}{n}\mathbf{L}(\mathcal{S}_n) \leqslant \delta_\epsilon(n) + \delta(\epsilon), \quad \text{and} \quad \frac{1}{n}\mathbf{U}(\mathcal{S}_n) \leqslant \delta_\epsilon(n) + \delta(\epsilon).$$

Hence, there exists a sequence of $(2^{nR}, n)$ key-distillation strategies $\{\mathcal{S}_n\}_{n\geqslant 1}$ with rate

$$R = \mathbb{H}(X|Z) - \mathbb{H}(X|Y) + \delta(\epsilon) = \mathbb{I}(X;Y) - \mathbb{I}(X;Z) + \delta(\epsilon)$$

and such that

$$\lim_{n\to\infty}\mathbf{P}_e(\mathcal{S}_n) = 0, \quad \lim_{n\to\infty}\frac{1}{n}\mathbf{L}(\mathcal{S}_n) \leqslant \delta(\epsilon), \quad \text{and} \quad \lim_{n\to\infty}\frac{1}{n}\mathbf{U}(\mathcal{S}_n) \leqslant \delta(\epsilon).$$

Since ϵ can be chosen arbitrarily small, we conclude that the secret-key capacity is at least $\mathbb{I}(X;Y) - \mathbb{I}(X;Z)$.

Note that, if $\mathbb{I}(X;Y) - \mathbb{I}(Y;Z) > 0$, we can reproduce the proof by swapping the roles of Alice and Bob and swapping X and Y in the equations. Therefore, the secret-key capacity is also at least $\mathbb{I}(Y;X) - \mathbb{I}(Y;Z)$.

Remark 4.5. *The existence of a public channel of unlimited capacity in the model is convenient because it allows us to focus solely on secrecy without having to account for the cost of communication and processing; however, this approach has a subtle drawback. In the real world, there exists no channel of unlimited capacity, and the public channel used in the model would be obtained through multiple uses of a side channel with finite capacity. Consequently, if a key-distillation strategy requires many rounds of communication over the public channel, the side channel would have to be used many times. Hence, the* effective *secret-key rate, obtained by normalizing the key size by the number of random realizations of the sources plus the number of uses of the side channel, may be much lower than what is predicted by the results obtained thus far. This motivates the study of key-distillation strategies with* rate-limited *public communication, that is, key-distillation strategies for which the messages exchanged over the public channel are subject to a rate constraint R_p. The construction of strategies based on Slepian–Wolf codes described above does not allow us to control precisely the rate of messages sent over the public channel. Actually, in* (4.16)*, we implicitly* required *the public rate R_p to be at least $\mathbb{H}(X|Y)$ (or $\mathbb{H}(Y|X)$ if the roles of Alice and Bob are swapped). The key idea to handle rate-limited public communication is to construct a Slepian–Wolf-based strategy that operates on a* quantized *version of X^n instead of on X^n directly. The quantization allows us to control the rate of public communication and to adjust it so that it falls below the rate constraint R_p. Specifically, one can combine the strategy described above with a Wyner–Ziv compression scheme, which can be thought of as a special case of vector quantization. We refer the reader to the bibliographical notes for further references.*

4.2.3 Upper bound for secret-key capacity

In this section, we show that $C_s^{SM} \leqslant \min(\mathbb{I}(X;Y), \mathbb{I}(X;Y|Z))$ with a converse argument. Let R be an achievable weak secret-key rate and let $\epsilon > 0$. For n sufficiently large, there exists a $(2^{nR}, n)$ key-distillation strategy $\mathcal{S}_n$ such that

$$\mathbf{P}_e(\mathcal{S}_n) \leqslant \epsilon, \quad \frac{1}{n}\mathbf{L}(\mathcal{S}_n) \leqslant \epsilon, \quad \text{and} \quad \frac{1}{n}\mathbf{U}(\mathcal{S}_n) \leqslant \epsilon.$$

For clarity, we drop the conditioning on the strategy $\mathcal{S}_n$ in all subsequent calculations. By virtue of Fano's inequality, we have

$$\frac{1}{n}\mathbb{H}\left(K|\hat{K}A^rB^rZ^n\right) \leqslant \frac{1}{n}\mathbb{H}\left(K|\hat{K}\right) \leqslant \delta(\mathbf{P}_e(\mathcal{S}_n)) \leqslant \delta(\epsilon).$$

First, we show that $R \leqslant \mathbb{I}(X;Y|Z) + \delta(\epsilon)$. Note that

$$
\begin{aligned}
R &\leqslant \frac{1}{n}\log\lceil 2^{nR}\rceil \\
&= \frac{1}{n}\mathbb{H}(K) + \frac{1}{n}\mathbf{U}(\mathcal{S}_n) \\
&\leqslant \frac{1}{n}\mathbb{H}(K) + \epsilon \\
&= \frac{1}{n}\mathbb{H}(K|A^rB^rZ^n) + \frac{1}{n}\mathbf{L}(\mathcal{S}_n) + \epsilon \\
&\leqslant \frac{1}{n}\mathbb{H}(K|A^rB^rZ^n) + \delta(\epsilon) \\
&= \frac{1}{n}\mathbb{I}\left(K;\hat{K}|A^rB^rZ^n\right) + \frac{1}{n}\mathbb{H}\left(K|\hat{K}A^rB^rZ^n\right) + \delta(\epsilon) \\
&\leqslant \frac{1}{n}\mathbb{I}\left(K;\hat{K}|A^rB^rZ^n\right) + \delta(\epsilon) \\
&\leqslant \frac{1}{n}\mathbb{I}(X^nR_X;Y^nR_Y|A^rB^rZ^n) + \delta(\epsilon),
\end{aligned}
\tag{4.20}
$$

where the last inequality follows from the data-processing inequality applied to the Markov chain $K \to X^nR_XB^r \to Y^nR_YA^r \to \hat{K}$. We upper bound $(1/n)\mathbb{I}(X^nR_X;Y^nR_Y|A^rB^rZ^n)$ by using the following lemma.

Lemma 4.2. *Let $r \in \mathbb{N}^*$. Let $S \in \mathcal{S}$, $T \in \mathcal{T}$, $U \in \mathcal{U}$, $V^r \in \mathcal{V}^r$, and $W^r \in \mathcal{W}^r$ be random vectors such that*

$$
\forall i \in [\![1,r]\!] \quad \begin{cases} V_i \text{ is a function of S and } W^{i-1}, \\ W_i \text{ is a function of T and } V^{i-1}. \end{cases}
$$

Then, $\mathbb{I}(S;T|V^rW^rU) \leqslant \mathbb{I}(S;T|U)$.

Proof. We upper bound $\mathbb{I}(S;T|V^rW^rU)$ as follows:

$$
\begin{aligned}
\mathbb{I}(S;T|V^rW^rU) &\leqslant \mathbb{I}\left(SV_r;T|V^{r-1}W^rU\right) \\
&\leqslant \mathbb{I}\left(SV_r;TW_r|V^{r-1}W^{r-1}U\right) \\
&= \mathbb{I}\left(S;TW_r|V^{r-1}W^{r-1}U\right) + \mathbb{I}\left(V_r;TW_r|SV^{r-1}W^{r-1}U\right) \\
&= \mathbb{I}\left(S;T|V^{r-1}W^{r-1}U\right) + \mathbb{I}\left(S;W_r|V^{r-1}W^{r-1}TU\right) \\
&\quad + \mathbb{I}\left(V_r;TW_r|SV^{r-1}W^{r-1}U\right).
\end{aligned}
\tag{4.21}
$$

Since V_r is a function of S and W^{r-1} and W_r is a function of T and V^{r-1}, note that

$$
\mathbb{I}\left(V_r;TW_r|SV^{r-1}W^{r-1}U\right) = 0 \quad \text{and} \quad \mathbb{I}\left(S;W_r|TV^{r-1}W^{r-1}U\right) = 0. \tag{4.22}
$$

On substituting (4.22) into (4.21), we obtain

$$
\mathbb{I}(S;T|V^rW^rU) \leqslant \mathbb{I}\left(S;T|V^{r-1}W^{r-1}U\right).
$$

By induction over r, we conclude that $\mathbb{I}(S;T|V^rW^rU) \leqslant \mathbb{I}(S;T|U)$. □

By using Lemma 4.2 with $\mathsf{S} \triangleq \mathsf{X}^n\mathsf{R}_\mathsf{X}$, $\mathsf{T} \triangleq \mathsf{Y}^n\mathsf{R}_\mathsf{Y}$, $\mathsf{U} \triangleq \mathsf{Z}^n$, $\mathsf{V}^r \triangleq \mathsf{A}^r$, and $\mathsf{W}^r \triangleq \mathsf{B}^r$, we obtain

$$\mathbb{I}(\mathsf{X}^n\mathsf{R}_\mathsf{X}; \mathsf{Y}^n\mathsf{R}_\mathsf{Y}|\mathsf{A}^r\mathsf{B}^r\mathsf{Z}^n) \leqslant \mathbb{I}(\mathsf{X}^n\mathsf{R}_\mathsf{X}; \mathsf{Y}^n\mathsf{R}_\mathsf{Y}|\mathsf{Z}^n).$$

Since R_X and R_Y are independent of the DMS $(\mathcal{XYZ}, p_{\mathsf{XYZ}})$, we have

$$\mathbb{I}(\mathsf{X}^n\mathsf{R}_\mathsf{X}; \mathsf{Y}^n\mathsf{R}_\mathsf{Y}|\mathsf{Z}^n) = \mathbb{I}(\mathsf{X}^n; \mathsf{Y}^n|\mathsf{Z}^n) = n\mathbb{I}(\mathsf{X}; \mathsf{Y}|\mathsf{Z});$$

therefore,

$$\mathbb{I}(\mathsf{X}^n\mathsf{R}_\mathsf{X}; \mathsf{Y}^n\mathsf{R}_\mathsf{Y}|\mathsf{A}^r\mathsf{B}^r\mathsf{Z}^n) \leqslant n\mathbb{I}(\mathsf{X}; \mathsf{Y}|\mathsf{Z}). \tag{4.23}$$

On substituting (4.23) into (4.20), we obtain

$$R \leqslant \mathbb{I}(\mathsf{X}; \mathsf{Y}|\mathsf{Z}) + \delta(\epsilon).$$

Finally, we show that $R \leqslant \mathbb{I}(\mathsf{X}; \mathsf{Y}) + \delta(\epsilon)$. Note that, by assumption,

$$\frac{1}{n}\mathbb{I}(\mathsf{K}; \mathsf{A}^r\mathsf{B}^r) \leqslant \frac{1}{n}\mathbb{I}(\mathsf{K}; \mathsf{A}^r\mathsf{B}^r\mathsf{Z}^n) = \frac{1}{n}\mathbf{L}(\mathcal{S}_n) \leqslant \delta(\epsilon).$$

Therefore, all the steps leading to (4.20) and (4.23) can be reapplied without conditioning on the variable Z^n, which yields $R \leqslant \mathbb{I}(\mathsf{X}; \mathsf{Y}) + \delta(\epsilon)$. Since $\epsilon > 0$ can be chosen arbitrarily small, we obtain the desired upper bound

$$C_\mathrm{s}^\mathrm{SM} \leqslant \min(\mathbb{I}(\mathsf{X}; \mathsf{Y}), \mathbb{I}(\mathsf{X}; \mathsf{Y}|\mathsf{Z})).$$

4.2.4 Alternative upper bounds for secret-key capacity

In general, the upper bound $C_\mathrm{s}^\mathrm{SM} \leqslant \mathbb{I}(\mathsf{X}; \mathsf{Y}|\mathsf{Z})$ established in Theorem 4.2 is loose, but the cause of this looseness is buried in the technical details of the proof; hence, it is worth developing an intuitive understanding of the bound before we try to improve it.

We start by showing that, for any source model, the quantity $\mathbb{I}(\mathsf{X}; \mathsf{Y}|\mathsf{Z})$ admits an operational interpretation: it is the secret-key capacity obtained by providing Bob with an explicit advantage over Eve. In fact, consider a source model with DMS $(\mathcal{XYZ}, p_{\mathsf{XYZ}})$ and secret-key capacity $C_\mathrm{s}^\mathrm{SM}(p_{\mathsf{XYZ}})$. Assume we provide an advantage to Bob by giving him access to Eve's observation Z, which creates a new source model in which Bob observes $\mathsf{Y}' = (\mathsf{YZ})$ instead of Y. Since a key-distillation strategy for the original source model remains a key-distillation strategy for the new source model, it holds that $C_\mathrm{s}^\mathrm{SM}(p_{\mathsf{XY}'\mathsf{Z}}) \geqslant C_\mathrm{s}^\mathrm{SM}(p_{\mathsf{XYZ}})$; however, because $\mathsf{X} \rightarrow \mathsf{Y}' \rightarrow \mathsf{Z}$ forms a Markov chain, Corollary 4.1 applies and

$$C_\mathrm{s}^\mathrm{SM}(p_{\mathsf{XY}'\mathsf{Z}}) = \mathbb{I}\big(\mathsf{X}; \mathsf{Y}'|\mathsf{Z}\big) = \mathbb{I}(\mathsf{X}; \mathsf{Y}|\mathsf{Z}).$$

On the basis of this operational interpretation, a natural approach to improve the bound $C_\mathrm{s}^\mathrm{SM}(p_{\mathsf{XYZ}}) \leqslant \mathbb{I}(\mathsf{X}; \mathsf{Y}|\mathsf{Z})$ is to reduce the advantage given to Bob.

A first possibility to mitigate Bob's advantage is to analyze more precisely how Eve could further process her observations. Specifically, consider a key-distillation strategy $\mathcal{S}_n$ for a source model with DMS $(\mathcal{XYZ}, p_{\mathsf{XYZ}})$, with key K and public messages $\mathsf{A}^r\mathsf{B}^r$. We allow Eve to send her observations Z^n through an arbitrary DMC $(\mathcal{Z}, p_{\bar{\mathsf{Z}}|\mathsf{Z}}, \bar{\mathcal{Z}})$, which results in a new source model with DMS $(\mathcal{XY}\bar{\mathcal{Z}}, p_{\mathsf{XY}\bar{\mathsf{Z}}})$. Because $\mathsf{KA}^r\mathsf{B}^r \rightarrow \mathsf{Z}^n \rightarrow \bar{\mathsf{Z}}^n$ forms a Markov chain, the data-processing inequality ensures that

$$\frac{1}{n}\mathbb{I}(\mathsf{K};\mathsf{A}^r\mathsf{B}^r\bar{\mathsf{Z}}^n|\mathcal{S}_n) \leqslant \frac{1}{n}\mathbb{I}(\mathsf{K};\mathsf{A}^r\mathsf{B}^r\mathsf{Z}^n|\mathcal{S}_n) = \frac{1}{n}\mathbf{L}(\mathcal{S}_n).$$

Hence, we have that K is also a secret key for the new source model and $C_\mathrm{s}^{\mathrm{SM}}(p_{\mathsf{XYZ}}) \leqslant C_\mathrm{s}^{\mathrm{SM}}(p_{\mathsf{XY}\bar{\mathsf{Z}}})$. By virtue of Theorem 4.1, we also have $C_\mathrm{s}^{\mathrm{SM}}(p_{\mathsf{XY}\bar{\mathsf{Z}}}) \leqslant \mathbb{I}(\mathsf{X};\mathsf{Y}|\bar{\mathsf{Z}})$. Since the DMC $(\mathcal{Z}, p_{\bar{\mathsf{Z}}|\mathsf{Z}}, \bar{\mathcal{Z}})$ is arbitrary, it must hold that

$$C_\mathrm{s}^{\mathrm{SM}}(p_{\mathsf{XYZ}}) \leqslant \inf_{p_{\bar{\mathsf{Z}}|\mathsf{Z}}} \mathbb{I}(\mathsf{X};\mathsf{Y}|\bar{\mathsf{Z}}).$$

This inequality motivates the definition of a new measure of information.

Definition 4.7. *For a DMS $(\mathcal{XYZ}, p_{\mathsf{XYZ}})$, the intrinsic conditional information between* X *and* Y *given* Z *is*

$$\mathbb{I}(\mathsf{X};\mathsf{Y}\downarrow\mathsf{Z}) \triangleq \inf_{p_{\bar{\mathsf{Z}}|\mathsf{Z}}} \mathbb{I}(\mathsf{X};\mathsf{Y}|\bar{\mathsf{Z}}).$$

Intuitively, the intrinsic conditional information measures the information between X and Y that remains after Eve has chosen the best memoryless processing of the observation Z. The following theorem shows that the intrinsic conditional information is an upper bound for the secret-key capacity that is at least as tight as Theorem 4.1.

Theorem 4.2 (Maurer). *The secret-key capacity of a source model with DMS $(\mathcal{XYZ}, p_{\mathsf{XYZ}})$ satisfies*

$$C_\mathrm{s}^{\mathrm{SM}} \leqslant \mathbb{I}(\mathsf{X};\mathsf{Y}\downarrow\mathsf{Z}) \leqslant \min(\mathbb{I}(\mathsf{X};\mathsf{Y}), \mathbb{I}(\mathsf{X};\mathsf{Y}|\mathsf{Z})).$$

Proof. The inequality $C_\mathrm{s}^{\mathrm{SM}} \leqslant \inf_{p_{\bar{\mathsf{Z}}|\mathsf{Z}}} \mathbb{I}(\mathsf{X};\mathsf{Y}|\bar{\mathsf{Z}})$ has already been proved in the paragraphs above. To establish the second inequality, note that $\mathbb{I}(\mathsf{X};\mathsf{Y}\downarrow\mathsf{Z}) \leqslant \mathbb{I}(\mathsf{X};\mathsf{Y}|\bar{\mathsf{Z}})$ for any choice of transition probabilities $p_{\bar{\mathsf{Z}}|\mathsf{Z}}$; therefore, we prove that $\mathbb{I}(\mathsf{X};\mathsf{Y}\downarrow\mathsf{Z}) \leqslant \min(\mathbb{I}(\mathsf{X};\mathsf{Y}), \mathbb{I}(\mathsf{X};\mathsf{Y}|\mathsf{Z}))$ by constructing specific DMCs $(\mathcal{Z}, p_{\bar{\mathsf{Z}}|\mathsf{Z}}, \bar{\mathcal{Z}})$ for which $\mathbb{I}(\mathsf{X};\mathsf{Y}|\bar{\mathsf{Z}})$ takes the values $\mathbb{I}(\mathsf{X};\mathsf{Y})$ and $\mathbb{I}(\mathsf{X};\mathsf{Y}|\mathsf{Z})$:

- if we set $\bar{\mathcal{Z}} = \mathcal{Z}$ and $p_{\bar{\mathsf{Z}}|\mathsf{Z}}(\bar{z}|z) = 1/|\mathcal{Z}|$ for all $\bar{z}$, then $\bar{\mathsf{Z}}$ is independent of X, Y, and Z so that $\mathbb{I}(\mathsf{X};\mathsf{Y}|\bar{\mathsf{Z}}) = \mathbb{I}(\mathsf{X};\mathsf{Y})$;
- if we set $p_{\bar{\mathsf{Z}}|\mathsf{Z}}(\bar{z}|z) = \mathbb{1}(\bar{z} - z)$, then $\mathsf{Z} = \bar{\mathsf{Z}}$ with probability one and $\mathbb{I}(\mathsf{X};\mathsf{Y}|\bar{\mathsf{Z}}) = \mathbb{I}(\mathsf{X};\mathsf{Y}|\mathsf{Z})$.

Therefore,

$$\mathbb{I}(\mathsf{X};\mathsf{Y}\downarrow\mathsf{Z}) \leqslant \min(\mathbb{I}(\mathsf{X};\mathsf{Y}), \mathbb{I}(\mathsf{X};\mathsf{Y}|\mathsf{Z})).$$ □

The following example shows that Theorem 4.2 is useful because $\mathbb{I}(\mathsf{X};\mathsf{Y}\downarrow\mathsf{Z})$ can be strictly tighter than $\mathbb{I}(\mathsf{X};\mathsf{Y}|\mathsf{Z})$.

Example 4.1. Let $\mathcal{X} = \mathcal{Y} = \{0, 1, 2, 3\}$ and let p_{XY} be defined in the table below.

X \ Y	0	1	2	3
0	1/8	1/8	0	0
1	1/8	1/8	0	0
2	0	0	1/4	0
3	0	0	0	1/4

Define Z as

$$\mathsf{Z} \triangleq \begin{cases} \mathsf{X} \oplus \mathsf{Y} & \text{if} \quad \mathsf{X} \in \{0, 1\}, \\ \mathsf{X} & \text{if} \quad \mathsf{X} \in \{2, 3\}. \end{cases}$$

One can verify that $\mathbb{I}(\mathsf{X}; \mathsf{Y}) = \frac{3}{2}$ and $\mathbb{I}(\mathsf{X}; \mathsf{Y}|\mathsf{Z}) = \frac{1}{2}$. The particularity of this DMS is that, for $\mathsf{X} \in \{0, 1\}$, individual knowledge of Z or Y does not resolve any uncertainty about X but joint knowledge of Z and Y determines X completely. Consider now the channel $(\mathcal{Z}, p_{\bar{\mathsf{Z}}|\mathsf{Z}}, \bar{\mathcal{Z}})$, such that

$$p_{\bar{\mathsf{Z}}|\mathsf{Z}}(0|0) = p_{\bar{\mathsf{Z}}|\mathsf{Z}}(0|1) = p_{\bar{\mathsf{Z}}|\mathsf{Z}}(1|0) = p_{\bar{\mathsf{Z}}|\mathsf{Z}}(1|1) = \frac{1}{2},$$

$$p_{\bar{\mathsf{Z}}|\mathsf{Z}}(2|2) = p_{\bar{\mathsf{Z}}|\mathsf{Z}}(3|3) = 1.$$

Notice that, if $\mathsf{Z} \in \{0, 1\}$, then $\bar{\mathsf{Z}}$ is obtained by sending Z through a BSC with cross-over probability $\frac{1}{2}$ and, therefore, $\bar{\mathsf{Z}}$ becomes independent of Z. As a result, for $\mathsf{X} \in \{0, 1\}$, knowledge of $\bar{\mathsf{Z}}$ still does not resolve any uncertainty about X, but knowledge of both $\bar{\mathsf{Z}}$ and Y does not resolve any uncertainty, either. Hence, $\mathbb{I}(\mathsf{X}; \mathsf{Y}|\bar{\mathsf{Z}}) = 0$ and, consequently, $\mathbb{I}(\mathsf{X}; \mathsf{Y} \downarrow \mathsf{Z}) = 0$ and $C_s^{\text{SM}} = 0$.

A second possibility to improve the result of Theorem 4.2 is not only to analyze how Eve could further process her observations but also to provide her with some side information represented by a correlated DMS $(\mathcal{U}, p_{\mathsf{U}})$; however, the side information must be introduced carefully if we want to retain an upper bound for $C_s^{\text{SM}}(p_{\mathsf{XYZ}})$ because, in general, the secret-key capacity is reduced if Eve has access to side information.

Proposition 4.1. *Consider a source model with DMS $(\mathcal{X}\mathcal{Y}\tilde{\mathcal{Z}}, p_{\mathsf{XY}\tilde{\mathsf{Z}}})$, in which the eavesdropper has access to the observations $\tilde{\mathsf{Z}} = (\mathsf{Z}, \mathsf{U}) \in \mathcal{Z} \times \mathcal{U}$. Then,*

$$C_s^{\text{SM}}(p_{\mathsf{XY}\tilde{\mathsf{Z}}}) \geqslant C_s^{\text{SM}}(p_{\mathsf{XYZ}}) - \mathbb{H}(\mathsf{U}).$$

Proof. We prove the inequality by constructing a key-distillation strategy for the DMS $(\mathcal{X}\mathcal{Y}\tilde{\mathcal{Z}}, p_{\mathsf{XY}\tilde{\mathsf{Z}}})$ from a key-distillation strategy for the DMS $(\mathcal{X}\mathcal{Y}\mathcal{Z}, p_{\mathsf{XYZ}})$.

Let R be an achievable weak secret-key rate for the DMS $(\mathcal{X}\mathcal{Y}\mathcal{Z}, p_{\mathsf{XYZ}})$. For any $\epsilon > 0$, there exists a $(2^{nR}, n)$ key-distillation strategy $\mathcal{S}_n$, such that

$$\mathbf{P}_e(\mathcal{S}_n) \leqslant \delta(\epsilon), \quad \frac{1}{n}\mathbf{L}(\mathcal{S}_n) \leqslant \delta(\epsilon), \quad \text{and} \quad \frac{1}{n}\mathbf{U}(\mathcal{S}_n) \leqslant \delta(\epsilon). \tag{4.24}$$

Using Fano's inequality, we obtain

$$\frac{1}{n}\mathbb{H}\left(\mathsf{K}|\hat{\mathsf{K}}\mathcal{S}_n\right) \leqslant \delta(\mathbf{P}_e(\mathcal{S}_n)) \leqslant \delta(\epsilon). \tag{4.25}$$

In addition,

$$\begin{aligned}
\frac{1}{n}\mathbb{I}(\mathsf{K};\mathsf{A}^r\mathsf{B}^r\mathsf{Z}^n\mathsf{U}^n|\mathcal{S}_n) &= \frac{1}{n}\mathbb{I}(\mathsf{K};\mathsf{A}^r\mathsf{B}^r\mathsf{Z}^n|\mathcal{S}_n) + \frac{1}{n}\mathbb{I}(\mathsf{K};\mathsf{U}^n|\mathsf{A}^r\mathsf{B}^r\mathsf{Z}^n\mathcal{S}_n)\\
&\leqslant \frac{1}{n}\mathbb{I}(\mathsf{K};\mathsf{A}^r\mathsf{B}^r\mathsf{Z}^n|\mathcal{S}_n) + \frac{1}{n}\mathbb{H}(\mathsf{U}^n)\\
&\leqslant \frac{1}{n}\mathbf{L}(\mathcal{S}_n) + \mathbb{H}(\mathsf{U})\\
&\leqslant \delta(\epsilon) + \mathbb{H}(\mathsf{U}).
\end{aligned} \tag{4.26}$$

We now turn our attention back to the DMS $\left(\mathcal{X}\mathcal{Y}\tilde{\mathcal{Z}}, p_{\mathsf{XY}\tilde{\mathsf{Z}}}\right)$. We construct a key-distillation strategy by first running m *independent* repetitions of the key-distillation strategy $\mathcal{S}_n$, from which Alice obtains i.i.d. sequences K^m, Bob obtains i.i.d. sequences $\hat{\mathsf{K}}^m$, and Eve observes A^{rm}, B^{rm}, Z^{nm}, and U^{nm}. Effectively, this creates a source model with DMS $(\mathcal{X}'\mathcal{Y}'\mathcal{Z}', p_{\mathsf{X}'\mathsf{Y}'\mathsf{Z}'})$ in which $\mathsf{X}' \triangleq \mathsf{K}$, $\mathsf{Y}' \triangleq \hat{\mathsf{K}}$, and $\mathsf{Z}' \triangleq \mathsf{A}^r\mathsf{B}^r\mathsf{Z}^n\mathsf{U}^n$. Since a key-distillation strategy for the DMS $(\mathcal{X}'\mathcal{Y}'\mathcal{Z}', p_{\mathsf{X}'\mathsf{Y}'\mathsf{Z}'})$ is a specific instance of key-distillation strategy for the DMS $\left(\mathcal{X}\mathcal{Y}\tilde{\mathcal{Z}}, p_{\mathsf{XY}\tilde{\mathsf{Z}}}\right)$, we have

$$C_s^{\text{SM}}(p_{\mathsf{XY}\tilde{\mathsf{Z}}}) \geqslant \frac{1}{n}C_s^{\text{SM}}(p_{\mathsf{X}'\mathsf{Y}'\mathsf{Z}'}), \tag{4.27}$$

where the normalization by n appears because each realization of the DMS $(\mathcal{X}'\mathcal{Y}'\mathcal{Z}', p_{\mathsf{X}'\mathsf{Y}'\mathsf{Z}'})$ relies on n realizations of the DMS $\left(\mathcal{X}\mathcal{Y}\tilde{\mathcal{Z}}, p_{\mathsf{XY}\tilde{\mathsf{Z}}}\right)$. Note that Theorem 4.1 guarantees that

$$C_s^{\text{SM}}(p_{\mathsf{X}'\mathsf{Y}'\mathsf{Z}'}) \geqslant \mathbb{I}(\mathsf{X}';\mathsf{Y}') - \mathbb{I}(\mathsf{X}';\mathsf{Z}').$$

Next, we use (4.24), (4.25), and (4.26) to lower bound $\mathbb{I}(\mathsf{X}';\mathsf{Y}') - \mathbb{I}(\mathsf{X}';\mathsf{Z}')$ as

$$\begin{aligned}
\mathbb{I}(\mathsf{X}';\mathsf{Y}') - \mathbb{I}(\mathsf{X}';\mathsf{Z}') &= \mathbb{I}\left(\mathsf{K};\hat{\mathsf{K}}|\mathcal{S}_n\right) - \mathbb{I}(\mathsf{K};\mathsf{A}^r\mathsf{B}^r\mathsf{Z}^n\mathsf{U}^n|\mathcal{S}_n)\\
&= \mathbb{H}(\mathsf{K}|\mathcal{S}_n) - \mathbb{H}\left(\mathsf{K}|\hat{\mathsf{K}}\mathcal{S}_n\right) - \mathbb{I}(\mathsf{K};\mathsf{A}^r\mathsf{B}^r\mathsf{Z}^n\mathsf{U}^n|\mathcal{S}_n)\\
&\geqslant nR - n\mathbb{H}(\mathsf{U}) - n\delta(\epsilon).
\end{aligned}$$

All in all, we obtain

$$C_s^{\text{SM}}(p_{\mathsf{XY}\tilde{\mathsf{Z}}}) \geqslant R - \mathbb{H}(\mathsf{U}) - \delta(\epsilon).$$

Since $\epsilon > 0$ can be arbitrarily small and R can be chosen arbitrarily close to $C_s^{\text{SM}}(p_{\mathsf{XYZ}})$, we get the desired result

$$C_s^{\text{SM}}(p_{\mathsf{XY}\tilde{\mathsf{Z}}}) \geqslant C_s^{\text{SM}}(p_{\mathsf{XYZ}}) - \mathbb{H}(\mathsf{U}). \qquad \square$$

Proposition 4.1 means that providing Eve with side information U reduces the secret-key capacity by at most $\mathbb{H}(\mathsf{U})$. This motivates the definition of another measure of information.

Definition 4.8. *For a DMS* $(\mathcal{XYZ}, p_{XYZ})$, *the reduced intrinsic conditional information between* X *and* Y *given* Z *is*

$$\mathbb{I}(X; Y \downarrow\downarrow Z) \triangleq \inf_{p_{U|XYZ}} (\mathbb{I}(X; Y \downarrow ZU) + \mathbb{H}(U)).$$

Intuitively, the reduced intrinsic conditional information measures the information between X and Y that remains after the best memoryless processing of Z and with the best memoryless side information U; however, the term $\mathbb{H}(U)$ is introduced to compensate for the decrease of $\mathbb{I}(X; Y \downarrow Z)$ caused by the side information. The next theorem shows that $\mathbb{I}(X; Y \downarrow\downarrow Z)$ is an upper bound for C_s^{SM} that is at least as good as $\mathbb{I}(X; Y \downarrow Z)$.

Theorem 4.3 (Renner and Wolf). *The secret-key capacity of a source model with DMS* $(\mathcal{XYZ}, p_{XYZ})$ *satisfies*

$$C_s^{SM} \leqslant \mathbb{I}(X; Y \downarrow\downarrow Z) \leqslant \mathbb{I}(X; Y \downarrow Z).$$

Proof. The inequality $\mathbb{I}(X; Y \downarrow\downarrow Z) \leqslant \mathbb{I}(X; Y \downarrow Z)$ follows on choosing $U = 0$ in the definition of $\mathbb{I}(X; Y \downarrow\downarrow Z)$. The inequality $C_s^{SM} \leqslant \mathbb{I}(X; Y \downarrow\downarrow Z)$ follows on noting that, for any U, Proposition 4.1 and Theorem 4.2 ensure that

$$C_s^{SM}(p_{XYZ}) \leqslant C_s^{SM}(p_{XY(ZU)}) + \mathbb{H}(U) \leqslant \mathbb{I}(X; Y \downarrow ZU) + \mathbb{H}(U)$$

and, therefore,

$$C_s^{SM}(p_{XYZ}) \leqslant \inf_{p_{U|XYZ}} (\mathbb{I}(X; Y \downarrow ZU) + \mathbb{H}(U)) = \mathbb{I}(X; Y \downarrow\downarrow Z).$$

□

As shown in the following example, the result of Theorem 4.3 is useful because $\mathbb{I}(X; Y \downarrow\downarrow Z)$ can provide a better bound than can $\mathbb{I}(X; Y \downarrow Z)$.

Example 4.2. Let $\mathcal{X} = \mathcal{Y} = \{0, 1, 2, 3\}$ and let us consider again the joint probability distribution p_{XY} defined in the table below.

X \ Y	0	1	2	3
0	1/8	1/8	0	0
1	1/8	1/8	0	0
2	0	0	1/4	0
3	0	0	0	1/4

Let Z be defined as

$$Z \triangleq \begin{cases} X \oplus Y & \text{if } X \in \{0, 1\}, \\ X \bmod 2 & \text{if } X \in \{2, 3\}. \end{cases}$$

In contrast to Example 4.1, knowledge of Z never fully resolves the uncertainty about X or Y. One can verify that $\mathbb{I}(X; Y) = \mathbb{I}(X; Y|Z) = \frac{3}{2}$. We now introduce the side information U as

$$U \triangleq \left\lfloor \frac{X}{2} \right\rfloor.$$

Let us consider the channel $\left(\mathcal{Z} \times \mathcal{U}, p_{\bar{Z}|ZU}, \{0, 1, 2\}\right)$ such that

$$p_{\bar{Z}|ZU}(2|0, 0) = p_{\bar{Z}|ZU}(2|1, 0) = p_{\bar{Z}|ZU}(0|0, 1) = p_{\bar{Z}|ZU}(1|1, 1).$$

One can check that $\mathbb{I}\left(X; Y|\bar{Z}\right) = 0$ and therefore $\mathbb{I}(X; Y \downarrow\downarrow Z) = 0$ as well.

To establish that $\mathbb{I}(X; Y \downarrow Z) > 0$, consider an arbitrary channel $\left(\mathcal{Z}, p_{\bar{Z}|Z}, \bar{\mathcal{Z}}\right)$ and consider an output symbol $\bar{z}^*$. Let

$$p \triangleq p_{\bar{Z}|Z}(\bar{z}^*|0) \quad \text{and} \quad q \triangleq p_{\bar{Z}|Z}(\bar{z}^*|1).$$

One can check that

$$p_{XY|Z}(0, 0|\bar{z}^*) = p_{XY|Z}(0, 1|\bar{z}^*) = p_{XY|Z}(1, 0|\bar{z}^*) = p_{XY|Z}(1, 1|\bar{z}^*) = \frac{p}{4(p+q)},$$

$$p_{XY|Z}(2, 2|\bar{z}^*) = p_{XY|Z}(3, 3|\bar{z}^*) = \frac{p}{2(p+q)},$$

and $\mathbb{I}\left(X; Y|\bar{Z} = \bar{z}^*\right) = \frac{3}{2}$. Consequently, $\mathbb{I}\left(X; Y|\bar{Z}\right) = \frac{3}{2}$ as well, and, since the channel $\left(\mathcal{Z}, p_{\bar{Z}|Z}, \bar{\mathcal{Z}}\right)$ was arbitrary, $\mathbb{I}(X; Y \downarrow Z) = \frac{3}{2}$.

Note that neither the intrinsic conditional information nor the reduced intrinsic conditional information really helps determine a generic expression for C_s^{SM}.

4.3 Sequential key distillation for the source model

The analysis of wiretap codes and key-distillation strategies is complex because messages or keys are subject to *simultaneous* reliability and secrecy constraints. Random-coding and random-binning arguments allow us to circumvent this difficulty and to establish achievability results, but they provide limited insight into the design of practical schemes. Even the proof in Section 4.2.2, which exploits a strong connection between key-distillation strategies and Slepian–Wolf codes, implicitly requires the key-distillation function and the public-message-encoding function to be designed *jointly*; this joint design makes it difficult to find such functions in practice. Hence, to further simplify the design of practical schemes, it is legitimate to wonder whether one could handle the reliability and secrecy requirements *independently*. This makes little sense for wiretap codes, but the idea is not totally contrived for key-distillation strategies because keys are random sequences that do not carry any information by themselves; there is a lot of leeway in the construction of key-distillation strategies and Alice and Bob are free to remove, combine, or shuffle their observations.

In this section, we show that, for a source model, it is indeed possible to design key-distillation strategies that handle reliability and secrecy independently. Such strategies, which we call *sequential key-distillation strategies* because they operate in sequential phases, play a particularly important role for three reasons:

- they incur no loss of optimality, since they can achieve all rates below the secret-key capacity;

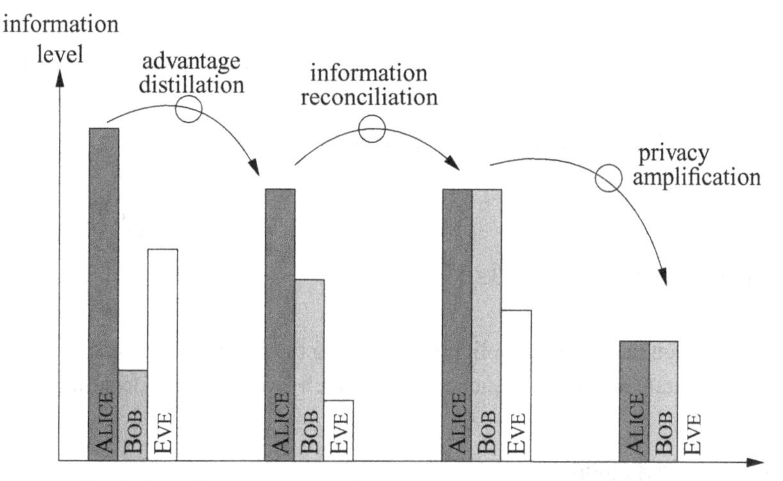

Figure 4.4 Evolution of information during the phases of a sequential key-generation strategy.

- they achieve strong secret-key rates, which allow us to prove $C_s^{SM} = \overline{C}_s^{SM}$;
- their analysis eventually leads to explicit and practical constructions.

Specifically, a sequential key-distillation strategy is a key-distillation strategy that operates in four successive phases.

1. *Randomness sharing.* Alice, Bob, and Eve observe n realizations of a DMS $(\mathcal{XYZ}, p_{XYZ})$.
2. *Advantage distillation.* If needed, Alice and Bob exchange messages over the public channel to process their observations and to "distill" observations for which they have an advantage over Eve.
3. *Information reconciliation.* Alice and Bob exchange messages over the public channel to process their observations and agree on a common bit sequence.
4. *Privacy amplification.* Alice and Bob publicly agree on a deterministic function they apply to their common sequence to generate a secret key.

Before we describe and analyze these phases precisely, it is useful to understand intuitively the role played by each of them in the key-distillation strategy. Figure 4.4 illustrates the evolution of each party's information about Alice's initial source observations during the different phases. The amount of information is represented qualitatively by the height of the bars in the figure. After the randomness-sharing phase, we assume that Eve has an advantage over Bob; hence, Bob's information is lower than Eve's information. During the advantage-distillation phase, Alice and Bob interact over the public channel to distill observations for which they have an advantage over Eve. Since the observations for which Eve has an advantage are discarded, Alice's information decreases; however, Bob's information now exceeds Eve's information. During the information-reconciliation phase, Alice provides Bob with side information that enables him to correct all the discrepancies between his sequence and Alice's; as a result, Bob's information increases to reach the level of Alice's, but, since the error-correction information is public, Eve's

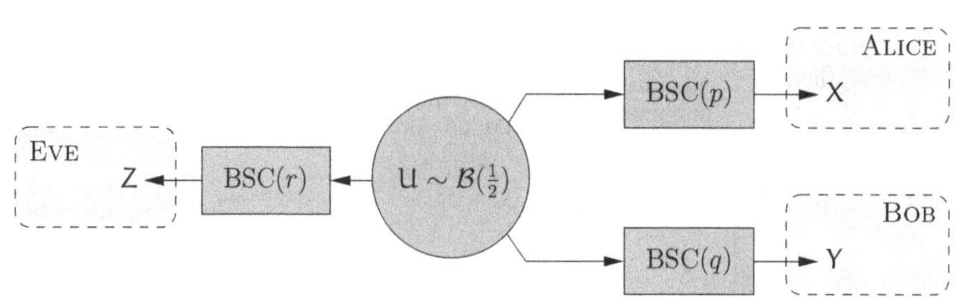

Figure 4.5 Satellite source model.

information increases as well. Finally, during the privacy-amplification phase, Alice and Bob generate a smaller sequence about which Eve has no information.

4.3.1 Advantage distillation

For some source models, the lower bound for the secret-key capacity provided by Theorem 4.1 is negative, and hence useless. Figure 4.5 illustrates such a source model, which is commonly called a "satellite" source model. It consists of a DMS $(\mathcal{U}, p_{\mathsf{U}})$ with $\mathsf{U} \sim \mathcal{B}(\frac{1}{2})$ that broadcasts sequences of bits to Alice, Bob, and Eve through independent binary symmetric channels with respective cross-over probabilities $p > 0$, $q > 0$, and $r > 0$. This source model can be thought of as a satellite transmitting to three base stations on Earth, one of which is an eavesdropper. It is assumed that Eve's cross-over probability r satisfies $r < p$ and $r < q$ so that

$$\mathbb{I}(\mathsf{X};\mathsf{Y}) < \mathbb{I}(\mathsf{X};\mathsf{Z}) \quad \text{and} \quad \mathbb{I}(\mathsf{X};\mathsf{Y}) < \mathbb{I}(\mathsf{Y};\mathsf{Z}).$$

In other words, Eve has an advantage over both Alice and Bob because the mutual information between Z and X and the mutual information between Z and Y are higher than that between X and Y.

The basic premise of advantage distillation is that Alice and Bob may reverse Eve's advantage by exchanging messages over the public channel. In fact, the mutual information $\mathbb{I}(\mathsf{X};\mathsf{Z})$ or $\mathbb{I}(\mathsf{Y};\mathsf{Z})$ measures only Eve's *average* advantage over Alice and Bob. Although $\mathbb{I}(\mathsf{X};\mathsf{Y}) < \mathbb{I}(\mathsf{X};\mathsf{Z})$ and $\mathbb{I}(\mathsf{X};\mathsf{Y}) < \mathbb{I}(\mathsf{Y};\mathsf{Z})$, there may exist some realizations of the DMS for which Eve's observations are loosely correlated to Alice and Bob's. One can think of an advantage-distillation protocol as a procedure to distill the realizations for which Alice and Bob have an advantage over Eve. Formally, an advantage-distillation protocol is defined as follows.

Definition 4.9. *An advantage-distillation protocol $\mathcal{D}_n$ for a source model with DMS $(\mathcal{XYZ}, p_{\mathsf{XYZ}})$ consists of*

- *two alphabets $\mathcal{X}'$ and $\mathcal{Y}'$;*
- *a source of local randomness $\left(\mathcal{R}_{\mathcal{X}}, p_{\mathsf{R_X}}\right)$ for Alice;*
- *a source of local randomness $\left(\mathcal{R}_{\mathcal{Y}}, p_{\mathsf{R_Y}}\right)$ for Bob;*
- *an integer $r \in \mathbb{N}^*$ that represents the number of rounds of communication;*
- *r encoding functions $f_i : \mathcal{X}^n \times \mathcal{B}^{i-1} \times \mathcal{R}_{\mathcal{X}} \to \mathcal{A}$ for $i \in [\![1, r]\!]$;*

- *r encoding functions* $g_i : \mathcal{Y}^n \times \mathcal{A}^{i-1} \times \mathcal{R}_\mathcal{Y} \to \mathcal{B}$ *for* $i \in [\![1, r]\!]$*;*
- *a function* $\theta_\text{a} : \mathcal{X}^n \times \mathcal{B}^r \times \mathcal{R}_\mathcal{X} \to \mathcal{X}'$*;*
- *a function* $\theta_\text{b} : \mathcal{Y}^n \times \mathcal{A}^r \times \mathcal{R}_\mathcal{Y} \to \mathcal{Y}'$*;*

and operates as follows:

- *Alice observes n realizations of the source* x^n *while Bob observes* y^n*;*
- *Alice generates a realization* r_x *of her source of local randomness while Bob generates* r_y *from his;*
- *in round* $i \in [\![1, r]\!]$*, Alice transmits* $a_i = f_i(x^n, b^{i-1}, r_x)$ *while Bob transmits* $b_i = g_i(y^n, a^{i-1}, r_y)$*;*
- *after round r, Alice distills* $x' = \theta_\text{a}(x^n, b^r, r_x)$ *while Bob distills* $y' = \theta_\text{b}(y^n, a^r, r_y)$*.*

By convention, $\mathcal{A}^0 \triangleq 0$ and $\mathcal{B}^0 \triangleq 0$. The number of rounds r and the sources of local randomness can be optimized during the design of the advantage-distillation protocol. By repeating an advantage-distillation protocol multiple times, Alice and Bob distill the realizations of a new DMS $(\mathcal{X}'\mathcal{Y}'\mathcal{Z}', p_{\text{X}'\text{Y}'\text{Z}'})$ with components X', Y', and $\text{Z}' \triangleq \text{Z}^n\text{A}^r\text{B}^r$. Ideally, this new DMS provides Alice and Bob with an advantage over Eve in the sense that

$$\mathbb{I}(\text{X}';\text{Y}') \geqslant \mathbb{I}(\text{X}';\text{Z}') \quad \text{or} \quad \mathbb{I}(\text{X}';\text{Y}') \geqslant \mathbb{I}(\text{Y}';\text{Z}').$$

Hence, it is natural to measure the performance of an advantage-distillation protocol in terms of the quantity

$$\mathbf{R}(\mathcal{D}_n) \triangleq \frac{1}{n}\max\left(\mathbb{I}(\text{X}';\text{Y}') - \mathbb{I}(\text{X}';\text{Z}'), \mathbb{I}(\text{X}';\text{Y}') - \mathbb{I}(\text{Y}';\text{Z}')\right),$$

which we call the *advantage-distillation rate*.[2] Notice that we have introduced a normalization by n to express the rate in bits per observation of the DMS $(\mathcal{XYZ}, p_{\text{XYZ}})$. The advantage-distillation rate captures an inherent trade-off in the design of an advantage-distillation protocol. On the one hand, Alice and Bob want to exchange messages to maximize $\mathbb{I}(\text{X}';\text{Y}')$, that is, they want to extract the observations for which their realizations are highly correlated; on the other hand, they must also minimize $\mathbb{I}(\text{X}';\text{Z}')$ or $\mathbb{I}(\text{Y}';\text{Z}')$, that is, they must choose their messages carefully in order to avoid revealing the values of their observations to Eve.

Definition 4.10. *An advantage-distillation rate R is achievable if there exists a sequence of advantage-distillation protocols* $\{\mathcal{D}_n\}_{n\geqslant 1}$ *such that*

$$\varliminf_{n\to\infty} \mathbf{R}(\mathcal{D}_n) \geqslant R.$$

Definition 4.11. *The advantage-distillation capacity* D^{SM} *of a source model with DMS* $(\mathcal{XYZ}, p_{\text{XYZ}})$ *is*

$$D^{\text{SM}} \triangleq \sup\{R : R \text{ is an achievable advantage-distillation rate}\}.$$

[2] Our definition allows an advantage-distillation rate to take negative values, which of course has little interest.

When required, we write $D^{\text{SM}}(p_{\text{XYZ}})$ in place of D^{SM} to specify explicitly the distribution of the DMS. We do not attempt to characterize the advantage distillation of a source model exactly; rather, we relate it to the secret-key capacity of the same source model.

Proposition 4.2 (Muramatsu). *For a source model with DMS $(\mathcal{X}\mathcal{Y}\mathcal{Z}, p_{\text{XYZ}})$,*

$$D^{\text{SM}} = C_{\text{s}}^{\text{SM}}.$$

Proof. A $(2^{nR}, n)$ key-distillation strategy for a source model with DMS $(\mathcal{X}\mathcal{Y}\mathcal{Z}, p_{\text{XYZ}})$ can be viewed as an advantage-distillation protocol for which $\text{X}' = \text{K}$, $\text{Y}' = \hat{\text{K}}$, and $\text{Z}' = \text{A}^r\text{B}^r\text{Z}^n$. If a secret-key rate R is achievable, there exists a sequence of key-distillation strategies $\{\mathcal{S}_n\}_{n\geqslant 1}$ such that

$$\lim_{n\to\infty} \mathbf{P}_{\text{e}}(\mathcal{S}_n) = 0, \quad \lim_{n\to\infty} \frac{1}{n}\mathbf{L}(\mathcal{S}_n) = 0, \quad \text{and} \quad \lim_{n\to\infty} \frac{1}{n}\mathbf{U}(\mathcal{S}_n) = 0.$$

By virtue of Fano's inequality,

$$\lim_{n\to\infty} \frac{1}{n}\mathbb{H}\left(\text{K}|\hat{\text{K}}\mathcal{S}_n\right) \leqslant \lim_{n\to\infty} \delta(\mathbf{P}_{\text{e}}(\mathcal{S}_n)) = 0.$$

and, from the definition of $\mathbf{L}(\mathcal{S}_n)$,

$$\lim_{n\to\infty} \frac{1}{n}\mathbb{I}(\text{K};\text{Z}^n\text{A}^r\text{B}^r|\mathcal{S}_n) = \lim_{n\to\infty} \frac{1}{n}\mathbf{L}(\mathcal{S}_n) = 0.$$

Hence, the sequence of advantage-distillation rate $\{\mathbf{R}(\mathcal{S}_n)\}_{n\geqslant 1}$ satisfies

$$\begin{aligned}
\lim_{n\to\infty} \mathbf{R}(\mathcal{S}_n) &= \lim_{n\to\infty} \left(\frac{1}{n}\mathbb{I}(\text{X}';\text{Y}') - \frac{1}{n}\mathbb{I}(\text{X}';\text{Z}')\right) \\
&= \lim_{n\to\infty} \left(\frac{1}{n}\mathbb{H}(\text{K}|\mathcal{S}_n) - \frac{1}{n}\mathbb{H}\left(\text{K}|\hat{\text{K}}\mathcal{S}_n\right) - \frac{1}{n}\mathbb{I}(\text{K};\text{Z}^n\text{A}^r\text{B}^r|\mathcal{S}_n)\right) \\
&= \lim_{n\to\infty} \left(\frac{1}{n}\log\lceil 2^{nR}\rceil - \frac{1}{n}\mathbf{U}(\mathcal{S}_n) - \frac{1}{n}\mathbb{H}\left(\text{K}|\hat{\text{K}}\mathcal{S}_n\right) - \frac{1}{n}\mathbb{I}(\text{K};\text{Z}^n\text{A}^r\text{B}^r|\mathcal{S}_n)\right) \\
&= R.
\end{aligned}$$

Therefore, $R \leqslant D^{\text{SM}}$; and, since R is an arbitrary achievable secret-key rate, $C_{\text{s}}^{\text{SM}} \leqslant D^{\text{SM}}$.

To prove the reverse inequality, notice that repeated application of an advantage-distillation protocol $\mathcal{D}_n$ creates a source model with DMS $(\mathcal{X}'\mathcal{Y}'\mathcal{Z}', p_{\text{X}'\text{Y}'\text{Z}'})$, whose secret-key capacity is $C_{\text{s}}^{\text{SM}}(p_{\text{X}'\text{Y}'\text{Z}'})$; by Theorem 4.1, $C_{\text{s}}^{\text{SM}}(p_{\text{X}'\text{Y}'\text{Z}'})$ satisfies

$$C_{\text{s}}^{\text{SM}}(p_{\text{X}'\text{Y}'\text{Z}'}) \geqslant \max\left(\mathbb{I}(\text{X}';\text{Y}') - \mathbb{I}(\text{X}';\text{Z}'), \mathbb{I}(\text{X}';\text{Y}') - \mathbb{I}(\text{Y}';\text{Z}')\right).$$

The secret-key capacity $C_{\text{s}}^{\text{SM}}(p_{\text{X}'\text{Y}'\text{Z}'})$ is expressed in bits per observation of the source $(\mathcal{X}'\mathcal{Y}'\mathcal{Z}', p_{\text{X}'\text{Y}'\text{Z}'})$; hence, the corresponding secret-key rate in bits per observation of the source $(\mathcal{X}\mathcal{Y}\mathcal{Z}, p_{\text{XYZ}})$ is $(1/n)C_{\text{s}}^{\text{SM}}(p_{\text{X}'\text{Y}'\text{Z}'})$ and, by definition, it cannot exceed $C_{\text{s}}^{\text{SM}}(p_{\text{XYZ}})$. Therefore,

$$\begin{aligned}
C_{\text{s}}^{\text{SM}}(p_{\text{XYZ}}) &\geqslant \frac{1}{n}C_{\text{s}}^{\text{SM}}(p_{\text{X}'\text{Y}'\text{Z}'}) \\
&\geqslant \frac{1}{n}\max\left(\mathbb{I}(\text{X}';\text{Y}') - \mathbb{I}(\text{X}';\text{Z}'), \mathbb{I}(\text{X}';\text{Y}') - \mathbb{I}(\text{Y}';\text{Z}')\right) \\
&= \mathbf{R}(\mathcal{D}_n).
\end{aligned}$$

Since the advantage-distillation rate $\mathbf{R}(\mathcal{D}_n)$ can be arbitrarily close to $D^{\text{SM}}(p_{\mathsf{XYZ}})$, we obtain

$$C_{\text{s}}^{\text{SM}}(p_{\mathsf{XYZ}}) \geqslant D^{\text{SM}}(p_{\mathsf{XYZ}}). \qquad \square$$

Proposition 4.2 does not provide an explicit characterization of the advantage-distillation capacity, but it shows that the secret-key capacity is equal to the maximum rate at which Alice and Bob can "generate an advantage" over Eve. This formalizes the intuition that the amount of secrecy that Alice and Bob can extract from their observations is related to the advantage they have over Eve. In addition, the proof shows that there is no loss of optimality in starting a key-distillation strategy with an advantage-distillation phase.

Unfortunately, there is no known generic procedure for designing advantage-distillation protocols because the distillation heavily depends on the specific statistics of the underlying DMS $(\mathcal{XYZ}, p_{\mathsf{XYZ}})$. Nevertheless, we illustrate the concept by analyzing a protocol for the satellite scenario of Figure 4.5. This protocol, which is called the "repetition" protocol, is due to Maurer and operates as follows.

1. Alice, Bob, and Eve observe m realizations of the DMS denoted by X^m, Y^m, and Z^m, respectively.
2. Alice generates a bit $\mathsf{V} \sim \mathcal{B}(\frac{1}{2})$.
3. Alice creates a vector $\overline{\mathsf{V}}^m \triangleq (\mathsf{V}, \dots, \mathsf{V})$ that consists of m repetitions of the same bit V, and transmits the bit-wise sum $\overline{\mathsf{V}}^m \oplus \mathsf{X}^m$ over the public channel.
4. Upon reception of $\overline{\mathsf{V}}^m \oplus \mathsf{X}^m$, Bob uses his observations Y^m to compute $\overline{\mathsf{V}}^m \oplus \mathsf{X}^m \oplus \mathsf{Y}^m$ and defines

$$\tilde{\mathsf{Y}} \triangleq \begin{cases} 0 & \text{if} \quad \overline{\mathsf{V}}^m \oplus \mathsf{X}^m \oplus \mathsf{Y}^m = (0, 0, \dots, 0), \\ 1 & \text{if} \quad \overline{\mathsf{V}}^m \oplus \mathsf{X}^m \oplus \mathsf{Y}^m = (1, 1, \dots, 1), \\ ? & \text{else.} \end{cases}$$

   Bob then sends Alice a message F over the public channel defined as

$$\mathsf{F} \triangleq \begin{cases} 1 & \text{if} \quad \tilde{\mathsf{Y}} \in \{0, 1\}, \\ 0 & \text{if} \quad \tilde{\mathsf{Y}} = ?. \end{cases}$$

   Notice that F carries information about $\tilde{\mathsf{Y}}$ but does not reveal its exact value.
5. Upon reception of F, Alice defines $\tilde{\mathsf{X}}$ as

$$\tilde{\mathsf{X}} \triangleq \begin{cases} \mathsf{V} & \text{if} \quad \mathsf{F} = 1, \\ ? & \text{if} \quad \mathsf{F} = 0. \end{cases}$$

By reiterating the repetition protocol multiple times, Alice and Bob effectively create a new DMS $(\mathcal{X}'\mathcal{Y}'\mathcal{Z}', p_{\mathsf{X}'\mathsf{Y}'\mathsf{Z}'})$ with components $\mathsf{X}' \triangleq \tilde{\mathsf{X}}\mathsf{F}$, $\mathsf{Y}' \triangleq \tilde{\mathsf{Y}}$, and $\mathsf{Z}' \triangleq (\mathsf{Z}^m, \overline{\mathsf{V}}^m \oplus \mathsf{X}^m, \mathsf{F})$. The protocol can be thought of as a *post-selection* procedure with a repetition code, by which Alice and Bob retain a bit only if their observations X^m and Y^m are highly correlated or anticorrelated. Since Eve is a passive eavesdropper and has no control over the post-selection, she cannot bias Alice or Bob towards selecting observations that would be favorable to her.

We now compute $\mathbb{I}(\mathsf{X}'; \mathsf{Y}')$ and $\mathbb{I}(\mathsf{X}'; \mathsf{Z}')$ obtained with the repetition protocol in order to characterize its advantage-distillation rate.

Proposition 4.3. *The mutual information between Alice and Bob after advantage distillation with the repetition protocol is*

$$\mathbb{I}(X';Y') = \mathbb{H}_b(\alpha) + \alpha(1 - \mathbb{H}_b(\beta)),$$

with

$$\alpha \triangleq (\bar{p}\bar{q} + pq)^m + (\bar{p}q + p\bar{q})^m, \qquad \beta \triangleq \frac{(\bar{p}q + p\bar{q})^m}{(\bar{p}\bar{q} + pq)^m + (\bar{p}q + p\bar{q})^m},$$

and $\bar{p} \triangleq (1-p)$ *and* $\bar{q} = (1-q)$.

Proof. The mutual information between Alice and Bob after the repetition protocol can be written

$$\mathbb{I}(X';Y') = \mathbb{I}(\tilde{X}F;\tilde{Y}) = \mathbb{I}(F;\tilde{Y}) + \mathbb{I}(\tilde{X};\tilde{Y}|F).$$

By construction, $\mathbb{H}(F|\tilde{Y}) = 0$ and $\mathbb{I}(\tilde{X};\tilde{Y}|F = 0) = 0$; therefore,

$$\mathbb{I}(X';Y') = \mathbb{H}(F) + \mathbb{P}[F = 1]\mathbb{I}(\tilde{X};\tilde{Y}|F = 1). \tag{4.28}$$

We compute each term on the right-hand side of (4.28) separately. By construction, $\mathbb{P}[F = 1]$ is the probability that Bob obtains $\tilde{Y} \in \{0, 1\}$, which can be computed explicitly as

$$\begin{aligned}\alpha \triangleq \mathbb{P}[F = 1] &= \mathbb{P}[X^m \oplus Y^m = 0 \quad \text{or} \quad X^m \oplus Y^m = 1] \\ &= (\bar{p}\bar{q} + pq)^m + (\bar{p}q + p\bar{q})^m.\end{aligned} \tag{4.29}$$

The probability that $\tilde{Y}$ differs from $\tilde{X}$ given $F = 1$ is simply the probability that $X^m \oplus Y^m = \bar{1}$ given $F = 1$; hence,

$$\beta \triangleq \mathbb{P}[\tilde{X} \neq \tilde{Y}|F = 1] = \frac{(\bar{p}q + p\bar{q})^m}{(\bar{p}\bar{q} + pq)^m + (\bar{p}q + p\bar{q})^m}. \tag{4.30}$$

The conditional probabilities $p_{\tilde{Y}|\tilde{X}F}$ are those of a BSC with cross-over probability $\mathbb{P}[\tilde{X} \neq \tilde{Y}|F = 1]$; therefore,

$$\mathbb{I}(\tilde{X};\tilde{Y}|F = 1) = 1 - \mathbb{H}_b(\beta). \tag{4.31}$$

On combining (4.29), (4.30), and (4.31) in (4.28), we obtain

$$\mathbb{I}(X';Y') = \mathbb{H}_b(\alpha) + \alpha(1 - \mathbb{H}_b(\beta)). \qquad \square$$

Proposition 4.4. *The mutual information between Alice and Eve after advantage distillation with the repetition protocol is*

$$\mathbb{I}(X';Z') = \mathbb{H}_b(\alpha) + \alpha\sum_{k=0}^{m}\binom{m}{k}p_k\left(1 - \mathbb{H}_b\left(\frac{p_k}{p_k + p_{m-k}}\right)\right),$$

with $\alpha \triangleq (\bar{p}\bar{q} + pq)^m + (\bar{p}q + p\bar{q})^m$ *and*

$$p_k \triangleq \frac{1}{\alpha}(\bar{p}\bar{q}r + pq\bar{r})^k(\bar{p}\bar{q}\bar{r} + pqr)^{m-k} + \frac{1}{\alpha}(\bar{p}qr + p\bar{q}\bar{r})^k(\bar{p}q\bar{r} + p\bar{q}r)^{m-k}.$$

Proof. Because Eve may not perform any hard decision on the bit V, evaluating Eve's probability of error[3] does not suffice to compute $\mathbb{I}(X'; Z')$. Nevertheless, the satellite source model and the repetition protocol are simple enough that we can compute $\mathbb{I}(X'; Z')$ in closed form. Using the chain rule,

$$\begin{aligned}\mathbb{I}(X'; Z') &= \mathbb{I}(\tilde{X}F; Z^m, \overline{V}^m \oplus X^m, F)\\ &= \mathbb{H}(\tilde{X}F) - \mathbb{H}(\tilde{X}F|Z^m, \overline{V}^m \oplus X^m, F)\\ &= \mathbb{H}(F) + \mathbb{H}(\tilde{X}|F) - \mathbb{H}(\tilde{X}|Z^m, \overline{V}^m \oplus X^m, F).\end{aligned} \tag{4.32}$$

Notice that the pair $(Z^m, \overline{V}^m \oplus X^m)$ uniquely determines $(Z^m, \overline{V}^m \oplus X^m \oplus Z^n)$ and vice versa, so that $\mathbb{H}(\tilde{X}|Z^m, \overline{V}^m \oplus X^m, F) = \mathbb{H}(\tilde{X}|Z^m, \overline{V}^m \oplus X^m \oplus Z^m, F)$. In addition, Z^m and X^m are observations of the same i.i.d. sequence U^m through indendent BSCs. Hence, we can write $Z^n = U^n \oplus E_z^m$ and $X^n = U^n \oplus E_x^m$, where E_z^m and E_x^m are the independent error patterns introduced by the BSCs. Consequently,

$$\begin{aligned}&\mathbb{H}(\tilde{X}|Z^m, \overline{V}^m \oplus X^m, F)\\ &\quad= \mathbb{H}(\tilde{X}|Z^m, \overline{V}^m \oplus X^m \oplus Z^m, F)\\ &\quad= \mathbb{H}(\tilde{X}|U^m \oplus E_z^m, \overline{V}^m \oplus E_x^m \oplus E_z^m, F)\\ &\quad= \mathbb{H}(\tilde{X}, U^m \oplus E_z^m|\overline{V}^m \oplus E_x^m \oplus E_z^m, F) - \mathbb{H}(U^m \oplus E_z^m|\overline{V}^m \oplus E_x^m \oplus E_z^m, F)\\ &\quad= \mathbb{H}(\tilde{X}|\overline{V}^m \oplus E_x^m \oplus E_z^m, F) + \mathbb{H}(U^m \oplus E_z^m|\overline{V}^m \oplus E_x^m \oplus E_z^m, F, \tilde{X})\\ &\quad\quad - \mathbb{H}(U^m \oplus E_z^m|\overline{V}^m \oplus E_x^m \oplus E_z^m, F).\end{aligned}$$

Since the sequence U^m is uniformly distributed, the crypto lemma ensures that

$$\mathbb{H}(U^m \oplus E_z^m|\overline{V}^m \oplus E_x^m \oplus E_z^m, F, \tilde{X}) = \mathbb{H}(U^m \oplus E_z^m|\overline{V}^m \oplus E_x^m \oplus E_z^m, F) = m,$$

and, therefore,

$$\mathbb{H}(\tilde{X}|Z^m, \overline{V}^m \oplus X^m, F) = \mathbb{H}(\tilde{X}|\overline{V}^m \oplus E_x^m \oplus E_z^m, F). \tag{4.33}$$

On substituting (4.33) into (4.32), we obtain

$$\begin{aligned}\mathbb{I}(X'; Z') &= \mathbb{H}(F) + \mathbb{H}(\tilde{X}|F) - \mathbb{H}(\tilde{X}|\overline{V}^m \oplus E_x^m \oplus E_z^m, F)\\ &= \mathbb{H}(F) + \mathbb{I}(\tilde{X}; \overline{V}^m \oplus E_x^m \oplus E_z^m|F)\\ &= \mathbb{H}(F) + \mathbb{P}[F = 1]\mathbb{I}(\tilde{X}; \overline{V}^m \oplus E_x^m \oplus E_z^m|F = 1),\end{aligned} \tag{4.34}$$

where we have used $\mathbb{I}(\tilde{X}; Z^m, \overline{V}^m \oplus X^m|F = 0) = 0$ to obtain the last equality. Given $F = 1$, note that $\tilde{X} = V$ and that the weight $W \triangleq w(\overline{V}^m \oplus E_x^m \oplus E_z^m)$ of the sequence $\overline{V}^m \oplus E_x^m \oplus E_z^m$ is a sufficient statistic for $\tilde{X}$ given $\overline{V}^m \oplus E_x^m \oplus E_z^m$. Hence,

$$\mathbb{I}(X'; Z') = \mathbb{H}(F) + \mathbb{P}[F = 1]\mathbb{I}(\tilde{X}; W|F = 1).$$

[3] If we were to compute the probability of error for Eve under maximum-likelihood decoding, Fano's inequality would yield only a *lower* bound for $\mathbb{I}(X'; Z')$.

Finally, all we need in order to compute $\mathbb{I}(\tilde{X}; W|F = 1)$ is the joint distribution $\mathbb{P}[W, \tilde{X}|F = 1]$. For any weight $k \in [\![0, m]\!]$,

$$\begin{aligned}\mathbb{P}\left[W = k, \tilde{X} = 0|F = 1\right] &= \mathbb{P}\left[w(\overline{V}^m \oplus E_x^m \oplus E_z^m) = k, \tilde{X} = 0|F = 1\right]\\ &= \mathbb{P}\left[w\left(E_x^m \oplus E_z^m\right) = k, V = 0|F = 1\right]\\ &= \mathbb{P}\left[w\left(E_x^m \oplus E_z^m\right) = k|F = 1\right]\mathbb{P}[V = 0].\end{aligned}$$

By construction, $\mathbb{P}[V = 0] = \frac{1}{2}$, and $\mathbb{P}\left[w\left(E_x^m \oplus E_z^m\right) = k|F = 1\right]$ can be written as

$$\begin{aligned}&\frac{\mathbb{P}\left[w\left(E_x^m \oplus E_z^m\right) = k, F = 1\right]}{\mathbb{P}[F = 1]}\\ &= \frac{1}{\alpha}\left(\mathbb{P}\left[w\left(E_x^m \oplus E_z^m\right) = k, X^m \oplus Y^m = 0\right] + \mathbb{P}\left[w\left(E_x^m \oplus E_z^m\right) = k, X^m \oplus Y^m = 1\right]\right)\\ &= \frac{1}{\alpha}\left(\binom{m}{k}(\bar{p}\bar{q}r + pq\bar{r})^k(\bar{p}\bar{q}\bar{r} + pqr)^{m-k} + \binom{m}{k}(\bar{p}qr + p\bar{q}\bar{r})^k(\bar{p}q\bar{r} + p\bar{q}r)^{m-k}\right).\end{aligned}$$

Upon defining p_k as

$$p_k \triangleq \frac{1}{\alpha}(\bar{p}\bar{q}r + pq\bar{r})^k(\bar{p}\bar{q}\bar{r} + pqr)^{m-k} + \frac{1}{\alpha}(\bar{p}qr + p\bar{q}\bar{r})^k(\bar{p}q\bar{r} + p\bar{q}r)^{m-k},$$

we obtain

$$\mathbb{P}\left[W = k, \tilde{X} = 0|F = 1\right] = \frac{1}{2}\binom{m}{k}p_k. \tag{4.35}$$

Similarly,

$$\mathbb{P}\left[W = k, \tilde{X} = 1|F = 1\right] = \frac{1}{2}\binom{m}{k}p_{m-k}, \tag{4.36}$$

and, on combining (4.35) and (4.36), we have

$$\mathbb{P}[W = k|F = 1] = \frac{1}{2}\binom{m}{k}(p_k + p_{m-k}).$$

We can now evaluate $\mathbb{I}(X'; W|F = 1)$ explicitly as

$$\begin{aligned}&\sum_{k=0}^{m}\left(\mathbb{P}\left[W = k, \tilde{X} = 0|F = 1\right]\log\left(\frac{\mathbb{P}\left[W = k|\tilde{X} = 0, F = 1\right]}{\mathbb{P}[W = k|F = 1]}\right)\right.\\ &\qquad\left. + \mathbb{P}\left[W = k, \tilde{X} = 1|F = 1\right]\log\left(\frac{\mathbb{P}\left[W = k|\tilde{X} = 1, F = 1\right]}{\mathbb{P}[W = k|F = 1]}\right)\right)\\ &= \sum_{k=0}^{m}\left(\frac{1}{2}\binom{m}{k}p_k\log\left(\frac{2p_k}{p_{m-k} + p_k}\right) + \frac{1}{2}\binom{m}{k}p_{m-k}\log\left(\frac{2p_{m-k}}{p_{m-k} + p_k}\right)\right)\\ &= \frac{1}{2}\sum_{k=0}^{m}\binom{m}{k}(p_k + p_{m-k}) + \frac{1}{2}\sum_{k=0}^{m}\binom{m}{k}(p_k + p_{m-k})\left(-\mathbb{H}_b\left(\frac{p_k}{p_k + p_{m-k}}\right)\right)\\ &= \sum_{k=0}^{m}\binom{m}{k}p_k\left(1 - \mathbb{H}_b\left(\frac{p_k}{p_k + p_{m-k}}\right)\right),\end{aligned}$$

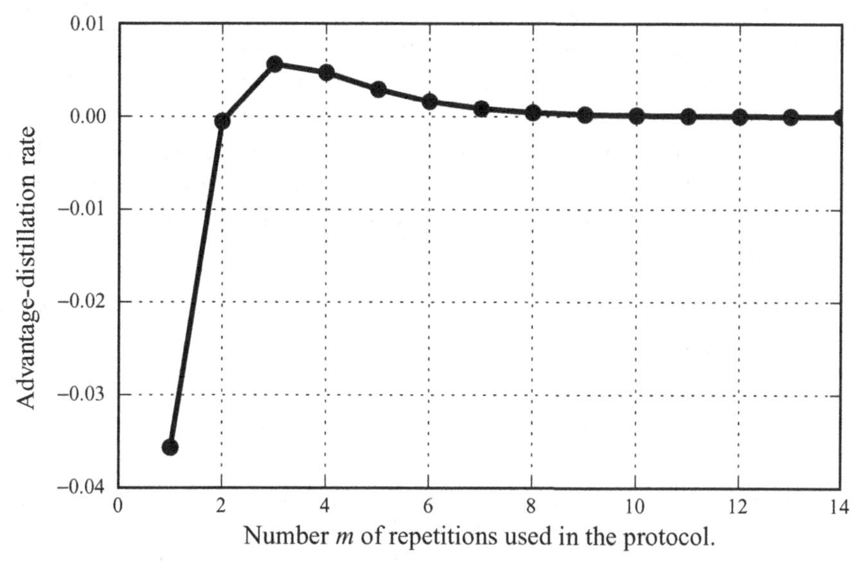

Figure 4.6 Advantage-distillation rate for different values of the repetition-protocol parameter m. The parameters of the satellite source are $p = q = 0.2$ and $r = 0.15$.

where the last equality follows because $\sum_{k=0}^{m} \binom{m}{k} p_k = \sum_{k=0}^{m} \binom{m}{k} p_{m-k}$ and

$$\mathbb{H}_b\left(\frac{p_k}{p_k + p_{m-k}}\right) = \mathbb{H}_b\left(1 - \frac{p_k}{p_k + p_{m-k}}\right) = \mathbb{H}_b\left(\frac{p_{m-k}}{p_k + p_{m-k}}\right).$$

On substituting (4.29) and (4.13) into (4.34), we obtain the desired result

$$\mathbb{I}(\mathrm{X}'; \mathrm{Z}') = \mathbb{H}_b(\alpha) + \alpha \sum_{k=0}^{m} \binom{m}{k} p_k \left(1 - \mathbb{H}_b\left(\frac{p_k}{p_k + p_{m-k}}\right)\right). \qquad \square$$

Figure 4.6 illustrates the advantage-distillation rate of the protocol for a satellite source with $p = q = 0.2$ and $r = 0.15$ for various values of the repetition parameter m. Note that, on choosing m large enough, the protocol achieves a strictly positive advantage-distillation rate; however, the protocol is inefficient and the advantage-distillation rates are quite low. For instance, with $p = q = 0.2$, $r = 0.15$, and $m = 3$, the advantage-distillation rate is on the order of 0.005 bits per observation of the DMS. This rate is relatively low because the post-selection with a repetition code is extremely wasteful of source observations. It is possible to improve these rates by using slightly better codes, and we refer the reader to the bibliographical notes at the end of this chapter for additional details.

4.3.2 Information reconciliation

After the advantage-distillation phase, Alice, Bob, and Eve, obtain the realizations of a DMS $(\mathcal{X}'\mathcal{Y}'\mathcal{Z}', p_{\mathrm{X}'\mathrm{Y}'\mathrm{Z}'})$. The objective of the information-reconciliation phase (reconciliation for short) is to allow Alice and Bob to agree on a common sequence S, but, at this stage, the sequence S is not subject to any secrecy constraint. To simplify the notation, we assume that the reconciliation protocol operates on the DMS $(\mathcal{XYZ}, p_{XYZ})$ instead of on the DMS $(\mathcal{X}'\mathcal{Y}'\mathcal{Z}', p_{X'Y'Z'})$ obtained with advantage distillation.

We place few restrictions on how reconciliation should be performed. The common sequence S could be a function of Alice and Bob's observations and of messages exchanged interactively over the public authenticated channel. Alice and Bob are also allowed to randomize their operations using sources of local randomness. Formally, a reconciliation protocol is defined as follows.

Definition 4.12. *A reconciliation protocol $\mathcal{R}_n$ for a source model with DMS $(\mathcal{XYZ}, p_{XYZ})$ consists of*

- *an alphabet $\mathcal{S} = [\![1, S]\!]$;*
- *a source of local randomness $(\mathcal{R}_\mathcal{X}, p_{R_X})$ for Alice;*
- *a source of local randomness $(\mathcal{R}_\mathcal{Y}, p_{R_Y})$ for Bob;*
- *an integer $r \in \mathbb{N}^*$ that represents the number of rounds of communication;*
- *r encoding functions $f_i : \mathcal{X}^n \times \mathcal{B}^{i-1} \times \mathcal{R}_\mathcal{X} \rightarrow \mathcal{A}$ for $i \in [\![1, r]\!]$;*
- *r encoding functions $g_i : \mathcal{Y}^n \times \mathcal{A}^{i-1} \times \mathcal{R}_\mathcal{Y} \rightarrow \mathcal{B}$ for $i \in [\![1, r]\!]$;*
- *a function $\eta_a : \mathcal{X}^n \times \mathcal{B}^r \times \mathcal{R}_\mathcal{X} \rightarrow \mathcal{S}$;*
- *a function $\eta_b : \mathcal{Y}^n \times \mathcal{A}^r \times \mathcal{R}_\mathcal{Y} \rightarrow \mathcal{S}$;*

and operates as follows:

- *Alice observes n realizations of the source x^n while Bob observes y^n;*
- *Alice generates a realization r_x of her source of local randomness while Bob generates r_y from his;*
- *in round $i \in [\![1, r]\!]$, Alice transmits $a_i = f_i(x^n, b^{i-1}, r_x)$ while Bob transmits $b_i = g_i(y^n, a^{i-1}, r_y)$;*
- *after round r, Alice computes $s = \eta_a(x^n, b^r, r_x)$ while Bob computes $\hat{s} = \eta_b(y^n, a^r, r_y)$.*

By convention, $\mathcal{A}^0 \triangleq 0$ and $\mathcal{B}_0 \triangleq 0$. The number of rounds r and the sources of local randomness can be optimized during the design of the reconciliation protocol. Note that the definition of a reconciliation protocol is slightly different from that of an advantage-distillation protocol because the objective is not the same. The goal of the reconciliation protocol is to guarantee that Alice and Bob agree on a common sequence; hence, the output alphabets of the functions η_a and η_b are the same. The goal of an advantage-distillation protocol is merely to generate a new source; hence, the output alphabets of the functions θ_a and θ_b in Definition 4.9 could be different.

The reliability performance of a reconciliation protocol is measured in terms of the average probability of error

$$\mathbf{P}_e(\mathcal{R}_n) \triangleq \mathbb{P}\left[S \neq \hat{S} \middle| \mathcal{R}_n\right].$$

In addition, since the common sequence S generated by a reconciliation protocol is eventually processed to generate a secret key, it is desirable that the protocol leaks as little information as possible over the public channel. Hence, a reasonable measure of performance is the difference between the entropy of the common sequence $\mathbb{H}(\mathsf{S})$ and the amount of public information $\mathbb{H}(\mathsf{A}^r\mathsf{B}^r)$ exchanged over the public channel. The quantity

$$\mathbf{R}(\mathcal{R}_n) \triangleq \frac{1}{n}(\mathbb{H}(\mathsf{S}|\mathcal{R}_n) - \mathbb{H}(\mathsf{A}^r\mathsf{B}^r|\mathcal{R}_n))$$

is called the *reconciliation rate* of a reconciliation protocol. The choice of this measure is fully justified when we discuss privacy amplification in Section 4.3.3.

Definition 4.13. *A reconciliation rate R is achievable if there exists a sequence of reconciliation protocols $\{\mathcal{R}_n\}_{n\geqslant 1}$ such that*

$$\lim_{n\to\infty} \mathbf{P}_e(\mathcal{R}_n) = 0 \quad \textit{and} \quad \underline{\lim}_{n\to\infty} \mathbf{R}(\mathcal{R}_n) \geqslant R.$$

Definition 4.14. *The reconciliation capacity R^{SM} of a DMS $(\mathcal{XY}, p_{\mathsf{XY}})$ is*

$$R^{\text{SM}} \triangleq \sup\{R : R \text{ is an achievable reconciliation rate}\}.$$

Proposition 4.5. *The reconciliation capacity of a source model with DMS $(\mathcal{XYZ}, p_{\mathsf{XYZ}})$ is*

$$R^{\text{SM}} = \mathbb{I}(\mathsf{X};\mathsf{Y}).$$

In addition the reconciliation rates below R^{SM} are achievable with one-way communication and without sources of local randomness.

Proof. We start by proving that the reconciliation capacity cannot exceed $\mathbb{I}(\mathsf{X};\mathsf{Y})$. Let R be an achievable reconciliation rate. By definition, for any $\epsilon > 0$, there exists a reconciliation protocol $\mathcal{R}_n$ such that

$$\mathbf{P}_e(\mathcal{R}_n) \leqslant \delta(\epsilon) \quad \text{and} \quad \mathbf{R}(\mathcal{R}_n) \geqslant R - \delta(\epsilon). \tag{4.37}$$

In the remainder of the proof, we omit the conditioning on $\mathcal{R}_n$ to simplify the notation. Fano's inequality guarantees that $(1/n)\mathbb{H}(\mathsf{S}|\hat{\mathsf{S}}) \leqslant \delta(\mathbf{P}_e(\mathcal{R}_n)) \leqslant \delta(\epsilon)$; therefore,

$$\begin{aligned}
R &\leqslant \mathbf{R}(\mathcal{R}_n) + \delta(\epsilon)\\
&= \frac{1}{n}(\mathbb{H}(\mathsf{S}) - \mathbb{H}(\mathsf{A}^r\mathsf{B}^r)) + \delta(\epsilon)\\
&\leqslant \frac{1}{n}\mathbb{H}(\mathsf{S}) - \frac{1}{n}\mathbb{H}(\mathsf{S}|\hat{\mathsf{S}}) - \frac{1}{n}\mathbb{H}(\mathsf{A}^r\mathsf{B}^r) + \delta(\epsilon)\\
&= \frac{1}{n}\mathbb{I}(\mathsf{S};\hat{\mathsf{S}}) - \frac{1}{n}\mathbb{H}(\mathsf{A}^r\mathsf{B}^r) + \delta(\epsilon).
\end{aligned} \tag{4.38}$$

Since $\mathsf{S} \rightarrow \mathsf{X}^n\mathsf{A}^r\mathsf{B}^r\mathsf{R}_\mathsf{X} \rightarrow \mathsf{Y}^n\mathsf{A}^r\mathsf{B}^r\mathsf{R}_\mathsf{Y} \rightarrow \hat{\mathsf{S}}$ forms a Markov chain, the data-processing inequality ensures that

$$\mathbb{I}(\mathsf{S};\hat{\mathsf{S}}) \leqslant \mathbb{I}(\mathsf{X}^n\mathsf{R}_\mathsf{X}\mathsf{A}^r\mathsf{B}^r;\mathsf{Y}^n\mathsf{R}_\mathsf{Y}\mathsf{A}^r\mathsf{B}^r). \tag{4.39}$$

We further expand $\mathbb{I}(X^n R_X A^r B^r; Y^n R_Y A^r B^r)$ as

$$\begin{aligned}
&\mathbb{I}(X^n R_X A^r B^r; Y^n R_Y A^r B^r)\\
&\quad = \mathbb{I}(X^n R_X; Y^n R_Y A^r B^r) + \mathbb{I}(A^r B^r; Y^n R_Y A^r B^r | X^n R_X)\\
&\quad = \mathbb{I}(X^n R_X; Y^n R_Y | A^r B^r) + \mathbb{I}(X^n R_X; A^r B^r) + \mathbb{I}(A^r B^r; Y^n R_Y A^r B^r | X^n R_X)\\
&\quad = \mathbb{I}(X^n R_X; Y^n R_Y | A^r B^r) + \mathbb{H}(A^r B^r) - \mathbb{H}(A^r B^r | X^n R_X)\\
&\qquad + \mathbb{I}(A^r B^r; Y^n R_Y A^r B^r | X^n R_X).
\end{aligned} \tag{4.40}$$

Note that

$$\mathbb{I}(A^r B^r; Y^n R_Y A^r B^r | X^n R_X) = \mathbb{H}(A^r B^r | X^n R_X). \tag{4.41}$$

By applying Lemma 4.2 with $S \triangleq X^n R_X$, $T \triangleq Y^n R_Y$, $V^r \triangleq A^r$, and $W^r \triangleq B^r$, we obtain

$$\mathbb{I}(X^n R_X; Y^n R_Y | A^r B^r) \leqslant \mathbb{I}(X^n R_X; Y^n R_Y).$$

Since the DMSs $(\mathcal{R}_\mathcal{X}, p_{R_X})$ and $(\mathcal{R}_\mathcal{Y}, p_{R_Y})$ are mutually independent and independent of the DMS $(\mathcal{XYZ}, p_{XYZ})$, we have

$$\mathbb{I}(X^n R_X; Y^n R_Y | A^r B^r) \leqslant n\mathbb{I}(X; Y). \tag{4.42}$$

On combining (4.41) and (4.42) in (4.40), we have

$$\mathbb{I}(X^n R_X A^r B^r; Y^n R_Y A^r B^r) \leqslant n\mathbb{I}(X; Y) + \mathbb{H}(A^r B^r), \tag{4.43}$$

and, using (4.43) in (4.39) and (4.38), we obtain

$$R \leqslant \mathbb{I}(X; Y) + \delta(\epsilon).$$

Since ϵ can be chosen arbitrarily small and R can be chosen arbitrarily close to R^{SM}, it must hold that $R^{SM} \leqslant \mathbb{I}(X; Y)$.

We now show that all reconciliation rates below R^{SM} are achievable. This result follows directly from Corollary 2.4. In fact, for any $\epsilon > 0$, Corollary 2.4 guarantees the existence of a $(2^{nR}, n)$ code $\mathcal{C}_n$ that compresses X^n into a message A at rate $R \leqslant \mathbb{H}(X|Y) + \delta(\epsilon)$ and such that X^n can be retrieved from Y^n and A with probability of error $\mathbf{P}_e(\mathcal{C}_n) \leqslant \delta(\epsilon)$. Such a code can be viewed as a reconciliation protocol $\mathcal{R}_n$ without sources of local randomness, for which $S = X^n$ and in which there is a single public message A of about $n\mathbb{H}(X|Y)$ bits exchanged over the public channel. The corresponding reconciliation rate is

$$\begin{aligned}
\mathbf{R}(\mathcal{R}_n) &= \frac{1}{n}(\mathbb{H}(S) - \mathbb{H}(A))\\
&\geqslant \frac{1}{n}\mathbb{H}(X^n) - R\\
&= \mathbb{I}(X; Y) - \delta(\epsilon).
\end{aligned}$$

Hence, all reconciliation rates $R < \mathbb{I}(X; Y)$ are achievable without sources of local randomness and with one-way communication. □

The achievability proof of Proposition 4.5 carries over directly if the roles of Alice and Bob are interchanged. By having Alice estimate Y^n and Bob send information over the public channel, Alice and Bob can recover a sequence of length $n\mathbb{H}(Y)$ while disclosing about $n\mathbb{H}(Y|X)$ bits. The rate of this reconciliation protocol is again on the order of $\mathbb{I}(X;Y)$. Reconciliation protocols for which $S = X^n$ are called *direct* reconciliation protocols, while those for which $S = Y^n$ are called *reverse* reconciliation protocols. Both direct and reverse reconciliation protocols can achieve the reconciliation capacity, but the keys that can be distilled subsequently might not be identical; this issue is discussed further in Section 4.3.3.

Although Proposition 4.5 ensures the existence of reconciliation protocols achieving reconciliation rates arbitrarily close to $\mathbb{I}(X;Y)$, this limit cannot be exactly attained. Any practical finite-length reconciliation protocol introduces an overhead and discloses strictly more than $n\mathbb{H}(X|Y)$ bits over the public channel. It is convenient to account for this overhead by defining the *efficiency* of a reconciliation protocol as follows.

Definition 4.15. *The efficiency of a reconciliation protocol $\mathcal{R}_n$ for a source model with DMS $(\mathcal{XYZ}, p_{XYZ})$ is*

$$\beta \triangleq \frac{\mathbb{H}(S) - r\log(|\mathcal{A}|\,|\mathcal{B}|)}{n\mathbb{I}(X;Y)}. \tag{4.44}$$

The quantity $r\log(|\mathcal{A}|\,|\mathcal{B}|)$ represents the number of bits required in order to describe all messages exchanged over the public channel. Note that $\beta \leqslant 1$ because

$$\begin{aligned}\frac{1}{n}(\mathbb{H}(S) - r\log(|\mathcal{A}|\,|\mathcal{B}|)) &\leqslant \frac{1}{n}(\mathbb{H}(S) - \mathbb{H}(A^rB^r))\\ &= \mathbf{R}(\mathcal{R}_n)\\ &\leqslant R^{\text{SM}}\\ &= \mathbb{I}(X;Y).\end{aligned}$$

In terms of efficiency, Proposition 4.5 states that there exist reconciliation protocols with efficiency arbitrarily close to one.

Remark 4.6. *With continuous correlated sources, lossless source coding with side information is not possible, since the discrepancies between continuous sources cannot be corrected exactly. Unlike traditional source coding problems, which can be analyzed in a rate-distortion framework, reconciliation* requires *Alice and Bob to agree on a common sequence for further processing. Consequently, the natural way of handling continuous sources is to quantize them to revert back to a discrete case. Assuming again that Alice's randomness X^n is chosen as the common sequence S, Alice can generate a quantized version X_d^n of X^n with a scalar quantizer. The upper bound of Proposition 4.5 applies even if Y^n is not quantized, and reconciliation rates can be no greater than $\mathbb{I}(X_d;Y)$. By Corollary 2.4, the upper bound can be approached with one-way communication only. Additionally, by choosing a fine enough quantizer, reconciliation rates can be made arbitrarily close to $\mathbb{I}(X;Y)$, and quantization incurs a negligible loss.*

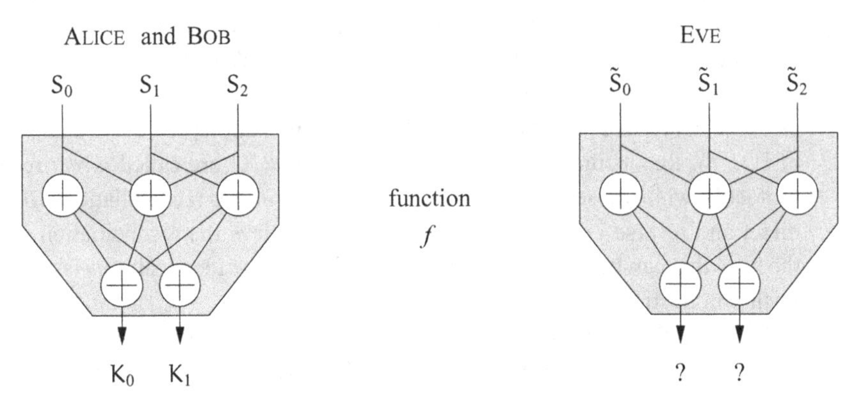

Figure 4.7 Principle of privacy amplification. Alice, Bob, and Eve apply a known function f to their respective sequences S and $\tilde{S}$. Because of the discrepancies between $\tilde{S}$ and S, the outputs $K = f(S)$ and $f(\tilde{S})$ are different. Eve cannot predict how her errors propagate and, for a well-chosen f, she obtains no information about K.

4.3.3 Privacy amplification

Privacy amplification is the final step of a sequential key-distillation strategy that allows Alice and Bob to distill a secret key. Specifically, the role of privacy amplification is to process the sequence S obtained by Alice and Bob after reconciliation to extract a shorter sequence of k bits that is provably unknown to Eve. Without loss of generality, we assume throughout this section that S is a binary sequence of n bits.

Before we analyze privacy amplification in detail, it is useful to develop an intuitive understanding of why and how this operation is possible. First, note that privacy amplification is straightforward in certain cases. For instance,

- if Alice and Bob know that Eve has no information about S, then the sequence S itself can be used as a key;
- if Alice and Bob know that Eve has access to S, then no secret key can be distilled;
- if Alice and Bob know that Eve has access to m bits of S then the remaining $n - m$ bits of S can be used as a secret key.

In general, privacy amplification is not so simple because Alice and Bob know a bound for Eve's information that cannot be tied to bits of S directly. Nevertheless, we show that this bound is all Alice and Bob need to *extract* bits from S about which Eve has little knowledge. Why this operation is possible can be understood intuitively as follows. For simplicity, assume that Eve computes her best estimate $\tilde{S}$ of S on the basis of her observations of the source and all messages exchanged over the public channel by the advantage-distillation and reconciliation protocols. Unless Eve's information about S is exactly $\mathbb{H}(S)$, her estimate $\tilde{S}$ differs from S in some positions. The key idea is that Eve cannot determine the *location* of the discrepancies; consequently, as illustrated in Figure 4.7, if Alice and Bob apply a deterministic transformation f to their sequence S to shuffle and remove some bits, Eve cannot predict how her errors propagate and affect her outcome $f(\tilde{S})$. We will show that there exist transformations that propagate Eve's errors so much that all possible outcomes of the transformation f become equally likely from Eve's perspective.

The precise analysis of privacy amplification is slightly involved because it does not rely on the Shannon entropy but on alternative measures called the *collision entropy* and the *min-entropy*. The detailed study of these entropies goes well beyond the scope of this book, and the following section presents only the properties that are useful in the context of secret-key distillation. We refer the interested reader to the bibliographical notes at the end of the chapter for further references.

Collision entropy and min-entropy

The *collision entropy* and the *min-entropy* are convenient metrics because they are tailored to the actual functions used for privacy amplification and because they are more sensitive than the Shannon entropy to deviations from uniform distributions. The latter property allows us to establish the achievability of strong secret-key rates rather than just weak secret-key rates. As an illustration, the following example exhibits a random variable that is not uniform but whose Shannon entropy is hardly distinguishable from that of a uniform random variable.

Example 4.3. Consider a random variable $\mathsf{K} \in [\![1, 2^k]\!]$ with probability distribution

$$\mathbb{P}[\mathsf{K} = 1] = 2^{-k/4} \quad \text{and} \quad \mathbb{P}[\mathsf{K} = i] = \frac{1 - 2^{-k/4}}{2^k - 1} \quad \text{for } i \neq 1.$$

In other words, K is approximately uniform because the realization $\mathsf{K} = 1$ has probability $2^{-k/4}$ while the others have probability $\sim 2^{-k}$. If k is large, this slight non-uniformity is not well captured by the Shannon entropy $\mathbb{H}(\mathsf{K})$. In fact,

$$\begin{aligned}
\mathbb{H}(\mathsf{K}) &= 2^{-k/4}\frac{k}{4} - (2^k - 1)\frac{1 - 2^{-k/4}}{2^k - 1}\log\left(\frac{1 - 2^{-k/4}}{2^k - 1}\right) \\
&= 2^{-k/4}\frac{k}{4} + \left(1 - 2^{-k/4}\right)\log\left(2^k - 1\right) - \left(1 - 2^{-k/4}\right)\log\left(1 - 2^{-k/4}\right) \\
&= k + 2^{-k/4}\left(\frac{k}{4} - k\right) + \left(1 - 2^{-k/4}\right)\log\left(\frac{1 - 2^{-k}}{1 - 2^{-k/4}}\right).
\end{aligned}$$

Therefore, $\lim_{k\to\infty}(1/k)\mathbb{H}(\mathsf{K}) = 1$, which obscures the fact that the random variable K is not exactly uniform.

Definition 4.16. *The collision entropy of a discrete random variable* $\mathsf{X} \in \mathcal{X}$ *is*

$$\mathbb{H}_c(\mathsf{X}) \triangleq -\log \mathbb{E}[p_{\mathsf{X}}(\mathsf{X})] = -\log\left(\sum_{x\in\mathcal{X}} p_{\mathsf{X}}(x)^2\right).$$

For two discrete random variables $\mathsf{X} \in \mathcal{X}$ *and* $\mathsf{Y} \in \mathcal{Y}$*, the conditional collision entropy of* X *given* Y *is*

$$\mathbb{H}_c(\mathsf{X}|\mathsf{Y}) \triangleq \sum_{y\in\mathcal{Y}} p_{\mathsf{Y}}(y)\,\mathbb{H}_c(\mathsf{X}|\mathsf{Y} = y).$$

The collision entropy bears such a name because it is a function of the collision probability $\sum_{x\in\mathcal{X}} p_X(x)^2$, which measures the probability of obtaining the same realization of a random variable twice in two independent experiments.

Proposition 4.6. *For any discrete random variable $X \in \mathcal{X}$, the collision entropy satisfies $\mathbb{H}(X) \geqslant \mathbb{H}_c(X) \geqslant 0$. If X is uniformly distributed over $\mathcal{X}$, then $\mathbb{H}(X) = \mathbb{H}_c(X) = \log|\mathcal{X}|$.*

Proof. Using Jensen's inequality and the convexity of the function $x \mapsto -\log x$, we obtain

$$\mathbb{H}(X) = \mathbb{E}_X[-\log p_X(X)] \geqslant -\log \mathbb{E}_X[p_X(X)] = \mathbb{H}_c(X).$$

In addition, since $p_X(x) \leqslant 1$ for all $x \in \mathcal{X}$, it holds that

$$\sum_{x\in\mathcal{X}} p_X(x)^2 \leqslant \sum_{x\in\mathcal{X}} p_X(x) = 1,$$

and, therefore, $\mathbb{H}_c(X) \geqslant 0$. If X is uniformly distributed, then $p_X(x) = 1/|\mathcal{X}|$ for all $x \in \mathcal{X}$ and we obtain $\mathbb{H}_c(X) = \log|\mathcal{X}|$ by direct calculation. □

Remark 4.7. *Many properties of the Shannon entropy $\mathbb{H}(X)$ do not hold for the collision entropy $\mathbb{H}_c(X)$. For instance, conditioning may* increase *the collision entropy; that is, some random variables are such that $\mathbb{H}_c(X|Y) > \mathbb{H}_c(X)$. In such cases, Y is called the* spoiling knowledge.

Definition 4.17. *The min-entropy of a discrete random variable $X \in \mathcal{X}$ is*

$$\mathbb{H}_\infty(X) = -\log\left(\max_{x\in\mathcal{X}} p_X(x)\right).$$

For two discrete random variables $X \in \mathcal{X}$ and $Y \in \mathcal{Y}$, the conditional min-entropy of X given Y is

$$\mathbb{H}_\infty(X|Y) \triangleq \sum_{y\in\mathcal{Y}} p_Y(y)\mathbb{H}_\infty(X|Y=y).$$

Proposition 4.7. *For any discrete random variable $X \in \mathcal{X}$, the min-entropy satisfies $\mathbb{H}_c(X) \geqslant \mathbb{H}_\infty(X) \geqslant 0$. If X is uniformly distributed over $\mathcal{X}$ then $\mathbb{H}_\infty(X) = \mathbb{H}_c(X) = \mathbb{H}(X) = \log|\mathcal{X}|$.*

Proof. Since $p_X(x) \leqslant 1$ for all $x \in \mathcal{X}$, it is obvious that $\mathbb{H}_\infty(X) \geqslant 0$. Also,

$$\begin{aligned}\mathbb{H}_c(X) = -\log\left(\sum_{x\in\mathcal{X}} p_X(x)^2\right) &\geqslant -\log\left(\sum_{x\in\mathcal{X}} p_X(x)\max_x p_X(x)\right)\\ &= -\log\left(\max_x p_X(x)\right) = \mathbb{H}_\infty(X).\end{aligned}$$

If X is uniformly distributed, we obtain $\mathbb{H}_\infty(X) = \log|\mathcal{X}|$ by direct calculation. □

To illustrate that the collision entropy $\mathbb{H}_c(X)$ and the min-entropy $\mathbb{H}_\infty(X)$ are indeed stronger measures of uniformity that $\mathbb{H}(X)$, we revisit Example 4.3.

Example 4.4. Consider the random variable $\mathsf{K} \in [\![1, 2^k]\!]$ with distribution

$$\mathbb{P}[\mathsf{K} = 1] = 2^{-k/4} \quad \text{and} \quad \mathbb{P}[\mathsf{K} = i] = \frac{1 - 2^{-k/4}}{2^k - 1} \quad \text{for } i \neq 1,$$

for which we showed that $\lim_{k\to\infty}(1/k)\mathbb{H}(\mathsf{K}) = 1$ in Example 4.3. The collision entropy of K is

$$\begin{aligned}\mathbb{H}_c(\mathsf{K}) &= -\log\left(2^{-k/2} + (2^k - 1)\left(\frac{1 - 2^{-k/4}}{2^k - 1}\right)^2\right)\\ &= \frac{k}{2} - \log\left(\frac{2^k + 2^{k/2} - 2 \times 2^{k/4}}{2^k - 1}\right).\end{aligned}$$

Hence, $\lim_{k\to\infty}(1/k)\mathbb{H}_c(\mathsf{K}) = \frac{1}{2}$.

For k large enough, the min-entropy of K is

$$\mathbb{H}_\infty(\mathsf{K}) = -\log 2^{-k/4} = \frac{k}{4}.$$

Therefore, $\lim_{k\to\infty}(1/k)\mathbb{H}_\infty(\mathsf{K}) = \frac{1}{4}$.

Remark 4.8. *The collision entropy, the min-entropy, and the Shannon entropy are special cases of the Rényi entropy. For a discrete random variable X, the Rényi entropy of order α is*

$$R_\alpha(\mathsf{X}) \triangleq \frac{1}{1-\alpha}\log\left(\sum_{x\in\mathcal{X}} p_\mathsf{X}(x)^\alpha\right).$$

One can check directly that

$$\mathbb{H}_c(\mathsf{X}) = R_2(\mathsf{X}), \quad \mathbb{H}_\infty(\mathsf{X}) = \lim_{\alpha\to\infty} R_\alpha(\mathsf{X}),$$

and, using l'Hôpital's rule,

$$\mathbb{H}(\mathsf{X}) = \lim_{\alpha\to 1} R_\alpha(\mathsf{X}).$$

Intuitively, the Shannon entropy, the collision entropy and the min-entropy of a random variable play the same role for discrete random variables as the arithmetic mean, geometric mean, and minimum value for a set of numbers.

Privacy amplification with hash functions

In this section, we introduce a generic privacy-amplification technique that exploits hash functions to distill a secret key. From the informal discussion at the beginning of Section 4.3.3, it should not be too surprising that hash functions play a role in privacy amplification. Hash functions are usually designed to produce significantly different outputs even when their inputs are quite similar, which is intuitively the sort of operation that we expect privacy amplification to perform. In what follows, we consider a specific class of hash functions called *universal families*.

Definition 4.18. *Given two finite sets $\mathcal{A}$ and $\mathcal{B}$, a family $\mathcal{G}$ of functions $g : \mathcal{A} \to \mathcal{B}$ is 2-universal (universal for short) if*

$$\forall x_1, x_2 \in \mathcal{A} \quad x_1 \neq x_2 \Rightarrow \mathbb{P}_{\mathsf{G}}[\mathsf{G}(x_1) = \mathsf{G}(x_2)] \leqslant \frac{1}{|\mathcal{B}|},$$

where G is the random variable that represents the choice of a function $g \in \mathcal{G}$ uniformly at random in $\mathcal{G}$.

Universal families of hash functions have been thoroughly studied and we provide two well-known examples of families without proving their universality.

Example 4.5. By identifying $\{0, 1\}^n$ with $\mathrm{GF}(2)^n$, we associate with any binary matrix $\mathbf{M} \in \mathrm{GF}(2)^{k\times n}$ a function

$$h_{\mathbf{M}} : \mathrm{GF}(2)^n \to \mathrm{GF}(2)^k : \mathbf{x} \mapsto \mathbf{Mx}.$$

The family of hash functions $\mathcal{H}_1 = \left\{h_{\mathbf{M}} : \mathbf{M} \in \mathrm{GF}(2)^{k\times n}\right\}$ is universal.

Example 4.6. By identifying $\{0, 1\}^n$ with $\mathrm{GF}(2^n)$, we associate with any element $y \in \mathrm{GF}(2^n)$ a function

$$h_y : \mathrm{GF}(2^n) \to \{0, 1\}^k : x \mapsto k \text{ bits of the product } xy.$$

The k bits are fixed but their position can be chosen arbitrarily. The family of hash functions $\mathcal{H}_2 = \left\{h_y : y \in \mathrm{GF}(2^n)\right\}$ is universal.

Remark 4.9. *Identifying a function in the family $\mathcal{H}_1$ requires nk bits, while identifying a function in the family $\mathcal{H}_2$ requires only n bits. This difference does not affect the operation of privacy amplification, but, in practice, it is often desirable to limit the amount of communication; hence, the family $\mathcal{H}_2$ would be preferred to the family $\mathcal{H}_1$.*

The usefulness of universal families of hash functions for privacy amplification is justified in the following theorem.

Theorem 4.4 (Bennett *et al.*). *Let $\mathsf{S} \in \{0, 1\}^n$ be the random variable that represents the common sequence shared by Alice and Bob, and let E be the random variable that represents the total knowledge about S available to Eve. Let e be a particular realization of E. If Alice and Bob know the conditional collision entropy $\mathbb{H}_c(\mathsf{S}|\mathsf{E} = e)$ to be at least some constant c, and if they choose $\mathsf{K} = \mathsf{G}(\mathsf{S})$ as their secret key, where G is a hash function chosen uniformly at random from a universal family of hash functions $\mathcal{G} : \{0, 1\}^n \to \{0, 1\}^k$, then*

$$\mathbb{H}(\mathsf{K}|\mathsf{G}, \mathsf{E} = e) \geqslant k - \frac{2^{k-c}}{\ln 2}.$$

Proof. Since $\mathbb{H}(\mathsf{K}|\mathsf{G}, \mathsf{E} = e) \geqslant \mathbb{H}_c(\mathsf{K}|\mathsf{G}, \mathsf{E} = e)$ by Proposition 4.6, it suffices to establish a lower bound for the collision entropy $\mathbb{H}_c(\mathsf{K}|\mathsf{G}, \mathsf{E} = e)$. Note that

$$\begin{aligned}
\mathbb{H}_c(\mathsf{K}|\mathsf{G}, \mathsf{E} = e) &= \sum_{g \in \mathcal{G}} p_{\mathsf{G}}(g) \mathbb{H}_c(\mathsf{K}|\mathsf{G} = g, \mathsf{E} = e) \\
&= \sum_{g \in \mathcal{G}} p_{\mathsf{G}}(g) \left(-\log \mathbb{E}_{\mathsf{K}|\mathsf{G}=g,\mathsf{E}=e} \left[p_{\mathsf{K}|\mathsf{GE}}(\mathsf{K}|g, e) \right] \right) \\
&\geqslant -\log \left(\sum_{g \in \mathcal{G}} p_{\mathsf{G}}(g) \mathbb{E}_{\mathsf{K}|\mathsf{G}=g,\mathsf{E}=e} \left[p_{\mathsf{K}|\mathsf{GE}}(\mathsf{K}|g, e) \right] \right),
\end{aligned} \tag{4.45}$$

where the last inequality follows from the convexity of the function $x \mapsto -\log x$ and Jensen's inequality. Now, let $\mathsf{S}_1 \in \{0, 1\}^n$ and $\mathsf{S}_2 \in \{0, 1\}^n$ be two random variables that are independent of each other and independent of G, which are distributed according to $p_{\mathsf{S}|\mathsf{E}=e}$. Then,

$$\begin{aligned}
\mathbb{P}[\mathsf{G}(\mathsf{S}_1) = \mathsf{G}(\mathsf{S}_2)|\mathsf{G} = g] &= \sum_{k \in \{0,1\}^k} p_{\mathsf{G}(\mathsf{S})|\mathsf{GE}}(k|g, e) p_{\mathsf{G}(\mathsf{S})|\mathsf{GE}}(k|g, e) \\
&= \mathbb{E}_{\mathsf{K}|\mathsf{G}=g,\mathsf{E}=e} \left[p_{\mathsf{K}|\mathsf{GE}}(\mathsf{K}|g, e) \right],
\end{aligned}$$

and we can rewrite inequality (4.45) as

$$\mathbb{H}_c(\mathsf{K}|\mathsf{G}, \mathsf{E} = e) \geqslant -\log \mathbb{P}[\mathsf{G}(\mathsf{S}_1) = \mathsf{G}(\mathsf{S}_2)]. \tag{4.46}$$

We now develop an upper bound for $\mathbb{P}[\mathsf{G}(\mathsf{S}_1) = \mathsf{G}(\mathsf{S}_2)]$. By the law of total probability,

$$\begin{aligned}
\mathbb{P}[\mathsf{G}(\mathsf{S}_1) = \mathsf{G}(\mathsf{S}_2)] &= \mathbb{P}[\mathsf{G}(\mathsf{S}_1) = \mathsf{G}(\mathsf{S}_2), \mathsf{S}_1 = \mathsf{S}_2] \mathbb{P}[\mathsf{S}_1 = \mathsf{S}_2] \\
&\quad + \mathbb{P}[\mathsf{G}(\mathsf{S}_1) = \mathsf{G}(\mathsf{S}_2), \mathsf{S}_1 \neq \mathsf{S}_2] \mathbb{P}[\mathsf{S}_1 \neq \mathsf{S}_2].
\end{aligned} \tag{4.47}$$

Note that

$$\mathbb{P}[\mathsf{G}(\mathsf{S}_1) = \mathsf{G}(\mathsf{S}_2)|\mathsf{E} = e, \mathsf{S}_1 = \mathsf{S}_2] \leqslant 1 \quad \text{and} \quad \mathbb{P}[\mathsf{S}_1 \neq \mathsf{S}_2|\mathsf{E} = e] \leqslant 1.$$

In addition, by virtue of the definition of the collision entropy,

$$\mathbb{P}[\mathsf{S}_1 = \mathsf{S}_2] = \sum_{s \in \{0,1\}^n} p_{\mathsf{S}|\mathsf{E}=e}(s|e)^2 = 2^{-\mathbb{H}_c(\mathsf{S}|\mathsf{E}=e)}.$$

Finally, because the hash function $\mathcal{G}$ is chosen in a universal family, it holds that

$$\mathbb{P}[\mathsf{G}(\mathsf{S}_1) = \mathsf{G}(\mathsf{S}_2)|\mathsf{S}_1 \neq \mathsf{S}_2] \leqslant 2^{-k}.$$

On substituting these inequalities into (4.47), we obtain

$$\mathbb{P}[\mathsf{G}(\mathsf{S}_1) = \mathsf{G}(\mathsf{S}_2)] \leqslant 2^{-\mathbb{H}_c(\mathsf{S}|\mathsf{E}=e)} + 2^{-k} \leqslant 2^{-k}(1 + 2^{k-c}), \tag{4.48}$$

where the last inequality follows from the assumption $\mathbb{H}_c(\mathsf{S}|\mathsf{E} = e) \geqslant c$. On substituting (4.48) into (4.46) and using the fact that $\ln(1 + x) \leqslant x$ for all $x > -1$, we obtain

$$\mathbb{H}_c(\mathsf{K}|\mathsf{G}, \mathsf{E} = e) \geqslant k - \frac{2^{k-c}}{\ln 2}.$$

□

Since $\mathbb{H}_c(\mathsf{K}|\mathsf{G}, \mathsf{E} = e) \leqslant k$ by definition, Theorem 4.4 states that it is possible to distill a secret key of k bits with hash functions, provided that their output size is small enough ($k \ll c$). The output sequence size depends directly on the a-priori uncertainty of the eavesdropper, which must be measured in terms of the collision entropy. Since this result is the essential tool that we use to analyze the achievable secret-key rates of sequential secret-key distillation strategies, it is worthwhile discussing it in detail. Note that, although the hash function G is chosen at random, the actual choice is known to Eve and this is reflected by the conditioning on G in the entropy; however, the theorem provides only a lower bound on $\mathbb{H}(\mathsf{K}|\mathsf{G}, \mathsf{E} = e)$, which is an *average* over all possible choices of hash functions in $\mathcal{G}$. Consequently, for a specific choice of $g \in \mathcal{G}$, the entropy $\mathbb{H}(\mathsf{K}|\mathsf{G} = g, \mathsf{E} = e)$ might be significantly different from k, even if $k \ll c$; luckily, this happens with negligible probability. Most importantly, Theorem 4.4 provides an *explicit* privacy-amplification technique because it shows that it suffices to choose a hash function at random in a universal family. This is in sharp contrast with the random-coding argument used in Section 4.2.2, which guarantees only the *existence* of a suitable key-distillation function. Finally, we emphasize that Theorem 4.4 bounds the entropy $\mathbb{H}(\mathsf{K}|\mathsf{G}, \mathsf{E} = e)$, which is a stronger result than a bound on the entropy rate $(1/n)\mathbb{H}(\mathsf{K}|\mathsf{G}, \mathsf{E} = e)$.

We now have all the tools to prove that sequential key-distillation strategies achieve strong secret-key rates.

Theorem 4.5. *Consider a source model with DMS $(\mathcal{X}\mathcal{Y}\mathcal{Z}, p_{\mathsf{XYZ}})$ and let $\beta \in [0, 1]$. All strong secret-key rates R_s that satisfy*

$$R_s < \beta\mathbb{I}(\mathsf{X};\mathsf{Y}) - \min(\mathbb{I}(\mathsf{X};\mathsf{Z}), \mathbb{I}(\mathsf{Y};\mathsf{Z}))$$

are achievable with sequential secret-key distillation strategies that consist of a reconciliation protocol with efficiency β and privacy amplification with a universal family of hash functions. Additionally, these rates are achievable with one-way communication.

Note that the secret-key rates given in Theorem 4.5 are achievable without advantage distillation. In addition, Theorem 4.5 shows that reconciliation efficiency acts as a penalty factor that reduces the information between Alice and Bob from $\mathbb{I}(\mathsf{X};\mathsf{Y})$ to $\beta\mathbb{I}(\mathsf{X};\mathsf{Z})$ but leaves the information leaked to the eavesdropper $\mathbb{I}(\mathsf{X};\mathsf{Z})$ or $\mathbb{I}(\mathsf{Y};\mathsf{Z})$ unchanged. This result is particularly relevant because any practical reconciliation protocol has an efficiency $\beta < 1$, which turns out to be one of the main limiting factors of secret-key rates. Although Proposition 4.5 guarantees the existence of reconciliation protocols with β arbitrarily close to unity, the design of efficient protocols can be challenging. The construction of practical yet efficient reconciliation protocols is discussed in greater detail in Chapter 6.

The proof of Theorem 4.5 is involved because we need to establish a lower bound for Eve's collision entropy about Alice and Bob's common sequence S before we can apply Theorem 4.4. There is no obvious bound since Eve's total knowledge consists not only of the observations of the source Z^n but also of the public messages A^r and B^r exchanged during the reconciliation phase.

Proof of Theorem 4.5. Fix an integer n and let $\epsilon > 0$. Let k be an integer to be determined later. We consider a sequential key-distillation strategy $\mathcal{S}_n$ that consists of

- a direct reconciliation protocol $\mathcal{R}_n$ with efficiency β, with which Alice sends messages A^r to Bob; Theorem 4.5 guarantees the existence of a protocol so that Bob's estimate $\hat{\mathsf{X}}^n$ of X^n satisfies $\mathbb{P}[\hat{\mathsf{X}}^n \neq \mathsf{X}^n|\mathcal{R}_n] \leqslant \delta_\epsilon(n)$;
- privacy amplification based on a universal family of hash functions with output size k, at the end of which Alice computes her key $\mathsf{K} = \mathsf{G}(\mathsf{X}^n)$, while Bob computes $\hat{\mathsf{K}} = \mathsf{G}(\hat{\mathsf{X}}^n)$.

Note that $\mathbb{P}[\hat{\mathsf{K}} \neq \mathsf{K}|\mathcal{S}_n] \leqslant \mathbb{P}[\hat{\mathsf{X}}^n \neq \mathsf{X}^n|\mathcal{R}_n] \leqslant \delta_\epsilon(n)$ and that the strategy uses only one-way communication from Alice to Bob. The total information available to Eve after reconciliation consists of her observation Z^n of the DMS $(\mathcal{XYZ}, p_{\mathsf{XYZ}})$, the public messages A^r exchanged during the reconciliation protocol, and the hash function G chosen for privacy amplification. The strategy $\mathcal{S}_n$ is also known to Eve, but we omit the conditioning on $\mathcal{S}_n$ in order to simplify the notation. We show that, for a suitable choice of the output size k,

$$k \geqslant \mathbb{H}(\mathsf{K}|\mathsf{Z}^n\mathsf{A}^r\mathsf{G}) \geqslant k - \delta_\epsilon(n).$$

This result will follow from Theorem 4.4, provided that we establish a lower bound for the collision entropy $\mathbb{H}_c(\mathsf{X}^n|\mathsf{Z}^n = z^n, \mathsf{A}^r = a^r)$. This is not straightforward because the collision entropy depends on the specific operation of the reconciliation protocol; nevertheless, we circumvent this difficulty in two steps.

First, we will relate $\mathbb{H}_c(\mathsf{X}^n|\mathsf{Z}^n = z^n, \mathsf{A}^r = a^r)$ to $\mathbb{H}_c(\mathsf{X}^n|\mathsf{Z}^n = z^n)$ by means of the following lemma, whose proof is relegated to the appendix at the end of this chapter.

Lemma 4.3 (Cachin). *Let $\mathsf{S} \in \mathcal{S}$ and $\mathsf{U} \in \mathcal{U}$ be two discrete random variables with joint distribution p_{SU}. For any $r > 0$, define the function $\chi : \mathcal{U} \rightarrow \{0, 1\}$ as*

$$\chi(u) \triangleq \begin{cases} 1 & \text{if } \mathbb{H}_c(\mathsf{S}) - \mathbb{H}_c(\mathsf{S}|u) \leqslant \log|\mathcal{U}| + 2r + 2, \\ 0 & \text{otherwise.} \end{cases}$$

Then, $\mathbb{P}_{\mathsf{U}}[\chi(\mathsf{U}) = 1] \geqslant 1 - 2^{-r}$.

Lemma 4.3 shows that, with high probability, the decrease in collision entropy caused by conditioning on U is bounded by $\log|\mathcal{U}| + 2r + 2$, which does not depend on the exact correlation between S and U.

Next, we will lower bound $\mathbb{H}_c(\mathsf{X}^n|\mathsf{Z}^n = z^n)$ by a term on the order of $n\mathbb{H}(\mathsf{X}|\mathsf{Z})$. Essentially, this result follows from the fact that, for n large enough, the realizations of the random source $(\mathsf{X}^n, \mathsf{Z}^n)$ are typical and almost uniformly distributed in the joint typical set. This idea is formalized in the following lemma, whose proof is again relegated to the appendix.

Lemma 4.4. *Consider a DMS $(\mathcal{XZ}, p_{\mathsf{XZ}})$ and define the random variable Θ as*

$$\Theta = \begin{cases} 1 & \text{if } (\mathsf{X}^n, \mathsf{Z}^n) \in \mathcal{T}_{2\epsilon}^n(\mathsf{XZ}) \text{ and } \mathsf{Z}^n \in \mathcal{T}_\epsilon^n(\mathsf{XZ}), \\ 0 & \text{otherwise.} \end{cases}$$

Then, for n sufficiently large, $\mathbb{P}[\Theta = 1] \geqslant 1 - 2^{-\sqrt{n}}$. *Moreover, if* $z^n \in \mathcal{T}_\epsilon^n(\mathsf{Z})$,

$$\mathbb{H}_c(\mathsf{X}^n|\mathsf{Z}^n = z^n, \Theta = 1) \geqslant \mathbb{H}_\infty(\mathsf{X}^n|\mathsf{Z}^n = z^n, \Theta = 1)$$
$$\geqslant n(\mathbb{H}(\mathsf{X}|\mathsf{Z}) - \delta(\epsilon)) - \delta_\epsilon(n).$$

To leverage the results of Lemma 4.3 and Lemma 4.4, we start by defining the random variables Υ (a function of A^r) and Θ (a function of X^n and Z^n) as follows:

$$\Upsilon \triangleq \begin{cases} 1 & \text{if} \quad \mathbb{H}_c(\mathsf{X}^n|\mathsf{Z}^n = z^n) - \mathbb{H}_c(\mathsf{X}^n|\mathsf{Z}^n = z^n, \mathsf{A}^r) \leqslant \log|\mathcal{A}|^r + 2\sqrt{n} + 2, \\ 0 & \text{otherwise;} \end{cases}$$

$$\Theta \triangleq \begin{cases} 1 & \text{if} \quad (\mathsf{X}^n, \mathsf{Z}^n) \in \mathcal{T}_{2\epsilon}^n(\mathsf{XZ}) \quad \text{and} \quad \mathsf{Z}^n \in \mathcal{T}_\epsilon^n(\mathsf{XZ}), \\ 0 & \text{otherwise.} \end{cases}$$

Lemma 4.3 guarantees that $\mathbb{P}[\Upsilon = 1] \geqslant 1 - 2^{-\sqrt{n}}$ and Lemma 4.4 ensures that $\mathbb{P}[\Theta = 1] \geqslant 1 - 2^{-\sqrt{n}}$; therefore, by the union bound,

$$\mathbb{P}[\Theta = 1, \Upsilon = 1] \geqslant 1 - 2 \cdot 2^{-\sqrt{n}}.$$

Consequently, we can lower bound $\mathbb{H}(\mathsf{K}|\mathsf{G}\mathsf{Z}^n\mathsf{A}^r)$ as

$$\begin{aligned} \mathbb{H}(\mathsf{K}|\mathsf{G}\mathsf{Z}^n\mathsf{A}^r) &\geqslant \mathbb{H}(\mathsf{K}|\mathsf{G}\mathsf{Z}^n\mathsf{A}^r\Theta\Upsilon) \\ &\geqslant \mathbb{P}[\Theta = 1, \Upsilon = 1]\mathbb{H}(\mathsf{K}|\mathsf{G}\mathsf{Z}^n\mathsf{A}^r, \Theta = 1, \Upsilon = 1) \\ &\geqslant \left(1 - 2 \cdot 2^{-\sqrt{n}}\right) \mathbb{H}(\mathsf{K}|\mathsf{G}\mathsf{Z}^n\mathsf{A}^r, \Theta = 1, \Upsilon = 1). \end{aligned} \tag{4.49}$$

To bound $\mathbb{H}(\mathsf{K}|\mathsf{G}\mathsf{Z}^n\mathsf{A}^r, \Theta = 1, \Upsilon = 1)$ with Theorem 4.4 it suffices to lower bound the collision entropy

$$\mathbb{H}_c(\mathsf{X}^n|\mathsf{G}, \mathsf{Z}^n = z^n, \mathsf{A}^r = a^r, \Theta = 1, \Upsilon = 1)$$

for any realization $z^n \in \mathcal{T}_\epsilon^n(\mathsf{Z})$ and a^r. By virtue of the definition of Υ,

$$\begin{aligned} &\mathbb{H}_c(\mathsf{X}^n|\mathsf{Z}^n = z^n, \mathsf{A}^r = a^r, \Theta = 1, \Upsilon = 1) \\ &\qquad \geqslant \mathbb{H}_c(\mathsf{X}^n|\mathsf{Z}^n = z^n, \Theta = 1) - \log|\mathcal{A}|^r - 2\sqrt{n} - 2. \end{aligned} \tag{4.50}$$

The quantity $\log|\mathcal{A}|^r$ represents the number of bits required to describe the messages exchanged during the reconciliation phase, which we can express in terms of the reconciliation efficiency β as

$$\log|\mathcal{A}|^r = \mathbb{H}(\mathsf{X}^n) - n\beta\mathbb{I}(\mathsf{X};\mathsf{Y}).$$

Therefore, we can rewrite (4.50) as

$$\begin{aligned} &\mathbb{H}_c(\mathsf{X}^n|\mathsf{Z}^n = z^n, \mathsf{A}^r = a^r, \Theta = 1, \Upsilon = 1) \\ &\qquad \geqslant \mathbb{H}_c(\mathsf{X}^n|\mathsf{Z}^n = z^n, \Theta = 1) - \mathbb{H}(\mathsf{X}^n) + n\beta\mathbb{I}(\mathsf{X};\mathsf{Y}) - 2\sqrt{n} - 2. \end{aligned} \tag{4.51}$$

Next, notice that Lemma 4.4 ensures

$$\mathbb{H}_c(\mathsf{X}^n|\mathsf{Z}^n = z^n, \Theta = 1) \geqslant n(\mathbb{H}(\mathsf{X}|\mathsf{Z}) - \delta(\epsilon)) - \delta_\epsilon(n). \tag{4.52}$$

Combining (4.51) and (4.52) yields

$$\begin{aligned}\mathbb{H}_c(X^n|Z^n = z^n, A^r = a^r, \Theta = 1, \Upsilon = 1) \\ &\geqslant n(\mathbb{H}(X|Z) - \delta(\epsilon)) - \mathbb{H}(X^n) + n\beta\mathbb{I}(X;Y) - 2\sqrt{n} - 2 - \delta_\epsilon(n) \\ &= n(\beta\mathbb{I}(X;Y) - \mathbb{I}(X;Z) - \delta(\epsilon)) - 2\sqrt{n} - 2 - \delta_\epsilon(n). \qquad (4.53)\end{aligned}$$

Hence, if we set the output size of the hash function k to be less than the lower bound in (4.53) by $\sqrt{n}$, say

$$k \triangleq \lfloor n(\beta\mathbb{I}(X;Y) - \mathbb{I}(X;Z) - \delta(\epsilon)) - 3\sqrt{n} - 2 - \delta_\epsilon(n)\rfloor, \qquad (4.54)$$

then Theorem 4.4 ensures that

$$\begin{aligned}\mathbb{H}(K|GZ^n = z^n, A^r = a^r, \Theta = 1, \Upsilon = 1) &\geqslant k - \frac{2^{-\sqrt{n}}}{\ln 2} \\ &= k - \delta(n). \qquad (4.55)\end{aligned}$$

On substituting (4.55) back into (4.49), we finally obtain

$$\mathbb{H}(K|GZ^nA^r) \geqslant \left(1 - 2\cdot 2^{-\sqrt{n}}\right)(k - \delta(n)) = k - \delta(n), \qquad (4.56)$$

and, consequently,

$$\mathbb{I}(K;GZ^nA^r) = \mathbb{H}(K) - \mathbb{H}(K|GZ^nA^r) \leqslant \delta(n).$$

Notice that the corresponding secret-key rate is

$$R \triangleq \frac{k}{n} = \beta\mathbb{I}(X;Y) - \mathbb{I}(X;Z) - \delta(\epsilon) - \delta(n). \qquad (4.57)$$

Hence, we have proved the existence of a $(2^{nR}, n)$ sequential key-distillation strategy $\mathcal{S}_n$ with rate given by (4.57) and based on a direct reconciliation protocol of efficiency β and privacy amplification with a universal family of hash functions such that

$$\mathbf{P}_e(\mathcal{S}_n) \leqslant \delta_\epsilon(n), \quad \mathbf{L}(\mathcal{S}_n) \leqslant \delta_\epsilon(n), \quad \text{and} \quad \mathbf{U}(\mathcal{S}_n) \leqslant \delta_\epsilon(n).$$

Hence, $\beta\mathbb{I}(X;Y) - \mathbb{I}(X;Z) - \delta(\epsilon)$ is an achievable strong secret-key rate. Since ϵ can be chosen arbitrarily small, all strong secret-key rates below $\beta\mathbb{I}(X;Y) - \mathbb{I}(X;Z)$ are achievable, as well.

The achievability of all strong secret-key rates below $\beta\mathbb{I}(X;Y) - \mathbb{I}(Y;Z)$ follows from the same arguments on reversing the roles of Alice and Bob and considering a reverse reconciliation protocol. □

Although both direct and reverse reconciliation protocols can operate close to the reconciliation capacity R^{SM}, the secret-key rates obtained after privacy amplification are different. In the proof of Theorem 4.5, the secret-key rates below $\beta\mathbb{I}(X;Y) - \mathbb{I}(X;Z)$ are achievable with a direct reconciliation protocol, whereas the secret-key rates below $\beta\mathbb{I}(X;Y) - \mathbb{I}(Y;Z)$ are achievable with a reverse reconciliation protocol. Intuitively, a direct reconciliation protocol uses Alice's observations as the reference to distill the key and the information leaked to Eve is therefore $\mathbb{I}(X;Z)$. In contrast, a reverse reconciliation uses Bob's observations as the reference and the information leaked to Eve is then $\mathbb{I}(Y;Z)$.

Remark 4.10. *Privacy amplification with a universal family of hash functions is a special case of privacy amplification with an almost universal family of hash functions. For $\gamma \geqslant 1$, a γ-almost universal family $\mathcal{G}$ of hash functions $g{:}\mathcal{A} \rightarrow \mathcal{B}$ is such that*

$$\forall x_1, x_2 \in \mathcal{A} \quad x_1 \neq x_2 \Rightarrow \mathbb{P}_{\mathrm{G}}[\mathrm{G}(x_1) = \mathrm{G}(x_2)] \leqslant \frac{\gamma}{|\mathcal{B}|},$$

where G *is the random variable that represents the choice of a function g uniformly at random in $\mathcal{G}$. It is possible to show that sequential key-distillation strategies based on γ-almost universal families of hash functions can achieve all strong secret-key rates below*

$$\beta\mathbb{I}(\mathrm{X};\mathrm{Y}) - \min(\mathbb{I}(\mathrm{X};\mathrm{Z}), \mathbb{I}(\mathrm{Y};\mathrm{Z})) - \log\gamma,$$

which is lower than the rate given in Theorem 4.5 if $\gamma > 1$. Nevertheless, almost-universal families of hash functions might be preferred to universal families of hash functions in practice because of their greater flexibility and lower complexity. A more detailed discussion of privacy amplification based on almost-universal families of hash functions can be found in the textbook of Van Assche [38].

Corollary 4.2. *The strong secret-key capacity $\overline{C}_{\mathrm{s}}^{\mathrm{SM}}$ of a source model with DMS $(\mathcal{XYZ}, p_{\mathrm{XYZ}})$ is equal to its weak secret-key capacity $C_{\mathrm{s}}^{\mathrm{SM}}$. In addition, all strong secret-key rates below $\overline{C}_{\mathrm{s}}^{\mathrm{SM}}$ are achievable with sequential key-distillation strategies.*

Proof. This result follows directly from Proposition 4.2 and Theorem 4.5. According to Proposition 4.2, an advantage-distillation protocol $\mathcal{D}_n$ can be used to transform n realizations of the original DMS $(\mathcal{XYZ}, p_{\mathrm{XYZ}})$ into a single realization of a new DMS $(\mathcal{X}'\mathcal{Y}'\mathcal{Z}', p_{\mathrm{X'Y'Z'}})$ for which

$$\mathbb{I}(\mathrm{X}';\mathrm{Y}') - \mathbb{I}(\mathrm{X}';\mathrm{Z}') > 0 \quad \text{or} \quad \mathbb{I}(\mathrm{X}';\mathrm{Y}') - \mathbb{I}(\mathrm{Y}';\mathrm{Z}') > 0.$$

For this new source and any $\epsilon > 0$, Theorem 4.5 ensures that the strong secret-key rate

$$R_{\mathrm{s}}' = \max\left(\mathbb{I}(\mathrm{X}';\mathrm{Y}') - \mathbb{I}(\mathrm{X}';\mathrm{Z}'), \mathbb{I}(\mathrm{X}';\mathrm{Y}') - \mathbb{I}(\mathrm{Y}';\mathrm{Z}')\right) - \delta(\epsilon)$$

is achievable, where R_{s}' is expressed in bits per observation of the DMS $(\mathcal{X}'\mathcal{Y}'\mathcal{Z}', p_{\mathrm{X'Y'Z'}})$. The rate in bits per observation of the original DMS $(\mathcal{XYZ}, p_{\mathrm{XYZ}})$ is then

$$\begin{aligned} R_{\mathrm{s}} &= \frac{1}{n}\max\left(\mathbb{I}(\mathrm{X}';\mathrm{Y}') - \mathbb{I}(\mathrm{X}';\mathrm{Z}'), \mathbb{I}(\mathrm{X}';\mathrm{Y}') - \mathbb{I}(\mathrm{Y}';\mathrm{Z}')\right) - \delta(\epsilon) \\ &= \mathbf{R}(\mathcal{D}_n) - \delta(\epsilon), \end{aligned}$$

which can be made as close as desired to the advantage-distillation capacity D^{SM}. By Proposition 4.2, $D^{\mathrm{SM}} = C_{\mathrm{s}}^{\mathrm{SM}}$ and, because R_{s} is a strong secret-key rate, we obtain $\overline{C}_{\mathrm{s}}^{\mathrm{SM}} \geqslant C_{\mathrm{s}}^{\mathrm{SM}}$ and hence $\overline{C}_{\mathrm{s}}^{\mathrm{SM}} = C_{\mathrm{s}}^{\mathrm{SM}}$. □

Theorem 4.5 and Corollary 4.2 have far-reaching implications. First, the fact that the strong secret-key capacity $\overline{C}_{\mathrm{s}}^{\mathrm{SM}}$ is equal to the weak secret-key capacity $C_{\mathrm{s}}^{\mathrm{SM}}$ is reassuring because it suggests that the fundamental limit of secret-key generation is fairly robust with respect to the actual choice of secrecy condition. Second, since sequential-key distillation strategies can achieve all rates below $C_{\mathrm{s}}^{\mathrm{SM}}$, we know that there is no loss of optimality in handling the reliability and secrecy requirements independently. Whether sequential key-distillation strategies can achieve secret-key rates close to C_s^{SM} depends on our ability to design advantage-distillation protocols, but at least we have an explicit procedure for privacy amplification, and we will see in Chapter 6 how to design efficient reconciliation protocols. Finally, the proof of Corollary 4.2 highlights the importance of two-way communications for secret-key agreement as a means to distill an advantage over the eavesdropper. Two-way communication is required for advantage distillation, although reconciliation and privacy amplification can be implemented with one-way communication only.

Privacy amplification with extractors

We conclude our study of privacy amplification by discussing an alternative to universal families of hash functions and by analyzing privacy amplification with a class of functions called *extractors*. In essence, the analysis and the results are identical to those obtained earlier, but they exploit the min-entropy in place of the collision entropy. The only difference between privacy amplification with hash functions and privacy amplification with extractors is the amount of communication over the public channel.

Definition 4.19. *A function $g : \{0, 1\}^n \times \{0, 1\}^d \to \{0, 1\}^k$ is called a (γ, ϵ)-extractor if, for any random variable $S \in \{0, 1\}^n$ with min-entropy $\mathbb{H}_\infty(S) \geqslant \gamma n$ and a random variable U_d uniformly distributed over $\{0, 1\}^d$, the variational distance between the random variable $(U_d, g(S, U_d)) \in \{0, 1\}^{d+k}$ and the random variable U_{d+k} with uniform distribution over $\{0, 1\}^{d+k}$ satisfies*

$$\mathbb{V}((U_d, g(S, U_d)), U_{d+k}) \leqslant \epsilon.$$

In other words, an extractor is a function that converts a sequence of n bits with arbitrary distribution into a sequence of k bits with almost uniform distribution using d bits of randomness as a catalyst. If $d \leqslant k$, that is the extractor outputs more uniform randomness than is used at the input, this operation can be thought of as a way of "extracting" uniform randomness from the random variable S. In practice, an extractor is useful if $d \ll k$, which means that little extra randomness is necessary for the extraction. The existence of such extractors is guaranteed by the following proposition, which we state without proof.

Proposition 4.8 (Vadhan). *For any $\epsilon > 0$, $\gamma \in (0, 1)$, there exists a (γ, ϵ)-extractor $g : \{0, 1\}^n \times \{0, 1\}^d \to \{0, 1\}^k$ with*

$$k = \gamma n - 2 \log\left(\frac{1}{\epsilon}\right) - O(1)$$

and

$$d = O\left(\left(\log\left(\frac{n}{\epsilon}\right)\right)^2 \log(\gamma n)\right).$$

In other words, for n large enough, there exist extractors that extract almost the entire min-entropy of the input S and require a comparatively *negligible* amount of uniform randomness. The existence of such extractors allows us to prove the counterpart of Theorem 4.4 with extractors in place of hash functions.

Theorem 4.6 (Maurer and Wolf). *Let $\mathsf{S} \in \{0, 1\}^n$ be the random variable that represents the common sequence shared by Alice and Bob, and let E be the random variable that represents the total knowledge about S available to Eve. Let e be a particular realization of E. If Alice and Bob know the conditional min-entropy $\mathbb{H}_\infty(\mathsf{S}|\mathsf{E} = e)$ to be at least γn for some $\gamma \in (0, 1)$, then there exists a function $g : \{0, 1\}^n \times \{0, 1\}^d \to \{0, 1\}^k$ with*

$$d \leqslant n\delta(n) \quad \text{and} \quad k \geqslant n(\gamma - \delta(n)),$$

such that, if U_d is a random variable with uniform distribution on $\{0, 1\}^d$ and Alice and Bob choose $\mathsf{K} = g(\mathsf{S}, \mathsf{U}_d)$ as their secret key, then

$$\mathbb{H}(\mathsf{K}|\mathsf{U}_d, \mathsf{E} = e) \geqslant k - \delta(n).$$

Proof. Let $\epsilon \triangleq 2^{-\sqrt{n}/\log n}$ and let $\mathsf{K}_e \in \{0, 1\}^k$ be the random variable with distribution

$$p_{\mathsf{K}_e} \triangleq p_{g(\mathsf{S},\mathsf{U}_d)|\mathsf{E}=e}.$$

According to Proposition 4.8, there exists a (γ, ϵ) extractor $g : \{0, 1\}^n \times \{0, 1\}^d \to \{0, 1\}^k$ with

$$d = O\left(\left(\log\left(\frac{n}{\epsilon}\right)\right)^2 \log(\gamma n)\right) = n\delta(n),$$

$$k = \gamma n - 2\log\left(\frac{1}{\epsilon}\right) - O(1) = n(\gamma - \delta(n)),$$

and such that $\mathbb{V}((\mathsf{U}_d, \mathsf{K}_e), \mathsf{U}_{d+k}) \leqslant 2^{-\sqrt{n}/\log n}$. Note that we can write the uniform distribution $p_{\mathsf{U}_{d+k}}$ over $\{0, 1\}^{d+k}$ as the product of the uniform distribution p_{U_d} over $\{0, 1\}^d$ with the uniform distribution p_{U_k} over $\{0, 1\}^k$; hence,

$$\begin{aligned}\mathbb{V}((\mathsf{U}_d, \mathsf{K}_e), \mathsf{U}_{d+k}) &= \sum_{u,s} \left|p_{\mathsf{U}_d}(u)p_{\mathsf{K}_e|\mathsf{U}_d}(s|u) - p_{\mathsf{U}_d}(u)p_{\mathsf{U}_k}(s)\right| \\ &= \sum_{u,s} p_{\mathsf{U}_d}(u)\left|p_{\mathsf{K}_e|\mathsf{U}_d}(s|u) - 2^{-k}\right| \\ &= \mathbb{E}_{\mathsf{U}_d}\left[\sum_s \left|p_{\mathsf{K}_e|\mathsf{U}_d}(s|\mathsf{U}_d) - 2^{-k}\right|\right].\end{aligned}$$

By Markov's inequality,

$$\begin{aligned}\mathbb{P}_{\mathsf{U}_d}\left[\sum_s \left|p_{\mathsf{K}_e|\mathsf{U}_d}(s|\mathsf{U}_d) - 2^{-k}\right| \geqslant 2^{-\sqrt{n}/(2\log n)}\right] &\leqslant \frac{\mathbb{E}_{\mathsf{U}_d}\left[\sum_s \left|p_{\mathsf{K}_e|\mathsf{U}_d}(s|\mathsf{U}_d) - 2^{-k}\right|\right]}{2^{-\sqrt{n}/(2\log n)}} \\ &\leqslant 2^{-\sqrt{n}/(2\log n)}.\end{aligned} \tag{4.58}$$

In other words, with high probability, the realization u_d of U_d is such that the variational distance between $p_{\mathsf{K}_e|\mathsf{U}_d}$ and a uniform distribution over $\{0, 1\}^k$ is small. Formally, if we define the random variable Θ function of U_d as

$$\Theta = \begin{cases} 1 & \text{if } \mathbb{V}((\mathsf{U}_d, \mathsf{K}_e), \mathsf{U}_{d+k}) \leqslant 2^{-\sqrt{n}/(2\log n)}, \\ 0 & \text{otherwise.} \end{cases}$$

then, by (4.58),

$$\mathbb{P}[\Theta = 1] \geqslant 1 - 2^{-\sqrt{n}/\log n}. \tag{4.59}$$

We now lower bound $\mathbb{H}(\mathsf{K}|\mathsf{U}_d, \mathsf{E} = e)$ as

$$\begin{aligned} \mathbb{H}(\mathsf{K}|\mathsf{U}_d, \mathsf{E} = e) &\geqslant \mathbb{H}(\mathsf{K}|\mathsf{U}_d, \Theta, \mathsf{E} = e) \\ &\geqslant \mathbb{P}[\Theta = 1]\mathbb{H}(\mathsf{K}_e|\mathsf{U}_d, \Theta = 1). \end{aligned} \tag{4.60}$$

Given $\Theta = 1$, the variational distance $\mathbb{V}(\mathsf{U}_d, \mathsf{K}_e; \mathsf{U}_{d+k})$ is less than $2^{-\sqrt{n}/\log n}$ and note that the function $x \mapsto x \log(2^k/x)$ is increasing for $x \in (0, 2^{k-1})$; therefore, for n large enough, we have, according to Proposition 2.1,

$$\begin{aligned} |\mathbb{H}(\mathsf{K}_e|\mathsf{U}_d, \Theta = 1) - k| &\leqslant 2^{-\sqrt{n}/\log n} \log\left(\frac{2^k}{2^{-\sqrt{n}/\log n}}\right) \\ &= 2^{-\sqrt{n}/\log n}\left(k + \frac{\sqrt{n}}{\log n}\right) \\ &= \delta(n). \end{aligned} \tag{4.61}$$

On combining (4.59) and (4.61) in (4.60), we finally obtain

$$\mathbb{H}(\mathsf{K}|\mathsf{U}_d, \mathsf{E} = e) \geqslant k - \delta(n),$$

which is the desired result. □

We can now establish the counterpart of Theorem 4.5.

Theorem 4.7. *Consider a source model with DMS $(\mathcal{X}\mathcal{Y}\mathcal{Z}, p_{\mathsf{XYZ}})$ and let $\beta \in [0, 1]$. All strong secret-key rates R_s that satisfy*

$$R_s < \beta\mathbb{I}(\mathsf{X}; \mathsf{Y}) - \min(\mathbb{I}(\mathsf{X}; \mathsf{Z}), \mathbb{I}(\mathsf{Y}; \mathsf{Z}))$$

are achievable with sequential secret-key distillation strategies that consist of a reconciliation protocol with efficiency β and privacy amplification with extractors. Additionally, these rates are achievable with one-way communication.

Sketch of proof. The proof of Theorem 4.7 follows exactly that of Theorem 4.5. The only difference is the use of extractors instead of hash functions, which requires the use of the min-entropy instead of the collision entropy. Notice that Lemma 4.4 already establishes a lower bound for the min-entropy, hence we need just the counterpart of Lemma 4.3 for the min-entropy. As shown in the appendix to this chapter, the following result holds.

Lemma 4.5. *Let $\mathsf{S} \in \mathcal{S}$ and $\mathsf{U} \in \mathcal{U}$ be two discrete random variables with joint distribution p_{SU}. For any $r > 0$, define the function $\chi : \mathcal{U} \to \{0, 1\}$ as*

$$\chi(u) \triangleq \begin{cases} 1 & \text{if } \mathbb{H}_\infty(\mathsf{S}) - \mathbb{H}_\infty(\mathsf{S}|u) \leqslant \log|\mathcal{U}| + r, \\ 0 & \text{otherwise.} \end{cases}$$

Then, $\mathbb{P}_{\mathsf{U}}[\chi(\mathsf{U}) = 1] \geqslant 1 - 2^{-r}$.

Essentially, Lemma 4.5 states that, with high probability, conditioning on U reduces the min-entropy by at most on the order of $\log|\mathcal{U}|$. This allows us to relate the min-entropy before and after reconciliation independently of the exact operation of the reconciliation protocol. □

4.4 Secret-key capacity of the channel model

In this section, we turn our attention to the channel model for secret-key agreement and study the secret-key capacity of a channel model. The secret-key capacity of a channel model is sometimes called the *secrecy capacity with public discussion*, because a channel model can be viewed as a WTC enhanced by a public channel between Alice and Bob. However, this denomination is slightly misleading because, in contrast to the secrecy capacity, the secrecy capacity with public discussion characterizes a secret-key rate, not a secure message rate. In this book, we restrict ourselves to the term secret-key capacity and it will be clear from the context whether this refers to a source model or a channel model.

Definition 4.20. *The weak secret-key capacity of a channel model with DMC* $(\mathcal{X}, p_{\text{YZ|X}}, \mathcal{Y}, \mathcal{Z})$ *is*

$$C_{\text{s}}^{\text{CM}} \triangleq \sup\{R : R \text{ is an achievable weak secret-key rate}\}.$$

Similarly, the strong secret-key capacity is

$$\overline{C}_{\text{s}}^{\text{CM}} \triangleq \sup\{R : R \text{ is an achievable strong secret-key rate}\}.$$

It follows from the definition of achievable rates that $C_{\text{s}}^{\text{CM}} \leqslant \overline{C}_{\text{s}}^{\text{CM}}$. We prove in Section 4.5 that $C_{\text{s}}^{\text{CM}} = \overline{C}_{\text{s}}^{\text{CM}}$, but until then we focus on the weak secret-key capacity. Note that the definition of key-distillation strategies for the channel model is so broad that an exact characterization of the secret-key capacity for a general channel model seems out of reach. Nevertheless, it is possible to develop upper and lower bounds similar to those developed in Theorem 4.1 for the secret-key capacity.

Theorem 4.8 (Ahlswede and Csiszár). *The secret-key capacity* C_{s}^{CM} *of a channel model satisfies*

$$\max\left(\max_{p_{\text{X}}}(\mathbb{I}(\text{X};\text{Y}) - \mathbb{I}(\text{X};\text{Z})), \max_{p_{\text{X}}}(\mathbb{I}(\text{X};\text{Y}) - \mathbb{I}(\text{Y};\text{Z}))\right) \leqslant C_{\text{s}}^{\text{CM}} \leqslant \max_{p_{\text{X}}} \min\left(\mathbb{I}(\text{X};\text{Y}), \mathbb{I}(\text{X};\text{Y}|\text{Z})\right).$$

Proof. We derive the lower bound by considering a specific key-distillation strategy, in which the the input X^n is chosen i.i.d. according to an arbitrary distribution p_{X}. For this choice of input, the channel model becomes a source model with DMS $(\mathcal{XYZ}, p_{\text{XYZ}})$

whose secret-key capacity is $C_s^{\text{SM}}(p_{XYZ})$. Therefore,

$$C_s^{\text{CM}} \geqslant \max_{p_X} C_s^{\text{SM}}(p_{XYZ})$$
$$\geqslant \max\left(\max_{p_X}(\mathbb{I}(X;Y) - \mathbb{I}(X;Z)), \max_{p_X}(\mathbb{I}(X;Y) - \mathbb{I}(Y;Z))\right),$$

where the second inequality follows from Theorem 4.1.

We establish the upper bound with a converse argument similar to that used in the proof of Theorem 4.1; the proof is more technical because the inputs of the channel depend on previously exchanged messages and past inputs. Let R be an achievable secret-key rate and let $\epsilon > 0$. For n sufficiently large, there exists a $(2^{nR}, n)$ key-distillation strategy $\mathcal{S}_n$ such that

$$\mathbf{P}_e(\mathcal{S}_n) \leqslant \delta(\epsilon), \quad \frac{1}{n}\mathbf{L}(\mathcal{S}_n) \leqslant \delta(\epsilon), \quad \text{and} \quad \frac{1}{n}\mathbf{U}(\mathcal{S}_n) \leqslant \delta(\epsilon). \tag{4.62}$$

In the following, we omit conditioning on $\mathcal{S}_n$ in order to simplify the notation. Fano's inequality combined with (4.62) ensures that

$$\frac{1}{n}\mathbb{H}(K|\hat{K}A^rB^rZ^n) \leqslant \frac{1}{n}\mathbb{H}(K|\hat{K}) \leqslant \delta(\mathbf{P}_e(\mathcal{S}_n)) \leqslant \delta(\epsilon).$$

We also introduce the random variables Λ_j for $j \in [\![0, n]\!]$, which represent all public messages exchanged *after* Y_j and Z_j have been received but *before* Y_{j+1} and Z_{j+1} are received:

$$\Lambda_0 \triangleq (A_1, \ldots, A_{i_1-1}, B_1, \ldots, B_{i_1-1}),$$
$$\Lambda_j \triangleq (A_{i_j+1}, \ldots, A_{i_{j+1}-1}, B_{i_j+1}, \ldots, B_{i_{j+1}-1}) \quad \text{for } j \in [\![1, n]\!].$$

With this definition, note that $A^rB^r = \Lambda_0\Lambda^n$. Following the steps used to prove (4.20) for the source model, we can show that

$$R \leqslant \frac{1}{n}\mathbb{I}(X^nR_X; Y^nR_Y|A^rB^rZ^n) + \delta(\epsilon)$$
$$= \frac{1}{n}\mathbb{I}(X^nR_X; Y^nR_Y|\Lambda_0\Lambda^nZ^n) + \delta(\epsilon). \tag{4.63}$$

However, in contrast to the case of the source model, $X_j = h_j(B^{i_j-1}, R_X)$ is a function of B^{i_j-1} and R_X; therefore, the inequality simplifies to

$$R \leqslant \frac{1}{n}\mathbb{I}(R_X; Y^nR_Y|\Lambda_0\Lambda^nZ^n) + \delta(\epsilon).$$

Next, we expand $\mathbb{I}(R_X; Y^nR_Y|\Lambda_0\Lambda^nZ^n)$ as

$$\mathbb{I}(R_X; Y^nR_Y|\Lambda_0\Lambda^nZ^n)$$
$$= \mathbb{I}(R_X; Y^nR_YZ^n\Lambda_0\Lambda^n) - \mathbb{I}(R_X; Z^n\Lambda_0\Lambda^n)$$
$$= \mathbb{I}(R_X; R_Y\Lambda_0) - \mathbb{I}(R_X; \Lambda_0) + \mathbb{I}(R_X; Y^nZ^n\Lambda^n|R_Y\Lambda_0) - \mathbb{I}(R_X; Z^n\Lambda^n|\Lambda_0),$$
$$= \mathbb{I}(R_X; R_Y|\Lambda_0) + \mathbb{I}(R_X; Y^nZ^n\Lambda^n|R_Y\Lambda_0) - \mathbb{I}(R_X; Z^n\Lambda^n|\Lambda_0) \tag{4.64}$$

and study each term separately.

Since $\Lambda_0 = A^{i_1-1}B^{i_1-1}$, we can apply Lemma 4.2 to $\mathbb{I}(R_X; R_Y|\Lambda_0)$ with $S \triangleq R_X$, $T \triangleq R_Y$, $U \triangleq 0$, $r \triangleq i_1 - 1$, $V_i \triangleq A_i$, and $W_i \triangleq B_i$ to obtain

$$\mathbb{I}(R_X; R_Y|\Lambda_0) \leqslant \mathbb{I}(R_X; R_Y) = 0. \tag{4.65}$$

Next, notice that

$$\begin{aligned}\mathbb{I}(R_X; Y^n Z^n \Lambda^n | R_Y \Lambda_0) &= \sum_{j=1}^{n} \mathbb{I}\left(R_X; Z_j Y_j \Lambda_j | R_Y \Lambda_0 Z^{j-1} Y^{j-1} \Lambda^{j-1}\right) \\ &= \sum_{j=1}^{n} \left(\mathbb{I}\left(R_X; Z_j Y_j | R_Y \Lambda_0 Z^{j-1} Y^{j-1} \Lambda^{j-1}\right)\right. \\ &\quad \left.+ \mathbb{I}\left(R_X; \Lambda_j | R_Y \Lambda_0 Z^j Y^j \Lambda^{j-1}\right)\right),\end{aligned} \tag{4.66}$$

and, similarly,

$$\mathbb{I}(R_X; Z^n \Lambda^n | \Lambda_0) = \sum_{j=1}^{n} \left(\mathbb{I}\left(R_X; Z_j | \Lambda_0 Z^{j-1} \Lambda^{j-1}\right) + \mathbb{I}\left(R_X; \Lambda_j | \Lambda_0 Z^j \Lambda^{j-1}\right)\right). \tag{4.67}$$

On substituting (4.65), (4.66), and (4.67) into (4.64), we obtain

$$\begin{aligned}&\mathbb{I}(R_X; Y^n R_Y | \Lambda_0 \Lambda^n Z^n) \\ &\quad\leqslant \sum_{j=1}^{n} \left(\mathbb{I}\left(R_X; Z_j Y_j | R_Y \Lambda_0 Z^{j-1} Y^{j-1} \Lambda^{j-1}\right) - \mathbb{I}\left(R_X; Z_j | \Lambda_0 Z^{j-1} \Lambda^{j-1}\right)\right) \\ &\qquad + \sum_{j=1}^{n} \left(\mathbb{I}\left(R_X; \Lambda_j | R_Y \Lambda_0 Z^j Y^j \Lambda^{j-1}\right) - \mathbb{I}\left(R_X; \Lambda_j | \Lambda_0 Z^j \Lambda^{j-1}\right)\right).\end{aligned} \tag{4.68}$$

We proceed to bound the terms in each sum separately. First,

$$\begin{aligned}&\mathbb{I}\left(R_X; Z_j Y_j | R_Y \Lambda_0 Z^{j-1} Y^{j-1} \Lambda^{j-1}\right) - \mathbb{I}\left(R_X; Z_j | \Lambda_0 Z^{j-1} \Lambda^{j-1}\right) \\ &\quad = \mathbb{H}\left(Z_j Y_j | R_Y \Lambda_0 Z^{j-1} Y^{j-1} \Lambda^{j-1}\right) - \mathbb{H}\left(Z_j Y_j | R_X R_Y \Lambda_0 Z^{j-1} Y^{j-1} \Lambda^{j-1}\right) \\ &\qquad - \mathbb{H}\left(Z_j | \Lambda_0 Z^{j-1} \Lambda^{j-1}\right) + \mathbb{H}\left(Z_j | R_X \Lambda_0 Z^{j-1} \Lambda^{j-1}\right).\end{aligned} \tag{4.69}$$

Since conditioning does not increase entropy, we have

$$\begin{aligned}&\mathbb{H}\left(Z_j Y_j | R_Y \Lambda_0 Z^{j-1} Y^{j-1} \Lambda^{j-1}\right) - \mathbb{H}\left(Z_j | \Lambda_0 Z^{j-1} \Lambda^{j-1}\right) \\ &\qquad\qquad \leqslant \mathbb{H}\left(Z_j Y_j | \Lambda_0 Z^{j-1} \Lambda^{j-1}\right) - \mathbb{H}\left(Z_j | \Lambda_0 Z^{j-1} \Lambda^{j-1}\right) \\ &\qquad\qquad = \mathbb{H}\left(Y_j | Z_j \Lambda_0 Z^{j-1} \Lambda^{j-1}\right) \\ &\qquad\qquad \leqslant \mathbb{H}\left(Y_j | Z_j\right).\end{aligned} \tag{4.70}$$

In addition, since $R_X R_Y \Lambda_0 Z^{j-1} Y^{j-1} \Lambda^{j-1} \to X_j \to Y_j Z_j$ forms a Markov chain and $X_j = h_j(B^{i_j-1}, R_X)$, we have

$$\mathbb{H}\left(Z_j Y_j | R_X R_Y \Lambda_0 Z^{j-1} Y^{j-1} \Lambda^{j-1}\right) = \mathbb{H}\left(Z_j Y_j | X_j\right) \tag{4.71}$$

and

$$\mathbb{H}\left(Z_j | R_X \Lambda_0 Z^{j-1} \Lambda^{j-1}\right) = \mathbb{H}\left(Z_j | X_j\right). \tag{4.72}$$

On substituting (4.70), (4.71), and (4.72) into (4.69), we obtain

$$\begin{aligned}\mathbb{I}(\mathsf{R_X};\mathsf{Z}_j\mathsf{Y}_j|\mathsf{R_Y}\Lambda_0\mathsf{Z}^{j-1}\mathsf{Y}^{j-1}\Lambda^{j-1}) - \mathbb{I}(\mathsf{R_X};\mathsf{Z}_j|\Lambda_0\mathsf{Z}^{j-1}\Lambda^{j-1}) \\ \leqslant \mathbb{H}(\mathsf{Y}_j|\mathsf{Z}_j) - \mathbb{H}(\mathsf{Z}_j\mathsf{Y}_j|\mathsf{X}_j) + \mathbb{H}(\mathsf{Z}_j|\mathsf{X}_j) \\ = \mathbb{I}(\mathsf{X}_j;\mathsf{Y}_j|\mathsf{Z}_j).\end{aligned} \tag{4.73}$$

We now turn our attention to the terms in the second sum of (4.68):

$$\begin{aligned}&\mathbb{I}(\mathsf{R_X};\Lambda_j|\mathsf{R_Y}\Lambda_0\mathsf{Z}^j\mathsf{Y}^j\Lambda^{j-1}) - \mathbb{I}(\mathsf{R_X};\Lambda_j|\Lambda_0\mathsf{Z}^j\Lambda^{j-1})\\ &\quad= \mathbb{I}(\mathsf{R_X};\Lambda_j\mathsf{R_Y}\mathsf{Y}^j|\Lambda_0\mathsf{Z}^j\Lambda^{j-1}) - \mathbb{I}(\mathsf{R_X};\mathsf{R_Y}\mathsf{Y}^j|\Lambda_0\mathsf{Z}^j\Lambda^{j-1}) - \mathbb{I}(\mathsf{R_X};\Lambda_j|\Lambda_0\mathsf{Z}^j\Lambda^{j-1})\\ &\quad= \mathbb{I}(\mathsf{R_X};\Lambda_j|\Lambda_0\mathsf{Z}^j\Lambda^{j-1}) + \mathbb{I}(\mathsf{R_X};\mathsf{R_Y}\mathsf{Y}^j|\Lambda_0\mathsf{Z}^j\Lambda^{j}) - \mathbb{I}(\mathsf{R_X};\mathsf{R_Y}\mathsf{Y}^j|\Lambda_0\mathsf{Z}^j\Lambda^{j-1})\\ &\qquad - \mathbb{I}(\mathsf{R_X};\Lambda_j|\Lambda_0\mathsf{Z}^j\Lambda^{j-1})\\ &\quad= \mathbb{I}(\mathsf{R_X};\mathsf{R_Y}\mathsf{Y}^j|\Lambda_0\mathsf{Z}^j\Lambda^{j}) - \mathbb{I}(\mathsf{R_X};\mathsf{R_Y}\mathsf{Y}^j|\Lambda_0\mathsf{Z}^j\Lambda^{j-1}).\end{aligned}$$

By applying again Lemma 4.2 with $\mathsf{S} \triangleq \mathsf{R_X}$, $\mathsf{T} \triangleq \mathsf{R_Y}\mathsf{Y}^j$, $\mathsf{U} \triangleq \Lambda_0\mathsf{Z}^j$, $r \triangleq j$, $\mathsf{V}_i \triangleq \mathsf{A}_i$, and $W_i \triangleq \mathsf{B}_i$, we obtain

$$\mathbb{I}(\mathsf{R_X};\mathsf{R_Y}\mathsf{Y}_j|\Lambda_0\mathsf{Z}^j\Lambda^{j}) \leqslant \mathbb{I}(\mathsf{R_X};\mathsf{R_Y}\mathsf{Y}_j|\Lambda_0\mathsf{Z}^j\Lambda^{j-1})$$

and, therefore,

$$\mathbb{I}(\mathsf{R_X};\Lambda_j|\mathsf{R_Y}\Lambda_0\mathsf{Z}^j\mathsf{Y}^j\Lambda^{j-1}) - \mathbb{I}(\mathsf{R_X};\Lambda_j|\Lambda_0\mathsf{Z}^j\Lambda^{j-1}) \leqslant 0. \tag{4.74}$$

On substituting (4.73) and (4.74) into (4.68), we finally obtain

$$\mathbb{I}(\mathsf{R_X};\mathsf{Y}^n\mathsf{R_Y}|\Lambda^n\mathsf{Z}^n) \leqslant \sum_{j=1}^{n}\mathbb{I}(\mathsf{X}_j;\mathsf{Y}_j|\mathsf{Z}_j) \leqslant n\max_{p_\mathsf{X}}\mathbb{I}(\mathsf{X};\mathsf{Y}|\mathsf{Z}). \tag{4.75}$$

Hence,

$$R \leqslant \sum_{j=1}^{n}\mathbb{I}(\mathsf{X}_j;\mathsf{Y}_j|\mathsf{Z}_j) + \delta(\epsilon).$$

It remains to show that $R \leqslant \sum_{j=1}^{n}\mathbb{I}(\mathsf{X}_j;\mathsf{Y}_j) + \delta(\epsilon)$. Since

$$\frac{1}{n}\mathbb{H}(\mathsf{K}|\hat{\mathsf{K}}\mathsf{A}^r\mathsf{B}^r) \leqslant \frac{1}{n}\mathbb{H}(\mathsf{K}|\hat{\mathsf{K}}) \leqslant \delta(\epsilon),$$

it suffices to reiterate all the steps leading to (4.75) without the conditioning on Z^n to obtain the desired result. □

Even though Theorem 4.8 does not characterize the secret-key capacity exactly, it provides simple bounds that do not depend on auxiliary random variables.

Example 4.7. For some channel models, the secret-key capacity is even equal to the secrecy capacity. The simplest example of such a situation is a channel model in which the DMC $(\mathcal{X}, p_{\mathsf{YZ}|\mathsf{X}}, \mathcal{Y}, \mathcal{Z})$ is physically degraded. In this case $\mathsf{X} \rightarrow \mathsf{Y} \rightarrow \mathsf{Z}$ forms a Markov chain and

$$C_\text{s}^\text{CM} = \max_{p_\mathsf{X}}\mathbb{I}(\mathsf{X};\mathsf{Y}|\mathsf{Z}) = C_\text{s}.$$

In other words, using a wiretap code is an optimal key-distillation strategy for a channel model in which the eavesdropper's channel is physically degraded with respect to the main channel.

Remark 4.11. *The converse technique used in the proof of Theorem 4.8 can also be used to derive simple bounds for secure rates over WTCs; in fact, the secret-key capacity is always an upper bound for the secrecy capacity since a wiretap code is a special case of key-distillation strategy for a channel model. As an application, we revisit the WTC with confidential rate-limited feedback, for which we computed achievable rates in Section 3.6.2. Following the steps used to establish* (4.20), *we can show that the secret-key capacity with public discussion of the WTC with confidential rate-limited feedback satisfies*

$$\frac{1}{n}\mathbb{H}(\mathsf{K}) \leqslant \frac{1}{n}\mathbb{I}(\mathsf{R_X F}^n; \mathsf{R_Y Y}^n|\mathsf{Z}^n \mathsf{A}^r \mathsf{B}^r) + \delta(\epsilon).$$

The only difference from (4.63) *is the presence of the term* F^n *representing the messages sent over the confidential feedback channel. Nevertheless, we have*

$$\begin{aligned}\frac{1}{n}\mathbb{I}(\mathsf{R_X F}^n; \mathsf{R_Y Y}^n|\mathsf{Z}^n \mathsf{A}^r \mathsf{B}^r) &= \frac{1}{n}\mathbb{I}(\mathsf{F}^n; \mathsf{R_Y Y}^n|\mathsf{A}^r \mathsf{B}^r \mathsf{Z}^n) + \frac{1}{n}\mathbb{I}(\mathsf{R_X}; \mathsf{R_Y Y}^n|\mathsf{A}^r \mathsf{B}^r \mathsf{F}^n \mathsf{Z}^n)\\ &\leqslant \frac{1}{n}\mathbb{H}(\mathsf{F}^n) + \frac{1}{n}\mathbb{I}(\mathsf{R_X}; \mathsf{R_Y Y}^n|\mathsf{A}^r \mathsf{B}^r \mathsf{F}^n \mathsf{Z}^n).\end{aligned}$$

The second term on the right-hand side is similar to (4.63)*, since the messages sent over the secure feedback channel appear along* A^r *and* B^r *and can now be interpreted as public messages. Following the same steps and using the fact that* $(1/n)\mathbb{H}(\mathsf{F}^n) \leqslant R_\text{f}$, *we obtain the upper bound*

$$\frac{1}{n}\mathbb{H}(\mathsf{K}) \leqslant R_\text{f} + \sum_{j=1}^{n}\mathbb{I}(\mathsf{X}_j; \mathsf{Y}_j|\mathsf{Z}_j).$$

This outer bound coincides with the achievable rate obtained in Proposition 3.9 if the channel is physically degraded.

4.5 Strong secrecy from weak secrecy

In this section, we develop a generic mathematical procedure by which to construct a scheme (wiretap code or key-distillation strategy) that guarantees a strong secrecy condition from a scheme that guarantees a weak secrecy condition. We show that this procedure entails no rate loss, which allows us to prove that $C_\text{s}^\text{CM} = \overline{C}_\text{s}^\text{CM}$ for a channel model and $C_\text{s} = \overline{C}_\text{s}$ for a wiretap channel and justifies a posteriori the use of a weak secrecy condition in previous chapters.

Proposition 4.9. *The strong secret-key capacity* $\overline{C}_\text{s}^\text{CM}$ *of a channel model with DMC* $(\mathcal{X}, p_{\mathsf{YZ}|\mathsf{X}}, \mathcal{Y}, \mathcal{Z})$ *is equal to its weak secret-key capacity* C_s^CM.

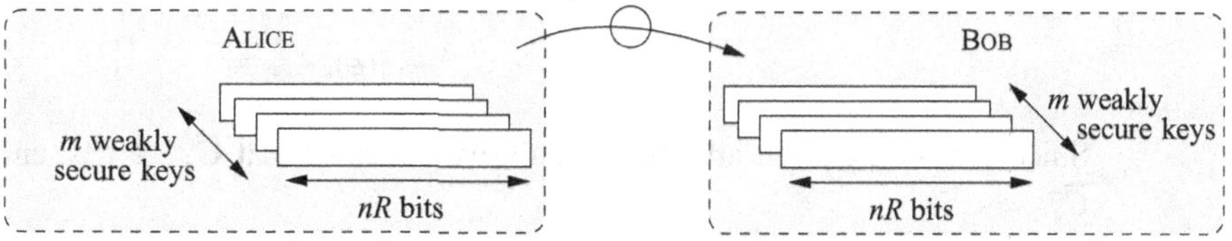

Figure 4.8 From weak to strong secrecy.

Proof. Consider a channel model with DMC $(\mathcal{X}, p_{\mathsf{YZ}|\mathsf{X}}, \mathcal{Y}, \mathcal{Z})$ and a $(2^{nR}, n)$ key-distillation strategy $\mathcal{S}_n$ that achieves weak secret-key rates. As illustrated in Figure 4.8, we construct a new strategy by

- using the key-distillation strategy $\mathcal{S}_n$ m times to generate m weakly secure keys;
- treating the weakly secure keys as the realizations of a DMS and distilling strong secret keys by means of information reconciliation and privacy amplification with extractors.

Note that the post-processing of the weakly secure keys is possible because keys are not meant to carry any information by themselves and do not need to be known ahead of time.

Formally, let $\epsilon > 0$. By definition, there exists a $(2^{nR}, n)$ key-distillation strategy $\mathcal{S}_n$ with rate $R \geqslant C_{\mathrm{s}}^{\mathrm{CM}} - \epsilon$ such that

$$\mathbf{P}_{\mathrm{e}}(\mathcal{S}_n) \leqslant \delta(\epsilon) \quad \text{and} \quad \frac{1}{n}\mathbf{L}(\mathcal{S}_n) \leqslant \delta(\epsilon).$$

Alice runs $\mathcal{S}_n$ m times to generate m independent keys. In each run $i \in [\![1, m]\!]$, Alice obtains a key K_i, Bob obtains a key $\hat{\mathsf{K}}_i$, and Eve obtains the observations Z_i^n together with public messages A_i^r and B_i^r. Effectively, the situation is as if Alice, Bob, and Eve observed m realizations of a DMS $(\mathcal{X}'\mathcal{Y}'\mathcal{Z}', p_{\mathsf{X}'\mathsf{Y}'\mathsf{Z}'})$ with

$$\mathsf{X}' \triangleq \mathsf{K}, \quad \mathsf{Y}' \triangleq \hat{\mathsf{K}}, \quad \text{and} \quad \mathsf{Z}' \triangleq \mathsf{Z}^n\mathsf{A}^r\mathsf{B}^r.$$

According to Theorem 4.7, Alice and Bob can distill a strong secret key $\overline{\mathsf{K}}$ of length

$$k \triangleq m(\mathbb{I}(\mathsf{X}';\mathsf{Y}') - \mathbb{I}(\mathsf{X}';\mathsf{Z}') - \delta(\epsilon))$$

and such that

$$\mathbb{H}(\overline{\mathsf{K}}|\mathsf{Z}'^m\mathcal{S}_n) \geqslant k - \delta_\epsilon(m)$$

by means of a one-way direct reconciliation protocol and privacy amplification with extractors. Note that

$$\begin{aligned}
\mathbb{I}(\mathsf{X}';\mathsf{Y}') - \mathbb{I}(\mathsf{X}';\mathsf{Z}') &= \mathbb{I}(\mathsf{K};\hat{\mathsf{K}}|\mathcal{S}_n) - \mathbb{I}(\mathsf{K};\mathsf{Z}^n\mathsf{A}^r\mathsf{B}^r|\mathcal{S}_n) \\
&= \mathbb{H}(\mathsf{K}) - \mathbb{H}(\mathsf{K}|\hat{\mathsf{K}}\mathcal{S}_n) - \mathbb{I}(\mathsf{K};\mathsf{Z}^n\mathsf{A}^r\mathsf{B}^r|\mathcal{S}_n) \\
&\geqslant nR - n\delta(\mathbf{P}_{\mathrm{e}}(\mathcal{S}_n)) - n\mathbf{L}(\mathcal{S}_n) \\
&\geqslant nC_{\mathrm{s}}^{\mathrm{CM}} - n\delta(\epsilon).
\end{aligned}$$

Therefore, the rate (in bits per use of the channel $(\mathcal{X}, p_{\mathrm{YZ|X}}, \mathcal{Y}, \mathcal{Z})$) at which the strong secret-key $\overline{\mathrm{K}}$ is generated is

$$\frac{k}{mn} \geqslant C_{\mathrm{s}}^{\mathrm{CM}} - \delta(\epsilon).$$

Since ϵ can be chosen arbitrarily small, we conclude that $\overline{C}_{\mathrm{s}}^{\mathrm{CM}} \geqslant C_{\mathrm{s}}^{\mathrm{CM}}$ and, therefore, $\overline{C}_{\mathrm{s}}^{\mathrm{CM}} = C_{\mathrm{s}}^{\mathrm{CM}}$. □

Proposition 4.10. *The strong secrecy capacity $\overline{C}_{\mathrm{s}}^{\mathrm{WT}}$ of a WTC $(\mathcal{X}, p_{\mathrm{YZ|X}}, \mathcal{Y}, \mathcal{Z})$ is equal to its weak secrecy capacity $C_{\mathrm{s}}^{\mathrm{WT}}$.*

Proof. The proof is similar to that of Proposition 4.9 but the objective is now to transmit messages (instead of generating keys) without relying on a public channel. Let $\epsilon > 0$. By definition, there exists a $(2^{nR}, n)$ code $\mathcal{C}_n$ with rate $R \geqslant C_{\mathrm{s}}^{\mathrm{WT}} - \epsilon$ such that

$$\mathbf{P}_{\mathrm{e}}(\mathcal{C}_n) \leqslant \delta(\epsilon) \quad \text{and} \quad \frac{1}{n}\mathbf{L}(\mathcal{C}_n) \leqslant \delta(\epsilon).$$

Alice uses the code $\mathcal{C}_n$ m times to transmit m independent messages. In each run $i \in [\![1, m]\!]$, Alice transmits a message M_i, Bob obtains a message $\hat{\mathrm{M}}_i$, and Eve obtains the observations Z_i^n. Again, the situation is as if Alice, Bob, and Eve observed m realizations of a DMS $(\mathcal{X}'\mathcal{Y}'\mathcal{Z}', p_{\mathrm{X'Y'Z'}})$ with

$$\mathrm{X}' \triangleq \mathrm{M}, \quad \mathrm{Y}' \triangleq \hat{\mathrm{M}}, \quad \text{and} \quad \mathrm{Z}' \triangleq \mathrm{Z}^n.$$

According to Theorem 4.7, Alice and Bob can distill a strong secret key $\overline{\mathrm{K}}$ of length

$$k \triangleq m(\mathbb{I}(\mathrm{X}';\mathrm{Y}') - \mathbb{I}(\mathrm{X}';\mathrm{Z}') - \delta(\epsilon))$$

and such that

$$\mathbb{H}(\overline{\mathrm{K}}|\mathrm{Z}'^m\mathcal{C}_n) \geqslant k - \delta_\epsilon(m)$$

by means of a one-way direct reconciliation protocol and privacy amplification with extractors. The reader can check that

$$\mathbb{I}(\mathrm{X}';\mathrm{Y}') - \mathbb{I}(\mathrm{X}';\mathrm{Z}') \geqslant nC_{\mathrm{s}}^{\mathrm{WT}} - n\delta(\epsilon).$$

However, because there is no public channel, the messages required in order to perform reconciliation and privacy amplification must be transmitted over the channel; we must also account for these additional channel uses in the calculation of the final key rate.

By Proposition 4.5, there exists a one-way reconciliation protocol that exchanges $m(\mathbb{H}(\mathrm{X}'|\mathrm{Y}') + \delta(\epsilon))$ bits of public messages and that guarantees $\mathbb{P}[\mathrm{X}' \neq \hat{\mathrm{X}}'] \leqslant \delta_\epsilon(m)$. By Theorem 4.6, there exists an extractor that requires the transmission of $m\delta(m)$ bits of uniform randomness in order to distill the key $\overline{\mathrm{K}}$. Alice can transmit these bits to Bob over the main channel $(\mathcal{X}, p_{\mathrm{Y|X}}, \mathcal{Y})$ with an error-correcting code of length m. Let C_{m} denote the capacity of the channel $(\mathcal{X}, p_{\mathrm{Y|X}}, \mathcal{Y})$. From Shannon's channel coding theorem, we know that there exists a code of rate $C_{\mathrm{m}} - \delta(\epsilon)$ that guarantees an average probability of error less than $\delta_\epsilon(m)$. Therefore, the transmission of these additional bits requires

$$m' \triangleq \left\lceil \frac{m(\mathbb{H}(\mathsf{X}'|\mathsf{Y}') + \delta(\epsilon)) + m\delta(m)}{C_{\mathrm{m}} - \delta(\epsilon)} \right\rceil$$

channel uses. By virtue of Fano's inequality,

$$\mathbb{H}(\mathsf{X}'|\mathsf{Y}') = \mathbb{H}(\mathsf{M}|\hat{\mathsf{M}}) \leqslant \mathbb{H}_{\mathrm{b}}(\mathbf{P}_{\mathrm{e}}(\mathcal{C}_n)) + \mathbf{P}_{\mathrm{e}}(\mathcal{C}_n) n C_{\mathrm{s}}^{\mathrm{WT}} = n\delta_\epsilon(n).$$

All in all, the strong secret key $\overline{\mathsf{K}}$ can be generated at a rate

$$\begin{aligned}\frac{k}{mn + m'} &\geqslant \frac{m(nC_{\mathrm{s}}^{\mathrm{WT}} - n\delta(\epsilon))}{mn + \dfrac{m(n\delta_\epsilon(n) + \delta(\epsilon)) + m\delta(m)}{C_{\mathrm{m}} - \delta(\epsilon)} + 1} \\ &= C_{\mathrm{s}}^{\mathrm{WT}} - \delta(\epsilon) - \delta_\epsilon(n, m).\end{aligned}$$

In other words, for n and m large enough, the transmission of reconciliation and privacy-amplification messages over the channel incurs a *negligible* rate penalty.

To conclude, it remains to show that the key $\overline{\mathsf{K}}$ can be interpreted as a message so that the key rate is a message rate. Notice that all communications over the channel are one-way; therefore, in principle, Alice could choose the final key $\overline{\mathsf{K}}$ ahead of time, "invert" the privacy-amplification process, and artificially split the transmission over mn channel uses. Hence, the final strong secret-key $\overline{\mathsf{K}}$ can be treated as a message M that satisfies a strong secrecy condition. Since ϵ can be arbitrarily small, we obtain $\overline{C}_{\mathrm{s}}^{\mathrm{WT}} \geqslant C_{\mathrm{s}}^{\mathrm{WT}}$ and thus $\overline{C}_{\mathrm{s}}^{\mathrm{WT}} = C_{\mathrm{s}}^{\mathrm{WT}}$. □

Remark 4.12. *Privacy amplification with extractors is critical to obtaining strong secrecy with a negligible rate penalty. It is possible to show that the minimum size of a universal family of hash functions $\mathcal{G} : \{0,1\}^{mn} \to \{0,1\}^k$ is 2^{mn-k}; therefore, the minimum number of bits to describe a randomly chosen hash function in the family is $mn - k$ and the number of channel uses required to transmit this choice would be*

$$\frac{mn - k}{C_{\mathrm{m}} - \delta(\epsilon)},$$

which incurs a non-negligible rate penalty.

Remark 4.13. *The mathematical procedure used to convert weakly secure codes in strongly secure codes is not really practical. Although the "inversion" of the process is conceptually feasible, it becomes quickly intractable as n and m become large. Hence, this procedure does not replace the construction of wiretap codes, which we discuss in Chapter 6.*

4.6 Conclusions and lessons learned

The results obtained in this chapter allow us to draw several crucial conclusions. First and foremost, the analysis of the secret-key capacity for source and channel models shows that *feedback improves secrecy*. We already knew from Section 3.6.2 that secure feedback is beneficial for secrecy, but this statement remains true even if the feedback is known to the eavesdropper. This suggests that the need for an advantage over the eavesdropper at the physical layer highlighted in Chapter 3 was largely the consequence of the restrictions imposed on the coding schemes.

Second, secret-key agreement seems much more practical than wiretap coding at this point. The sequential key-distillation strategies based on advantage distillation, information reconciliation, and privacy amplification described in Section 4.3 handle the reliability and secrecy requirements *independently*, which leads to effective ways of distilling secret keys. Nevertheless, note that the fundamental limits of secret-key agreement from source models or channel models are not as well understood as those of secure communication over wiretap channels. This state of affairs is partially explained by the fact that two-way communications seem to be an essential ingredient of key-distillation strategies, and they are significantly harder to analyze than one-way communications. One should also note the tight connection between key distillation and source coding with side information.

Finally, the study of secret-key agreement allows us to develop a generic mathematical procedure by which to strengthen secrecy results, which justifies a posteriori the relevance of the fundamental limits derived with a weak secrecy criterion in Chapter 3 and the first sections of this chapter. In some sense, strong secrecy comes "for free" but the reader should keep in mind that, although the fundamental limits of secure communication and secret-key generation remain the same, the coding schemes achieving strong secrecy might be quite different from those achieving weak secrecy.

4.7 Appendix

Proof of Lemma 4.1

Proof. The result will follow from Chebyshev's inequality, provided that we first establish an upper bound for $\mathrm{Var}\left(p_{\mathrm{K}_{\mathrm{S}_n}}(1)\right) = \mathbb{E}_{\mathrm{S}_n}\left[(p_{\mathrm{K}_{\mathrm{S}_n}}(1))^2\right] - \mathbb{E}_{\mathrm{S}_n}\left[(p_{\mathrm{K}_{\mathrm{S}_n}}(1))\right]^2$. By virtue of the definition of $\mathrm{K}_{\mathcal{S}_n}$, we can write $p_{\mathrm{K}_{\mathcal{S}_n}}(1)$ as

$$p_{\mathrm{K}_{\mathcal{S}_n}}(1) = \sum_{x^n \in \mathcal{T}_\epsilon^n(\mathrm{X})} \frac{p_{\mathrm{X}^n}(x^n)}{\mathbb{P}[\Xi = 1]} \mathbb{1}(\kappa_{\mathrm{a}}(x^n) = 1),$$

where κ_{a} is the key-distillation function used by Alice in the strategy $\mathcal{S}_n$. Hence,

$$\begin{aligned}\mathbb{E}_{\mathrm{S}_n}\left[p_{\mathrm{K}_{\mathrm{S}_n}}(1)\right] &= \sum_{x^n \in \mathcal{T}_\epsilon^n(\mathrm{X})} \frac{p_{\mathrm{X}^n}(x^n)}{\mathbb{P}[\Xi = 1]} \mathbb{E}_{\mathrm{S}_n}[\mathbb{1}(\kappa_{\mathrm{a}}(x^n) = 1)]. \\ &= \sum_{x^n \in \mathcal{T}_\epsilon^n(\mathrm{X})} \frac{p_{\mathrm{X}^n}(x^n)}{\mathbb{P}[\Xi = 1]} \frac{1}{\lceil 2^{nR} \rceil} \\ &= \frac{1}{\lceil 2^{nR} \rceil}.\end{aligned}$$

Similarly,

$$\mathbb{E}_{S_n}\left[(p_{K|S_n}(1))^2\right]$$
$$= \mathbb{E}_{S_n}\left[\left(\sum_{x^n \in \mathcal{T}_\epsilon^n(X)} \frac{p_{X^n}(x^n)}{\mathbb{P}[\Xi = 1]} \mathbb{1}(\kappa_a(x^n) = 1)\right)^2\right]$$
$$\leqslant \mathbb{E}_{S_n}\left[\sum_{x^n \in \mathcal{T}_\epsilon^n(X)} \left(\frac{p_{X^n}(x^n)}{\mathbb{P}[\Xi = 1]}\right)^2 \mathbb{1}(\kappa_a(x^n) = 1)\right.$$
$$\left. + \sum_{x^n \in \mathcal{T}_\epsilon^n(X)} \sum_{x'^n \in \mathcal{T}_\epsilon^n(X)} \frac{p_{X^n}(x^n)\, p_{X^n}(x'^n)}{\mathbb{P}[\Xi = 1]^2} \mathbb{1}(\kappa_a(x^n) = 1)\, \mathbb{1}\left(\kappa_a(x'^n) = 1\right)\right].$$

Hence,

$$\mathbb{E}_{S_n}\left[(p_{K S_n}(1))^2\right]$$
$$\leqslant \sum_{x^n \in \mathcal{T}_\epsilon^n(X)} \left(\frac{p_{X^n}(x^n)}{\mathbb{P}[\Xi = 1]}\right)^2 \mathbb{E}_{S_n}[\mathbb{1}(\kappa_a(x^n) = 1)]$$
$$+ \sum_{x^n \in \mathcal{T}_\epsilon^n(X)} \sum_{x'^n \in \mathcal{T}_\epsilon^n(X)} \frac{p_{X^n}(x^n) p_{X^n}(x'^n)}{\mathbb{P}[\Xi = 1]^2} \mathbb{E}_{S_n}\left[\mathbb{1}(\kappa_a(x^n) = 1)\, \mathbb{1}\left(\kappa_a(x'^n) = 1\right)\right].$$

Using the AEP, we bound the first term on the right-hand side as

$$\sum_{x^n \in \mathcal{T}_\epsilon^n(X)} \left(\frac{p_{X^n}(x^n)}{\mathbb{P}[\Xi = 1]}\right)^2 \mathbb{E}_{S_n}[\mathbb{1}(\kappa_a(x^n) = 1)] \leqslant \frac{2^{-n(\mathbb{H}(X) - \delta(\epsilon))}}{\lceil 2^{nR} \rceil (1 - \delta_\epsilon(n))}.$$

Similarly, we bound the second term on the right-hand side as

$$\sum_{x^n \in \mathcal{T}_\epsilon^n(X)} \sum_{x'^n \in \mathcal{T}_\epsilon^n(X)} \frac{p_{X^n}(x^n) p_{X^n}(x'^n)}{\mathbb{P}[\Xi = 1]^2} \mathbb{E}_{S_n}\left[\mathbb{1}(\kappa_a(x^n) = 1)\, \mathbb{1}\left(\kappa_a(x'^n) = 1\right)\right] = \left(\frac{1}{\lceil 2^{nR} \rceil}\right)^2.$$

Therefore, we obtain the following bound for $\mathrm{Var}\left(p_{K S_n}(1)\right)$:

$$\mathrm{Var}\left(p_{K S_n}(1)\right) = \mathbb{E}_{S_n}\left[(p_{K S_n}(1))^2\right] - \mathbb{E}_{S_n}\left[(p_{K S_n}(1))\right]^2 \leqslant \frac{1}{2^{nR}} \frac{2^{-n(\mathbb{H}(X) - \delta(\epsilon))}}{1 - \delta_\epsilon(n)}.$$

By virtue of Chebyshev's inequality,

$$\mathbb{P}_{S_n}\left[\left|p_{K S_n}(1) - 2^{-nR}\right| > \epsilon 2^{-nR}\right] \leqslant \frac{\mathrm{Var}\left(p_{K S_n}(1)\right)}{\epsilon^2 2^{-2nR}}$$
$$\leqslant \frac{1}{\epsilon^2} \frac{2^{-n(\mathbb{H}(X) - R - \delta(\epsilon))}}{1 - \delta_\epsilon(n)}.$$

If $R < \mathbb{H}(X) - \delta(\epsilon)$ then $\mathbb{P}_{S_n}\left[\left|p_{K S_n}(1) - 2^{-nR}\right| > \epsilon 2^{-nR}\right] \leqslant \delta_\epsilon(n)$. □

Proof of Lemma 4.3

Proof. Let $t \triangleq r + 1$. We show that $\mathbb{P}_{\mathrm{U}}[\chi(\mathrm{U}) = 0] \leqslant 2^{-r}$ by developing an upper bound as follows:

$$\begin{aligned}
\mathbb{P}_{\mathrm{U}}[\chi(\mathrm{U}) = 0] &= \mathbb{P}_{\mathrm{U}}[\mathbb{H}_c(\mathrm{S}) - \log|\mathcal{U}| - 2t - \mathbb{H}_c(\mathrm{S}|\mathrm{U}) > 0] \\
&= \mathbb{P}_{\mathrm{U}}[\mathbb{H}_c(\mathrm{S}) + \log p_{\mathrm{U}}(u) - \mathbb{H}_c(\mathrm{S}|\mathrm{U}) - t - \log p_{\mathrm{U}}(u) - \log|\mathcal{U}|] \\
&\quad - t > 0 \\
&\leqslant \mathbb{P}_{\mathrm{U}}[\mathbb{H}_c(\mathrm{S}) + \log p_{\mathrm{U}}(u) - \mathbb{H}_c(\mathrm{S}|\mathrm{U}) - t \geqslant 0] \\
&\quad + \mathbb{P}_{\mathrm{U}}[-\log p_{\mathrm{U}}(u) - \log|\mathcal{U}| - t \geqslant 0].
\end{aligned}$$

We introduce the random variables

$$\mathrm{X}_{\mathrm{U}} \triangleq 2^{\log p_{\mathrm{U}}(\mathrm{U}) - \mathbb{H}_c(\mathrm{S}|\mathrm{U}) + \mathbb{H}_c(\mathrm{S})} \quad \text{and} \quad \mathrm{Y}_{\mathrm{U}} \triangleq -\log|\mathcal{U}| - \log p_{\mathrm{U}}(\mathrm{U})$$

so that we obtain

$$\mathbb{P}_{\mathrm{U}}[\chi(\mathrm{U}) = 0] \leqslant \mathbb{P}\left[\mathrm{X}_{\mathrm{U}} \geqslant 2^t\right] + \mathbb{P}[\mathrm{Y}_{\mathrm{U}} \geqslant t], \tag{4.76}$$

and we upper bound each term on the right-hand side of (4.76) separately.

First, note that

$$\mathbb{P}[\mathrm{Y}_{\mathrm{U}} \geqslant t] = \mathbb{P}\left[p_{\mathrm{U}}(\mathrm{U}) < \frac{2^{-t}}{|\mathcal{U}|}\right] = \sum_{u \in \mathcal{U}: p_{\mathrm{U}}(u) < 2^{-t}/|\mathcal{U}|} p_{\mathrm{U}}(u) \leqslant 2^{-t}. \tag{4.77}$$

Next, we develop an upper bound $\mathbb{P}\left[\mathrm{X}_{\mathrm{U}} \geqslant 2^t\right]$. Since X_{U} is positive, we establish it by upper bounding $\mathbb{E}_{\mathrm{U}}[\mathrm{X}_{\mathrm{U}}]$ and using Markov's inequality. We start by writing the collision entropy $\mathbb{H}_c(\mathrm{SU})$ as

$$\begin{aligned}
\mathbb{H}_c(\mathrm{SU}) &= -\log\left(\sum_{s \in \mathcal{S}} \sum_{u \in \mathcal{U}} (p_{\mathrm{SU}}(s, u))^2\right) \\
&= -\log\left(\sum_{u \in \mathcal{U}} (p_{\mathrm{U}}(u))^2 \sum_{s \in \mathcal{S}} \left(p_{\mathrm{S}|\mathrm{U}}(s|u)\right)^2\right).
\end{aligned}$$

Note that

$$\sum_{s \in \mathcal{S}} \left(p_{\mathrm{S}|\mathrm{U}}(s|u)\right)^2 = 2^{-\mathbb{H}_c(\mathrm{S}|\mathrm{U}=u)} \quad \text{and} \quad p_{\mathrm{U}}(u) = 2^{\log p_{\mathrm{U}}(u)};$$

therefore,

$$\mathbb{H}_c(\mathrm{SU}) = -\log\left(\sum_{u \in \mathcal{U}} p_{\mathrm{U}}(u) 2^{\log p_{\mathrm{U}}(u) - \mathbb{H}_c(\mathrm{S}|\mathrm{U}=u)}\right),$$

which we can rewrite as

$$2^{-\mathbb{H}_c(\mathrm{SU})} = \mathbb{E}_{\mathrm{U}}\left[2^{\log p_{\mathrm{U}}(\mathrm{U}) - \mathbb{H}_c(\mathrm{S}|\mathrm{U})}\right].$$

Hence,

$$\mathbb{E}_{\mathsf{U}}[\mathsf{X}_{\mathsf{U}}] = 2^{-\mathbb{H}_c(\mathsf{SU})+\mathbb{H}_c(\mathsf{S})}.$$

The reader can easily check that $\mathbb{H}_c(\mathsf{S}) \leqslant \mathbb{H}_c(\mathsf{SU})$ and, therefore, $\mathbb{E}_{\mathsf{U}}[\mathsf{X}_{\mathsf{U}}] \leqslant 1$. Hence, by virtue of Markov's inequality,

$$\mathbb{P}_{\mathsf{U}}\left[\mathsf{X}_{\mathsf{U}} \geqslant 2^t\right] \leqslant \frac{\mathbb{E}_{\mathsf{U}}[\mathsf{U}]}{2^t} \leqslant 2^{-t}. \tag{4.78}$$

On substituting (4.77) and (4.78) into (4.76), we finally obtain

$$\mathbb{P}_{\mathsf{U}}[\chi(\mathsf{U}) = 0] \leqslant 2^{-t} + 2^{-t} = 2^{-(t-1)} = 2^{-r}. \qquad \square$$

Proof of Lemma 4.4

Proof. Using a strengthened version of Theorem 2.1 and Corollary 2.1 [6] we obtain

$$\mathbb{P}\left[\mathsf{Z}^n \in \mathcal{T}_\epsilon^n(\mathsf{Z})\right] \geqslant 1 - \frac{2^{-\sqrt{n}}}{2} \quad \text{and} \quad \mathbb{P}\left[(\mathsf{X}^n, \mathsf{Z}^n) \in \mathcal{T}_{2\epsilon}^n(\mathsf{XZ})\right] \geqslant 1 - \frac{2^{-\sqrt{n}}}{2}$$

for n sufficiently large. Hence, by the union bound, $\mathbb{P}[\Theta = 1] \geqslant 1 - 2^{-\sqrt{n}}$. In addition, for $z^n \in \mathcal{T}_\epsilon^n(\mathsf{Z})$, Theorem 2.2 guarantees that $\mathbb{P}_{\mathsf{X}^n|\mathsf{Z}^n}\left[\mathsf{X}^n \in \mathcal{T}_{2\epsilon}^n(\mathsf{XZ}|z^n)|\mathsf{Z}^n = z^n\right] \geqslant 1 - 2^{-\sqrt{n}}$.

To obtain the remaining part of the lemma, it suffices to establish a lower bound for the min-entropy $\mathbb{H}_\infty(\mathsf{X}^n|\mathsf{Z}^n = z^n, \Theta = 1)$ with $z^n \in \mathcal{T}_\epsilon^n(\mathsf{Z})$ because Proposition 4.7 already proves that

$$\mathbb{H}_c(\mathsf{X}^n|\mathsf{Z}^n = z^n, \Theta = 1) \geqslant \mathbb{H}_\infty(\mathsf{X}^n|\mathsf{Z}^n = z^n, \Theta = 1).$$

By definition,

$$\mathbb{H}_\infty(\mathsf{X}^n|\mathsf{Z}^n = z^n, \Theta = 1) = -\log \max_{x^n} p_{\mathsf{X}^n|\mathsf{Z}^n\Theta}(x^n|z^n, 1).$$

By Bayes' rule, we obtain for all $x^n \in \mathcal{X}^n$

$$\begin{aligned} p_{\mathsf{X}^n|\mathsf{Z}^n\Theta}(x^n|z^n, 1) &= \frac{\mathbb{P}[\Theta = 1|\mathsf{X}^n = x^n, \mathsf{Z}^n = z^n]p_{\mathsf{X}^n|\mathsf{Z}^n}(x^n|z^n)}{\mathbb{P}[\Theta = 1|\mathsf{Z}^n = z^n]} \\ &= \frac{\mathbb{P}[\Theta = 1|\mathsf{X}^n = x^n, \mathsf{Z}^n = z^n]p_{\mathsf{X}^n|\mathsf{Z}^n}(x^n|z^n)}{\mathbb{P}\left[\mathsf{X}^n \in \mathcal{T}_{2\epsilon}^n(\mathsf{XZ}|z^n)|\mathsf{Z}^n = z^n\right]} \\ &\leqslant \frac{2^{-n(\mathbb{H}(\mathsf{X}|\mathsf{Z})-\delta(\epsilon))}}{1 - \delta_\epsilon(n)}, \end{aligned}$$

where the last inequality follows from $p_{\mathsf{X}^n|\mathsf{Z}^n}(x^n|z^n) \leqslant 2^{-n(\mathbb{H}(\mathsf{X}|\mathsf{Z})-\delta(\epsilon))}$ by Theorem 2.1 if $\mathbb{P}[\Theta = 1|\mathsf{X}^n = x^n, \mathsf{Z}^n = z^n] > 0$. Hence,

$$\begin{aligned} \mathbb{H}_\infty(\mathsf{X}^n|\mathsf{Z}^n = z^n, \Theta = 1) &\geqslant n(\mathbb{H}(\mathsf{X}|\mathsf{Z}) - \delta(\epsilon)) + \log(1 - \delta_\epsilon(n)) \\ &= n(\mathbb{H}(\mathsf{X}|\mathsf{Z}) - \delta(\epsilon)) - \delta_\epsilon(n). \end{aligned} \qquad \square$$

Proof of Lemma 4.5

Proof. We show that $\mathbb{P}_U[\chi(U) = 0] \leqslant 2^{-r}$. Note that $\mathbb{H}_\infty(S|U = u)$ satisfies

$$\begin{aligned}\mathbb{H}_\infty(S|U = u) &= -\log \max_s p_{S|U}(s|u) \\ &= -\log \max_s p_{SU}(s, u) + \log p_U(u) \\ &\geqslant -\log \max_s p_S(s) + \log p_U(u) \\ &= \mathbb{H}_\infty(S) + \log p_U(u),\end{aligned}$$

where the inequality follows because $\forall (s, u) \in \mathcal{S} \times \mathcal{U}\ p_{SU}(s, u) \leqslant p_S(s)$. Hence,

$$\begin{aligned}\mathbb{P}_U[\chi(U) = 0] &= \mathbb{P}_U[\mathbb{H}_\infty(S) > \mathbb{H}_\infty(S|U = u) + r + \log|\mathcal{U}|] \\ &\leqslant \mathbb{P}[\log p_U(U) < -r - \log|\mathcal{U}|] \\ &= \mathbb{P}\left[p_U(U) < \frac{2^{-r}}{|\mathcal{U}|}\right] \\ &= \sum_{u \in \mathcal{U}: p_U(u) < 2^{-r}/|\mathcal{U}|} p_U(u) \\ &\leqslant 2^{-r}.\end{aligned}$$

□

4.8 Bibliographical notes

Since the design of sequential key-distillation strategies was largely motivated by applications to quantum key distribution, several survey papers and textbooks discuss secret-key agreement in this context. A high-level presentation of quantum cryptography can be found in the textbook of Nielsen and Chuang [39]. The textbook of Van Assche [38] provides a comprehensive discussion of secret-key distillation and its applications to continuous-variable quantum key distribution. The survey paper of Gisin, Ribordy, Tittel, and Zbinden [40] also discusses secret-key agreement for quantum key distribution and focuses on practical system implementation. A more recent survey of secret-key agreement by Maurer, Renner, and Wolf is also available [41].

The fundamental limits of secret-key agreement by public discussion from common randomness were first investigated by Maurer [42] and Ahlswede and Csiszár [43]. The bounds for secret-key capacity presented in this chapter are the simplest known bounds for a general source or channel model; nevertheless, alternative bounds exist, such as those derived by Gohari and Anantharam [44, 45]. The alternative upper bounds for the secret-key capacity presented in this chapter are those obtained by Maurer and Wolf [46] and Renner and Wolf [47].

When restrictions are placed on public communication, it is sometimes possible to characterize the secret-key capacity exactly. For instance, Ahlswede and Csiszár characterized the forward secret-key capacity of source and channel models, defined as the secret-key capacity when public communication is limited to a single transmission from Alice to Bob [43]. The secret-key capacity of a source model with rate-limited and one-way communication was characterized by Csiszár and Narayan [48] as the consequence of more general results on secret-key agreement with a helper; these results were used subsequently by Khisti, Diggavi, and Wornell [49] as well as Prabhakaran and Ramchandran [50] to analyze secret-key agreement without a public channel. A closed-form expression for the secret-key capacity of a Gaussian source model with rate-limited public communication was established by Watanabe and Oohama [51].

The concept of advantage distillation and the "satellite" source model were introduced by Maurer [42] to show that key-distillation strategies with two-way communication are, in general, more powerful than key-distillation strategies with one-way communication. Maurer's repetition protocol suffices to illustrate the principle of advantage distillation but it achieves a low advantage-distillation rate. Alternative protocols with higher advantage-distillation rates have been studied, such as the "bit-pair iteration" protocol of Gander and Maurer [52]. Liu, Van Tilborg, and Van Dijk also extended and optimized the bit-pair iteration protocol to perform advantage distillation and reconciliation jointly and further increase the advantage-distillation rate [53]. Although the satellite source model is based on a binary DMS $(\mathcal{XYZ}, p_{XYZ})$, the ideas behind the design of advantage protocols generalize to continuous sources. For instance, Naito, Watanabe, Matsumoto, and Uyematsu proposed an advantage-distillation protocol based on a post-selection procedure for a Gaussian satellite model [54]. The notion of advantage-distillation capacity and its relation to secret-key capacity were studied by Muramatsu, Yoshimura, and Davis [55].

Information-reconciliation protocols for secret-key agreement were first introduced by Brassard and Salvail [56] for discrete random variables. In essence, the original reconciliation protocols are simple error-correcting codes that rely on two-way communication to exchange parity checks and identify error locations. Examples of such protocols include CASCADE, proposed by Brassard and Salvail [56], and WINNOW, developed by Buttler, Lamoreaux, Torgerson, Nickel, Donahue, and Peterson [57]. More efficient protocols based on powerful error-control codes can be designed at the cost of increased complexity. For instance, Elkouss, Leverrier, Alléaume, and Boutros optimized a reconciliation protocol based on low-density parity-check codes [58]. Reconciliation protocols for continuous random variables were developed susbsequently by Van Assche, Cardinal, and Cerf [59], Nguyen, Van Assche, and Cerf [60], Bloch, Thangaraj, McLaughlin, and Merolla [61], and Ye, Reznik, and Shah [62].

The Rényi entropy, from which the collision entropy and the min-entropy are derived, was introduced and studied by Rényi [63]. Privacy amplification with universal families of hash functions was analyzed by Bennett, Brassard, Crépeau, and Maurer [64]. The examples of universal families provided in this chapter are due to Carter and Wegman [65, 66], who introduced these functions in the context of authentication. Stinson generalized the work of Carter and Wegman and proved that the minimum size of a universal family of hash functions $\mathcal{G} : \{0, 1\}^n \rightarrow \{0, 1\}^k$ is 2^{n-k} [67]. A thorough discussion of the implementation and properties of privacy amplification with almost universal families of hash functions can be found in the textbook of Van Assche [38]. Privacy amplification with extractors was studied by Maurer and Wolf [68], on the basis of the existence of extractors requiring only a small amount of input randomness as established by Vadhan [69]. The link between reconciliation and privacy amplification was developed by Cachin and Maurer [70, 71]. Although privacy amplification is often described explicitly in terms of universal families of hash functions or extractors, other classes of functions may, in principle, perform the same function. For instance, Muramatsu proved the existence of key-distillation strategies based on low-density parity-check codes [72].

The generic procedure to create strongly secure keys from weakly secure ones presented in this chapter is due to Maurer and Wolf [68]. An alternative (perhaps more direct) approach to strong secrecy was developed by Csiszár [24] for discrete random variables, and was extended to continuous correlations by Nitinawarat [73].

The authentication of the public channel is a crucial assumption used to derive all of the results in this chapter. The problem of secret-key agreement over unauthenticated public channels is significantly more involved; nevertheless, Maurer and Wolf showed that the secret-key capacity remains the same, provided that the DMS of the source model satisfies a condition called the *simulatability* condition [74]. In general, deciding whether a source satisfies the simulatability condition is not straightforward; however, Maurer and Wolf developed an efficient algorithm to check a necessary condition for simulatability [75] and studied privacy amplification protocols for key distillation over unauthenticated channels [76].

There have been a few experimental demonstrations of key-distillation strategies. For instance, Imai, Kobara, and Morozov gathered experimental data to assess the possibility of key agreement with a directional antenna [77]. Ye, Mathur, Reznik, Shah, Trappe, and Mandayam [78] and Chen and Jensen [79] also implemented several key-distillation strategies relying on the reciprocity of fading channels. Conceptually, the key-distillation strategies for wireless channels are similar to those implemented for continuous-variable quantum key distribution, see for instance the quantum key-distribution systems proposed by Grosshans, Van Assche, Wenger, Tualle-Brouri, Cerf, and Grangier [80] and Lodewyck, Bloch, García-Patrón, Fossier, Karpov, Diamanti, Debuisschert, Cerf, Tualle-Brouri, McLaughlin, and Grangier [81]. Chabanne and Fumaroli also analyzed a low-cost key-distillation strategy for RFID tags [82].

5 Security limits of Gaussian and wireless channels

This chapter extends the results obtained in Chapter 3 and Chapter 4 for discrete memoryless channels and sources to Gaussian channels and wireless channels, for which numerical applications provide insight beyond that of the general formula in Theorem 3.3. Gaussian channels are of particular importance, not only because the secrecy capacity admits a simple, intuitive, and easily computable expression but also because they provide a reasonable approximation of the physical layer encountered in many practical systems. The analysis of Gaussian channels also lays the foundations for the study of wireless channels.

The application of physical-layer security paradigms to wireless channels is perhaps one of the most promising research directions in physical-layer security. While wireline systems offer some security, because the transmission medium is confined, wireless systems are intrinsically susceptible to eavesdropping since all transmissions are broadcast over the air and overheard by neighboring devices. Other users can be viewed as potential eavesdroppers if they are not the intended recipients of a message. However, as seen in earlier chapters, the randomness present at the physical layer can be harnessed to provide security, and randomness is a resource that abounds in a wireless medium. For instance, we show that fading can be exploited opportunistically to guarantee secrecy even if an eavesdropper obtains on average a higher signal-to-noise ratio than a legitimate receiver.

We start this chapter with a detailed study of Gaussian channels and sources, including multiple-input multiple-output channels (Section 5.1.2). We then move on to wireless channels, and we analyze the fundamental limits of secure communications for ergodic fading (Section 5.2.1), block fading (Section 5.2.2), and quasi-static fading (Section 5.2.3).

5.1 Gaussian channels and sources

5.1.1 Gaussian broadcast channel with confidential messages

Communication over a (real) Gaussian broadcast channel with confidential messages (Gaussian BCC for short) is illustrated in Figure 5.1. This channel model is a specific instance of a BCC in which the codewords transmitted by Alice are corrupted by additive Gaussian noise. Specifically, the relationships between the inputs and outputs

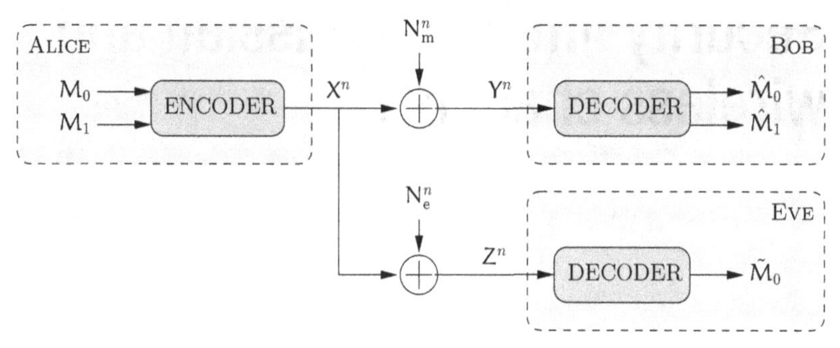

Figure 5.1 Communication over a Gaussian BCC.

of the channel are given by

$$Y_i = X_i + N_{m,i} \quad \text{and} \quad Z_i = X_i + N_{e,i},$$

where the noise processes $\{N_{m,i}\}_{i\geqslant 1}$ and $\{N_{e,i}\}_{i\geqslant 1}$ are i.i.d. and

$$N_{m,i} \sim \mathcal{N}(0, \sigma_m^2) \qquad \text{and} \qquad N_{e,i} \sim \mathcal{N}(0, \sigma_e^2).$$

The statistics of $N_{m,i}$ and $N_{e,i}$ are assumed known to the transmitter, the receiver, and the eavesdropper prior to transmission. The input of the channel is also subject to an average power constraint

$$\frac{1}{n}\sum_{i=1}^{n} \mathbb{E}\left[X_i^2\right] \leqslant P.$$

Definitions 3.6 and 3.7 for codes, achievable rates, and the secrecy capacity then apply readily to the Gaussian BCC. The key property of the Gaussian BCC that makes it more amenable to study than the general BCC is that either the eavesdropper's channel is stochastically degraded with respect to the main channel or the main channel is stochastically degraded with respect to the eavesdropper's channel. In fact, if $\sigma_e^2 \geqslant \sigma_m^2$, the marginal probabilities $p_{Y|X}$ and $p_{Z|X}$ are the same as those of the channel characterized by

$$Y_i = X_i + N_{m,i} \quad \text{and} \quad Z_i = Y_i + N_i' \qquad \text{with } N_i' \sim \mathcal{N}(0, \sigma_e^2 - \sigma_m^2).$$

Similarly, if $\sigma_e^2 < \sigma_m^2$, the marginal probabilities $p_{Y|X}$ and $p_{Z|X}$ are the same as those of the channel characterized by

$$Z_i = X_i + N_{e,i} \quad \text{and} \quad Y_i = Z_i + N_i'' \qquad \text{with } N_i'' \sim \mathcal{N}(0, \sigma_m^2 - \sigma_e^2).$$

In the latter case, the secrecy capacity is zero according to Proposition 3.4.

Theorem 5.1 (Liang *et al.*). *The secrecy-capacity region of the Gaussian BCC is*

$$\mathcal{C}^{\text{GBCC}} = \bigcup_{\beta\in[0,1]} \left\{ (R_0, R_1)\colon \begin{array}{l} R_0 \leqslant \min\left(\frac{1}{2}\log\left(1 + \frac{(1-\beta)P}{\sigma_m^2 + \beta P}\right), \frac{1}{2}\log\left(1 + \frac{(1-\beta)P}{\sigma_e^2 + \beta P}\right)\right) \\ R_1 \leqslant \left(\frac{1}{2}\log\left(1 + \frac{\beta P}{\sigma_m^2}\right) - \frac{1}{2}\log\left(1 + \frac{\beta P}{\sigma_e^2}\right)\right)^+ \end{array} \right\}.$$

Proof. If we treat the Gaussian BCC as a special case of the BCC studied in Section 3.5, then the achievability of $\mathcal{C}^{\text{GBCC}}$ follows from Theorem 3.3 with the following choice of random variables:

$$\mathsf{U} \sim \mathcal{N}(0, (1-\beta)P), \quad \mathsf{V} \triangleq \mathsf{U} + \mathsf{X}' \text{ with } \mathsf{X}' \sim \mathcal{N}(0, \beta P) \quad \text{and} \quad \mathsf{X} \triangleq \mathsf{V}.$$

To make the proof rigorous, we would need to modify the proof of Section 3.5.2 appropriately to take into account the continuous nature of the Gaussian BCC and the power constraint. This can be done by noting that strongly typical sequences can be replaced by weakly typical sequences in the random-coding argument to handle continuous distributions.[1] Then, the input power constraint can be dealt with by introducing an error event that accounts for the violation of the constraint as done in [3, Chapter 9].

For the converse part of the proof, note that all the steps in Section 3.5.3 up to (3.57) involve "basic" properties of mutual information (the chain rule and positivity) that hold irrespective of the continuous or discrete nature of the channel. Therefore, if a rate pair (R_0, R_1) is achievable for the Gaussian BCC, it must hold for any $\epsilon > 0$ that

$$R_0 \leqslant \min\left(\frac{1}{n}\sum_{i=1}^{n}\mathbb{I}(\mathsf{M}_0\tilde{\mathsf{Z}}^{i+1}\mathsf{Y}^{i-1};\mathsf{Y}_i), \frac{1}{n}\sum_{i=1}^{n}\mathbb{I}(\mathsf{M}_0\tilde{\mathsf{Z}}^{i+1}\mathsf{Y}^{i-1};\mathsf{Z}_i)\right) + \delta(\epsilon),$$

$$R_1 \leqslant \frac{1}{n}\sum_{i=1}^{n}\left(\mathbb{I}(\mathsf{M}_1;\mathsf{Y}_i|\mathsf{M}_0\mathsf{Y}^{i-1}\tilde{\mathsf{Z}}^{i+1}) - \mathbb{I}(\mathsf{M}_1;\mathsf{Z}_i|\mathsf{M}_0\mathsf{Y}^{i-1}\tilde{\mathsf{Z}}^{i+1})\right) + \delta(\epsilon), \tag{5.1}$$

where we have used $\mathsf{Y}^{i-1} \triangleq (\mathsf{Y}_1 \dots \mathsf{Y}_{i-1})$, and $\tilde{\mathsf{Z}}^{i+1} = (\mathsf{Z}_{i+1} \dots \mathsf{Z}_n)$. Next, we introduce the random variables $\mathsf{U}_i \triangleq \mathsf{Y}^{i-1}\tilde{\mathsf{Z}}^{i+1}\mathsf{M}_0$ and $\mathsf{V}_i \triangleq \mathsf{U}_i\mathsf{M}_1$. One can verify that the joint distribution of $\mathsf{U}_i, \mathsf{V}_i, \mathsf{X}_i, \mathsf{Y}_i$, and Z_i satisfies

$$\forall (u, v, x, y, z) \in \mathcal{U} \times \mathcal{V} \times \mathcal{X} \times \mathcal{Y} \times \mathcal{Z}$$
$$p_{\mathsf{U}_i}(u)p_{\mathsf{V}_i|\mathsf{U}_i}(v|u)p_{\mathsf{X}_i|\mathsf{V}_i}(x|v)p_{\mathsf{YZ}|\mathsf{X}}(y, z|x),$$

where $p_{\mathsf{YZ}|\mathsf{X}}$ are the transition probabilities of the Gaussian BCC. On substituting these random variables into (5.1), we obtain

$$R_0 \leqslant \min\left(\frac{1}{n}\sum_{i=1}^{n}\mathbb{I}(\mathsf{U}_i;\mathsf{Y}_i), \frac{1}{n}\sum_{i=1}^{n}\mathbb{I}(\mathsf{U}_i;\mathsf{Z}_i)\right) + \delta(\epsilon), \tag{5.2}$$

$$R_1 \leqslant \frac{1}{n}\sum_{i=1}^{n}(\mathbb{I}(\mathsf{V}_i;\mathsf{Y}_i|\mathsf{U}_i) - \mathbb{I}(\mathsf{V}_i;\mathsf{Z}_i|\mathsf{U}_i)) + \delta(\epsilon). \tag{5.3}$$

It remains to upper bound (5.2) and (5.3) with terms that depend on the power constraint P. We first assume that $\sigma_{\text{m}}^2 \leqslant \sigma_{\text{e}}^2$ so that the eavesdropper's channel is stochastically degraded with respect to the main channel. We expand $(1/n)\sum_{i=1}^{n}\mathbb{I}(\mathsf{U}_i;\mathsf{Y}_i)$ in terms of the differential entropy as

$$\frac{1}{n}\sum_{i=1}^{n}\mathbb{I}(\mathsf{U}_i;\mathsf{Y}_i) = \frac{1}{n}\sum_{i=1}^{n}\mathbb{h}(\mathsf{Y}_i) - \frac{1}{n}\sum_{i=1}^{n}\mathbb{h}(\mathsf{Y}_i|\mathsf{U}_i), \tag{5.4}$$

[1] Note that the use of weakly typical sequences limits us to bounds on the probability of error of the form $\mathbf{P}_{\text{e}}(\mathcal{C}_n) \leqslant \delta(\epsilon)$ instead of $\mathbf{P}_{\text{e}}(\mathcal{C}_n) \leqslant \delta_\epsilon(n)$ for DMCs; however, this has no effect on the secrecy-capacity region.

and we bound each sum separately. Notice that $\mathbb{E}\left[\mathsf{Y}_i^2\right] = \mathbb{E}\left[\mathsf{X}_i^2\right] + \sigma_{\mathrm{m}}^2$ since $\mathsf{Y}_i = \mathsf{X}_i + \mathsf{N}_{\mathrm{m},i}$ and X_i is independent of $\mathsf{N}_{\mathrm{m},i}$. The differential entropy of Y_i is upper bounded by the entropy of a Gaussian random variable with the same variance; therefore,

$$\frac{1}{n}\sum_{i=1}^{n} \mathbb{h}(\mathsf{Y}_i) \leqslant \frac{1}{n}\sum_{i=1}^{n}\frac{1}{2}\log\left(2\pi e\left(\mathbb{E}\left[\mathsf{X}_i^2\right] + \sigma_{\mathrm{m}}^2\right)\right).$$

Since $x \mapsto \log(2\pi e x)$ is a concave function of x, we have, by application of Jensen's inequality,

$$\frac{1}{n}\sum_{i=1}^{n}\frac{1}{2}\log\left(2\pi e\left(\mathbb{E}\left[\mathsf{X}_i^2\right] + \sigma_{\mathrm{m}}^2\right)\right) \leqslant \frac{1}{2}\log\left(2\pi e\left(\frac{1}{n}\sum_{i=1}^{n}\mathbb{E}\left[\mathsf{X}_i^2\right] + \sigma_{\mathrm{m}}^2\right)\right).$$

On setting $Q \triangleq (1/n)\sum_{i=1}^{n}\mathbb{E}\left[\mathsf{X}_i^2\right]$, we finally obtain

$$\frac{1}{n}\sum_{i=1}^{n} \mathbb{h}(\mathsf{Y}_i) \leqslant \frac{1}{2}\log\left(2\pi e\left(Q + \sigma_{\mathrm{m}}^2\right)\right). \tag{5.5}$$

To bound the second sum $(1/n)\sum_{i=1}^{n}\mathbb{h}(\mathsf{Y}_i|\mathsf{U}_i)$, notice that

$$\frac{1}{n}\sum_{i=1}^{n} \mathbb{h}(\mathsf{Y}_i|\mathsf{U}_i) \leqslant \frac{1}{n}\sum_{i=1}^{n} \mathbb{h}(\mathsf{Y}_i) \leqslant \frac{1}{2}\log\left(2\pi e\left(Q + \sigma_{\mathrm{m}}^2\right)\right).$$

Moreover, because $\mathsf{U}_i \rightarrow \mathsf{X}_i \rightarrow \mathsf{Y}_i\mathsf{Z}_i$ forms a Markov chain, we have

$$\frac{1}{n}\sum_{i=1}^{n} \mathbb{h}(\mathsf{Y}_i|\mathsf{U}_i) \geqslant \frac{1}{n}\sum_{i=1}^{n} \mathbb{h}(\mathsf{Y}_i|\mathsf{X}_i\mathsf{U}_i) = \frac{1}{n}\sum_{i=1}^{n} \mathbb{h}(\mathsf{Y}_i|\mathsf{X}_i) = \frac{1}{2}\log\left(2\pi\sigma_{\mathrm{m}}^2\right).$$

Since $x \mapsto \frac{1}{2}\log\left(2\pi e\left(xQ + \sigma_{\mathrm{m}}^2\right)\right)$ is a continuous function on the interval $[0, 1]$, the intermediate-value theorem ensures the existence of $\beta \in [0, 1]$ such that

$$\frac{1}{n}\sum_{i=1}^{n} \mathbb{h}(\mathsf{Y}_i|\mathsf{U}_i) = \frac{1}{2}\log\left(2\pi e\left(\beta Q + \sigma_{\mathrm{m}}^2\right)\right). \tag{5.6}$$

On substituting (5.5) and (5.6) into (5.4), we obtain

$$\begin{aligned}\frac{1}{n}\sum_{i=1}^{n} \mathbb{I}(\mathsf{U}_i;\mathsf{Y}_i) &\leqslant \frac{1}{2}\log\left(2\pi e\left(Q + \sigma_{\mathrm{m}}^2\right)\right) - \frac{1}{2}\log\left(2\pi e\left(\beta Q + \sigma_{\mathrm{m}}^2\right)\right)\\ &= \frac{1}{2}\log\left(1 + \frac{(1-\beta)Q}{\sigma_{\mathrm{m}}^2}\right).\end{aligned} \tag{5.7}$$

We now need to upper bound $(1/n)\sum_{i=1}^{n}\mathbb{I}(\mathsf{Z}_i;\mathsf{U}_i)$. If we follow the same steps as above with Z_i in place of Y_i, we obtain the upper bound

$$\frac{1}{n}\sum_{i=1}^{n} \mathbb{I}(\mathsf{U}_i;\mathsf{Z}_i) \leqslant \frac{1}{2}\log\left(1 + \frac{(1-\beta')Q}{\sigma_{\mathrm{m}}^2}\right),$$

with $\beta' \in [0, 1]$. Unfortunately, this upper bound is not really useful because β' is a priori different from β; we need an alternative technique to show that $(1/n)\sum_{i=1}^{n}\mathbb{I}(\mathsf{U}_i;\mathsf{Z}_i)$ can be upper bounded with the *same* parameters β and Q as $(1/n)\sum_{i=1}^{n}\mathbb{I}(\mathsf{U}_i;\mathsf{Y}_i)$. The key tool that allows us to do so is the entropy–power inequality introduced in Lemma 2.14.

Note that we can repeat the steps leading to (5.5) with Z_i in place of Y_i to obtain

$$\frac{1}{n}\sum_{i=1}^{n} \mathbb{h}(Z_i) \leqslant \frac{1}{2}\log\big(2\pi e\big(Q+\sigma_e^2\big)\big); \tag{5.8}$$

therefore, we need to develop a lower bound for $(1/n)\sum_{i=1}^{n}\mathbb{h}(Z_i|U_i)$ as a function of β and Q. Since we have assumed that the eavesdropper's channel is stochastically degraded with respect to the main channel, we can write $Z_i = Y_i + N_i'$ with $N_i' \sim \mathcal{N}(0, \sigma_e^2 - \sigma_m^2)$. Applying the entropy–power inequality to the random variable Z_i conditioned on $U_i = u_i$, we have

$$\begin{aligned}
\mathbb{h}(Z_i|U_i = u_i) &= \mathbb{h}(Y_i + N_i'|U_i = u_i)\\
&\geqslant \frac{1}{2}\log\Big(2^{2\mathbb{h}(Y_i|U_i=u_i)} + 2^{2\mathbb{h}(N_i'|U_i=u_i)}\Big)\\
&= \frac{1}{2}\log\big(2^{2\mathbb{h}(Y_i|U_i=u_i)} + 2\pi e\big(\sigma_e^2 - \sigma_m^2\big)\big).
\end{aligned}$$

Therefore,

$$\begin{aligned}
\frac{1}{n}\sum_{i=1}^{n}\mathbb{h}(Z_i|U_i) &= \frac{1}{n}\sum_{i=1}^{n}\mathbb{E}_{U_i}[\mathbb{h}(Z_i|U_i)]\\
&\geqslant \frac{1}{2n}\sum_{i=1}^{n}\mathbb{E}_{U_i}\big[\log\big(2^{2\mathbb{h}(Y_i|U_i)} + 2\pi e\big(\sigma_e^2 - \sigma_m^2\big)\big)\big]\\
&\overset{(a)}{\geqslant} \frac{1}{2n}\sum_{i=1}^{n}\log\big(2^{2\mathbb{E}_{U_i}[\mathbb{h}(Y_i|U_i)]} + 2\pi e\big(\sigma_e^2 - \sigma_m^2\big)\big)\\
&= \frac{1}{2n}\sum_{i=1}^{n}\log\big(2^{2\mathbb{h}(Y_i|U_i)} + 2\pi e\big(\sigma_e^2 - \sigma_m^2\big)\big)\\
&\overset{(b)}{\geqslant} \frac{1}{2}\log\Big(2^{2(1/n)\sum_{i=1}^{n}\mathbb{h}(Y_i|U_i)} + 2\pi e\big(\sigma_e^2 - \sigma_m^2\big)\Big)\\
&\overset{(c)}{=} \frac{1}{2}\log\big(2\pi e\big(\beta Q + \sigma_m^2\big) + 2\pi e\big(\sigma_e^2 - \sigma_m^2\big)\big)\\
&= \frac{1}{2}\log\big(2\pi e\big(\beta Q + \sigma_e^2\big)\big),
\end{aligned} \tag{5.9}$$

where both (a) and (b) follow from the convexity of the function $x \mapsto \log(2^x + c)$ for $c \in \mathbb{R}_+$ and Jensen's inequality while (c) follows from (5.6). Hence,

$$\begin{aligned}
\frac{1}{n}\sum_{i=1}^{n}\mathbb{I}(U_i;Z_i) &= \frac{1}{n}\sum_{i=1}^{n}(\mathbb{h}(Z_i) - \mathbb{h}(Z_i|U_i))\\
&\leqslant \frac{1}{2}\log\big(2\pi e\big(Q+\sigma_e^2\big)\big) - \frac{1}{2}\log\big(2\pi e\big(\beta Q + \sigma_e^2\big)\big)\\
&= \frac{1}{2}\log\left(1 + \frac{(1-\beta)Q}{\sigma_e^2}\right),
\end{aligned} \tag{5.10}$$

where the inequality follows from (5.8) and (5.9). On substituting (5.7) and (5.10) into (5.2), we obtain

$$R_0 \leqslant \min\left(\frac{1}{2}\log\left(1+\frac{(1-\beta)Q}{\sigma_\text{m}^2}\right), \frac{1}{2}\log\left(1+\frac{(1-\beta)Q}{\sigma_\text{e}^2}\right)\right) + \delta(\epsilon). \tag{5.11}$$

We now develop an upper bound for R_1 as a function of the same parameters Q and β starting from (5.3). First, we eliminate the auxiliary random variable V_i by introducing the random variable X_i as follows:

$$\begin{aligned}
&\frac{1}{n}\sum_{i=1}^{n}(\mathbb{I}(\mathsf{V}_i;\mathsf{Y}_i|\mathsf{U}_i) - \mathbb{I}(\mathsf{V}_i;\mathsf{Z}_i|\mathsf{U}_i)) \\
&\quad = \frac{1}{n}\sum_{i=1}^{n}(\mathbb{I}(\mathsf{V}_i\mathsf{X}_i;\mathsf{Y}_i|\mathsf{U}_i) - \mathbb{I}(\mathsf{X}_i;\mathsf{Y}_i|\mathsf{U}_i\mathsf{V}_i) - \mathbb{I}(\mathsf{V}_i\mathsf{X}_i;\mathsf{Z}_i|\mathsf{U}_i) + \mathbb{I}(\mathsf{X}_i;\mathsf{Z}_i|\mathsf{U}_i\mathsf{V}_i)) \\
&\quad \overset{(a)}{=} \frac{1}{n}\sum_{i=1}^{n}(\mathbb{I}(\mathsf{X}_i;\mathsf{Y}_i|\mathsf{U}_i) - \mathbb{I}(\mathsf{X}_i;\mathsf{Z}_i|\mathsf{U}_i) - \mathbb{I}(\mathsf{X}_i;\mathsf{Y}_i|\mathsf{U}_i\mathsf{V}_i) + \mathbb{I}(\mathsf{X}_i;\mathsf{Z}_i|\mathsf{U}_i\mathsf{V}_i)) \\
&\quad = \frac{1}{n}\sum_{i=1}^{n}(\mathbb{I}(\mathsf{X}_i;\mathsf{Y}_i|\mathsf{U}_i) - \mathbb{I}(\mathsf{X}_i;\mathsf{Z}_i|\mathsf{U}_i) - \mathbb{I}(\mathsf{X}_i;\mathsf{Y}_i\mathsf{Z}_i|\mathsf{U}_i\mathsf{V}_i) + \mathbb{I}(\mathsf{X}_i;\mathsf{Z}_i|\mathsf{Y}_i\mathsf{U}_i\mathsf{V}_i) \\
&\qquad\qquad + \mathbb{I}(\mathsf{X}_i;\mathsf{Z}_i|\mathsf{U}_i\mathsf{V}_i)) \\
&\quad \overset{(b)}{=} \frac{1}{n}\sum_{i=1}^{n}(\mathbb{I}(\mathsf{X}_i;\mathsf{Y}_i|\mathsf{U}_i) - \mathbb{I}(\mathsf{X}_i;\mathsf{Z}_i|\mathsf{U}_i) - \mathbb{I}(\mathsf{X}_i;\mathsf{Y}_i\mathsf{Z}_i|\mathsf{U}_i\mathsf{V}_i) + \mathbb{I}(\mathsf{X}_i;\mathsf{Z}_i|\mathsf{U}_i\mathsf{V}_i)) \\
&\quad \overset{(c)}{\leqslant} \frac{1}{n}\sum_{i=1}^{n}(\mathbb{I}(\mathsf{X}_i;\mathsf{Y}_i|\mathsf{U}_i) - \mathbb{I}(\mathsf{X}_i;\mathsf{Z}_i|\mathsf{U}_i)),
\end{aligned}$$

where (a) follows from $\mathbb{I}(\mathsf{V}_i;\mathsf{Z}_i|\mathsf{U}_i\mathsf{X}_i) = \mathbb{I}(\mathsf{V}_i;\mathsf{Y}_i|\mathsf{U}_i\mathsf{X}_i) = 0$ since $\mathsf{U}_i \to \mathsf{V}_i \to \mathsf{X}_i \to \mathsf{Y}_i\mathsf{Z}_i$ forms a Markov chain, (b) follows from $\mathbb{I}(\mathsf{X}_i;\mathsf{Z}_i|\mathsf{U}_i\mathsf{V}_i\mathsf{Y}_i) = 0$ since Z_i is stochastically degraded with respect to Y_i, and (c) follows from $\mathbb{I}(\mathsf{X}_i;\mathsf{Z}_i|\mathsf{U}_i\mathsf{V}_i) \leqslant \mathbb{I}(\mathsf{X}_i;\mathsf{Z}_i\mathsf{Y}_i|\mathsf{U}_i\mathsf{V}_i)$. Next, we use (5.6) and (5.9) to introduce β and Q as follows:

$$\begin{aligned}
&\frac{1}{n}\sum_{i=1}^{n}(\mathbb{I}(\mathsf{X}_i;\mathsf{Y}_i|\mathsf{U}_i) - \mathbb{I}(\mathsf{X}_i;\mathsf{Z}_i|\mathsf{U}_i)) \\
&\quad = \frac{1}{n}\sum_{i=1}^{n}(\mathbb{h}(\mathsf{Y}_i|\mathsf{U}_i) - \mathbb{h}(\mathsf{Y}_i|\mathsf{X}_i\mathsf{U}_i) - \mathbb{h}(\mathsf{Z}_i|\mathsf{U}_i) + \mathbb{h}(\mathsf{Z}_i|\mathsf{X}_i\mathsf{U}_i)) \\
&\quad \leqslant \frac{1}{2}\log\left(2\pi e(\beta Q + \sigma_\text{m}^2)\right) - \frac{1}{2}\log\left(2\pi e\sigma_\text{m}^2\right) \\
&\qquad - \frac{1}{2}\log\left(2\pi \text{e}(\beta Q + \sigma_\text{e}^2)\right) + \frac{1}{2}\log\left(2\pi e\sigma_\text{e}^2\right) \\
&\quad = \frac{1}{2}\log\left(1+\frac{\beta Q}{\sigma_\text{m}^2}\right) - \frac{1}{2}\log\left(1+\frac{\beta Q}{\sigma_\text{e}^2}\right).
\end{aligned} \tag{5.12}$$

By substituting (5.12) into (5.3), we obtain the desired upper bound for R_1:

$$R_1 \leqslant \frac{1}{2}\log\left(1+\frac{\beta Q}{\sigma_{\mathrm{m}}^2}\right) - \frac{1}{2}\log\left(1+\frac{\beta Q}{\sigma_{\mathrm{e}}^2}\right) + \delta(\epsilon). \tag{5.13}$$

If $\sigma_{\mathrm{e}}^2 \leqslant \sigma_{\mathrm{m}}^2$, then the main channel is stochastically degraded with respect to the eavesdropper's channel and $R_1 = 0$ by virtue of Proposition 3.4. By swapping the roles of Y_i and Z_i in the proof, the reader can verify that (5.11) still holds. We combine the two cases $\sigma_{\mathrm{e}}^2 \leqslant \sigma_{\mathrm{m}}^2$ and $\sigma_{\mathrm{e}}^2 \geqslant \sigma_{\mathrm{m}}^2$ by writing

$$R_0 \leqslant \min\left(\frac{1}{2}\log\left(1+\frac{(1-\beta)Q}{\sigma_{\mathrm{m}}^2}\right), \frac{1}{2}\log\left(1+\frac{(1-\beta)Q}{\sigma_{\mathrm{e}}^2}\right)\right) + \delta(\epsilon),$$

$$R_1 \leqslant \left(\frac{1}{2}\log\left(1+\frac{\beta Q}{\sigma_{\mathrm{m}}^2}\right) - \frac{1}{2}\log\left(1+\frac{\beta Q}{\sigma_{\mathrm{e}}^2}\right)\right)^+ + \delta(\epsilon).$$

To conclude the proof, notice that

$$Q \mapsto \min\left(\frac{1}{2}\log\left(1+\frac{(1-\beta)Q}{\sigma_{\mathrm{m}}^2}\right), \frac{1}{2}\log\left(1+\frac{(1-\beta)Q}{\sigma_{\mathrm{e}}^2}\right)\right)$$

and

$$Q \mapsto \left(\frac{1}{2}\log\left(1+\frac{\beta Q}{\sigma_{\mathrm{m}}^2}\right) - \frac{1}{2}\log\left(1+\frac{\beta Q}{\sigma_{\mathrm{e}}^2}\right)\right)^+$$

are increasing functions of Q and, by definition, $Q = (1/n)\sum_{i=1}^n \mathbb{E}\left[\mathsf{X}_i^2\right] \leqslant P$. Additionally, ϵ can be chosen arbitrarily small; therefore, it must hold that

$$R_0 \leqslant \min\left(\frac{1}{2}\log\left(1+\frac{(1-\beta)P}{\sigma_{\mathrm{m}}^2}\right), \frac{1}{2}\log\left(1+\frac{(1-\beta)P}{\sigma_{\mathrm{e}}^2}\right)\right),$$

$$R_1 \leqslant \left(\frac{1}{2}\log\left(1+\frac{\beta P}{\sigma_{\mathrm{m}}^2}\right) - \frac{1}{2}\log\left(1+\frac{\beta P}{\sigma_{\mathrm{e}}^2}\right)\right)^+. \qquad \square$$

In contrast to the general BCC, the capacity region of the Gaussian BCC does not require the introduction of a prefix channel. In fact, the proof of Theorem 5.1 shows that the choice $\mathsf{X} = \mathsf{V}$ is optimal. The typical shape of the region $\mathcal{C}^{\text{GBCC}}$ is illustrated in Figure 5.2, together with the capacity region of the same Gaussian broadcast channel *without* confidential messages. It may seem that communicating securely inflicts a strong rate penalty and that a significant portion of the available capacity has to be sacrificed to confuse the eavesdropper; however, this is again somewhat misleading because the achievability proof shows that it is possible to transmit an additional individual message to the legitimate receiver. On specializing the results of Section 3.6.1 and assuming $\sigma_{\mathrm{m}}^2 \leqslant \sigma_{\mathrm{e}}^2$, we see that it is actually possible to transmit three messages over a Gaussian broadcast channel with confidential messages:

(1) a common message to both Bob and Eve at rate

$$R_0 = \min\left(\frac{1}{2}\log\left(1+\frac{(1-\beta)P}{\sigma_{\mathrm{m}}^2+\beta P}\right), \frac{1}{2}\log\left(1+\frac{(1-\beta)P}{\sigma_{\mathrm{e}}^2+\beta P}\right)\right);$$

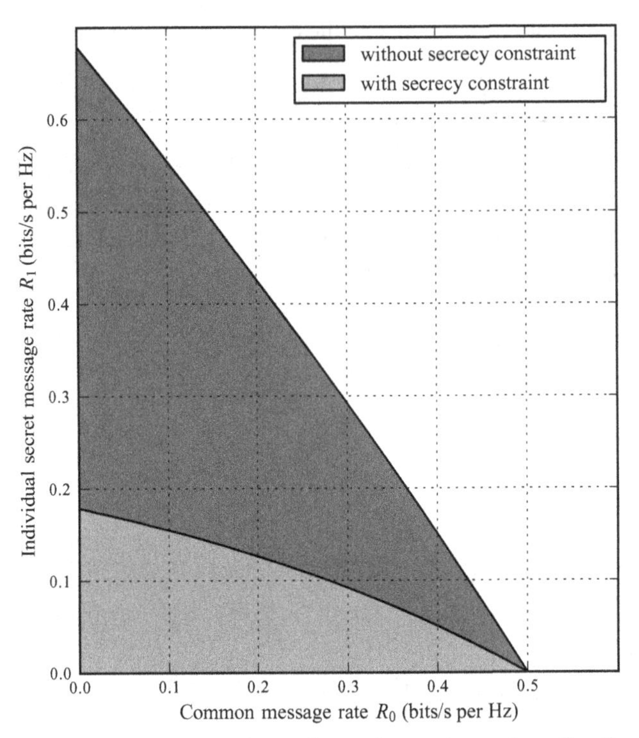

Figure 5.2 Secrecy-capacity region and capacity region of a Gaussian broadcast channel with $\sigma_m = 0.8$, $\sigma_e = 1$, and $P = 1$. The light gray region is the secrecy-capacity region, whereas the darker gray region is the capacity region without secrecy constraints.

(2) a confidential message to Bob at rate

$$R_1 = \frac{1}{2}\log\left(1 + \frac{\beta P}{\sigma_m^2}\right) - \frac{1}{2}\log\left(1 + \frac{\beta P}{\sigma_e^2}\right);$$

(3) a public message to Bob with no guaranteed secrecy at rate

$$R_d = \frac{1}{2}\log\left(1 + \frac{\beta P}{\sigma_e^2}\right).$$

Without any common message sent to Eve ($\beta = 1$ and $R_0 = 0$), the total rate effectively available to communicate with Bob is $R_1 + R_d = \frac{1}{2}\log\left(1 + P/\sigma_m^2\right)$, which is the capacity of the main channel.

The secrecy capacity of the Gaussian WTC is obtained by specializing Theorem 5.1 to $\beta = 1$ ($R_0 = 0$).

Corollary 5.1 (Leung-Yan-Cheong and Hellman). *The secrecy capacity of the Gaussian WTC is*

$$C_s = \left(\frac{1}{2}\log\left(1 + \frac{P}{\sigma_m^2}\right) - \frac{1}{2}\log\left(1 + \frac{P}{\sigma_e^2}\right)\right)^+ \triangleq (C_m - C_e)^+,$$

where $C_m \triangleq \frac{1}{2}\log\left(1 + P/\sigma_m^2\right)$ is the capacity of the main channel and $C_e \triangleq 1 + P/\sigma_e^2$ is that of the eavesdropper's channel.

The expression for the secrecy capacity of the Gaussian WTC implies that secure communication is possible if and only if the legitimate receiver has a better signal-to-noise ratio (SNR) than that of the eavesdropper. In practice, this is likely to happen if the eavesdropper is located farther away from the transmitter than the legitimate receiver and receives attenuated signals. Near-field communication is a good example of such a situation, but this requires the eavesdropper to have a disadvantage at the physical layer itself. Also notice that, unlike the channel capacity, the secrecy capacity does not grow unbounded as $P \to \infty$. Taking the limit in Corollary 5.1, we obtain

$$\lim_{P\to\infty} C_{\mathrm{s}}(P) = \left(\frac{1}{2}\log\left(\frac{\sigma_{\mathrm{e}}^2}{\sigma_{\mathrm{m}}^2}\right)\right)^+ .$$

Therefore, increasing the power results in only marginal secrecy gains beyond a certain point.

Remark 5.1. *All of the results above extend to the* complex *Gaussian WTC, for which the noise sources are complex and circularly symmetric, that is* $\mathrm{N}_{\mathrm{m},i} \sim \mathcal{CN}(0, \sigma_{\mathrm{m}}^2)$ *and* $\mathrm{N}_{\mathrm{e},i} \sim \mathcal{CN}(0, \sigma_{\mathrm{e}}^2)$*, and can account for constant (and known) multiplicative coefficients* $h_{\mathrm{m}} \in \mathbb{C}$ *and* $h_{\mathrm{e}} \in \mathbb{C}$ *in the main channel and in the eavesdropper's channel, respectively. By noting that a complex Gaussian WTC is equivalent to two parallel real Gaussian WTCs with power constraint* $P/2$ *(and half the noise variance), and that a multiplicative coefficient induces a scaling of the received SNR, the secrecy capacity follows directly from the previous analysis and we have*

$$C_{\mathrm{s}} = \left(\log\left(1 + \frac{|h_{\mathrm{m}}|^2 P}{\sigma_{\mathrm{m}}^2}\right) - \log\left(1 + \frac{|h_{\mathrm{e}}|^2 P}{\sigma_{\mathrm{e}}^2}\right)\right)^+ .$$

Remark 5.2. *Suppose that the eavesdropper's noise is known to be Gaussian, but the variance is known only to satisfy* $\sigma_{\mathrm{e}}^2 \geqslant \sigma_0^2$ *for some fixed* σ_0^2*. One can check that a set of Gaussian channels with noise variance* $\sigma_{\mathrm{e}}^2 \geqslant \sigma_0^2$ *forms a class of stochastically degraded channels, as introduced in Definition 3.10. Proposition 3.3 guarantees that a wiretap code designed for an eavesdropper's noise variance* σ_0^2 *will also ensure secrecy if the actual variance is* $\sigma_{\mathrm{e}}^2 \geqslant \sigma_0^2$*.*

5.1.2 Multiple-input multiple-output Gaussian wiretap channel

Generalizing the results obtained in the previous section to a multiple-input multiple-output (MIMO) situation is not merely useful for the sake of completeness; it also allows us to study the effect of spatial dimensionality and collusion of eavesdroppers on secure communications rates. The MIMO wiretap channel[2] is illustrated in Figure 5.3. The numbers of antennas used by the transmitter, receiver, and eavesdropper are denoted n_{t}, n_{r}, and n_{e}, respectively. Notice that the model does not distinguish between a single eavesdropper with multiple antennas and a set of multiple eavesdroppers who collude

[2] This model is also called the multiple-input multiple-output multiple-eavesdropper (MIMOME) channel to emphasize that all parties have multiple antennas.

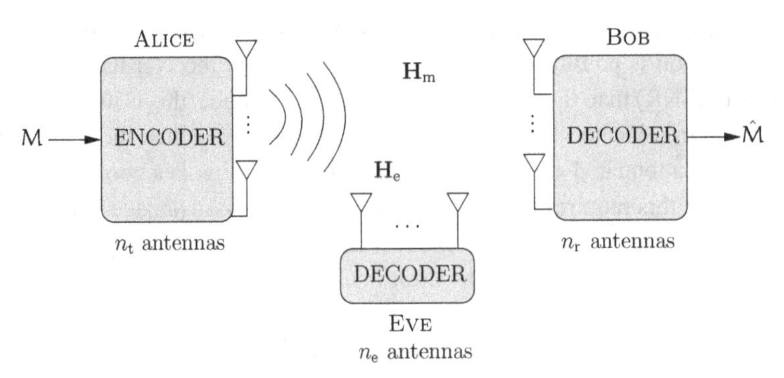

Figure 5.3 Communication over a MIMO Gaussian WTC.

and process their measurements jointly. In practice, there is a physical limit to the number of useful collocated antennas that one can deploy; therefore, a collusion of eavesdroppers is likely to be more powerful than a single eavesdropper with multiple antennas.

For a Gaussian MIMO wiretap channel (Gaussian MIMO WTC for short), the relationships between the inputs and outputs of the channel at each time i are

$$\mathrm{Y}_i^{n_\mathrm{r}} = \mathbf{H}_\mathrm{m}\mathrm{X}_i^{n_\mathrm{t}} + \mathrm{N}_{\mathrm{m},i} \quad \text{and} \quad \mathrm{Z}_i^{n_\mathrm{e}} = \mathbf{H}_\mathrm{e}\mathrm{X}_i^{n_\mathrm{t}} + \mathrm{N}_{\mathrm{e},i},$$

where $\mathrm{X}_i^{n_\mathrm{t}} \in \mathbb{C}^{n_\mathrm{t}\times 1}$ is the channel input vector, $\mathbf{H}_\mathrm{m} \in \mathbb{C}^{n_\mathrm{r}\times n_\mathrm{t}}$ and $\mathbf{H}_\mathrm{e} \in \mathbb{C}^{n_\mathrm{e}\times n_\mathrm{t}}$ are deterministic complex matrices, $\mathrm{Y}_i^{n_\mathrm{r}} \in \mathbb{C}^{n_\mathrm{r}\times 1}$ is the legitimate receiver's observation vector, and $\mathrm{Z}_i^{n_\mathrm{e}} \in \mathbb{C}^{n_\mathrm{e}\times 1}$ is the eavesdropper's observation vector. The channel matrices $\mathbf{H}_\mathrm{m}$ and $\mathbf{H}_\mathrm{e}$ are fixed for the entire transmission and known to all three terminals. The noise processes $\{\mathrm{N}_{\mathrm{m},i}\}_{i\geqslant 1}$ and $\{\mathrm{N}_{\mathrm{e},i}\}_{i\geqslant 1}$ are i.i.d.; at each time i the vectors $\mathrm{N}_{\mathrm{m},i} \in \mathbb{C}^{n_\mathrm{r}\times 1}$ and $\mathrm{N}_{\mathrm{e},i} \in \mathbb{C}^{n_\mathrm{e}\times 1}$ are circularly symmetric complex Gaussian random vectors with covariance matrices $\mathbf{K}_\mathrm{m} = \sigma_\mathrm{m}^2\mathbf{I}_{n_\mathrm{r}}$ and $\mathbf{K}_\mathrm{e} = \sigma_\mathrm{e}^2\mathbf{I}_{n_\mathrm{e}}$, respectively, where $\mathbf{I}_n$ is the identity matrix of dimension n. The channel input is also subject to the long-term average power constraint $(1/n)\sum_{i=1}^n \mathbb{E}\left[\left\|\mathrm{X}_i^{n_\mathrm{t}}\right\|^2\right] \leqslant P$.

Theorem 5.2 (Khisti and Wornell, Oggier and Hassibi, Liu and Shamai). *The secrecy capacity of the Gaussian MIMO WTC is*

$$C_\mathrm{s}^{\mathrm{MIMO}} = \max\left(\log\left|\mathbf{I}_{n_\mathrm{r}} + \frac{1}{\sigma_\mathrm{m}^2}\mathbf{H}_\mathrm{m}\mathbf{K}_\mathrm{X}\mathbf{H}_\mathrm{m}^\dagger\right| - \log\left|\mathbf{I}_{n_\mathrm{e}} + \frac{1}{\sigma_\mathrm{e}^2}\mathbf{H}_\mathrm{e}\mathbf{K}_\mathrm{X}\mathbf{H}_\mathrm{e}^\dagger\right|\right),$$

where the maximization is over all positive semi-definite matrices $\mathbf{K}_\mathrm{X}$ *such that* $\mathrm{tr}(\mathbf{K}_\mathrm{X}) \leqslant P$.

The expression for $C_\mathrm{s}^{\mathrm{MIMO}}$ is the natural generalization of the scalar case obtained in Corollary 5.1 which we could have expected; however, the proof of this result is significantly more involved. On the one hand, the achievability of rates below $C_\mathrm{s}^{\mathrm{MIMO}}$ follows from Corollary 3.4, which can be shown to hold for continuous vector channels with multiple inputs and outputs. Choosing an $n_\mathrm{t} \times n_\mathrm{t}$ positive semi-definite matrix $\mathbf{K}$ such that $\mathrm{tr}(\mathbf{K}) = P$ and substituting the random variables

$$\mathrm{V} \sim \mathcal{CN}(0, \mathbf{K}) \quad \text{and} \quad \mathrm{X} \triangleq \mathrm{V} \tag{5.14}$$

into Corollary 3.4 yields the desired result. On the other hand, proving that the choice of random variables in (5.14) is optimal is arduous because, in general and in contrast to the scalar Gaussian WTC, the eavesdropper's channel is not stochastically degraded with respect to the main channel. We refer the reader to the bibliographical notes at the end of this chapter for references to various proofs.

The maximization over covariance matrices subject to a trace constraint in the expression for C_s^{MIMO} makes it difficult to develop much intuition from Theorem 5.2 directly. Nevertheless, it is possible to develop a necessary and sufficient condition for $C_s^{\text{MIMO}} = 0$ that admits a more intuitive interpretation.

Proposition 5.1 (Khisti and Wornell). *The secrecy capacity of the Gaussian MIMO WTC is zero if and only if* $\lambda_{\max}(\mathbf{H}_w, \mathbf{H}_e) \leqslant 1$, *where*

$$\lambda_{\max}(\mathbf{H}_w, \mathbf{H}_e) \triangleq \sup_{\mathbf{v}\in\mathbb{C}^{n_t}} \frac{\sigma_e}{\sigma_m} \frac{\|\mathbf{H}_m\mathbf{v}\|}{\|\mathbf{H}_e\mathbf{v}\|}.$$

Sketch of proof. The kernel of a matrix $\mathbf{H}$ is $\text{Ker}(\mathbf{H}) \triangleq \{\mathbf{v} : \mathbf{H}\mathbf{v} = \mathbf{0}\}$. If $\text{Ker}(\mathbf{H}_e) \cap \text{Ker}(\mathbf{H}_m)^{\perp} \neq \emptyset$, there exists a vector $\mathbf{v}$ such that $\|\mathbf{H}_m\mathbf{v}\| > 0$ and $\|\mathbf{H}_e\mathbf{v}\| = 0$. In this case, $\lambda_{\max}$ is undefined and the transmitter can communicate securely by beamforming his signal in the direction of $\mathbf{v}$, which is unheard by the eavesdropper. Notice that this strategy does not require a wiretap code and beamforming is sufficient to secure communications.

If $\text{Ker}(\mathbf{H}_e) \cap \text{Ker}(\mathbf{H}_m)^{\perp} = \emptyset$, then beamforming is not sufficient to secure communications. Nevertheless, if $\lambda_{\max}(\mathbf{H}_w, \mathbf{H}_e) > 1$, then there exists $\mathbf{v}$ with $\|\mathbf{v}\| = 1$ such that $\|\mathbf{H}_m\mathbf{v}\|/\sigma_m > \|\mathbf{H}_e\mathbf{v}\|/\sigma_e$; in other words, even though the eavesdropper overhears all signals, there exists (at least) one direction in which the legitimate receiver benefits from a higher gain than the eavesdropper. Substituting the random variables

$$\mathsf{V} \sim \mathcal{CN}(0, P\mathbf{v}\mathbf{v}^{\mathsf{T}}) \quad \text{and} \quad \mathsf{X} \triangleq \mathsf{V}$$

into Corollary 3.4 shows that

$$C_s \geqslant \log\left|\mathbf{I}_{n_r} + \frac{P}{\sigma_m^2}\mathbf{H}_m\mathbf{v}\mathbf{v}^{\dagger}\mathbf{H}_m^{\dagger}\right| - \log\left|\mathbf{I}_{n_r} + \frac{P}{\sigma_e^2}\mathbf{H}_e\mathbf{v}\mathbf{v}^{\dagger}\mathbf{H}_e^{\dagger}\right|. \tag{5.15}$$

Using the identity $\log|\mathbf{I} + \mathbf{AB}| = \log|\mathbf{I} + \mathbf{BA}|$, we can rewrite (5.15) as

$$C_s \geqslant \log\left(1 + \frac{P}{\sigma_m^2}\|\mathbf{H}_m\mathbf{v}\|^2\right) - \log\left(1 + \frac{P}{\sigma_e^2}\|\mathbf{H}_e\mathbf{v}\|^2\right),$$

and the right-hand side is strictly positive since $\|\mathbf{H}_m\mathbf{v}\|/\sigma_m > \|\mathbf{H}_e\mathbf{v}\|/\sigma_e$.

If $\lambda_{\max}(\mathbf{H}_w, \mathbf{H}_e) \leqslant 1$, it is also possible to show that $C_s^{\text{MIMO}} = 0$. The proof hinges on a closed-form expression for $C_s^{\text{MIMO}} = 0$ in the high-SNR regime obtained using a generalized singular-value decomposition of $\mathbf{H}_m$ and $\mathbf{H}_e$. We refer the reader to [83, 84] for details of the proof. □

As expected, Proposition 5.1 confirms that secure communication is possible if the transmitter can beamform his signals in such a direction that the eavesdropper does not overhear. Perhaps more interestingly, Proposition 5.1 also shows that the secrecy capacity is strictly positive as long as the transmitter can beamform his signals in a

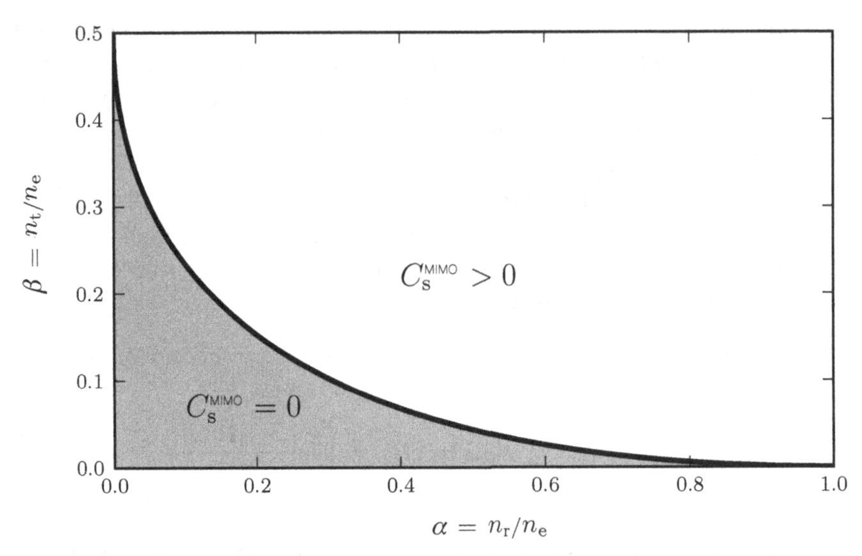

Figure 5.4 Condition for zero secrecy capacity in the limit of a large number of antennas. The elements of $\mathbf{H}_m$ and $\mathbf{H}_e$ are assumed to be generated i.i.d. according to the distribution $\mathcal{CN}(0, 1)$ and $\sigma_m = \sigma_e = 1$.

direction for which the eavesdropper obtains a lower SNR than does the legitimate receiver. In other words, the combination of coding and beamforming is more powerful than beamforming alone. Without additional assumptions about the specific structure of $\mathbf{H}_m$ and $\mathbf{H}_e$, little can be said regarding the existence of a secure beamforming direction. Nevertheless, if the entries of $\mathbf{H}_m$ and $\mathbf{H}_e$ are generated i.i.d. according to $\mathcal{CN}(0, 1)$ and if their realizations are known to Alice, Bob, and Eve, the behavior of C_s^{MIMO} can be further analyzed using tools from random-matrix theory.

Proposition 5.2 (Khisti and Wornell). *Suppose that $\sigma_m = \sigma_e = 1$ and n_r, n_t, and n_e go to infinity while the ratios $\alpha \triangleq n_r/n_e$ and $\beta \triangleq n_t/n_e$ are kept fixed. Then, the secrecy capacity converges almost surely to zero if and only if*

$$0 \leqslant \beta \leqslant \frac{1}{2}, \quad 0 \leqslant \alpha \leqslant 1, \quad \textit{and} \quad \alpha \leqslant \left(1 - \sqrt{2\beta}\right)^2.$$

The proof of Proposition 5.2 can be found in [83, 84]. Proposition 5.2 allows us to relate the possibility of secure communication directly to the number of antennas deployed by Alice, Bob, and Eve. As expected, and as illustrated in Figure 5.4, the secrecy capacity is positive as long as Eve does not deploy too many antennas compared with the numbers deployed by Alice and Bob. For instance, if $\alpha = 0$, which corresponds to a single receive antenna for Bob, the secrecy capacity is positive if Eve has fewer than twice as many antennas as Alice. For $\beta = 0$, which corresponds to a single transmit antenna for Alice, the secrecy capacity is positive provided that Eve has fewer antennas than Bob. This leads to the pessimistic conclusion that little can be done against an all-powerful eavesdropper who is able to deploy many antennas; however, one can perhaps draw a more optimistic conclusion and argue that Alice and Bob can mitigate the impact of colluding eavesdroppers by deploying multiple transmit and receive antennas.

We conclude this section on the MIMO Gaussian WTC with a brief discussion of a suboptimal strategy that sheds light on the choice of the input covariance matrix $\mathbf{K}_X$ in Theorem 5.2. Let $r = \text{rk}(\mathbf{H}_m)$ and consider the compact singular-value decomposition $\mathbf{H}_m = \mathbf{U}_m\Lambda_m\mathbf{V}_m^\dagger$, in which $\mathbf{U}_m \in \mathbb{C}^{n_r \times r}$ and $\mathbf{V}_m \in \mathbb{C}^{n_t \times r}$ have unitary columns, and $\Lambda_m \in \mathbb{C}^{r\times r}$ is a diagonal matrix with non-zero diagonal terms. We construct a unitary matrix $\mathbf{V} \in \mathbb{C}^{n_t \times n_t}$, by appending appropriate column vectors to $\mathbf{V}_m$, and we let

$$\mathbf{V} \triangleq (\mathbf{V}_m\ \mathbf{V}_n) \quad \text{with} \quad \mathbf{V}_m \triangleq (\mathbf{v}_1 \dots \mathbf{v}_r) \quad \text{and} \quad \mathbf{V}_n \triangleq \left(\mathbf{v}_{r+1} \dots \mathbf{v}_{n_t}\right).$$

This decomposition allows us to interpret the channel to the legitimate receiver as n_t parallel channels, of which only the first r can effectively be used for communication. Alice simultaneously transmits r symbols $b_1, \dots, b_r$ by sending the vector

$$\mathbf{x} = \sum_{j=1}^{r} b_j\mathbf{v}_j + \sum_{j=r+1}^{n_t} n_j\mathbf{v}_j,$$

where $\{n_j\}_{j=r+1}^{n_t}$ are dummy noise symbols. These dummy symbols do not affect Bob's received signal because they lie in the kernel of $\mathbf{H}_m$, but they are mixed with the useful symbols in Eve's received signal. This scheme is called an *artificial noise transmission* strategy since it consists essentially of transmitting information in the direction of the non-zero singular values of $\mathbf{H}_m$, and sending noise in all other directions to harm the eavesdropper. A simple way of ensuring that the power constraint $(1/n)\sum_{i=1}^{n}\mathbb{E}\left[\|X_i^{n_t}\|^2\right] \leqslant P$ is satisfied is to allocate the same average power P/n_t to all n_t sub-channels. In this case, achievable secure communication rates are given by the following proposition.

Proposition 5.3 (Khisti *et al.*). *The artificial noise transmission strategy achieves all the secure rates R_s such that*

$$R_s < \log\left|\mathbf{I}_r + \frac{P}{\sigma_m^2 n_t}\Lambda_m\Lambda_m^\dagger\right| + \log\left|\mathbf{V}_m^\dagger\left(\mathbf{I} + \frac{P}{\sigma_e^2 n_t}\mathbf{H}_e^\dagger\mathbf{H}_e\right)^{-1}\mathbf{V}_m\right|$$

Proof. The secure communication rates are obtained by substituting the random variables

$$\mathsf{V} \triangleq \sum_{j=1}^{r}\mathsf{B}_j\mathbf{v}_j \quad \text{and} \quad \mathsf{X} = \mathsf{V} + \sum_{j=r+1}^{n_t}\mathsf{N}_j\mathbf{v}_j$$

into Corollary 3.4, where the random variables $\{\mathsf{B}_j\}_{j=1}^{r}$ and $\{\mathsf{N}_j\}_{j=r+1}^{n_t}$ are i.i.d. and drawn according to $\mathcal{CN}(0, P/n_t)$. Then,

$$\begin{aligned}
\mathbb{I}(\mathsf{V};\mathsf{Y}) &= \mathbb{h}(\mathsf{Y}) - \mathbb{h}(\mathsf{Y}|\mathsf{X}) && (5.16)\\
&= \log\left|\sigma_m^2\mathbf{I} + \frac{P}{n_t}\mathbf{H}_m\mathbf{V}\mathbf{V}^\dagger\mathbf{H}_m^\dagger\right| - \log\left|\sigma_m^2\mathbf{I}\right| \\
&= \log\left|\mathbf{I} + \frac{P}{\sigma_m^2 n_t}\mathbf{U}_m\Lambda_m\mathbf{V}_m^\dagger\mathbf{V}\mathbf{V}^\dagger\mathbf{V}_m\Lambda_m^\dagger\mathbf{U}_m^\dagger\right| \\
&= \log\left|\mathbf{I} + \frac{P}{\sigma_m^2 n_t}\Lambda_m\Lambda_m^\dagger\right|, && (5.17)
\end{aligned}$$

where we have used $\mathbf{V}_{\mathrm{m}}^{\dagger}\mathbf{V}_{\mathrm{m}} = \mathbf{I}_{\mathrm{r}}$, $\mathbf{U}_{\mathrm{m}}^{\dagger}\mathbf{U}_{\mathrm{m}} = \mathbf{I}_{\mathrm{r}}$, and $\log|\mathbf{I} + \mathbf{AB}| = \log|\mathbf{I} + \mathbf{BA}|$ to obtain the last equality. Similarly,

$$\begin{aligned}
\mathbb{I}(\mathsf{V}; \mathsf{Z}) &= \mathbb{h}(\mathsf{Z}) - \mathbb{h}(\mathsf{Z}|\mathsf{V}) \\
&= \log\left|\sigma_{\mathrm{e}}^2\mathbf{I} + \frac{P}{n_{\mathrm{t}}}\mathbf{H}_{\mathrm{e}}\mathbf{V}\mathbf{V}^{\dagger}\mathbf{H}_{\mathrm{e}}^{\dagger}\right| - \log\left|\sigma_{\mathrm{e}}^2\mathbf{I} + \frac{P}{n_{\mathrm{t}}}\mathbf{H}_{\mathrm{e}}\mathbf{V}_{\mathrm{n}}\mathbf{V}_{\mathrm{n}}^{\dagger}\mathbf{H}_{\mathrm{e}}^{\dagger}\right| \\
&= \log\left|\mathbf{I} + \frac{P}{\sigma_{\mathrm{e}}^2 n_{\mathrm{t}}}\mathbf{H}_{\mathrm{e}}\mathbf{H}_{\mathrm{e}}^{\dagger}\right| - \log\left|\mathbf{I} + \frac{P}{\sigma_{\mathrm{e}}^2 n_{\mathrm{t}}}\mathbf{V}_{\mathrm{n}}\mathbf{V}_{\mathrm{n}}^{\dagger}\mathbf{H}_{\mathrm{e}}^{\dagger}\mathbf{H}_{\mathrm{e}}\right|.
\end{aligned}$$

Since $\mathbf{V}$ is unitary, $\mathbf{V}_{\mathrm{n}}\mathbf{V}_{\mathrm{n}}^{\dagger} + \mathbf{V}_{\mathrm{m}}\mathbf{V}_{\mathrm{m}}^{\dagger} = \mathbf{I}$; hence,

$$\begin{aligned}
&\log\left|\mathbf{I} + \frac{P}{\sigma_{\mathrm{e}}^2 n_{\mathrm{t}}}\mathbf{V}_{\mathrm{n}}\mathbf{V}_{\mathrm{n}}^{\dagger}\mathbf{H}_{\mathrm{e}}^{\dagger}\mathbf{H}_{\mathrm{e}}\right| \\
&\quad= \log\left|\mathbf{I} + \frac{P}{\sigma_{\mathrm{e}}^2 n_{\mathrm{t}}}(\mathbf{I} - \mathbf{V}_{\mathrm{m}}\mathbf{V}_{\mathrm{m}}^{\dagger})\mathbf{H}_{\mathrm{e}}^{\dagger}\mathbf{H}_{\mathrm{e}}\right| \\
&\quad= \log\left|\left(\mathbf{I} + \frac{P}{\sigma_{\mathrm{e}}^2 n_{\mathrm{t}}}\mathbf{H}_{\mathrm{e}}^{\dagger}\mathbf{H}_{\mathrm{e}}\right)\left(\mathbf{I} - \frac{P}{\sigma_{\mathrm{e}}^2 n_{\mathrm{t}}}\left(\mathbf{I} + \frac{P}{\sigma_{\mathrm{e}}^2 n_{\mathrm{t}}}\mathbf{H}_{\mathrm{e}}^{\dagger}\mathbf{H}_{\mathrm{e}}\right)^{-1}\mathbf{V}_{\mathrm{m}}\mathbf{V}_{\mathrm{m}}^{\dagger}\mathbf{H}_{\mathrm{e}}^{\dagger}\mathbf{H}_{\mathrm{e}}\right)\right|.
\end{aligned}$$

Therefore,

$$\begin{aligned}
\mathbb{I}(\mathsf{V}; \mathsf{Z}) &= -\log\left|\mathbf{I} - \frac{P}{\sigma_{\mathrm{e}}^2 n_{\mathrm{t}}}\left(\mathbf{I} + \frac{P}{\sigma_{\mathrm{e}}^2 n_{\mathrm{t}}}\mathbf{H}_{\mathrm{e}}^{\dagger}\mathbf{H}_{\mathrm{e}}\right)^{-1}\mathbf{V}_{\mathrm{m}}\mathbf{V}_{\mathrm{m}}^{\dagger}\mathbf{H}_{\mathrm{e}}^{\dagger}\mathbf{H}_{\mathrm{e}}\right| \\
&= -\log\left|\mathbf{I} - \frac{P}{\sigma_{\mathrm{e}}^2 n_{\mathrm{t}}}\mathbf{V}_{\mathrm{m}}^{\dagger}\mathbf{H}_{\mathrm{e}}^{\dagger}\mathbf{H}_{\mathrm{e}}\left(\mathbf{I} + \frac{P}{\sigma_{\mathrm{e}}^2 n_{\mathrm{t}}}\mathbf{H}_{\mathrm{e}}^{\dagger}\mathbf{H}_{\mathrm{e}}\right)^{-1}\mathbf{V}_{\mathrm{m}}\right| \\
&= -\log\left|\mathbf{V}_{\mathrm{m}}^{\dagger}\left(\mathbf{I} + \frac{P}{\sigma_{\mathrm{e}}^2 n_{\mathrm{t}}}\mathbf{H}_{\mathrm{e}}^{\dagger}\mathbf{H}_{\mathrm{e}}\right)^{-1}\mathbf{V}_{\mathrm{m}}\right|. \qquad (5.18)
\end{aligned}$$

By Corollary 3.4, all the rates $R_{\mathrm{s}} < \mathbb{I}(\mathsf{V}; \mathsf{Y}) - \mathbb{I}(\mathsf{V}; \mathsf{Z})$ with $\mathbb{I}(\mathsf{V}; \mathsf{Y})$ given by (5.17) and $\mathbb{I}(\mathsf{V}; \mathsf{Z})$ given by (5.18) are achievable. □

The idea of introducing artificial noise into the system to hinder the eavesdropper is a powerful concept that will reappear in Chapter 8 for multi-user secure communication systems.

Remark 5.3. *Although the signaling used in the artificial noise transmission strategy relies solely on knowledge of* $\mathbf{H}_{\mathrm{m}}$ *and does not exploit knowledge of the eavesdropper's channel* $\mathbf{H}_{\mathrm{e}}$*, notice that knowledge of* $\mathbf{H}_{\mathrm{e}}$ *is* required *in order to design the wiretap code and select the secure communication rate appropriately. Hence, the artificial noise operates in only a* semi-blind *fashion.*

5.1.3 Gaussian source model

A Gaussian source model for secret-key agreement consists of a memoryless source $(\mathcal{XYZ}, p_{\mathsf{XYZ}})$ whose components are jointly Gaussian with zero mean. The distribution is entirely characterized by the second-order moments

$$\mathbb{E}\left[X^2\right] = \sigma_X^2, \quad \mathbb{E}\left[Y^2\right] = \sigma_Y^2, \quad \mathbb{E}\left[Z^2\right] = \sigma_Z^2,$$
$$\mathbb{E}[XY] = \rho_1\sigma_X\sigma_Y, \quad \mathbb{E}[XZ] = \rho_2\sigma_X\sigma_Z, \quad \mathbb{E}[YZ] = \rho_3\sigma_Y\sigma_Z,$$

where ρ_1, ρ_2, and ρ_3 are the correlation coefficients of the source components. The definitions of key-distillation strategies, achievable key rates, and the secret-key capacity are those used for discrete memoryless sources. A closer look at the proof of the upper bound for the secret-key capacity in Section 4.2.1 shows that the derivation does not rely on the discrete nature of the source $(\mathcal{XYZ}, p_{XYZ})$; however, the achievability proof based on a conceptual WTC relies on the crypto lemma, which does not apply to Gaussian random variables. Nevertheless, we show in this section that the lower bound is still valid for a Gaussian source model.

Proposition 5.4. *For a Gaussian source model,*

$$\mathbb{I}(X;Y) - \min(\mathbb{I}(X;Z), \mathbb{I}(Y;Z)) \leqslant C_s^{SM} \leqslant \min(\mathbb{I}(X;Y), \mathbb{I}(X,Y|Z)).$$

Proof. The upper bound follows from the same steps as in the proof of Theorem 4.1, and we need only show that the lower bound holds. To do so, we construct a conceptual WTC as in Section 4.2.1 but we use this time the addition over real numbers. Specifically, to send a symbol $U \in \mathbb{R}$ independent of the DMS to Bob, Alice observes a realization X of the DMS and transmits $U + X$ over the public channel, where + denotes the usual addition over $\mathbb{R}$. This creates a conceptual memoryless WTC, for which Bob observes $(U + X, Y)$ and Eve observes $(U + X, Z)$. From the results of Chapter 3, we know that the secrecy capacity is at least

$$\mathbb{I}(U;Y,U+X) - \mathbb{I}(U;Z,U+X),$$

where the distribution p_U can be chosen arbitrarily; here, we choose $U \sim \mathcal{N}(0, P)$ for some $P > 0$. Using the chain rule of mutual information repeatedly, we obtain

$$\begin{aligned}
&\mathbb{I}(U;Y,U+X) - \mathbb{I}(U;Z,U+X)\\
&\quad = \mathbb{I}(U;Y) + \mathbb{I}(U;U+X|Y) - \mathbb{I}(U;Z) - \mathbb{I}(U;U+X|Z)\\
&\quad = \mathbb{h}(U|Z) - \mathbb{h}(U|Y) + \mathbb{h}(U+X|Y) - \mathbb{h}(U+X|UY) - \mathbb{h}(U+X|Z)\\
&\qquad + \mathbb{h}(U+X|UZ)\\
&\quad \stackrel{(a)}{=} \mathbb{h}(U|Z) - \mathbb{h}(U|Y) + \mathbb{h}(U+X|Y) - \mathbb{h}(X|Y) - \mathbb{h}(U+X|Z) + \mathbb{h}(X|Z). \qquad (5.19)
\end{aligned}$$

Equality (a) follows because U is independent of (X, Y), which implies that

$$\mathbb{h}(U+X|UY) = \mathbb{h}(X|UY) = \mathbb{h}(X|Y) \quad \text{and} \quad \mathbb{h}(U+X|UZ) = \mathbb{h}(X|Z).$$

Now,

$$\begin{aligned}
\mathbb{h}(U|Z) - \mathbb{h}(U+X|Z) &\leqslant \mathbb{h}(U|Z) - \mathbb{h}(U+X|Z,X)\\
&= \mathbb{h}(U|Z) - \mathbb{h}(U|Z,X)\\
&= 0,
\end{aligned}$$

where the last equality follows again from the independence of U and (X, Z). Also,

$$\mathbb{h}(\mathsf{U}|\mathsf{Z}) - \mathbb{h}(\mathsf{U}+\mathsf{X}|\mathsf{Z}) \geqslant \mathbb{h}(\mathsf{U}|\mathsf{Z}) - \mathbb{h}(\mathsf{U}+\mathsf{X}) \geqslant \log\left(\frac{P}{P+\sigma_{\mathsf{x}}^2}\right),$$

where the last inequality follows from $\mathbb{h}(\mathsf{U}|\mathsf{Z}) = \mathbb{h}(\mathsf{U}) = \log(2\pi e P)$ and the bound $\mathbb{h}(\mathsf{U}+\mathsf{X}) \leqslant \log(2\pi e(P+\sigma_{\mathsf{x}}^2))$. Because all communication takes place over a public noiseless channel (of infinite capacity), P can be arbitrarily large,[3] and, for any $\epsilon > 0$, we can choose P such that

$$|\mathbb{h}(\mathsf{U}|\mathsf{Z}) - \mathbb{h}(\mathsf{U}+\mathsf{X}|\mathsf{Z})| \leqslant \frac{\epsilon}{2}. \tag{5.20}$$

Repeating the same argument with Y instead of Z shows that for P large enough we have

$$|\mathbb{h}(\mathsf{U}|\mathsf{Y}) - \mathbb{h}(\mathsf{U}+\mathsf{X}|\mathsf{Y})| \leqslant \frac{\epsilon}{2}. \tag{5.21}$$

On combining (5.19), (5.20), and (5.21), we obtain

$$\begin{aligned}\mathbb{I}(\mathsf{U};\mathsf{Y},\mathsf{U}+\mathsf{X}) - \mathbb{I}(\mathsf{U};\mathsf{Z},\mathsf{U}+\mathsf{X}) &\geqslant \mathbb{h}(\mathsf{X}|\mathsf{Z}) - \mathbb{h}(\mathsf{X}|\mathsf{Y}) - \epsilon \\ &= \mathbb{I}(\mathsf{X};\mathsf{Y}) - \mathbb{I}(\mathsf{X};\mathsf{Z}) - \epsilon.\end{aligned}$$

Since $\epsilon > 0$ can be chosen arbitrarily small, we must have

$$C_{\mathrm{s}}^{\mathrm{SM}} \geqslant \mathbb{I}(\mathsf{X};\mathsf{Y}) - \mathbb{I}(\mathsf{X};\mathsf{Z}).$$

Similarly, by interchanging the roles of Y and Z, we can show that

$$C_{\mathrm{s}}^{\mathrm{SM}} \geqslant \mathbb{I}(\mathsf{X};\mathsf{Y}) - \mathbb{I}(\mathsf{Y};\mathsf{Z}).$$

□

Remark 5.4. *There is no loss of generality by restricting our analysis to a centered Gaussian source model. If X, Y, and Z have non-zero means μ_1, μ_1, and μ_3, one can simply consider the centered random variables $\mathsf{X}' \triangleq \mathsf{X} - \mu_1$, $\mathsf{Y}' \triangleq \mathsf{Y} - \mu_2$, and $\mathsf{Z}' \triangleq \mathsf{Z} - \mu_3$ and note that the bounds on the secret-key capacity remain unchanged.*

The bounds given by Proposition 5.4 can be computed explicitly in terms of the parameters ρ_1, ρ_2, and ρ_3.

Corollary 5.2. *The secret-key capacity of a Gaussian source model satisfies*

$$\begin{aligned}&\max\left(\frac{1}{2}\log\left(\frac{1-\rho_2^2}{1-\rho_1^2}\right), \frac{1}{2}\log\left(\frac{1-\rho_3^2}{1-\rho_1^2}\right)\right) \\ &\quad\leqslant C_{\mathrm{s}}^{\mathrm{SM}} \leqslant \min\left(\frac{1}{2}\log\left(\frac{1}{1-\rho_1^2}\right), \frac{1}{2}\log\left(\frac{(1-\rho_2^2)(1-\rho_3^2)}{1+2\rho_1\rho_2\rho_3-\rho_1^2-\rho_2^2-\rho_3^2}\right)\right).\end{aligned}$$

[3] Note that our ability to choose P as large as desired is a mathematical convenience rather than a realistic solution. In practice, even public communication would be subject to a power constraint and thus to a rate constraint.

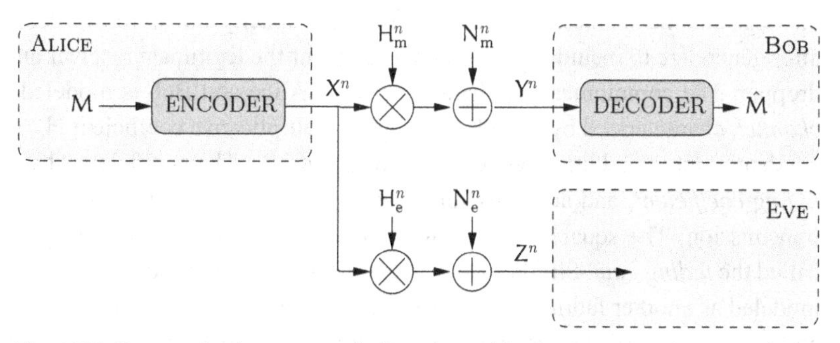

Figure 5.5 Communication over a wireless channel in the presence of an eavesdropper.

Remark 5.5. *Note that if $\rho_2 \geqslant \rho_1$ and $\rho_3 \geqslant \rho_1$ the lower bound obtained is negative and not really useful. Nevertheless, it is sometimes still possible to obtain a positive secret-key rate using an advantage-distillation protocol, as discussed in Section 4.3.1.*

Example 5.1. An interesting instance of a Gaussian source model is one in which $\mathsf{X} \sim \mathcal{N}(0, P)$ is a Gaussian random variable transmitted over a Gaussian WTC, such that

$$\mathsf{Y} = \mathsf{X} + \mathsf{N}_\mathrm{m} \quad \text{and} \quad \mathsf{Z} = \mathsf{X} + \mathsf{N}_\mathrm{e}$$

with $\mathsf{N}_\mathrm{m} \sim \mathcal{N}(0, \sigma_\mathrm{m}^2)$ and $\mathsf{N}_\mathrm{e} \sim \mathcal{N}(0, \sigma_\mathrm{e}^2)$. In this case, Corollary 4.1 and Proposition 5.4 apply and $C_\mathrm{s}^\mathsf{SM} = \mathbb{I}(\mathsf{X};\mathsf{Y}) - \mathbb{I}(\mathsf{Y};\mathsf{Z})$, which can be computed explicitly as

$$C_\mathrm{s}^\mathsf{SM} = \frac{1}{2}\log\left(1 + \frac{P\sigma_\mathrm{e}^2}{(P + \sigma_\mathrm{e}^2)\sigma_\mathrm{m}^2}\right).$$

Note that C_s^SM is positive even if $\sigma_\mathrm{e}^2 < \sigma_\mathrm{m}^2$. In contrast, the secrecy capacity of the same Gaussian WTC given in Theorem 5.1 is

$$C_\mathrm{s} = \left(\frac{1}{2}\log\left(1 + \frac{P}{\sigma_\mathrm{m}^2}\right) - \frac{1}{2}\log\left(1 + \frac{P}{\sigma_\mathrm{e}^2}\right)\right)^+,$$

which is zero if $\sigma_\mathrm{e}^2 < \sigma_\mathrm{m}^2$. Hence, the impossibility of secure communication over a Gaussian WTC when the eavesdropper has a higher SNR than that of the legitimate receiver is *solely* due to the restrictions placed on the communication scheme. In reality, as long as the eavesdropper obtains a different observation, the legitimate parties always have an advantage over the eavesdropper, and they can distill a secret key.

5.2 Wireless channels

The general channel model we use to investigate secure wireless communications is illustrated in Figure 5.5. For simplicity, we focus on the transmission of a single secure message and the characterization of the secrecy capacity, but all results described thereafter generalize to include a common message for the legitimate receiver and the eavesdropper. The communication channel between Alice and Bob is modeled as a *fading channel*, characterized by a random complex multiplicative coefficient H_m and an independent complex additive white Gaussian noise N_m. The coefficient H_m is called the *fading coefficient*, and accounts for the multipath interference occurring in a wireless transmission. The square of the magnitude of the fading coefficient, $\mathsf{G}_\text{m} \triangleq |\mathsf{H}_\text{m}|^2$, is called the *fading gain*. Similarly, the channel between Alice and the eavesdropper Eve is modeled as another fading channel with fading coefficient H_e, fading gain $\mathsf{G}_\text{e} \triangleq |\mathsf{H}_\text{e}|^2$, and additive white Gaussian noise N_e. In a continuous-time model, the time interval during which the fading coefficients remain almost constant is called a *coherence interval*; with a slight abuse of terminology, we call a realization (h_m, h_e) of the fading coefficients a coherence interval as well.

The relationships between inputs and outputs for each channel use i are given by

$$\mathsf{Y}_i = \mathsf{H}_{\text{m},i}\mathsf{X}_i + \mathsf{N}_{\text{m},i} \quad \text{and} \quad \mathsf{Z}_i = \mathsf{H}_{\text{e},i}\mathsf{X}_i + \mathsf{N}_{\text{e},i},$$

where $\mathsf{H}_{\text{m},i}$, $\mathsf{N}_{\text{m},i}$, $\mathsf{H}_{\text{e},i}$, and $\mathsf{N}_{\text{e},i}$ are mutually independent. The input of the channel is also subject to the power constraint

$$\frac{1}{n}\sum_{i=1}^{n}\mathbb{E}\left[\mathsf{X}_i^2\right] \leqslant P.$$

The noise processes $\left\{\mathsf{N}_{\text{m},i}\right\}_{i\geqslant 1}$ and $\left\{\mathsf{N}_{\text{e},i}\right\}_{i\geqslant 1}$ are i.i.d. complex Gaussian with

$$\mathsf{N}_{\text{m},i} \sim \mathcal{CN}(0,\sigma_\text{m}^2) \quad \text{and} \quad \mathsf{N}_{\text{e},i} \sim \mathcal{CN}(0,\sigma_\text{e}^2).$$

Different types of fading can be modeled by choosing the statistics of the fading coefficients $\left\{\mathsf{H}_{\text{m},i}\right\}_{i\geqslant 1}$ and $\left\{\mathsf{H}_{\text{e},i}\right\}_{i\geqslant 1}$ appropriately. In the remainder of the section, we focus on three standard fading models.

- **Ergodic-fading model:** this model characterizes a situation in which the duration of a coherence interval is on the order of the time required to send a single symbol. The processes $\left\{\mathsf{H}_{\text{m},i}\right\}_{i\geqslant 1}$ and $\left\{\mathsf{H}_{\text{e},i}\right\}_{i\geqslant 1}$ are mutually independent and i.i.d.; fading coefficients change at every channel use and a codeword experiences many fading realizations.
- **Block-fading model:** in this model, a codeword experiences many fading realizations; however, the time required to send a single symbol is much smaller than the duration of a coherence interval. The processes $\left\{\mathsf{H}_{\text{m},i}\right\}_{i\geqslant 1}$ and $\left\{\mathsf{H}_{\text{e},i}\right\}_{i\geqslant 1}$ are again mutually independent and i.i.d., but change every N channel uses; N is assumed to be sufficiently large for asymptotic coding results to hold within each coherence interval.
- **Quasi-static fading model:** this model differs fundamentally from the previous ones in that fading coefficients are assumed to remain *constant* over the transmission of an entire codeword, but change independently and randomly from one codeword to another. The processes $\left\{\mathsf{H}_{\text{m},i}\right\}_{i\geqslant 1}$ and $\left\{\mathsf{H}_{\text{e},i}\right\}_{i\geqslant 1}$ are mutually independent and i.i.d., characterizing a situation in which fading variations are on the order of the time required to send an entire codeword.

Remark 5.6. *We assume that the fading processes* $\{\mathsf{H}_{\mathrm{m},i}\}_{i\geqslant 1}$ *and* $\{\mathsf{H}_{\mathrm{e},i}\}_{i\geqslant 1}$ *are mutually independent and i.i.d to simplify the analysis; however, all results described thereafter generalize to the situation in which the processes are correlated, stationary, and ergodic.*

For all three models, we illustrate the results numerically by considering the special case of i.i.d. *Rayleigh fading*, for which $\{\mathsf{H}_{\mathrm{m},i}\}_{i\geqslant 1}$ and $\{\mathsf{H}_{\mathrm{e},i}\}_{i\geqslant 1}$ are mutually independent i.i.d. complex Gaussian processes with

$$\mathsf{H}_{\mathrm{m},i} \sim \mathcal{CN}(0, \alpha_{\mathrm{m}}^2) \quad \text{and} \quad \mathsf{H}_{\mathrm{e},i} \sim \mathcal{CN}(0, \alpha_{\mathrm{e}}^2).$$

In this case, the fading gains $\mathsf{G}_{\mathrm{m},i} \triangleq |\mathsf{H}_{\mathrm{m},i}|^2$ and $\mathsf{G}_{\mathrm{e},i} = |\mathsf{H}_{\mathrm{e},i}|^2$ are exponentially distributed with respective means

$$\mu_{\mathrm{m}} \triangleq \mathbb{E}[\mathsf{G}_{\mathrm{m},i}] = \alpha_{\mathrm{m}}^2 \quad \text{and} \quad \mu_{\mathrm{e}} \triangleq \mathbb{E}[\mathsf{G}_{\mathrm{e},i}] = \alpha_{\mathrm{e}}^2.$$

The statistics of the noise $\mathsf{N}_{\mathrm{m},i}$ and $\mathsf{N}_{\mathrm{e},i}$ are assumed known to Alice, Bob, and Eve, prior to transmission. Bob has at least instantaneous access to the fading coefficient $h_{\mathrm{m},i}$ and is able to detect symbols *coherently*. In addition, Eve has instantaneous access to both of the fading coefficients $h_{\mathrm{m},i}$ and $h_{\mathrm{e},i}$ so that the information leakage is always implicitly defined as

$$\mathbf{L}(\mathcal{C}_n) \triangleq \frac{1}{n}\mathbb{I}(\mathsf{M}; \mathsf{Z}^n | \mathsf{H}_{\mathrm{m}}^n \mathsf{H}_{\mathrm{e}}^n \mathcal{C}_n),$$

where $\mathsf{H}_{\mathrm{m}}^n$ and $\mathsf{H}_{\mathrm{e}}^n$ are the sequences of fading coefficients for the main and eavesdropper channel and $\mathcal{C}_n$ is the code used by Alice. Although providing the channel state information of the main channel to the eavesdropper is a pessimistic assumption, this assumption is required in the proofs. We will see that whether or not Alice has instantaneous access to the coefficients $h_{\mathrm{m},i}$ and $h_{\mathrm{e},i}$ has a significant impact on achievable communication rates, and several situations are considered in the next sections.

5.2.1 Ergodic-fading channels

For ergodic-fading channels, the processes $\{\mathsf{H}_{\mathrm{m},i}\}_{i\geqslant 1}$ and $\{\mathsf{H}_{\mathrm{e},i}\}_{i\geqslant 1}$ are mutually independent and i.i.d. We first assume that Alice, Bob, and Eve have full *channel state information* (CSI); that is, they all have access *instantaneously* to the realizations of the fading coefficients $(h_{\mathrm{m},i}, h_{\mathrm{e},i})$. In addition, a symbol sequence X^n is allowed to experience (infinitely) many fading realizations as the blocklength n goes to infinity; the average power constraint $(1/n)\sum_{i=1}^n \mathbb{E}[\mathsf{X}_i^2] \leqslant P$ is understood as a *long-term* constraint so that the power can be adjusted depending on the current fading realization.

Theorem 5.3 (Liang *et al.*). *With full CSI, the secrecy capacity of an ergodic fading wireless channel is*

$$C_{\mathrm{s}} = \max_{\gamma} \mathbb{E}_{\mathsf{G}_{\mathrm{m}}\mathsf{G}_{\mathrm{e}}}\left[\log\left(1 + \frac{\gamma(\mathsf{G}_{\mathrm{m}}, \mathsf{G}_{\mathrm{e}})\mathsf{G}_{\mathrm{m}}}{\sigma_{\mathrm{m}}^2}\right) - \log\left(1 + \frac{\gamma(\mathsf{G}_{\mathrm{m}}, \mathsf{G}_{\mathrm{e}})\mathsf{G}_{\mathrm{e}}}{\sigma_{\mathrm{e}}^2}\right)\right],$$

where $\gamma : \mathbb{R}_+^2 \to \mathbb{R}_+$ *is subject to the constraint* $\mathbb{E}[\gamma(\mathsf{G}_{\mathrm{m}}, \mathsf{G}_{\mathrm{e}})] \leqslant P$.

Proof. The key idea for the achievability part of the proof is that knowledge of the CSI allows Alice, Bob, and Eve to demultiplex the ergodic-fading WTC into a set of *parallel and time-invariant* Gaussian WTCs. Specifically, this transformation can be done as follows. We partition the range of $\mathrm{G_m}$ into k intervals $[g_{\mathrm{m},i}, g_{\mathrm{m},i+1})$ with $i \in [\![1, k]\!]$. Similarly, we partition the range of fading gains $\mathrm{G_e}$ into k intervals $[g_{\mathrm{e},j}, g_{\mathrm{e},j+1})$ with $j \in [\![1, k]\!]$. For simplicity, we first assume that the fading gains are bounded (that is, $g_{\mathrm{m},k+1} < \infty$ and $g_{\mathrm{e},k+1} < \infty$) and let

$$p_i \triangleq \mathbb{P}\left[\mathrm{G_m} \in [g_{\mathrm{m},i}, g_{\mathrm{m},i+1})\right] \quad \text{and} \quad q_j \triangleq \mathbb{P}\left[\mathrm{G_e} \in [g_{\mathrm{e},j}, g_{\mathrm{e},j+1})\right].$$

For each pair of indices (i, j), Alice and Bob publicly agree on a transmit power γ_{ij} and on a wiretap code $\mathcal{C}_n^{ij}$ of length n designed to operate on a Gaussian WTC with main channel gain $g_{\mathrm{m},i}$ and eavesdropper's channel gain $g_{\mathrm{e},j+1}$. The set of transmit powers $\{\gamma_{ij}\}_{k,k}$ is also chosen such that $\sum_{i=1}^k \sum_{j=1}^k p_i q_j \gamma_{ij} \leqslant P$. If we define

$$C_{ij} \triangleq \left(\log\left(1 + \frac{g_{\mathrm{m},i}\gamma_{ij}}{\sigma_\mathrm{m}^2}\right) - \log\left(1 + \frac{g_{\mathrm{e},j+1}\gamma_{ij}}{\sigma_\mathrm{e}^2}\right)\right)^+,$$

then, for any $\epsilon > 0$, Corollary 5.1 ensures the existence of a $(2^{nR_{ij}}, n)$ code $\mathcal{C}_n^{ij}$ such that

$$R_{ij} \geqslant C_{ij} - \epsilon, \quad \frac{1}{n}\mathbf{L}(\mathcal{C}_n^{ij}) \leqslant \delta(\epsilon), \quad \text{and} \quad \mathbf{P}_\mathrm{e}(\mathcal{C}_n^{ij}) \leqslant \delta(\epsilon).$$

Note that a Gaussian channel with gain $g_{\mathrm{m},i}$ is stochastically degraded with respect to any Gaussian channel with gain $g \in \left[g_{\mathrm{m},i}, g_{\mathrm{m},i+1}\right)$; therefore $\mathcal{C}_n^{ij}$ also guarantees that $\mathbf{P}_\mathrm{e}(\mathcal{C}_n^{ij}) \leqslant \delta(\epsilon)$ for a main channel gain $g \in \left[g_{\mathrm{m},i}, g_{\mathrm{m},i+1}\right)$. Similarly, a Gaussian channel with gain $g \in \left[g_{\mathrm{e},j}, g_{\mathrm{e},j+1}\right)$ is stochastically degraded with respect to a Gaussian channel with gain $g_{\mathrm{e},j+1}$; therefore, by Proposition 3.3, $\mathcal{C}_n^{ij}$ guarantees that $\mathbf{L}(\mathcal{C}_n^{ij}) \leqslant \delta(\epsilon)$ for an eavesdropper's channel gain $g \in \left[g_{\mathrm{e},j}, g_{\mathrm{e},j+1}\right)$.

Since all fading coefficients are available to the transmitter and receivers, the ergodic-fading WTC can be demultiplexed into k^2 independent, time-invariant, Gaussian WTCs. The set of codes $\{\mathcal{C}_n^{ij}\}_{k,k}$ for the demultiplexed channels can be viewed as a single code $\mathcal{C}_n$ for the ergodic-fading channel, whose rate R_s is the sum of secure rates R_{ij} achieved over each channel weighted by the probability $p_i q_j$ that the code $\mathcal{C}_n^{ij}$ is used. Therefore,

$$\begin{aligned} R_\mathrm{s} &= \sum_{i=1}^k \sum_{j=1}^k p_i q_j R_{ij} \\ &\geqslant \sum_{i=1}^k \sum_{j=1}^k p_i q_j \left(\log\left(1 + \frac{g_{\mathrm{m},i}\gamma_{ij}}{\sigma_\mathrm{m}^2}\right) - \log\left(1 + \frac{g_{\mathrm{e},j+1}\gamma_{ij}}{\sigma_\mathrm{e}^2}\right)\right)^+ - \epsilon \end{aligned}$$

subject to the power constraint

$$\sum_{i=1}^k \sum_{j=1}^k p_i q_j \gamma_{ij} \leqslant P.$$

Additionally,

$$\frac{1}{n}\mathbf{L}(\mathcal{C}_n) = \sum_{i=1}^{k}\sum_{j=1}^{k} p_i q_j \frac{1}{n}\mathbf{L}(\mathcal{C}_n^{ij}) \leqslant \delta(\epsilon)$$

and

$$\mathbf{P}_e(\mathcal{C}_n) = \sum_{i=1}^{k}\sum_{j=1}^{k} p_i q_j \mathbf{P}_e(\mathcal{C}_n^{ij}) \leqslant \delta(\epsilon).$$

Note that k can be chosen arbitrarily large and ϵ can be chosen arbitrarily small. Hence, the ergodicity of the channel ensures that all rates R_s such that

$$R_s < \mathbb{E}_{G_m G_e}\left[\left(\log\left(1 + \frac{G_m \gamma(G_m, G_e)}{\sigma_m^2}\right) - \log\left(1 + \frac{G_e \gamma(G_m, G_e)}{\sigma_e^2}\right)\right)^+\right], \quad (5.22)$$

with $\gamma : \mathbb{R}^2 \to \mathbb{R}^+$ a power-allocation function such that $\mathbb{E}[\gamma(G_m, G_e)] \leqslant P$, are achievable full secrecy rates. One can further increase the upper bound in (5.22) by optimizing over the power allocations γ such that $\mathbb{E}[\gamma(G_m, G_e)] \leqslant P$. Note that the maximization over γ also allows us to drop the operator $(\cdot)^+$, which yields the expression in Theorem 5.3.

If the range of fading gain is unbounded, we define an arbitrary but finite threshold g_{max} beyond which Alice and Bob do not communicate. The multiplexing scheme developed above applies directly for fading gains below g_{max}; however, there is a rate penalty because $\mathbb{P}[G_m > g_{max} \text{ or } G_e > g_{max}] > 0$ and Alice and Bob do not communicate for a fraction of fading realizations. Nevertheless, the penalty can be made as small as desired by choosing g_{max} large enough.

We omit the converse part of the proof of the theorem, which can be obtained from a converse argument for parallel Gaussian WTCs. The ideas behind the proof are similar to those used in Section 3.5.3, with the necessary modifications to account for parallel channels. We refer the interested reader to [85] for more details. □

The power allocation γ which maximizes C_s in Theorem 5.3 can be characterized exactly.

Proposition 5.5. *The power allocation $\gamma^* : \mathbb{R}_+^2 \to \mathbb{R}_+$ that achieves C_s in Theorem 5.3 is defined as follows:*

- *if $u > 0$ and $v = 0$, then*

$$\gamma^*(u, v) = \left(\frac{1}{\lambda} - \frac{1}{u}\right)^+;$$

- *if*

$$\frac{u}{\sigma_m^2} > \frac{v}{\sigma_e^2} > 0,$$

then

$$\gamma^*(u, v) = \frac{1}{2}\left(-\left(\frac{\sigma_m^2}{u} + \frac{\sigma_e^2}{v}\right) + \sqrt{\left(\frac{\sigma_e^2}{v} - \frac{\sigma_m^2}{u}\right)\left(\frac{4}{\lambda} + \frac{\sigma_m^2}{u} - \frac{\sigma_e^2}{v}\right)}\right)^+;$$

- $\gamma^*(u, v) = 0$ *otherwise;*

with $\lambda > 0$ *such that* $\mathbb{E}[\gamma^*(G_m, G_e)] = P$.

Proof. For simplicity, we assume that the fading realizations take a finite number of values and we accept the fact that the proof extends to an infinite number of values. For any $(u, v) \in \mathbb{R}_+^2$, we define the function

$$f_{uv} : \gamma \mapsto \log\left(1 + \frac{u\gamma}{\sigma_m^2}\right) - \log\left(1 + \frac{v\gamma}{\sigma_e^2}\right).$$

If $u/\sigma_m^2 < v/\sigma_e^2$, the function f_{uv} takes negative values; hence, without loss of optimality, we can set

$$\gamma^*(u, v) = 0 \quad \text{if} \quad \frac{u}{\sigma_m^2} < \frac{v}{\sigma_e^2}.$$

If $u/\sigma_m^2 \geqslant v/\sigma_e^2$, the function f_{uv} is concave in γ; consequently, the secrecy capacity, which is a weighted sum of functions f_{uv}, is concave, as well. Therefore, the optimal power allocation γ^* can be obtained by forming the Lagrangian

$$\begin{aligned}\mathcal{L} \triangleq &\sum_u \sum_v \log\left(1 + \frac{\gamma(u, v)u}{\sigma_m^2}\right) p_{G_m}(u) p_{G_e}(v) \\ &- \sum_u \sum_v \log\left(1 + \frac{\gamma(u, v)v}{\sigma_e^2}\right) p_{G_m}(u) p_{G_e}(v) \\ &- \lambda \sum_u \sum_v \gamma(u, v) p_{G_m}(u) p_{G_e}(v),\end{aligned}$$

and finding $\gamma(u, v) \geqslant 0$ maximizing $\mathcal{L}$ for each (u, v) such that $u/\sigma_m^2 \geqslant v/\sigma_e^2$.

- If $v = 0$ and $u > 0$, the derivative of $\mathcal{L}$ with respect to $\gamma(u, v)$ is

$$\frac{\partial \mathcal{L}}{\partial \gamma(u, v)} = \frac{u}{\sigma_m^2 + \gamma(u, v)u} p_{G_m}(u) p_{G_e}(v) - \lambda p_{G_m}(u) p_{G_e}(v).$$

Therefore,

$$\frac{\partial \mathcal{L}}{\partial \gamma(u, v)} = 0 \Leftrightarrow \gamma(u, v) = \frac{1}{\lambda} - \frac{\sigma_m^2}{u}.$$

- If $u/\sigma_m^2 > v/\sigma_e^2 > 0$, we obtain

$$\begin{aligned}\frac{\partial \mathcal{L}}{\partial \gamma(u, v)} = &\frac{\sigma_e^2 u - \sigma_m^2 v}{(\sigma_m^2 + \gamma(u, v)u)(\sigma_e^2 + \gamma(u, v)v)} p_{G_m}(u)\, p_{G_e}(v) \\ &- \lambda p_{G_m}(u) p_{G_e}(v).\end{aligned}$$

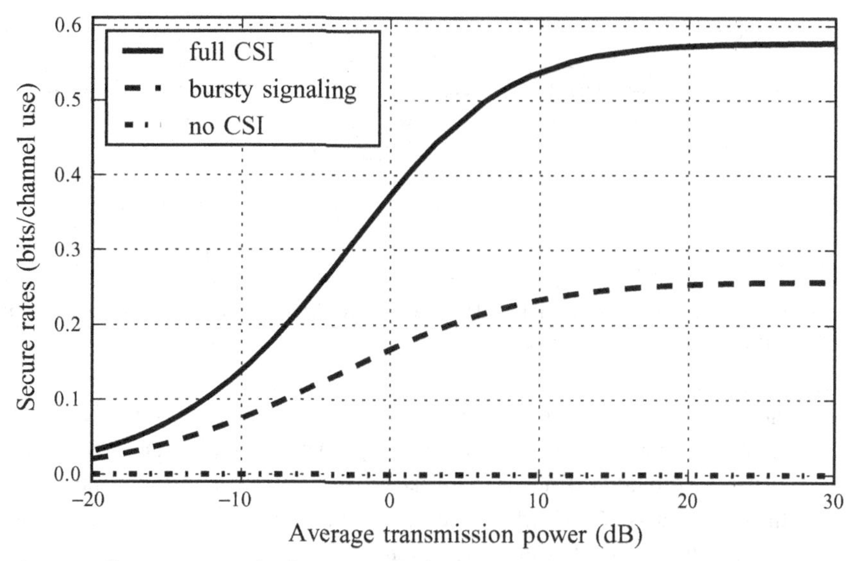

Figure 5.6 Secure communication rates over the Rayleigh fading channel for $\mu_m = 1$, $\mu_e = 2$, and $\sigma_m^2 = \sigma_e^2 = 1$, and for different knowledge of the CSI. The bursty signaling strategy is based only on knowledge of the CSI for the main channel.

Therefore,

$$\frac{\partial \mathcal{L}}{\partial \gamma(u,v)} = 0 \Leftrightarrow \frac{\sigma_e^2 u - \sigma_m^2 v}{(\sigma_m^2 + \gamma(u,v)u)(\sigma_e^2 + \gamma(u,v)v)} - \lambda = 0,$$

and, consequently,

$$\gamma(u,v) = \frac{1}{2}\left(-\left(\frac{\sigma_m^2}{u} + \frac{\sigma_e^2}{v}\right) + \sqrt{\left(\frac{\sigma_e^2}{v} - \frac{\sigma_m^2}{u}\right)\left(\frac{4}{\lambda} + \frac{\sigma_e^2}{v} - \frac{\sigma_m^2}{u}\right)}\right)^+ . \qquad \square$$

Note that the optimal power-allocation strategy for the secrecy capacity with full CSI depends on the fading statistics only through the parameter λ. In addition, the fact that $\gamma^*(g_m, g_e) = 0$ if $g_m/\sigma_m^2 \leqslant g_e/\sigma_e^2$ is consistent with the intuition that no power should be allocated when the eavesdropper obtains a better instantaneous SNR than does the legitimate receiver.

Theorem 5.3 is illustrated in the case of Rayleigh fading in Figure 5.6. Even in this case, there is no closed-form expression for the secrecy capacity; nevertheless, since $\gamma^*(g_m, g_e) \to \infty$ as $P \to \infty$ for all (g_m, g_e) such that $g_m/\sigma_m^2 > g_e/\sigma_e^2$, we can approximate C_s in the limit of high power as follows. If $\mathbb{P}\left[\mathrm{G}_m/\sigma_m^2 > \mathrm{G}_e/\sigma_e^2\right] > 0$, then

$$\lim_{P\to\infty} C_s(P) = \mathbb{E}_{\mathrm{G}_m/\sigma_m^2 > \mathrm{G}_e/\sigma_e^2}\left[\log\left(\frac{\sigma_e^2}{\sigma_m^2}\frac{\mathrm{G}_m}{\mathrm{G}_e}\right)\right] = \log\left(1 + \frac{\sigma_e^2}{\sigma_m^2}\frac{\mu_m}{\mu_e}\right),$$

which depends only on the ratio of the average SNR at the receiver $\mu_m P/\sigma_m^2$ and that at the eavesdropper $\mu_e P/\sigma_e^2$. Notice that, as P goes to infinity, $C_s(P)$ is strictly positive even if, on average, the eavesdropper has a better channel than does the legitimate receiver. A

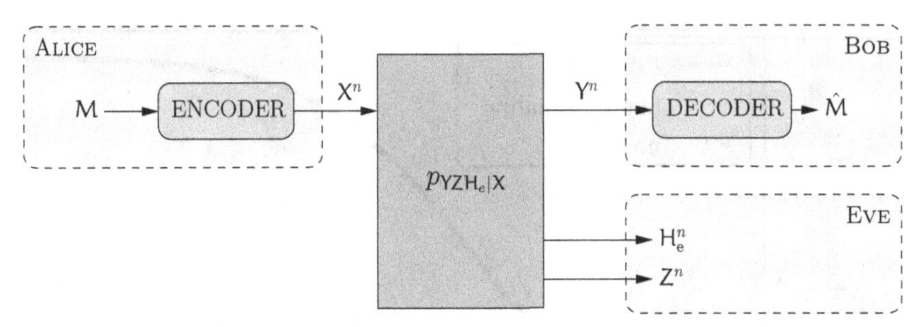

Figure 5.7 Equivalent channel model without eavesdropper's channel state information at the transmitter and legitimate receiver.

closer look at the formula for the secrecy capacity with full CSI given in Theorem 5.3 shows that the secrecy capacity is strictly positive for *any* transmit power and channel statistics, provided that $\mathbb{P}\left[\text{G}_\text{m}/\sigma_\text{m}^2 > \text{G}_\text{e}/\sigma_\text{e}^2\right] > 0$. This result contrasts sharply with the Gaussian WTC, for which secure communication is impossible if the eavesdropper has a better channel. Hence, the *fading affecting wireless channels is beneficial for security*. Nevertheless, our result relies on the demultiplexing of the ergodic-fading WTC, which requires the fading coefficients of both channels to be known at the transmitter. The optimal power allocation derived in Proposition 5.5 requires Alice to allocate *opportunistically* more power during the coherence intervals (h_m, h_e) for which Eve has a lower SNR than does Bob.

For more realistic applications, it is desirable to relax the full CSI assumption and, in particular, to evaluate secure communication rates *without* knowledge of the eavesdropper's fading coefficient h_e at the transmitter.

Proposition 5.6. *With CSI for the main channel but without CSI for the eavesdropper's channel, all rates R_s such that*

$$R_\text{s} < \max_{\gamma} \mathbb{E}_{\text{G}_\text{m}\text{G}_\text{e}}\left[\log\left(1 + \frac{\gamma(\text{G}_\text{m})\text{G}_\text{m}}{\sigma_\text{m}^2}\right) - \log\left(1 + \frac{\gamma(\text{G}_\text{m})\text{G}_\text{e}}{\sigma_\text{e}^2}\right)\right],$$

where $\gamma : \mathbb{R}_+ \to \mathbb{R}_+$ is subject to the constraint $\mathbb{E}[\gamma(\text{G}_\text{m})] \leqslant P$, are achievable full secrecy rates.

Proof. The demultiplexing scheme used to prove Theorem 5.3 cannot be used directly because the transmitter and legitimate receiver do not know the eavesdropper's CSI. Nevertheless, it is still possible to demultiplex the channel based on the main channel CSI, and one can include the fading coefficients affecting the eavesdropper's channel in the channel statistics. As illustrated in Figure 5.7, the eavesdropper's knowledge of her channel coefficients can be taken into account by treating H_e as a second output for the eavesdropper's channel. The range of G_m is partitioned into k intervals $[g_{\text{m},i}, g_{\text{m},i+1})$ with $i \in [\![1, k]\!]$. For simplicity, we assume again that the fading gain is bounded ($g_{\text{m},k+1} < \infty$) and let

$$p_i \triangleq \mathbb{P}\left[\text{G}_\text{m} \in [g_{\text{m},i}, g_{\text{m},i+1})\right].$$

For each index $i \in [\![1, k]\!]$, Alice and Bob publicly agree on a transmit power γ_i and on a wiretap code $\mathcal{C}_n^i$ designed to operate on a WTC with transition probabilities $p^{(i)}_{\mathsf{YZH_e|X}}$, such that the marginal $p^{(i)}_{\mathsf{Y|X}}$ corresponds to a Gaussian channel with known constant channel gain $g_{\mathrm{m},i}$ while $p^{(i)}_{\mathsf{ZH_e|X}}$ corresponds to a fading eavesdropper channel with i.i.d. fading coefficient $\mathsf{H_e}$ treated as a second output for the eavesdropper. Since the fading coefficient $\mathsf{H_e}$ is independent of the input, note that

$$\forall (x, z, h_\mathrm{e}) \quad p^{(i)}_{\mathsf{ZH_e|X}}(z, h_\mathrm{e}|x) \triangleq p_{\mathsf{Z|H_eX}}(z|h_\mathrm{e}, x)\, p_{\mathsf{H_e}}(h_\mathrm{e}).$$

The set of transmit powers $\{\gamma_i\}_k$ is also chosen such that $\sum_{i=1}^k p_i \gamma_i \leqslant P$. We can apply Theorem 3.4 to this channel and, for any $\epsilon > 0$ and input distribution p_{X}, this shows the existence of a wiretap code $\mathcal{C}_n^i$ of length n with rate

$$R_i \geqslant \mathbb{I}(\mathsf{X};\mathsf{Y}) - \mathbb{I}(\mathsf{X};\mathsf{ZH_e}) - \epsilon,$$

such that $(1/n)\mathbf{L}(\mathcal{C}_n^i) \leqslant \delta(\epsilon)$ and $\mathbf{P}_\mathrm{e}(\mathcal{C}_n^i) \leqslant \delta(\epsilon)$. In particular, for the specific choice $\mathsf{X} \sim \mathcal{CN}(0, \gamma_i)$, we obtain

$$\mathbb{I}(\mathsf{X};\mathsf{Y}) = \log\left(1 + \frac{g_{\mathrm{m},i}\gamma_i}{\sigma_\mathrm{m}^2}\right)$$

and

$$\mathbb{I}(\mathsf{X};\mathsf{ZH_e}) = \mathbb{I}(\mathsf{X};\mathsf{H_e}) + \mathbb{I}(\mathsf{X};\mathsf{Z}|\mathsf{H_e}) = \mathbb{E}_{\mathsf{G_e}}\left[\log\left(1 + \frac{\mathsf{G_e}\gamma_i}{\sigma_\mathrm{e}^2}\right)\right],$$

where we have used $\mathbb{I}(\mathsf{X};\mathsf{H_e}) = 0$ since $\mathsf{H_e}$ is independent of the channel input X by assumption. Therefore,

$$R_i \geqslant \log\left(1 + \frac{g_{\mathrm{m},i}\gamma_i}{\sigma_\mathrm{m}^2}\right) - \mathbb{E}_{\mathsf{G_e}}\left[\log\left(1 + \frac{\mathsf{G_e}\gamma_i}{\sigma_\mathrm{e}^2}\right)\right] - \epsilon.$$

Because the fading coefficient $\mathsf{H_m}$ is known by the transmitter and receivers, this ergodic-fading WTC can be demultiplexed into k WTCs with time-invariant Gaussian main channel and fading eavesdropper's channel as in Figure 5.7. The set of codes $\{\mathcal{C}_n^i\}_k$ for the demultiplexed channels can be viewed as a single code $\mathcal{C}_n$ for the ergodic-fading channel, whose rate is

$$\begin{aligned} R_\mathrm{s} &= \sum_{i=1}^k p_i R_i \\ &= \sum_{i=1}^k p_i \left(\log\left(1 + \frac{g_{\mathrm{m},i}\gamma_i}{\sigma_\mathrm{m}^2}\right) - \mathbb{E}_{\mathsf{G_e}}\left[\log\left(1 + \frac{\mathsf{G_e}\gamma_i}{\sigma_\mathrm{e}^2}\right)\right]\right) - \epsilon, \end{aligned}$$

and subject to the constraint

$$\sum_{i=1}^k p_i \gamma_i \leqslant P.$$

In addition,

$$\frac{1}{n}\mathbf{L}(\mathcal{C}_n) = \sum_{i=1}^{k} p_i \frac{1}{n}\mathbf{L}(\mathcal{C}_n^i) \leqslant \delta(\epsilon)$$

and

$$\mathbf{P}_e(\mathcal{C}_n) = \sum_{i=1}^{k} p_i \mathbf{P}_e(\mathcal{C}_n^i) \leqslant \delta(\epsilon).$$

Note that k can be chosen arbitrarily large and ϵ can be chosen arbitrarily small. Hence, the ergodicity of the channel guarantees that all rates R_s such that

$$R_s < \mathbb{E}_{G_m G_e}\left[\log\left(1 + \frac{G_m \gamma(G_m)}{\sigma_e^2}\right) - \log\left(1 + \frac{G_e \gamma(G_m)}{\sigma_e^2}\right)\right] \tag{5.23}$$

with $\mathbb{E}[\gamma(G_m, G_e)] \leqslant P$ are achievable full secrecy rates. Finally, we can improve the upper bound in (5.23) by optimizing over all power allocations γ satisfying the constraint $\mathbb{E}_{G_m}[\gamma(G_m)] \leqslant P$. □

Although the achievable rates in Theorem 5.3 and Proposition 5.6 differ only in the arguments of the power-allocation function γ, this similarity is misleading because the underlying codes are fundamentally different. In Theorem 5.3, all parties have access to full CSI about the channels, and the code is composed of independent wiretap codes for Gaussian WTCs that are multiplexed to adapt to the time-varying fading gains. However, in Proposition 5.6, the code is composed of independent wiretap codes that are interleaved to adapt to the main channel fading gain only and whose codewords spread over many different realizations of the eavesdropper's channel gain.

The optimal power allocation $\gamma : \mathbb{R}_+ \to \mathbb{R}_+$ for Proposition 5.6 cannot be derived exactly because the objective function

$$f_u : \gamma \mapsto \log\left(1 + \frac{\gamma u}{\sigma_m^2}\right) - \mathbb{E}\left[\log\left(1 + \frac{\gamma G_e}{\sigma_e^2}\right)\right]$$

is not concave in γ. A Lagrangian maximization as in Proposition 5.5 would allow us to compute achievable full secrecy rates, but, in general, γ does not admit a closed-form expression. Instead, we consider a simple *bursty signaling*[4] strategy, in which the transmitter selects a threshold $\tau > 0$ and allocates power as

$$\gamma(u) = \begin{cases} P_\tau \triangleq P/\mathbb{P}[G_m > \tau] & \text{if} \quad u > \tau, \\ 0 & \text{otherwise.} \end{cases}$$

For Rayleigh fading, we can compute the bound in Theorem 5.6 in closed form in terms of the exponential-integral function

$$E_1 : x \to \int_x^\infty \frac{e^{-y}}{y}\, dy.$$

[4] Bursty signaling is also called "on–off" power control.

Evaluating (5.23) explicitly for bursty signaling over a wireless channel with i.i.d. Rayleigh fading shows that all secure rates R_s such that

$$R_s < \exp\left(-\frac{\tau}{\mu_m}\right)\log\left(1+\frac{\tau P_\tau}{\sigma_m^2}\right) + \exp\left(\frac{\sigma_m^2}{\mu_m P_\tau}\right)E_1\left(\frac{\sigma_m^2}{\mu_m P_\tau}+\frac{\tau}{\mu_m}\right)\log(e)$$
$$- \exp\left(\frac{\sigma_e^2}{\mu_e P_\tau}-\frac{\tau}{\mu_m}\right)E_1\left(\frac{\sigma_e^2}{\mu_e P_\tau}\right)\log(e).$$

are achievable. As illustrated in Figure 5.6, the lack of knowledge about the eavesdropper's channel has a detrimental effect on secure communication rates.

If assumptions are further relaxed, and no CSI is available at the transmitter, then power cannot be allocated to avoid harmful situations in which the eavesdropper has a higher SNR than that of the legitimate receiver. In particular, if $\mathbb{E}[G_e]/\sigma_e^2 \geqslant \mathbb{E}[G_m]/\sigma_m^2$ and the eavesdropper obtains a better average SNR than does the legitimate receiver, the secrecy capacity without any CSI is zero.

5.2.2 Block-fading channels

It is important to realize that the conclusions drawn regarding the effect of CSI depend on the fading statistics considered; different fading models lead to slightly different conclusions. In this section, we consider a *block-fading* model, for which the coherence interval is sufficiently long that coding can also be performed *within* the interval. Specifically, the processes $\{H_{m,i}\}_{i\geqslant 1}$ and $\{H_{e,i}\}_{i\geqslant 1}$ are i.i.d., but for each realization $(h_{m,i}, h_{e,i})$ the relationships between channel inputs and outputs are

$$\begin{cases} Y_{i,j} = h_{m,i}X_{i,j} + N_{m,i,j}, \\ Z_{i,j} = h_{e,i}X_{i,j} + N_{e,i,j}, \end{cases} \quad \text{for} \quad j \in [\![1, N]\!],$$

where N is assumed to be sufficiently large for asymptotic coding results to hold. If the transmitter and receivers have CSI about all channels, then the demultiplexing and power-allocation scheme used for the ergodic-fading WTC can be used, and the secrecy capacity is given again by Theorem 5.3 with the optimal power allocation of Proposition 5.5. The situation is quite different without knowledge about the eavesdropper's fading at the transmitter. In fact, for the ergodic-fading model considered in Section 5.2.1, the transmitter is allowed a single channel use per coherence interval; consequently, the information leaked to the eavesdropper can be arbitrarily large. In contrast, for the block-fading model, the transmitter can *code* within each coherence interval and the information leaked to the eavesdropper *cannot exceed* the information communicated to the legitimate receiver. Specifically, we let X^N represent a coded sequence chosen at random in the transmitter's codebook and sent during one coherence interval, and we let Z^N denote the corresponding eavesdropper's observation. Then, it holds that

$$\mathbb{I}(X^N; Z^N) \leqslant \mathbb{H}(X^N) < \infty,$$

because X^N takes a *finite* number of values.

Theorem 5.4 (Gopala *et al.*). *The secrecy capacity of a block-fading WTC with CSI about the main channel but no CSI about the eavesdropper's channel is*

$$C_s = \max_{\gamma} \mathbb{E}_{\mathsf{G_m G_e}}\left[\left(\log\left(1+\frac{\gamma(\mathsf{G_m})\mathsf{G_m}}{\sigma_m^2}\right) - \log\left(1+\frac{\gamma(\mathsf{G_m})\mathsf{G_e}}{\sigma_e^2}\right)\right)^+\right],$$

subject to the constraint $\mathbb{E}[\gamma(\mathsf{G_m})] \leqslant P$.

Proof. We provide only the achievability part of the proof, and refer the reader to [86] for details regarding the converse. The key ideas behind the code construction are to code within each coherence interval in order to bound the information leaked to the eavesdropper and to spread the codewords over many realizations of the eavesdropper's fading gain. The proof is greatly simplified by noting that the block-fading channel can be treated as an ergodic-fading channel with a *vector* input X^N and *vector* outputs Y^N and Z^N such that

$$\mathsf{Y}^N = \mathsf{H_m}\mathsf{X}^N + \mathsf{N}_m^N \quad \text{and} \quad \mathsf{Z}^N = \mathsf{H_e}\mathsf{X}^N + \mathsf{N}_e^N.$$

Therefore, we can use the same approach as in the proof of Proposition 5.6.

The range of $\mathsf{G_m}$ is partitioned into k intervals $[g_{m,i}, g_{m,i+1})$ with $i \in [\![1,k]\!]$. We assume the fading gain to be bounded ($g_{m,k+1} < \infty$), and we let

$$p_i \triangleq \mathbb{P}\left[\mathsf{G_m} \in [g_{m,i}, g_{m,i+1})\right].$$

For each index $i \in [\![1,k]\!]$, Alice and Bob publicly agree on a transmit power γ_i and on a wiretap code $\mathcal{C}_n^i$ of length n designed to operate on a *vector* WTC with transition probabilities $p^{(i)}_{\mathsf{Y}^N\mathsf{Z}^N\mathsf{H_e}|\mathsf{X}^N}$. Note that the marginal $p^{(i)}_{\mathsf{Y}^N|\mathsf{X}^N}$ is such that

$$\forall (y^N, x^N) \quad p^{(i)}_{\mathsf{Y}^N|\mathsf{X}^N}(y^N|x^N) = \prod_{i=1}^{N} p_{\mathsf{Y}|\mathsf{X}}(y_i|x_i),$$

where $p^{(i)}_{\mathsf{Y}|\mathsf{X}}$ corresponds to a Gaussian channel with known constant fading coefficient $h_{m,i}$. Similarly, the marginal $p^{(i)}_{\mathsf{Z}^N\mathsf{H_e}|\mathsf{X}^N}$ is such that

$$\forall (z^N, h_e, x^N) \quad p^{(i)}_{\mathsf{Z}^N\mathsf{H_e}|\mathsf{X}^N}(z^N, h_e|x^N) = \left(\prod_{i=1}^{N} p_{\mathsf{Z}|\mathsf{H_e}\mathsf{X}}(z_i, |h_e, x_i)\right) p_{\mathsf{H_e}}(h_e),$$

where $p_{\mathsf{Z}|\mathsf{H_e}\mathsf{X}}$ corresponds to a fading eavesdropper channel with fading coefficient $\mathsf{H_e}$ available to the eavesdropper. The set of transmit powers is also chosen such that $\sum_{i=1}^k p_i\gamma_i \leqslant P$. By Theorem 3.4, for any $\epsilon > 0$ and input distribution p_{X^N}, there exists a wiretap code $\mathcal{C}_n^i$ of length n with rate

$$R_i \geqslant \mathbb{I}(\mathsf{X}^N;\mathsf{Y}^N) - \mathbb{I}(\mathsf{X}^N;\mathsf{Z}^N\mathsf{H_e}) - \epsilon, \tag{5.24}$$

measured in bits per *vector* channel use and such that $(1/n)\mathbf{L}(\mathcal{C}_n^i) \leqslant \delta(\epsilon)$ and $\mathbf{P}_e(\mathcal{C}_n^i) \leqslant \delta(\epsilon)$. We are free to optimize the distribution of X^N as long as the power constraint $\sum_{j=1}^N \mathbb{E}\left[\mathsf{X}_j^2\right] \leqslant \gamma_i$ is satisfied; in particular, we can choose X^N to represent the codewords chosen uniformly at random in a codebook such that

$$\log\left(1+\frac{g_{\mathrm{m},i}\gamma_i}{\sigma_{\mathrm{m}}^2}\right)-\epsilon \leqslant \mathbb{H}(\mathsf{X}^N) \leqslant \log\left(1+\frac{g_{\mathrm{m},i}\gamma_i}{\sigma_{\mathrm{m}}^2}\right),$$

and whose probability of error over a Gaussian channel with gain $g_{\mathrm{m},i}$ is at most $\delta(\epsilon)$. The existence of this code is ensured directly by the channel coding theorem if N is large enough. Since X^N represents a codeword in a codebook, it follows from Fano's inequality that

$$\mathbb{H}(\mathsf{X}^N|\mathsf{Y}^N) \leqslant N\delta(\epsilon).$$

Therefore,

$$\mathbb{I}(\mathsf{X}^N;\mathsf{Y}^N) = \mathbb{H}(\mathsf{X}^N) - \mathbb{H}(\mathsf{X}^N|\mathsf{Y}^N) \geqslant N\log\left(1+\frac{g_{\mathrm{m},i}\gamma_i}{\sigma_{\mathrm{m}}^2}\right) - N\delta(\epsilon). \tag{5.25}$$

Note that $\mathbb{I}(\mathsf{X}^N;\mathsf{H}_\mathrm{e}) = 0$ because the fading coefficient H_e is independent of the input and that the channel is memoryless; therefore,

$$\begin{aligned}\mathbb{I}(\mathsf{X}^N;\mathsf{Z}^N\mathsf{H}_\mathrm{e}) &= \mathbb{I}(\mathsf{X}^N;\mathsf{Z}^N|\mathsf{H}_\mathrm{e})\\ &= \mathbb{E}_{\mathsf{H}_\mathrm{e}}\left[\mathbb{I}(\mathsf{X}^N;\mathsf{Z}^N|\mathsf{H}_\mathrm{e})\right]\\ &\leqslant N\mathbb{E}_{\mathsf{G}_\mathrm{e}}\left[\log\left(1+\frac{\gamma_i\mathsf{G}_\mathrm{e}}{\sigma_\mathrm{e}^2}\right)\right].\end{aligned} \tag{5.26}$$

Finally, note that the following trivial upper bound holds:

$$\forall h_\mathrm{e}\quad \mathbb{I}(\mathsf{X}^N;\mathsf{Z}^N|\mathsf{H}_\mathrm{e}=h_\mathrm{e}) \leqslant \mathbb{H}(\mathsf{X}^N) \leqslant N\log\left(1+\frac{g_{\mathrm{m},i}\gamma_i}{\sigma_{\mathrm{m}}^2}\right). \tag{5.27}$$

On combining (5.25), (5.26), and (5.27) in (5.24) we obtain

$$R_i \geqslant N\mathbb{E}_{\mathsf{G}_\mathrm{e}}\left[\left(\log\left(1+\frac{g_{\mathrm{m},i}\gamma_i}{\sigma_{\mathrm{m}}^2}\right)-\log\left(1+\frac{\gamma_i\mathsf{G}_\mathrm{e}}{\sigma_\mathrm{e}^2}\right)\right)^+\right] - N\delta(\epsilon).$$

Since the fading coefficient H_m is known by the transmitter and receivers, the channel can be demultiplexed into k vector input WTCs. The set of codes $\{\mathcal{C}_n^i\}_k$ for the demultiplexed vector channels can be viewed as a single code $\mathcal{C}_n$ for the block-ergodic fading channel, whose rate in bits per channel use is

$$\begin{aligned}R_\mathrm{s} &= \frac{1}{N}\sum_{i=1}^{k} p_i R_i\\ &\geqslant \sum_{i=1}^{k} p_i\mathbb{E}_{\mathsf{G}_\mathrm{e}}\left[\left(\log\left(1+\frac{g_{\mathrm{m},i}\gamma_i}{\sigma_{\mathrm{m}}^2}\right)-\log\left(1+\frac{\gamma_i\mathsf{G}_\mathrm{e}}{\sigma_\mathrm{e}^2}\right)\right)^+\right] - \delta(\epsilon),\end{aligned}$$

and subject to the constraint $\sum_{i=1}^{k} p_i\gamma_i \leqslant P$. In addition,

$$\frac{1}{nN}\mathbf{L}(\mathcal{C}_n) = \frac{1}{N}\sum_{i=1}^{k} p_i\frac{1}{n}\mathbf{L}(\mathcal{C}_n^i) \leqslant \delta(\epsilon)$$

and

$$\mathbf{P}_e(\mathcal{C}_n) = \sum_{i=1}^{k} p_i \mathbf{P}_e(\mathcal{C}_n^i) \leqslant \delta(\epsilon).$$

Note that k can be chosen arbitrarily large and ϵ can be chosen arbitrarily small. Hence, the ergodicity of the channel guarantees that all rates R_s, such that

$$R_s < \mathbb{E}_{G_m G_e}\left[\left(\log\left(1 + \frac{G_m \gamma(G_m)}{\sigma_m^2}\right) - \log\left(1 + \frac{G_e \gamma(G_m)}{\sigma_e^2}\right)\right)^+\right],$$

with $\mathbb{E}_{G_m}[\gamma(G_m)] \leqslant P$, are achievable full secrecy rates (in bits per channel use). □

Theorem 5.4 differs from Corollary 5.6 only by the presence of the operator $(\cdot)^+$ in the expectation, which appears because the information leaked to the eavesdropper within each coherence interval is bounded. The formula for the secrecy capacity highlights again that fading is beneficial for security and, in contrast to ergodic fading, the lack of knowledge about the eavesdropper's CSI at the transmitter seems to incur a lesser penalty for block-ergodic fading. For Rayleigh fading, the upper bound in Theorem 5.4 can be computed in closed form for the bursty signaling strategy defined in Section 5.2.1. Bursty signaling over wireless channels with i.i.d. Rayleigh fading can achieve all rates R_s such that

$$\begin{aligned} R_s < {} & \exp\left(-\frac{\tau}{\mu_m}\right)\log\left(1 + \frac{\tau P_\tau}{\sigma_m^2}\right) + \exp\left(\frac{\sigma_m^2}{\mu_m P_\tau}\right) E_1\left(\frac{\tau}{\mu_m} + \frac{\sigma_m^2}{\mu_m P_\tau}\right)\log(e) \\ & + \exp\left(\frac{\sigma_e^2}{\mu_e P_\tau} - \frac{\tau}{\mu_m}\right)\left(E_1\left(\frac{\sigma_e^2 \tau}{\mu_e \sigma_m^2} + \frac{\sigma_e^2}{\mu_e P_\tau}\right) - E_1\left(\frac{\sigma_e^2}{\mu_e P_\tau}\right)\right)\log(e) \\ & - \exp\left(\frac{\sigma_e^2}{\mu_e P_\tau} - \frac{\sigma_m^2}{\mu_m P_\tau}\right) E_1\left(\left(\frac{1}{\mu_m} + \frac{\sigma_e^2}{\mu_e \sigma_m^2}\right)\left(\tau + \frac{\sigma_m^2}{P_\tau}\right)\right)\log(e). \end{aligned} \tag{5.28}$$

Interestingly, if P goes to infinity and τ goes to zero, (5.28) becomes

$$R_s < \log\left(1 + \frac{\sigma_e^2}{\sigma_m^2}\frac{\mu_m}{\mu_e}\right).$$

Note that the right-hand side is the secrecy capacity with full CSI as P goes to infinity. Therefore, for the block-fading model, the bursty signaling strategy approaches the secrecy capacity in the limit of large power. This is illustrated in Figure 5.8, which shows the secrecy capacity of a block-fading model with perfect knowledge of all fading coefficients and the secure rate achievable with the bursty signaling strategy for different values of the power P.

5.2.3 Quasi-static fading channels

In this last section, we consider the situation in which the fading coefficients $\{H_{m,i}^n\}_{i\geqslant 1}$ and $\{H_{e,i}^n\}_{i\geqslant 1}$ remain constant over the transmission of an entire codeword and

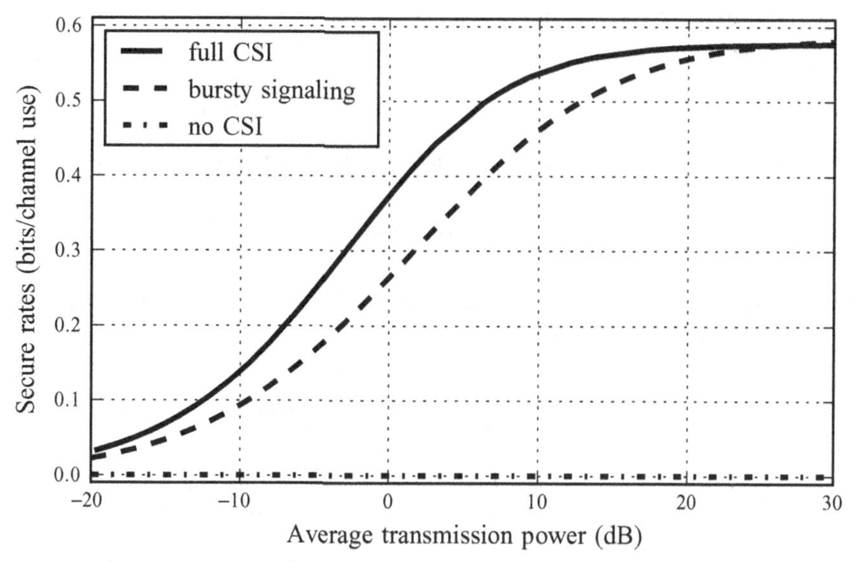

Figure 5.8 Secure communication rates over the Rayleigh block-fading channel with parameters $\mu_{\rm m} = 1$, $\mu_{\rm e} = 2$, and $\sigma_{\rm m}^2 = \sigma_{\rm e}^2 = 1$.

change independently at random from one codeword to another. This contrasts with the ergodic-fading and block-fading models, in which every transmitted codeword experiences many fading realizations during transmission. This model is often called a *quasi-static* fading model, and, for each coherence interval characterized by fading realizations $(h_{\rm m}, h_{\rm e})$, the model reduces to a Gaussian WTC defined by

$$\mathrm{Y}_i = h_{\rm m}\mathrm{X}_i + \mathrm{N}_{{\rm m},i} \quad \text{and} \quad \mathrm{Z}_i = h_{\rm e}\mathrm{X}_i + \mathrm{N}_{{\rm e},i}.$$

The input is subject to a power constraint $(1/n)\sum_{i=1}^{n}\mathbb{E}\left[\mathrm{X}_i^2\right] \leqslant P$, which is interpreted as a short-term constraint and must be satisfied within each coherence interval. Again, this contrasts with the long-term power constraint we used for the ergodic-fading and block-fading models, and the short-term constraint of the quasi-static model prevents the transmitter from allocating power opportunistically depending on the fading gain; nevertheless, the transmitter can still adapt its coding rate to the realization of the fading coefficients. While we have seen in previous sections that the possibility of secure communications with ergodic fading is determined by the *average* fading realization, we will see that secure communications over quasi-static channels are determined by the *instantaneous* one.

If the transmitter, the legitimate receiver, and the eavesdropper have perfect knowledge of the instantaneous realizations of the fading coefficients $(h_{\rm m}, h_{\rm e})$, the wiretap code used for each realization of the fading can be chosen opportunistically. The aggregate secure communications rate achievable over a long period of time is then given by the following theorem.

Theorem 5.5 (Barros and Rodrigues). *With full CSI, the average secrecy capacity of a quasi-static wiretap channel is*

$$C_s^{\text{avg}} = \mathbb{E}_{G_m G_e}\left[C_s^{\text{inst}}(G_m, G_e)\right],$$

where $C_s^{\text{inst}}(g_m, g_e)$ *is the* instantaneous secrecy capacity, *defined as*

$$C_s^{\text{inst}}(g_m, g_e) \triangleq \left(\log\left(1 + \frac{g_m P}{\sigma_m^2}\right) - \log\left(1 + \frac{g_e P}{\sigma_e^2}\right)\right)^+.$$

In the case of i.i.d. Rayleigh fading, C_s^{avg} can be computed explicitly using the exponential-integral function as

$$\begin{aligned} C_s^{\text{avg}} &= \exp\left(\frac{\sigma_m^2}{\mu_m P}\right) \mathrm{E}_1\left(\frac{\sigma_m^2}{\mu_m P}\right) \log(e) \\ &\quad - \exp\left(\frac{\sigma_m^2}{\mu_m P} + \frac{\sigma_e^2}{\mu_e P}\right) \mathrm{E}_1\left(\frac{\sigma_m^2}{\mu_m P} + \frac{\sigma_e^2}{\mu_e P}\right) \log(e), \end{aligned}$$

and one can check that

$$\lim_{P\to\infty} C_s^{\text{avg}}(P) = \log\left(1 + \frac{\sigma_e^2}{\sigma_m^2}\frac{\mu_m}{\mu_e}\right).$$

If the transmitter knows the fading coefficient h_m of the main channel but does not know the fading coefficient h_e of the eavesdropper's channel, then the average secrecy capacity for a quasi-static fading model is zero, no matter what the statistics of the channels are. In fact, since a codeword experiences a single realization of the fading gain, the probability of the eavesdropper obtaining a better instantaneous SNR for an entire codeword is always strictly positive, and no coding can guarantee secrecy. Nevertheless, one can still obtain insight into the security of wireless communications by taking a probabilistic view of security.

If we assume that the transmitter knows h_m, then the rate of the code used within each coherence interval can be adapted to guarantee reliability; however, without knowledge of h_e, the transmitter can use only a wiretap code *targeted* for a predefined secure communication rate R. Whenever the realization g_e is such that $R < C_s^{\text{inst}}(g_m, g_e)$, it follows from Remark 5.2 that the message is transmitted securely; however, if $R > C_s^{\text{inst}}(g_m, g_e)$, then some information is leaked to the eavesdropper. This behavior can be characterized by using the notion of *the outage probability of the secrecy capacity*.

Definition 5.1. *If the transmitter knows the fading coefficient of the main channel, the outage probability of the secrecy capacity is defined as*

$$P_{\text{out}}(R) \triangleq \mathbb{P}_{G_m G_e}\left[C_s^{\text{inst}}(G_m, G_e) < R\right].$$

For Rayleigh fading, $P_{\text{out}}(R)$ takes the closed-form expression

$$P_{\text{out}}(R) = 1 - \frac{\mu_m}{\mu_m + 2^R \mu_e \sigma_m^2/\sigma_e^2} \exp\left(-\sigma_m^2 \frac{2^R - 1}{\mu_m P}\right).$$

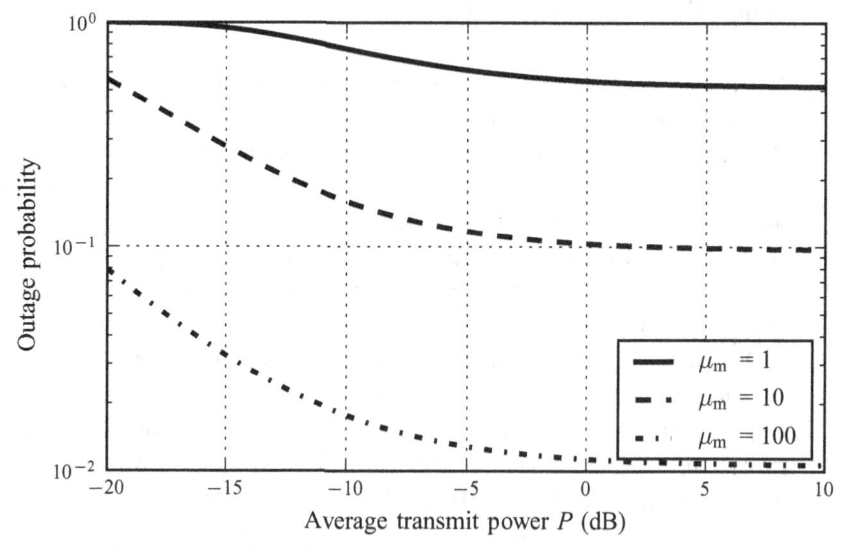

Figure 5.9 Outage probability of the secrecy capacity for various values of μ_m ($R = 0.1$, $\mu_e = 1$, $\sigma_m^2 = \sigma_e^2 = 1$).

Figure 5.9 illustrates the behavior of $P_{out}(R)$ as a function of the transmit power P for various values of the average fading gain μ_m. The channel statistics must give Alice and Bob a clear advantage over Eve in order to achieve low values of the outage probability. The relevance of the outage approach is also very much application-dependent, insofar as even leaking information with a low probability might sometimes be unacceptable.

Note that $P_{out}(R)$ is a decreasing function of P and cannot be reduced by decreasing the transmission power. However, reducing the targeted secure transmission rate R reduces the outage probability, and the minimum outage probability is obtained as R goes to zero, that is

$$P_{out}(0) = \frac{\mu_e}{\mu_e + \mu_m \sigma_m^2/\sigma_e^2},$$

which is always strictly positive. As expected, no matter what the transmitter does, information is leaked to the eavesdropper. Despite this seemingly disappointing result, it is worth noting that the outage probability is a pessimistic metric, which does not discriminate between events for which $R \gg C_s$ and events for which R exceeds C_s by a small amount. In addition, if the fading realizations are independent, outage events are independent of each other as well; as a result, a security leakage at a given time instant does not necessarily hinder security at later times.

Remark 5.7. *If the transmitter does not have any CSI, both reliability and security need to be assessed in terms of outage. The definition of the outage probability can be modified to*

$$P_{out} \triangleq \mathbb{P}_{G_m G_e}\left[C_s^{inst}(G_m, G_e) < R, C_m^{inst}(G_m) < R\right],$$

where $C_m^{inst}(g_m) \triangleq \log\left(1 + g_m P/\sigma_m^2\right)$ *is the instantaneous main channel capacity.*

5.3 Conclusions and lessons learned

The secrecy capacity of a Gaussian WTC admits a simple characterization as the difference between the main channel capacity and the eavesdropper's channel capacity. Consequently, secure communication over a Gaussian WTC is possible if and only if the legitimate receiver obtains a higher SNR than does the eavesdropper. This result is somewhat disappointing because it seems to limit the scope of applications. However, as shown by the analysis of Gaussian source models, this limitation can be overcome by considering more powerful communication schemes exploiting feedback, such as secret-key agreement schemes.

Our analysis of the MIMO Gaussian WTC leads to a severe conclusion: little can be done against the collusion of many eavesdroppers. Nevertheless, the MIMO model may be overly pessimistic because it ignores the communication requirements of the eavesdroppers. Most often, the bandwidth of eavesdroppers will be limited, which is likely to mitigate the detrimental impact of a collusion. On a more positive note, our study shows that coding for secrecy is, in general, more powerful than beamforming alone.

Perhaps surprisingly, the fluctuations of received SNR induced by fading in wireless transmissions are beneficial for security. If the instantaneous fading realizations can be accurately estimated by the transmitter, transmit power can be allocated opportunistically to the fading realizations for which the eavesdropper obtains a lower instantaneous SNR than that of the legitimate receiver. As a result, strictly positive secure communication rates are achievable even if, on average, the eavesdropper obtains a better SNR than that of the legitimate receiver. For some fading models, this is possible even if the transmitter does not have access to the eavesdropper's instantaneous fading realization.

5.4 Bibliographical notes

The secrecy capacity of the physically degraded Gaussian wiretap channel was established by Leung-Yan-Cheong and Hellman [87], and its generalization to the Gaussian BCC is due to Liang, Poor, and Shamai [85]. For the Gaussian WTC, an elegant converse proof based on the relation between MMSE and mutual information has been proposed by Bustin, Liu, Poor, and Shamai [88]. The secrecy capacity of the MIMOME channel was characterized by Khisti and Wornell [83, 84], Oggier and Hassibi [89], and Liu and Shamai [90], using different techniques (see also an earlier result of Shafiee, Liu, and Ulukus [91]). Although the general form of the covariance matrix $\mathbf{K}_X$ maximizing C_s^{MIMO} is not known, the optimal signaling strategies are known in certain regimes. At high SNR, Khisti and Wornell have shown that it is sufficient to transmit in (all) the directions in which the legitimate receiver has a better SNR than that of the eavesdropper. At low SNR, Gursoy has shown that it is sufficient to transmit in the direction of the largest eigenvalue of the matrix $\mathbf{H}_m^\dagger \mathbf{H}_m - (\sigma_m^2/\sigma_e^2)\mathbf{H}_e^\dagger \mathbf{H}_e$ [92]. The benefit of transmitting artificial noise in the null space of the main channel to impair the eavesdropper's observation was first suggested by Negi and Goel [93, 94] and subsequently analyzed by Khisti, Wornell, Wiesel, and Eldar [95].

The beneficial role of fading for secure communications was highlighted by Barros and Rodrigues [96] for the quasi-static fading wireless channel. The secrecy-capacity region of ergodic-fading channels with perfect channel state information (CSI) about all channels was characterized by Liang, Poor, and Shamai [85]. Bursty signaling was shown to mitigate the absence of information about the eavesdropper's fading by Li, Yates, and Trappe [97] for i.i.d. Rayleigh fading. The secrecy capacity of block-fading channels with no information about the eavesdropper's channels was established by Gopala, Lai, and El Gamal [86], and the near optimality of bursty signaling was proved for i.i.d. Rayleigh block-fading in the high-power regime. A closed-form expression for the secrecy capacity of wireless channels with correlated fading in the limit of high power can be found in [98]. The security of frequency-selective channels has been studied by Koyluoglu, El Gamal, Lai, and Poor [99] and Kobayashi, Debbah, and Shamai [100]. Bloch and Laneman investigated the impact of imperfect CSI on achievable secure rates for general WTCs [101] and showed that little CSI is needed to enable secure communication over wireless channels. Studies of secure wireless communications in terms of the outage probability can be found in [96] for quasi-static fading channels and in [85] for ergodic-fading channels. Tang, Liu, Spasojević, and Poor also analyzed the outage probability of secure hybrid-ARQ schemes [102]. A layered broadcast coding approach, which operates both in ergodic and in non-ergodic fading, has been investigated by Tang, Poor, Liu, and Spasojević [103] as well as Liang, Lai, Poor, and Shamai [104].

Although this chapter primarily focused on the problem of secure communication over wireless channels, many results extend to secret-key agreement over wireless channels. Bloch, Barros, Rodrigues, and McLaughlin investigated key-distillation strategies for quasi-static fading channels [105], and Wong, Bloch, and Shea studied the secret-key capacity of MIMO ergodic channels [106]. Wilson, Tse, and Sholtz [107] also performed an extensive analysis of secret sharing using the reciprocity of an ultrawideband channel. All these studies highlight again the beneficial role of fading for secrecy. In addition, since the channel fading coefficients can always be included in the statistics of the source used for key distillation, key-distillation strategies are often less sensitive to the availability of CSI than wiretap codes.

Practical constructions of wiretap codes for Gaussian and wireless communications remain elusive, but Bloch, Barros, Rodrigues, and McLaughlin [105] proposed a pragmatic approach to secure wireless communications that is based on key-agreement techniques over quasi-static fading channels, which was shown to incur little loss of secure rate in certain regimes.

Part III

Coding and system aspects

6 Coding for secrecy

In this chapter, we discuss the construction of practical codes for secrecy. The design of codes for the wiretap channel turns out to be surprisingly difficult, and this area of information-theoretic security is still largely in its infancy. To some extent, the major obstacles in the road to secrecy capacity are similar to those that lay in the path to channel capacity: the random-coding arguments used to establish the secrecy capacity do not provide explicit code constructions. However, the design of wiretap codes is further impaired by the absence of a simple metric, such as a bit error rate, which could be evaluated numerically. Unlike codes designed for reliable communication, whose performance is eventually assessed by plotting a bit-error-rate curve, we cannot simulate an eavesdropper with unlimited computational power; hence, wiretap codes must possess enough structure to be provably secure. For certain channels, such as binary erasure wiretap channels, the information-theoretic secrecy constraint can be recast in terms of an algebraic property for a code-generator matrix. Most of the chapter focuses on such cases since this algebraic view of secrecy simplifies the analysis considerably.

As seen in Chapter 4, the design of secret-key distillation strategies is a somewhat easier problem insofar as reliability and security can be handled separately by means of information reconciliation and privacy amplification. Essentially, the construction of coding schemes for key agreement reduces to the design of Slepian–Wolf-like codes for information reconciliation, which can be done efficiently with low-density parity-check (LDPC) codes or turbo-codes.

We start this chapter by clarifying the connection between secrecy and capacity-achieving codes (Section 6.1), which was used implicitly in Chapter 3 and Chapter 4, to highlight the insight that can be gained from the information-theoretic proofs. We then briefly recall some fundamental properties of linear codes and LDPC codes (Section 6.2), and we use these codes as building blocks for the construction of wiretap codes over the binary erasure wiretap channel (Section 6.3) and efficient Slepian–Wolf codes for information reconciliation in secret-key distillation strategies (Section 6.4 and Section 6.5). We conclude with a discussion of secure communication over wiretap channels using secret-key distillation strategies (Section 6.6).

6.1 Secrecy and capacity-achieving codes

A natural approach by which to construct practical wiretap codes is to mimic the code structure used in the achievability proofs of Theorem 3.2 and Theorem 3.3. Specifically, we established the existence of codes for a WTC $(\mathcal{X}, p_{\mathrm{YZ|X}}, \mathcal{Y}, \mathcal{Z})$ by partitioning a codebook with $\lceil 2^{nR} \rceil \lceil 2^{nR_\mathrm{d}} \rceil$ codewords into $\lceil 2^{nR} \rceil$ bins of $\lceil 2^{nR_\mathrm{d}} \rceil$ codewords each. The $\lceil 2^{nR} \rceil \lceil 2^{nR_\mathrm{d}} \rceil$ codewords were chosen so that the legitimate receiver could decode reliably. In addition, the bins were constructed so that an eavesdropper knowing which bin is used could decode reliably, as well.

Each bin of codewords can be thought of as a subcode of a "mother code," which is known in coding theory as a *nested code* structure. More importantly, a closer look at the proofs shows that these subcodes are implicitly capacity-achieving codes for the eavesdropper's channel, since the rate R_d of each subcode is chosen in (3.24) such that

$$R_\mathrm{d} = \mathbb{I}(\mathrm{X};\mathrm{Z}) - \delta(\epsilon) \quad \text{for some small } \epsilon > 0.$$

This condition is somewhat buried in the technical details and it is worth clarifying the connection between secrecy and capacity-achieving codes with a more direct proof.

Consider a WTC $(\mathcal{X}, p_{\mathrm{YZ|X}}, \mathcal{Y}, \mathcal{Z})$, and let $\mathcal{C}$ be a code of length n with $\lceil 2^{nR} \rceil$ disjoint subcodes $\{\mathcal{C}_i\}_{\lceil 2^{nR} \rceil}$ such that

$$\mathcal{C} = \bigcup_{i=1}^{\lceil 2^{nR} \rceil} \mathcal{C}_i.$$

For simplicity, we assume that $\mathcal{C}$ guarantees reliable communication over the main channel and analyze only its secrecy properties. Following the stochastic encoding suggested by the proof in Section 3.4.1, a message $m \in [\![1, 2^{nR}]\!]$ is sent by transmitting a codeword chosen uniformly at random in the subcode $\mathcal{C}_\mathrm{m}$. The following theorem provides a *sufficient condition* for this coding scheme to guarantee secrecy with respect to the eavesdropper.

Theorem 6.1 (Thangaraj *et al.*). *If each subcode in the set* $\{\mathcal{C}_i\}_{\lceil 2^{nR} \rceil}$ *stems from a sequence of capacity-achieving codes over the eavesdropper's channel as n goes to infinity, then*

$$\lim_{n\to\infty} \frac{1}{n}\mathbb{I}(\mathrm{M};\mathrm{Z}^n) = 0.$$

Proof. Let C_e denote the capacity of the eavesdropper's channel. If each subcode in $\{\mathcal{C}_i\}_{\lceil 2^{nR} \rceil}$ stems from a sequence of capacity-achieving codes for the eavesdropper's channel then, for any $\epsilon > 0$, there exists n large enough that

$$\forall i \in [\![1, 2^{nR}]\!] \quad \frac{1}{n}\mathbb{I}(\mathrm{X}^n;\mathrm{Z}^n|\mathrm{M}=i) \geqslant C_\mathrm{e} - \epsilon.$$

Consequently, $(1/n)\mathbb{I}(\mathrm{X}^n;\mathrm{Z}^n|\mathrm{M}) \geqslant C_\mathrm{e} - \epsilon$ as well. We now expand the mutual information $\mathbb{I}(\mathrm{M};\mathrm{Z}^n)$ as

$$\begin{aligned}\mathbb{I}(\mathrm{M};\mathrm{Z}^n) &= \mathbb{I}(\mathrm{Z}^n;\mathrm{X}^n\mathrm{M}) - \mathbb{I}(\mathrm{X}^n;\mathrm{Z}^n|\mathrm{M})\\ &= \mathbb{I}(\mathrm{X}^n;\mathrm{Z}^n) + \mathbb{I}(\mathrm{M};\mathrm{Z}^n|\mathrm{X}^n) - \mathbb{I}(\mathrm{X}^n;\mathrm{Z}^n|\mathrm{M}).\end{aligned}$$

Note that $\mathbb{I}(\mathsf{M}; \mathsf{Z}^n|\mathsf{X}^n) = 0$ since $\mathsf{M} \rightarrow \mathsf{X}^n \rightarrow \mathsf{Z}^n$ forms a Markov chain. In addition, $\mathbb{I}(\mathsf{X}^n; \mathsf{Z}^n) \leqslant nC_\text{e}$ since the eavesdropper's channel is memoryless. Therefore,

$$\begin{aligned}\frac{1}{n}\mathbb{I}(\mathsf{M}; \mathsf{Z}^n) &= \frac{1}{n}\mathbb{I}(\mathsf{X}^n; \mathsf{Z}^n) - \frac{1}{n}\mathbb{I}(\mathsf{X}^n; \mathsf{Z}^n|\mathsf{M})\\ &\leqslant C_\text{e} - (C_\text{e} - \epsilon)\\ &= \epsilon.\end{aligned}$$

□

Theorem 6.1 naturally suggests a code-design methodology based on nested codes and capacity-achieving codes over the eavesdropper's channel. Unfortunately, practical families of capacity-achieving codes are known for only a few channels, such as LDPC codes for binary erasure channels and polar codes for binary input symmetric channels; even for these channels, constructing a nested code with capacity-achieving subcodes remains a challenging task. Despite this pessimistic observation, note that the use of capacity-achieving codes for the eavesdropper's channel is merely a *sufficient condition* for secrecy, which leaves open the possibility that alternative approaches might turn out to be more successful. For instance, the code constructions for binary erasure wiretap channels presented in Section 6.3 are based on a somewhat different methodology.

Remark 6.1. *The connection between secrecy and capacity-achieving codes also holds for secret-key distillation strategies. In fact, for the secret-key distillation strategies based on Slepian–Wolf codes analyzed in Section 4.2.2, the number of bins in* (4.16) *was chosen arbitrarily close to the fundamental limit of source coding with side information. Nevertheless, the alternative approach based on sequential key-distillation circumvents this issue, and provides a design methodology that does not depend on Slepian–Wolf codes achieving the fundamental limits of source coding with side information.*

6.2 Low-density parity-check codes

Low-density parity-check (LDPC) codes constitute a family of graph-based block codes, whose performance approaches the fundamental limits of channel coding or source coding when the block length is large, and which can also be decoded efficiently with an iterative algorithm. Since we use LDPC codes extensively in the remainder of this chapter, we devote this section to a brief review (without proofs) of binary LDPC codes and their properties. We refer the interested reader to the textbook by Richardson and Urbanke [108] for a comprehensive and in-depth exposition on the subject.

6.2.1 Binary linear block codes and LDPC codes

Before discussing LDPC codes, we review some basics of binary linear block codes; in particular, the notions of dual code and coset code will be useful for secrecy codes.

Definition 6.1. *A binary* $(n, n-k)$ *block code is a set* $\mathcal{C} \subseteq \text{GF}(2)^n$ *of cardinality*[1] $|\mathcal{C}| = 2^{n-k}$. *The elements of* $\mathcal{C}$ *are called codewords. Associated with the code is a bijective mapping between* $\text{GF}(2)^{n-k}$ *and* $\mathcal{C}$, *which is called an encoder. The elements of* $\text{GF}(2)^{n-k}$ *are called messages. An* $(n, n-k)$ *code is linear if* $\mathcal{C}$ *is an* $(n-k)$*-dimensional subspace of* $\text{GF}(2)^n$. *The rate of a code* $\mathcal{C}$ *is defined as* $R \triangleq (n-k)/n$.

A linear code $\mathcal{C}$ is represented concisely by a matrix $\mathbf{G} \in \text{GF}(2)^{n-k\times n}$, called the *generator matrix*, whose rows form a basis of $\mathcal{C}$. An encoder can then be described by the matrix operation

$$\mathbf{m} \mapsto \mathbf{G}^{\mathsf{T}}\mathbf{m}.$$

A generator matrix $\mathbf{G}$ specifies a code completely, but notice that $\mathbf{G}$ is not unique (any basis of $\mathcal{C}$ can be used to construct $\mathbf{G}$). Different generator matrices define different encoders.

Definition 6.2. *The dual of an* $(n, n-k)$ *linear code* $\mathcal{C}$ *is the set* $\mathcal{C}^{\perp}$ *defined as*

$$\mathcal{C}^{\perp} \triangleq \left\{ \mathbf{c} \in \text{GF}(2)^n : \forall \mathbf{x} \in \mathcal{C} \quad \sum_{i=1}^{n} c_i x_i = 0 \right\}.$$

In other words, $\mathcal{C}^{\perp}$ *contains all vectors of* $\text{GF}(2)^n$ *that are orthogonal to* $\mathcal{C}$.

The reader can check that $\mathcal{C}^{\perp}$ is actually an (n, k) linear code. A generator matrix of $\mathcal{C}^{\perp}$ is denoted by a matrix $\mathbf{H} \in \text{GF}(2)^{k\times n}$ and is called the *parity-check matrix* of $\mathcal{C}$. Note that $\mathbf{H}$ satisfies $\mathbf{G}\mathbf{H}^{\mathsf{T}} = \mathbf{0}$ and that all codewords $\mathbf{x} \in \mathcal{C}$ must satisfy the parity-check equations $\mathbf{H}\mathbf{x} = \mathbf{0}$.

Definition 6.3. *For a linear* $(n, n-k)$ *code* $\mathcal{C}$ *with parity-check matrix* $\mathbf{H}$ *and for* $\mathbf{s} \in \text{GF}(2)^k$, *the set*

$$\mathcal{C}(\mathbf{s}) \triangleq \{\mathbf{x} \in \text{GF}(2)^n : \mathbf{H}\mathbf{x} = \mathbf{s}\}$$

is called the coset code of $\mathcal{C}$ *with syndrome* $\mathbf{s} \in \text{GF}(2)^k$. *In particular,* $\mathcal{C} = \mathcal{C}(\mathbf{0})$.

A coset code is also described by a translation of the original code. In fact, if $\mathbf{x}' \in \text{GF}(2)^n$ is such that $\mathbf{H}\mathbf{x}' = \mathbf{s}$, then

$$\mathcal{C}(\mathbf{s}) = \{\mathbf{x}' \oplus \mathbf{x} : \mathbf{x} \in \mathcal{C}\}.$$

A sequence $\mathbf{x}' \in \mathcal{C}(\mathbf{s})$ with minimum weight is called a *coset leader* of $\mathcal{C}(\mathbf{s})$. It is possible to show that an $(n, n-k)$ code has 2^k disjoint cosets, which form a partition of $\text{GF}(2)^n$.

Binary LDPC codes are a special class of binary linear codes, characterized by a sparse parity-check matrix $\mathbf{H}$, which contains a much smaller number of ones than zeros. In other words, the parity-check equations defining the code involve only a small number of bits. Rather than specifying the LDPC code in terms of its parity-check matrix, it is convenient to use a graphical representation of $\mathbf{H}$ called the *Tanner graph*. The Tanner

[1] The usual convention is to consider (n, k) block codes so that the number of codewords is 2^k rather than 2^{n-k}; nevertheless, this alternative convention simplifies our notation later on.

$$\mathbf{H} = \begin{pmatrix} 1&1&1&1&0&1&1&0&0&0 \\ 0&0&1&1&1&1&1&1&0&0 \\ 0&1&0&1&0&1&0&1&1&1 \\ 1&0&1&0&1&0&0&1&1&1 \\ 1&1&0&0&1&0&1&0&1&1 \end{pmatrix}$$

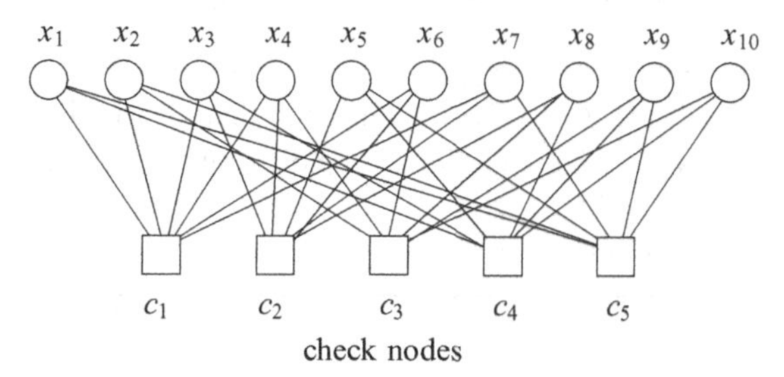

Figure 6.1 A parity-check matrix and its corresponding Tanner graph for a code with blocklength $n = 10$.

graph of $\mathbf{H} \in \mathrm{GF}(2)^{k \times n}$ is a bipartite graph with n *variable nodes* and k *check nodes* connected by *edges*. Each variable node represents a bit x_i in a codeword, while each check node represents a parity-check equation satisfied by the codewords. Specifically, letting $\mathbf{H} = \left(h_{ji}\right)_{kn}$, the jth check node represents the equation

$$c_j = \bigoplus_{i=1}^{n} x_i h_{ji}.$$

An edge connects variable node x_i to check node c_j if and only if x_i is involved in the jth parity-check equation, that is $h_{ji} = 1$. The *degree* of a node is defined as the number of edges incident to it. As an example, Figure 6.1 illustrates a parity-check matrix and its corresponding Tanner graph for a binary linear code of length $n = 10$, in which all variable nodes have degree 3 and all check nodes have degree 6.

Given a Tanner graph, it is possible to compute its *variable-node edge-degree distribution* $\{\lambda_i\}_{i \geqslant 1}$, in which λ_i is the fraction of edges incident on a variable node with degree i. Similarly, the *check-node edge-degree distribution* is $\{\rho_j\}_{j \geqslant 1}$, in which ρ_j is the fraction of edges incident on a check node with degree j. These edge-degree distributions (degree distributions for short) are often represented in compact form by the following polynomials in x:

$$\lambda(x) = \sum_{i \geqslant 1} \lambda_i x^{i-1} \quad \text{and} \quad \rho(x) = \sum_{j \geqslant 1} \rho_j x^{j-1}.$$

The rate of the code is directly related to the edge-degree distribution by

$$R = 1 - \frac{\int_0^1 \rho(x)\mathrm{d}x}{\int_0^1 \lambda(x)\mathrm{d}x}.$$

Note that a parity-check matrix $\mathbf{H}$ specifies a unique Tanner graph and thus a unique degree distribution, whereas a degree distribution specifies only an *ensemble* of codes with the same rate R (for instance, all permutations of nodes in a graph have the same degree distribution). Fortunately, for large block lengths, all codes within a given ensemble have roughly the same decoding performance; hence, LDPC codes are often specified by their degree distributions $(\lambda(x), \rho(x))$ alone.

An LDPC code is called *regular* if all variable nodes have the same degree, and all check nodes have the same degree. Otherwise, it is called *irregular*.

Example 6.1. A rate-$\frac{1}{2}$ regular $(3, 6)$ LDPC code is such that all variable nodes have degree 3, and all check nodes have degree 6. Its degree distributions are simply

$$\lambda(x) = x^2 \quad \text{and} \quad \rho(x) = x^5.$$

Example 6.2. The following irregular degree distributions correspond to another rate-$\frac{1}{2}$ LDPC code:

$$\lambda(x) = 0.106\,257x + 0.486\,659x^2 + 0.010\,390x^{10} + 0.396\,694x^{19},$$
$$\rho(x) = 0.5x^7 + 0.5x^8.$$

6.2.2 Message-passing decoding algorithm

Let $\mathcal{C}$ be an $(n, n-k)$ LDPC code with parity-check matrix $\mathbf{H} \in \text{GF}(2)^{k\times n}$. Consider a codeword $\mathbf{x} = (x_1, \dots, x_n)^\mathsf{T} \in \mathcal{C}$ whose bits are transmitted over a binary-input memoryless channel $(\{0, 1\}, p_{\mathrm{Y}|\mathrm{X}}(y|x), \mathcal{Y})$. Let $\mathbf{y} = (y_1, \dots, y_n)^\mathsf{T}$ denote the vector of received symbols. The success of LDPC codes stems from the existence of a computationally efficient algorithm to approximate the a-posteriori log-likelihood ratios (LLRs)

$$\lambda_i = \log\left(\frac{\mathbb{P}[\mathrm{X}_i = 0|\mathbf{y}]}{\mathbb{P}[\mathrm{X}_i = 1|\mathbf{y}]}\right) \quad \text{for } i \in [\![1, n]\!].$$

The sign of λ_i provides the most likely value of the bit x_i, while the magnitude $|\lambda_i|$ provides a measure of the reliability associated with the decision. For instance, if $|\lambda_i| = 0$, the bit x_i is equally likely to be zero or one; in contrast, if $|\lambda_i| = \infty$, there is no uncertainty regarding the value of x_i.

For $i \in [\![1, n]\!]$, we let $\mathcal{N}(i)$ denote the indices of check nodes connected to the variable node x_i in the Tanner graph; the set can be obtained from the parity-check matrix $\mathbf{H} = (h_{ji})_{kn}$ as

$$\mathcal{N}(i) \triangleq \{j : h_{ji} = 1\}.$$

Table 6.1 Belief-propagation algorithm

□ **Initialization.**

► For each $i \in [\![1, n]\!]$ and for each $j \in \mathcal{N}(i)$

$$u_{ji}^{(0)} = v_{ij}^{(0)} = 0.$$

► For each $i \in [\![1, n]\!]$

$$\lambda_i^{\text{INT}} = \log\left(\frac{p_{Y|X}(y_i|0)}{p_{Y|X}(y_i|1)}\right).$$

□ **Iterations.** For each iteration $l \in [\![1, l_{\max}]\!]$

► For each $i \in [\![1, n]\!]$ and for each $j \in \mathcal{N}(i)$

$$v_{ij}^{(l)} = \lambda_i^{\text{INT}} + \sum_{m \in \mathcal{N}(i) \backslash j} u_{mi}^{(l-1)}.$$

► For each $j \in [\![1, k]\!]$ and for each $i \in \mathcal{M}(j)$

$$u_{ji}^{(l)} = 2 \tanh^{-1}\left(\prod_{m \in \mathcal{M}(j) \backslash i} \tanh\left(\frac{v_{mj}^{(l)}}{2}\right)\right).$$

□ **Extrinsic information**. For all $i \in [\![1, n]\!]$

$$\lambda_i^{\text{EXT}} = \sum_{m \in \mathcal{N}(i)} u_{mi}^{(l_{\max})}.$$

□ **Hard decisions.** For all $i \in [\![1, n]\!]$

$$\hat{x}_i = \frac{1}{2}\left(1 - \text{sign}\left(\lambda_i^{\text{INT}} + \lambda_i^{\text{EXT}}\right)\right).$$

Similarly, for $j \in [\![1, k]\!]$, we let $\mathcal{M}(j)$ denote the indices of variable nodes connected to the check node c_j in the Tanner graph; that is,

$$\mathcal{M}(j) \triangleq \{i : h_{ji} = 1\}.$$

The LLRs $\{\lambda_i\}_n$ can be approximated using the iterative algorithm described in Table 6.1. This algorithm is called the "belief-propagation" algorithm or the "sum–product" algorithm.

As illustrated in Figure 6.2, the belief-propagation algorithm belongs to the class of "message-passing" algorithms, since the quantities $v_{ij}^{(l)}$ and $u_{ji}^{(l)}$, which are updated at each iteration l, can be understood as "messages" exchanged between the variable nodes and check nodes along the edges of the Tanner graph. The final hard decision for each bit x_i is based on two terms: λ_i^{INT}, which is called the *intrinsic information* (or intrinsic LLR) because it depends only on the current observation y_i; and λ_i^{EXT}, which is called the *extrinsic information* (or extrinsic LLR) because it contains the information about x_i provided by other observations. The usefulness of the algorithm is justified by the following result.

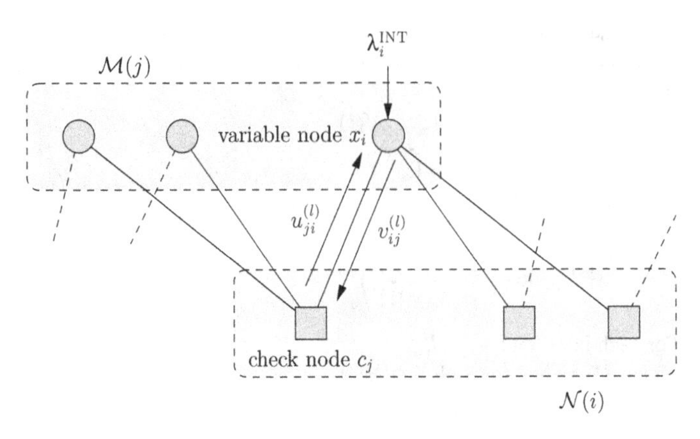

Figure 6.2 Illustration of message-passing behavior for the belief-propagation algorithm.

Theorem 6.2. *If the Tanner graph of an LDPC code does not contain cycles, then the values* $\left\{\lambda_i^{\text{INT}} + \lambda_i^{\text{EXT}}\right\}_n$ *computed by the message-passing algorithm converge to the true a-posteriori LLRs* $\{\lambda_i\}_n$. *The hard decisions are then equivalent to bit-wise maximum a-posteriori (MAP) estimations.*

In practice, even if the Tanner graph contains a few cycles, the message-passing algorithm performs reasonably well. The complexity of the algorithm is linear in the number of edges, and is particularly useful for codes with sparse Tanner graphs, such as LDPC codes.

6.2.3 Properties of LDPC codes under message-passing decoding

Definition 6.4. *Consider a set of binary-input memoryless channels that are all characterized by a parameter* α. *The set of channels is ordered if* $\alpha_1 < \alpha_2$ *implies that the channel with parameter* α_2 *is stochastically degraded with respect to the channel with parameter* α_1.

Definition 6.5. *A binary-input memoryless channel* $\left(\{\pm 1\}, p_{\mathsf{Y}|\mathsf{X}}(y|x), \mathcal{Y}\right)$ *is output-symmetric if the transition probabilities are such that*

$$\forall y \in \mathcal{Y} \quad p_{\mathsf{Y}|\mathsf{X}}(y|-1) = p_{\mathsf{Y}|\mathsf{X}}(-y|1).$$

Examples of ordered families of binary-input symmetric-output channels include binary symmetric channels with cross-over probability p and binary erasure channels with erasure probability ϵ (after relabeling of the inputs), and binary-input additive white Gaussian noise channels with noise variance σ^2 under the same input power constraint.

Theorem 6.3 (Richardson and Urbanke). *Consider an LDPC code of length n chosen uniformly at random from the ensemble of codes with degree distributions* $(\lambda(x), \rho(x))$ *and used over an ordered family of binary-input symmetric-output channels with parameter* α. *Then, the following results hold.*

1. ***Convergence to the cycle-free case:*** *as n goes to infinity, the Tanner graph becomes cycle-free.*
2. ***Concentration around the average:*** *as n goes to infinity, the error-decoding capability of the code under message-passing decoding converges to the average error-decoding capability of the code ensemble.*
3. ***Threshold behavior:*** *there exists a channel parameter α^*, called the* threshold, *such that the bit error probability goes to zero as n goes to infinity if and only if $\alpha < \alpha^*$.*

Theorem 6.3 simplifies the design of LDPC codes tremendously because it states that it is sufficient to analyze the average performance of an ensemble of LDPC code with given degree distributions $(\lambda(x), \rho(x))$ rather than focus on an individual code. For large length n, the probability of error with belief-propagation decoding of most codes in the ensemble is close to the probability of error averaged over the ensemble; hence, to construct good codes with rate R, it suffices to optimize the degree distributions $(\lambda(x), \rho(x))$ so as to maximize the threshold α^*. This optimization is, in general, non-convex in the degree distribution, but it can be numerically solved by combining an efficient algorithm to compute the threshold of given degree distributions $(\lambda(x), \rho(x))$, called density evolution, and a heuristic genetic optimization algorithm, called differential evolution.

Example 6.3. Consider the family of binary erasure channels with parameter α, the erasure probability. The threshold of a regular (3, 6) LDPC code is $\alpha^* \approx 0.42$. The threshold of a rate-$\frac{1}{2}$ irregular code with distribution as in Example 6.2 is $\alpha^* \approx 0.4741$.

Note that Theorem 6.3 characterizes a threshold for the asymptotic behavior of the *bit error probability*. This result can be refined and, as stated in the following theorem, one can show that the same threshold also characterizes the behavior of the *block error probability* for some LDPC ensembles.

Theorem 6.4 (Jin and Richardson). *If the degree distributions $(\lambda(x), \rho(x))$ of an LDPC ensemble do not contain any variable nodes of degree 2, then the bit error probability threshold is also the block error probability threshold.*

LDPC ensembles with high thresholds usually have a high fraction of nodes of degree 2, and it may seem that Theorem 6.4 prevents us from obtaining high thresholds for the block error probability. However, it is possible to strengthen Theorem 6.4 to accomodate some fraction of degree-2 nodes. The presentation of this result goes beyond the scope of this chapter and we refer the interested reader to [109] for more details.

6.3 Secrecy codes for the binary erasure wiretap channel

In this section, we restrict our attention to the binary erasure wiretap channel illustrated in Figure 6.3, in which the main channel is noiseless while the eavesdropper's channel is a BEC with erasure probability ϵ. From Corollary 3.1, the secrecy capacity of this

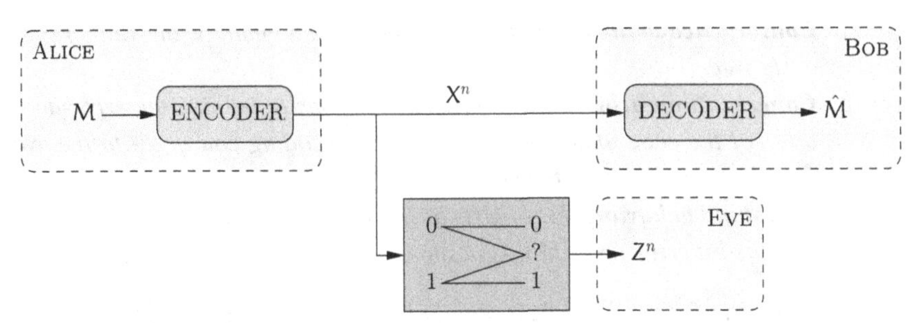

Figure 6.3 Binary erasure wiretap channel.

channel is $C_s = 1 - (1 - \epsilon) = \epsilon$. Following the observations made in Section 6.1, we investigate a code construction based on nested codes. However, since any codeword sent by the transmitter is received without errors by the legitimate receiver, the construction is much simpler than in the general case: for any set of disjoint subcodes $\{\mathcal{C}_i\}$, the mother code $\mathcal{C} = \bigcup_i \mathcal{C}_i$ is always a reliable code for the (noiseless) main channel.

A set of subcodes that leads to a particularly simple stochastic encoder consists of an $(n, n-k)$ binary linear code $\mathcal{C}_0$ and its cosets. For this choice of subcodes, the mother code $\mathcal{C}$ is

$$\mathcal{C} = \bigcup_{\mathbf{s} \in \mathrm{GF}(2)^k} \mathcal{C}_0(\mathbf{s}) = \mathrm{GF}(2)^n.$$

The corresponding stochastic encoding procedure is called *coset coding*, and it consists of encoding a message $\mathbf{m} \in \mathrm{GF}(2)^k$ by selecting a codeword uniformly at random in the coset code $\mathcal{C}_0(\mathbf{m})$ of $\mathcal{C}_0$ with syndrome $\mathbf{m}$. The following proposition shows that the encoding and decoding operations in coset coding can be implemented efficiently with matrix multiplications.

Proposition 6.1. *Let $\mathcal{C}_0$ be an $(n, n-k)$ binary linear code. Then there exists a generator matrix $\mathbf{G} \in \mathrm{GF}(2)^{n-k \times n}$ and a parity-check matrix $\mathbf{H} \in \mathrm{GF}(2)^{k \times n}$ for $\mathcal{C}_0$ and a matrix $\mathbf{G}' \in \mathrm{GF}(2)^{k \times n}$ such that*

- *the encoder maps a message $\mathbf{m}$ to a codeword as*

$$\mathbf{m} \mapsto (\mathbf{G}'^{\mathsf{T}} \;\; \mathbf{G}^{\mathsf{T}}) \begin{pmatrix} \mathbf{m} \\ \mathbf{v} \end{pmatrix},$$

  where the vector $\mathbf{v} \in \mathrm{GF}(2)^{n-k}$ is chosen uniformly at random;
- *the decoder maps a codeword $\mathbf{x}$ to a message as*

$$\mathbf{x} \mapsto \mathbf{Hx}.$$

Proof. Let $\{\mathbf{g}_i\}_{n-k}$ be a basis of $\mathcal{C}_0$ and let $\mathbf{G}$ be a generator matrix whose rows are the vectors $\{\mathbf{g}_i^{\mathsf{T}}\}_{n-k}$. The set $\{\mathbf{g}_i\}_{n-k}$, which is linearly independent, can be completed by a linearly independent set $\{\mathbf{h}_i\}_k$ to obtain a basis of $\mathrm{GF}(2)^n$. Let $\mathbf{G}'$ be a matrix whose rows are the vectors $\{\mathbf{h}_i^{\mathsf{T}}\}_k$.

Let us now consider the encoding of a message $\mathbf{m} \in \mathrm{GF}(2)^k$ as

$$\mathbf{m} \mapsto (\mathbf{G}'^{\mathsf{T}} \ \mathbf{G}^{\mathsf{T}}) \begin{pmatrix} \mathbf{m} \\ \mathbf{v} \end{pmatrix} = \sum_{i=1}^{k} m_i \mathbf{h}_i + \sum_{i=1}^{n-k} v_i \mathbf{g}_i, \tag{6.1}$$

where $\mathbf{v} \in \mathrm{GF}(2)^{n-k}$ is chosen uniformly at random. The term $\sum_{i=1}^{n-k} v_i \mathbf{g}_i$ corresponds to the choice of a codeword uniformly at random in $\mathcal{C}_0$; therefore, the operation (6.1) is equivalent to coset coding if we can prove that two different messages $\mathbf{m}$ and $\mathbf{m}'$ generate sequences in different cosets of $\mathcal{C}_0$. Assume that two messages $\mathbf{m}$ and $\mathbf{m}'$ are encoded as sequences $\mathbf{x}$ and $\mathbf{x}'$ in the same coset with coset leader $\mathbf{e} \in \mathrm{GF}(2)^n$. Then, there exist codewords $\mathbf{c}_1, \mathbf{c}_2 \in \mathcal{C}_0$ such that

$$\mathbf{x} = \mathbf{e} + \mathbf{c}_1 \quad \text{and} \quad \mathbf{x}' = \mathbf{e} + \mathbf{c}_2.$$

Consequently, $\mathbf{x} + \mathbf{x}' = \mathbf{c}_1 + \mathbf{c}_2 \in \mathcal{C}_0$, which is impossible unless $\mathbf{m} = \mathbf{m}'$.

It remains to prove that there exists a parity-check matrix $\mathbf{H}$ such that $\mathbf{Hx} = \mathbf{m}$. Let $\mathbf{H}'$ be an arbitrary parity-check matrix of $\mathcal{C}_0$. In general, if $\mathbf{x}$ is obtained from $\mathbf{m}$ according to (6.1), then $\mathbf{H}'\mathbf{x} \neq \mathbf{m}$. However, the application $\mathbf{m} \mapsto \mathbf{H}'\mathbf{x}$ is injective (and hence bijective); hence, there exists an invertible matrix $\mathbf{A} \in \mathrm{GF}(2)^{k \times k}$ such that

$$\mathbf{AH}'\mathbf{x} = \mathbf{m}.$$

Note that $\mathbf{H} = \mathbf{AH}'$ is another parity-check matrix for $\mathcal{C}_0$ and is such that $\mathbf{Hx} = \mathbf{m}$. □

6.3.1 Algebraic secrecy criterion

Since coset coding can be defined in terms of the parity-check matrix and generator matrix of a linear code $\mathcal{C}_0$, the algebraic structure of the linear code is likely to have a critical effect on secrecy. Hence, to clarify this connection and simplify the analysis, it is convenient to develop an *algebraic secrecy criterion* equivalent to the original information-theoretic secrecy criterion for coset coding.

Consider an eavesdropper's observation $\mathbf{z}$ with μ unerased bits in positions $(i_1, \ldots, i_\mu)$. If a sequence $\mathbf{x} \in \mathrm{GF}(2)^n$ is such that

$$(\mathbf{x}_{i_1}, \ldots, \mathbf{x}_{i_\mu}) = (\mathbf{z}_{i_1}, \ldots, \mathbf{z}_{i_\mu}),$$

then the sequence is said to be *consistent* with $\mathbf{z}$. If a coset of $\mathcal{C}_0$ contains at least one sequence $\mathbf{x}$ that is consistent with $\mathbf{z}$, then the coset itself is said to be consistent with $\mathbf{z}$. The total number of cosets of $\mathcal{C}_0$ consistent with $\mathbf{z}$ is denoted by $N(\mathcal{C}_0, \mathbf{z})$.

Lemma 6.1. *Let $\mathbf{z}$ be an eavesdropper's observation at the output of the BEC. Then all cosets of $\mathcal{C}_0$ that are consistent with $\mathbf{z}$ contain the same number of sequences consistent with $\mathbf{z}$.*

Proof. Let $(i_1, \ldots, i_\mu)$ denote the set of unerased positions of $\mathbf{z}$. Let $\mathbf{G} \in \mathrm{GF}(2)^{n-k \times n}$ be a generator matrix of $\mathcal{C}_0$, and let $\mathbf{g}_i$ denote the ith column of $\mathbf{G}$. We define the matrix $\mathbf{G}_\mu$ as

$$\mathbf{G}_\mu \triangleq \left(\mathbf{g}_{i_1} \ \mathbf{g}_{i_2} \cdots \mathbf{g}_{i_\mu}\right).$$

Let $\mathcal{C}_1$ be a coset of $\mathcal{C}_0$ consistent with $\mathbf{z}$. Define $\mathfrak{C}_1(\mathbf{z}) \triangleq \{\mathbf{x} \in \mathcal{C}_1 : \mathbf{x} \text{ consistent with } \mathbf{z}\}$ and let $\mathbf{x}_1 \in \mathfrak{C}_1(\mathbf{z})$. We show that $|\mathfrak{C}_1(\mathbf{z})| = \left|\text{Ker}(\mathbf{G}_\mu)\right|$.

- For any $\mathbf{m} \in \text{Ker}(\mathbf{G}_\mu)$, the vector $\mathbf{G}^\mathsf{T}\mathbf{m}$ is in $\mathcal{C}_0$ and contains zeros in positions $(i_1, \dots, i_\mu)$. Therefore, for any $\mathbf{m} \in \text{Ker}(\mathbf{G}_\mu)$, $\mathbf{x}_1 + \mathbf{G}^\mathsf{T}\mathbf{m} \in \mathfrak{C}_1(\mathbf{z})$ and
$$\left\{\mathbf{x}_1 + \mathbf{G}^\mathsf{T}\mathbf{m} : \mathbf{m} \in \text{Ker}(\mathbf{G}_\mu)\right\} \subseteq \mathfrak{C}_1(\mathbf{z}).$$
- Now, assume $\mathbf{x}_1' \in \mathfrak{C}_1(\mathbf{z})$. Then $\mathbf{x}_1' + \mathbf{x}_1 \in \mathcal{C}_0$ and contains zeros in positions $(i_1, \dots, i_\mu)$. Hence, there exists $\mathbf{m} \in \text{Ker}(\mathbf{G}_\mu)$ such that $\mathbf{x}_1' + \mathbf{x}_1 = \mathbf{G}^\mathsf{T}\mathbf{m}$. Therefore, $\mathbf{x}_1' \in \left\{\mathbf{x}_1 + \mathbf{G}^\mathsf{T}\mathbf{m} : \mathbf{m} \in \text{Ker}(\mathbf{G}_\mu)\right\}$ and
$$\mathfrak{C}_1(\mathbf{z}) \subseteq \left\{\mathbf{x}_1 + \mathbf{G}^\mathsf{T}\mathbf{m} : \mathbf{m} \in \text{Ker}(\mathbf{G}_\mu)\right\}.$$

Hence, $\mathfrak{C}_1(\mathbf{z}) = \left\{\mathbf{x}_1 + \mathbf{G}^\mathsf{T}\mathbf{m} : \mathbf{m} \in \text{Ker}(\mathbf{G}_\mu)\right\}$ and $|\mathfrak{C}_1(\mathbf{z})| = \left|\text{Ker}(\mathbf{G}_\mu)\right|$. Therefore, any coset of $\mathcal{C}_0$ consistent with $\mathbf{z}$ contains exactly $\left|\text{Ker}(\mathbf{G}_\mu)\right|$ sequences consistent with $\mathbf{z}$. □

Proposition 6.2. *Let* $\mathbf{z}$ *be an eavesdropper's observation at the output of the BEC. Then the eavesdropper's uncertainty about* M *given his observation* $\mathbf{z}$ *is*
$$\mathbb{H}(\mathsf{M}|\mathbf{z}) = \log N(\mathcal{C}_0, \mathbf{z}).$$

Proof. Let X^n be the random variable that represents the sequence sent over the channel. Then,
$$\begin{aligned}\mathbb{H}(\mathsf{M}|\mathbf{z}) &= \mathbb{H}(\mathsf{M}\mathsf{X}^n|\mathbf{z}) - \mathbb{H}(\mathsf{X}^n|\mathsf{M}\mathbf{z})\\ &= \mathbb{H}(\mathsf{X}^n|\mathbf{z}) - \mathbb{H}(\mathsf{X}^n|\mathsf{M}\mathbf{z}).\end{aligned}$$

The term $\mathbb{H}(\mathsf{X}^n|\mathbf{z})$ is the uncertainty in the codeword that was sent given the observation $\mathbf{z}$. By virtue of the definition of coset coding, all codewords are used with equal probability; therefore, $\mathbb{H}(\mathsf{X}^n|\mathbf{z}) = \log N$, where N is the number of sequences that are consistent with $\mathbf{z}$. Now,
$$\mathbb{H}(\mathsf{X}^n|\mathsf{M}\mathbf{z}) = \sum_m \mathbb{H}(\mathsf{X}^n|\mathsf{M} = m, \mathbf{z}) p_{\mathsf{M}|\mathsf{Z}}(m|\mathbf{z}),$$
and the term $\mathbb{H}(\mathsf{X}^n|\mathsf{M} = m, \mathbf{z})$ is the uncertainty in the sequence that was sent given $\mathbf{z}$ and knowing the coset m that was used. By definition, all codewords are used with equal probability and, by Lemma 6.1, all cosets consistent with $\mathbf{z}$ contain the same number of sequences consistent with $\mathbf{z}$; hence, $\mathbb{H}(\mathsf{X}^n|\mathsf{M} = m, \mathbf{z}) = \log N_\mathrm{c}$, where N_c is the number of sequences consistent with $\mathbf{z}$ in a coset consistent with $\mathbf{z}$. Therefore,
$$\mathbb{H}(\mathsf{M}|\mathbf{z}) = \log N - \log N_\mathrm{c} = \log\left(\frac{N}{N_\mathrm{c}}\right) = \log N(\mathcal{C}_0, \mathbf{z}).$$ □

The number of cosets is 2^k, therefore $N(\mathcal{C}_0, \mathbf{z}) \leqslant 2^k$. If $N(\mathcal{C}_0, \mathbf{z}) = 2^k$, then all cosets are consistent with $\mathbf{z}$ and we say that $\mathbf{z}$ is *secured* by the code $\mathcal{C}_0$. The following theorem provides a necessary and sufficient condition for an observation $\mathbf{z}$ to be secured by $\mathcal{C}_0$.

Proposition 6.3 (Ozarow and Wyner). *Let* $\mathcal{C}_0$ *be an* $(n, n-k)$ *binary linear code with generator matrix* $\mathbf{G}$*, and let* $\mathbf{g}_i$ *denote the* i*th column of* $\mathbf{G}$*. Let* $\mathbf{z}$ *be an observation of* *the eavesdropper with μ unerased bits in positions $(i_1, \dots, i_\mu)$. Then $\mathbf{z}$ is secured by $\mathcal{C}_0$ if and only if the matrix*

$$\mathbf{G}_\mu \triangleq (\mathbf{g}_{i_1}\ \mathbf{g}_{i_2} \dots \mathbf{g}_{i_\mu})$$

has rank μ.

Proof. Assume that $\mathbf{G}_\mu$ has rank μ. Then, by definition, the code $\mathcal{C}_0$ has codewords with all possible sequences of $\mathrm{GF}(2)^\mu$ in the μ unerased positions. Since cosets are obtained by translating $\mathcal{C}_0$, all cosets also have codewords with all possible binary sequences in the μ unerased positions. Therefore, $N(\mathcal{C}_0, \mathbf{z}) = 2^k$.

Now, assume that $\mathbf{G}_\mu$ has rank strictly less than μ. Then, there exists at least one sequence of μ bits $\mathbf{c}_\mu$ that does not appear in any codeword of $\mathcal{C}_0$ in the μ unerased positions. For any sequence $\mathbf{x}' \in \mathrm{GF}(2)^n$, we let $\mathbf{x}'_\mu$ denote the bits of $\mathbf{x}'$ in the μ unerased positions. Since the cosets form a partition of $\mathrm{GF}(2)^n$, there exists a sequence $\mathbf{x}'$ in a coset $\mathcal{C}'$ such that $\mathbf{x}'_\mu \oplus \mathbf{c}_\mu$ has the same value as $\mathbf{z}$ in the unerased positions. Since $\mathbf{c}_\mu$ does not appear in any codeword of $\mathcal{C}_0$, the coset $\mathcal{C}'$ is not consistent with $\mathbf{z}$; therefore $N(\mathcal{C}_0, \mathbf{z}) < 2^k$. □

As an immediate consequence of Proposition 6.3, we obtain a necessary and sufficient condition for communication in perfect secrecy with respect to an eavesdropper who observes any set of μ unerased bits.

Corollary 6.1. *Let $\mathcal{C}_0$ be an $(n, n-k)$ binary linear code with generator matrix* $\mathbf{G}$. *Coset coding with $\mathcal{C}_0$ guarantees perfect secrecy against an eavesdropper who observes* any *set of μ unerased bits if and only if all submatrices of* $\mathbf{G}$ *with μ columns have rank μ.*

Proof. If all submatrices of $\mathbf{G}$ with μ columns have rank μ, then any observation $\mathbf{z}$ with μ unerased positions is secured by $\mathcal{C}_0$ and $\mathbb{H}(\mathsf{M}|\mathbf{z}) = k$; therefore,

$$\begin{aligned}\mathbb{I}(\mathsf{M}; \mathsf{Z}^n) &= \mathbb{H}(\mathsf{M}) - \mathbb{H}(\mathsf{M}|\mathsf{Z}^n)\\ &= k - \sum_{\mathbf{z}} p_{\mathsf{Z}^n}(\mathbf{z})\mathbb{H}(\mathsf{M}|\mathbf{z})\\ &= 0.\end{aligned}$$

Conversely, if there exists a submatrix of $\mathbf{G}$ with μ columns and rank less than μ, then there exists an observation $\mathbf{z}'$ that is not secured by $\mathcal{C}_0$ and such that $\mathbb{H}(\mathsf{M}|\mathbf{z}') < k$. Therefore,

$$\begin{aligned}\mathbb{I}(\mathsf{M}; \mathsf{Z}^n) &= \mathbb{H}(\mathsf{M}) - \mathbb{H}(\mathsf{M}|\mathsf{Z}^n)\\ &= k - \sum_{\mathbf{z}} p_{\mathsf{Z}^n}(\mathbf{z})\mathbb{H}(\mathsf{M}|\mathbf{z})\\ &> 0.\end{aligned}$$

□

Remark 6.2. *A binary wiretap channel model in which the eavesdropper is known to access no more than μ of n transmitted bits is called a* wiretap channel of type II.

This model differs from the binary erasure wiretap channel of Figure 6.3 in that the eavesdropper can in principle choose which μ bits are observed. This model was extensively studied by Ozarow and Wyner, and we discuss it in the context of network coding in Chapter 9.

6.3.2 Coset coding with dual of LDPC codes

We now go back to the binary erasure wiretap channel of Figure 6.3. In general, we cannot guarantee that the eavesdropper's observation contains a fixed number E of unerased symbols with probability one; however, by Chebyshev's inequality,

$$\forall \beta > 0 \quad \mathbb{P}\left[\left|\frac{\mathsf{E}}{n} - (1-\epsilon)\right| > \beta\right] \leqslant \frac{\operatorname{Var}\mathsf{E}}{n^2\beta^2} = \frac{\epsilon(1-\epsilon)}{n\beta^2}.$$

In other words, the fraction of unerased positions is arbitrarily close to $(1-\epsilon)$ with high probability as n becomes large. Consequently, we will be able to leverage the results of Section 6.3.1 if we can find a suitable matrix $\mathbf{G}$ for coset coding such that the observations of the eavesdropper are secured with high probability. It turns out that *parity-check* matrices of LDPC codes satisfy the desired condition. In fact, the threshold property of decoding under message-passing can be interpreted as follows.

Lemma 6.2. *Let $\mathbf{H}$ be the parity-check matrix of a length-n LDPC code selected uniformly at random in an ensemble whose block error probability threshold under belief-propagation decoding for the erasure channel is α^*. Form a submatrix $\mathbf{H}'$ of $\mathbf{H}$ by selecting each column of $\mathbf{H}$ with probability $\alpha < \alpha^*$. Then,*

$$\mathbb{P}\left[\operatorname{rk}(\mathbf{H}') = \alpha n\right] = 1 - \delta(n).$$

Proof. By Theorem 6.3 and Theorem 6.4, with probability $1-\delta(n)$, $\mathbf{H}$ is such that the block error probability under message-passing decoding vanishes as n goes to infinity if $\alpha < \alpha^*$. In other words, if the erased bits in a given observation $\mathbf{z}$ are treated as unknown, the equation $\mathbf{Hz} = \mathbf{0}$ has a unique solution. Without loss of generality, we assume that the first bits of $\mathbf{z}$ are erased and rewrite the equation $\mathbf{Hz} = \mathbf{0}$ as $(\mathbf{H}'\ \mathbf{H}'')\mathbf{z} = \mathbf{0}$, where $\mathbf{H}'$ corresponds to the erased position of $\mathbf{z}$. This equation has a unique solution if and only if $\mathbf{H}'$ has full column rank αn. □

Theorem 6.5 (Thangaraj *et al.*). *Let C_0 be an $(n, n-k)$ binary LDPC code selected uniformly at random in an ensemble with erasure threshold α^* and let $C_0^\perp$ be its dual. Then, as n goes to infinity, coset coding with $C_0^\perp$ and its cosets guarantees (weak) secrecy at rate $R = k/n$ over any binary erasure wiretap channel with erasure probability $\epsilon > 1-\alpha^*$.*

Proof. By definition, a generator matrix $\mathbf{G}$ of $C_0^\perp$ is a parity-check matrix $\mathbf{H}$ of C_0. Therefore, by Lemma 6.2, if $1-\epsilon < \alpha^*$, any submatrix formed by selecting columns of $\mathbf{G}$ with probability $1-\epsilon$ has full column rank with probability $1-\delta(n)$; hence, by Proposition 6.3, any observation Z^n of the eavesdropper is secured by $C_0^\perp$ with probability

$1 - \delta(n)$. Equivalently, if we let Q be the random variable defined as

$$\mathrm{Q} \triangleq \begin{cases} 1 & \text{if } \mathrm{Z}^n \text{ is secured by } \mathcal{C}_0^\perp, \\ 0 & \text{else,} \end{cases}$$

then $\mathbb{P}[\mathrm{Q} = 1] \geqslant 1 - \delta(n)$. Consequently,

$$\begin{aligned}
\frac{1}{n}\mathbb{I}(\mathrm{M}; \mathrm{Z}^n) &\leqslant \frac{1}{n}\mathbb{I}(\mathrm{M}; \mathrm{Z}^n\mathrm{Q}) \\
&\leqslant \frac{1}{n}\left(\mathbb{H}(\mathrm{M}) - \mathbb{H}(\mathrm{M}|\mathrm{Z}^n\mathrm{Q})\right) \\
&= \frac{1}{n}\left(\mathbb{H}(\mathrm{M}) - \mathbb{H}(\mathrm{M}|\mathrm{Z}^n\mathrm{Q} = 1)\mathbb{P}[\mathrm{Q} = 1] - \mathbb{H}(\mathrm{M}|\mathrm{Z}^n\mathrm{Q} = 0)\mathbb{P}[\mathrm{Q} = 0]\right) \\
&\leqslant \frac{1}{n}\left(\mathbb{H}(\mathrm{M}) - \mathbb{H}(\mathrm{M})(1 - \delta(n))\right) \\
&\leqslant \frac{\mathbb{H}(\mathrm{M})}{n}\delta(n) \\
&\leqslant \delta(n).
\end{aligned}$$

In other words, coset coding with $\mathcal{C}_0^\perp$ guarantees weak secrecy at rate k/n. □

Note that the analysis of coset coding with the dual of LDPC codes does not rely on any capacity-achieving property; our proof relies solely on the concentration and threshold properties established by Theorem 6.3 and Theorem 6.4. Nevertheless, as illustrated by the following examples, the price paid is that we cannot achieve rates arbitrarily close to the secrecy capacity.

Example 6.4. A rate-$\frac{1}{2}$ (3, 6) regular LDPC code, with erasure threshold $\alpha^* \approx 0.42$, can be used for secure communication over any binary erasure wiretap channel with erasure probability $\epsilon > 1 - \alpha^* = 0.58$. The communication rate is 0.5, which is at most 86% of the secrecy capacity.

6.3.3 Degrading erasure channels

As shown in Proposition 3.3, a wiretap code designed for a specific eavesdropper's channel $(\mathcal{X}, p_{\mathrm{Z}|\mathrm{X}}, \mathcal{Z})$ can be used over any other eavesdropper's channel that is stochastically degraded with respect to $(\mathcal{X}, p_{\mathrm{Z}|\mathrm{X}}, \mathcal{Z})$. Codes designed for an erasure eavesdropper's channel are therefore useful over a much broader class of channels. Nevertheless, rates are then bounded strictly below the secrecy capacity since the full characteristics of the eavesdropper's channel are not necessarily exploited. The following proposition shows that all binary-input channels are stochastically degraded with respect to some binary erasure channels.

Proposition 6.4. *A memoryless binary-input channel* $(\{0, 1\}, p_{\mathrm{Y}|\mathrm{X}}, \mathcal{Y})$ *(the alphabet* $\mathcal{Y}$ *may be continuous or finite) is stochastically degraded with respect to an erasure* *channel with erasure probability*

$$\epsilon = \int_{\mathcal{Y}} \left(\min_{u\in\{0,1\}} p_{Y|X}(y|u) \right) dy.$$

Proof. Since $\int_{\mathcal{Y}} p_{Y|X}(y|x)dy = 1$ for any x and $p_{Y|X}(y|x) \geqslant 0$ for any x and y, it is clear that $\epsilon \in [0, 1]$. Let $(\{0, 1\}, p_{Z|X}, \{0, 1, ?\})$ be an erasure channel with erasure probability ϵ defined as above, that is

$$\forall x \in \{0, 1\} \qquad p_{Z|X}(?|x) = \epsilon \quad \text{and} \quad p_{Z|X}(x|x) = 1 - \epsilon.$$

Consider the channel $(\{0, 1, ?\}, \mathcal{Y}, p_{Y|Z}(y|z))$ such that

$$p_{Y|Z}(y|?) = \frac{1}{\epsilon} \min_{u\in\{0,1\}} p_{Y|X}(y|u),$$

$$p_{Y|Z}(y|z) = \frac{1}{1-\epsilon} \left(p_{Y|X}(y|z) - \min_{u\in\{0,1\}} p_{Y|X}(y|u) \right) \quad \text{if } z \in \{0, 1\}.$$

One can check that these are valid transition probabilities with the value of ϵ above. In addition, for any $(x, y) \in \{0, 1\} \times \mathcal{Y}$,

$$\begin{aligned}\sum_{z\in\{0,1,?\}} p_{Y|Z}(y|z)p_{Z|X}(z|x) &= p_{Y|Z}(y|?)\,\epsilon + \sum_{z\in\{0,1\}} p_{Y|Z}(y|z)\,(1-\epsilon)\mathbb{1}(z-x) \\ &= \min_{u\in\{0,1\}} p_{Y|X}(y|u) + p_{Y|X}(y|x) - \min_{u\in\{0,1\}} p_{Y|X}(y|u) \\ &= p_{Y|X}(y|x).\end{aligned}$$

Therefore, the channel $\left(\{0, 1\}, p_{Y|X}, \mathcal{Y}\right)$ is stochastically degraded with respect to a binary erasure channel with probability ϵ. □

The following examples are direct applications of Proposition 6.4 and Proposition 3.3.

Example 6.5. Consider a binary symmetric wiretap channel, in which the main channel is noiseless and the eavesdropper's channel is binary symmetric with cross-over probability $p < \frac{1}{2}$. A code designed for an erasure wiretap channel with erasure probability $\epsilon^* = 2p$ could achieve a secure rate ϵ^*. Figure 6.4 shows the secrecy capacity $C_s = \mathbb{H}_b(p)$ and the achievable rates as a function of the cross-over probability p. For $p = 0.29$, $\epsilon^* = 0.58$, and we can use the irregular code of Example 6.1 and its cosets for secure communication. Note that the rate of secure communication is 0.5 bits per channel use, compared with the secrecy capacity $C_s \approx 0.86$ bits per channel use.

Example 6.6. Consider a Gaussian wiretap channel in which the main channel is noiseless and the eavesdropper's channel is an AWGN channel with noise variance σ^2. Let us restrict our attention to binary inputs $x \in \{-1; +1\}$. A code designed for an erasure wiretap channel with erasure probability

$$\epsilon^* = \text{erfc}\left(\frac{1}{\sqrt{2\sigma^2}}\right)$$

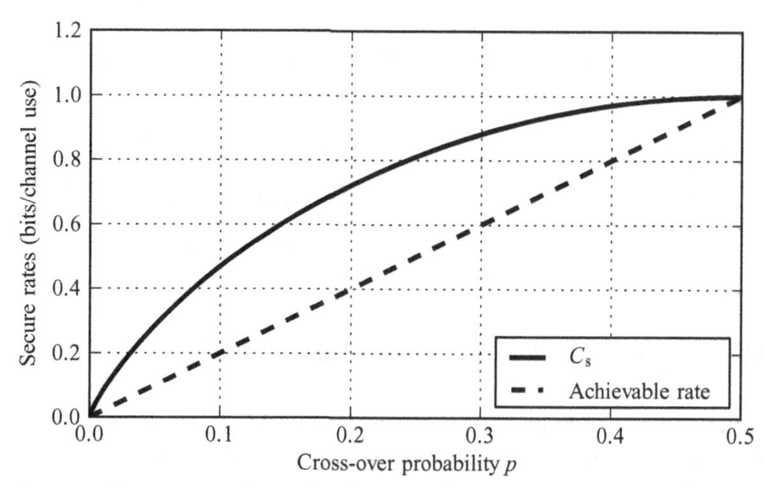

Figure 6.4 Secrecy capacity and achievable rates for a WTC with noiseless main channel and BSP(p) eavesdropper's channel.

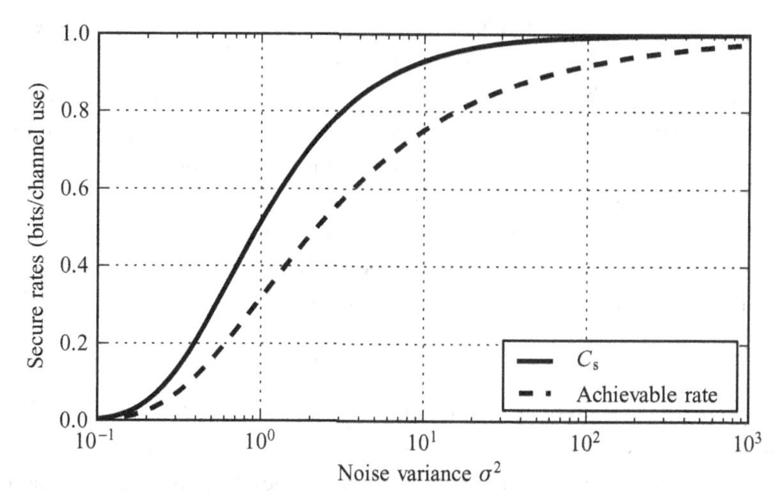

Figure 6.5 Secrecy capacity and achievable rates for a WTC with noiseless main channel and binary-input Gaussian eavesdropper's channel.

will achieve a secure rate ϵ^*. Figure 6.5 shows the secrecy capacity of the binary-input Gaussian wiretap channel and the achievable rates as a function of the eavesdropper's channel variance σ^2. For $\sigma^2 = 3.28$, $\epsilon^* \approx 0.58$, and we can again use the irregular code of Example 6.1. The communication rate is 0.5 bits per channel use, compared with the secrecy capacity $C_s \approx 0.81$ bits per channel use.

6.4 Reconciliation of binary memoryless sources

As seen in Section 4.3, one can generate secret keys from a DMS by performing information reconciliation followed by privacy amplification. Hash functions for privacy

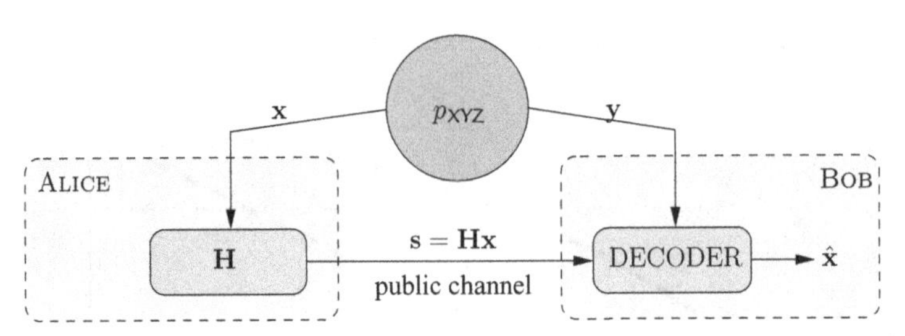

Figure 6.6 Source coding with side information with syndromes of linear codes.

amplification are already known (see for instance Example 4.5 and Example 4.6), and the only missing piece for implementing a complete key-distillation strategy is an efficient information-reconciliation protocol. For clarity, this section focuses solely on binary memoryless sources and the extension to arbitrary discrete sources and continuous sources is relegated to Section 6.5.

Since Proposition 4.5 shows that, without loss of optimality, the reconciliation of discrete random variables can be treated as a problem of source coding with side information, we need only design an encoder for a binary memoryless source X so that a receiver, who has access to a correlated binary memoryless source Y, retrieves X with arbitrarily small probability of error. The Slepian–Wolf theorem (Theorem 2.10) guarantees the existence of codes compressing at a rate arbitrarily close to $\mathbb{H}(X|Y)$ coded bits, but does not provide explicit constructions. As illustrated in Figure 6.6, one can build source encoders from well-chosen linear codes. Given the parity-check matrix $\mathbf{H} \in \text{GF}(2)^{k\times n}$ of a linear code, a vector of n observations $\mathbf{x}$ is encoded by computing the syndrome $\mathbf{s} = \mathbf{Hx}$. Upon reception of $\mathbf{s}$, the decoder can minimize its probability of error by looking for the sequence $\mathbf{x}$ that maximizes the a-posteriori probability

$$\mathbb{P}[\mathbf{y}|\mathbf{x}, \mathbf{s} = \mathbf{Hx}].$$

This procedure is equivalent to maximum a-posteriori (MAP) estimation of $\mathbf{x}$ within the coset code with syndrome $\mathbf{s}$. One can show that there exist good linear codes such that the probability of error can be made as small as desired, provided that the number of syndrome bits k is at least $n\mathbb{H}(X|Y)$. Note that the code rate of the linear code used for syndrome coding is $1 - k/n$, while the compression rate is k/n; in the remainder of this chapter, we explicitly specify which rate is being considered in order to avoid confusion.

In practice, the computational complexity required to perform MAP decoding is prohibitive, but we can use LDPC codes and adapt the message-passing algorithm described in Table 6.1 to obtain a suboptimal yet efficient algorithm. Essentially, given a syndrome $\mathbf{s}$ and a sequence of observations $\mathbf{y}$, the key idea is to slightly modify the intrinsic LLRs to account for the value of the syndrome and the a-priori distribution of X. To describe the modified message-passing algorithm, we introduce the following notation. The n-bit vector observed by the encoder is denoted $\mathbf{x} = (x_1 \dots x_n)^\intercal$, while the n-bit vector available to the decoder as side information is denoted $\mathbf{y} = (y_1 \dots y_n)^\intercal$. The decoder receives the syndrome vector $\mathbf{s} = (s_1 \dots s_k)^\intercal$, whose entries correspond to the values of check nodes $\{c_1, \dots, c_k\}$ in the Tanner graph of the LDPC code. The set of

Table 6.2 Belief-propagation algorithm for source coding with side information

☐ **Initialization.**

► For each $i \in [\![1, n]\!]$ and for each $j \in \mathcal{N}(i)$

$$v_{ij}^{(0)} = u_{ji}^{(0)} = 0.$$

► For each $i \in [\![1, n]\!]$

$$\lambda_i^{\text{INT}} = \log\left(\frac{p_{XY}(0, y_i)}{p_{XY}(1, y_i)}\right).$$

☐ **Iterations**. For each iteration $l \in [\![1, l_{\max}]\!]$

► For each $i \in [\![1, n]\!]$ and for each $j \in \mathcal{N}(i)$

$$v_{ij}^{(l)} = \lambda_i^{\text{INT}} + \sum_{m \in \mathcal{N}(i)\setminus j} u_{mj}^{(l-1)}.$$

► For each $j \in [\![1, k]\!]$ and $i \in \mathcal{M}(j)$

$$u_{ji}^{(l)} = 2\tanh^{-1}\left((1 - 2s_j) \prod_{m \in \mathcal{M}(j)\setminus i} \tanh\left(\frac{v_{mj}^{(l)}}{2}\right)\right).$$

☐ **Extrinsic information**. For all $i \in [\![1, n]\!]$

$$\lambda_i^{\text{EXT}} = \sum_{m \in \mathcal{N}(i)} u_{mi}^{(l_{\max})}.$$

☐ **Hard decisions.** For all $i \in [\![1, n]\!]$

$$\hat{x}_i = \frac{1}{2}\left(1 - \text{sign}\left(\lambda_i^{\text{INT}} + \lambda_i^{\text{EXT}}\right)\right).$$

check-node indices connected to a variable node x_i is denoted by $\mathcal{N}(i)$, while the set of variable-node indices connected to a check node c_j is denoted by $\mathcal{M}(j)$. The algorithm is described in Table 6.2.

Compared with the algorithm in Table 6.1, note that the initialization of the LLRs is now based on the joint distribution $p_{XY}(x, y)$ and that the jth syndrome value s_j affects the sign of the messages $u_{ji}^{(l)}$.

Example 6.7. To illustrate the performance of the algorithm, we consider a uniform binary memoryless source X, and Y is obtained by sending X through a binary symmetric channel with cross-over probability p. In principle, one can reconstruct the source X^n, provided that it is compressed at a rate of at least $\mathbb{H}_b(p)$. For a code of rate $\frac{1}{2}$, the efficiency is $\beta = 1/(2(1 - \mathbb{H}_b(p)))$. Figure 6.7 shows the bit-error-rate versus efficiency performance of syndrome coding using LDPC codes of length 10^4 with the degree distributions given in Example 6.1 and Example 6.2. These non-optimized degree distributions already provide over 80% efficiency at an error rate of 10^{-5}. Longer codes with optimized degree distributions can easily achieve over 90% efficiency.

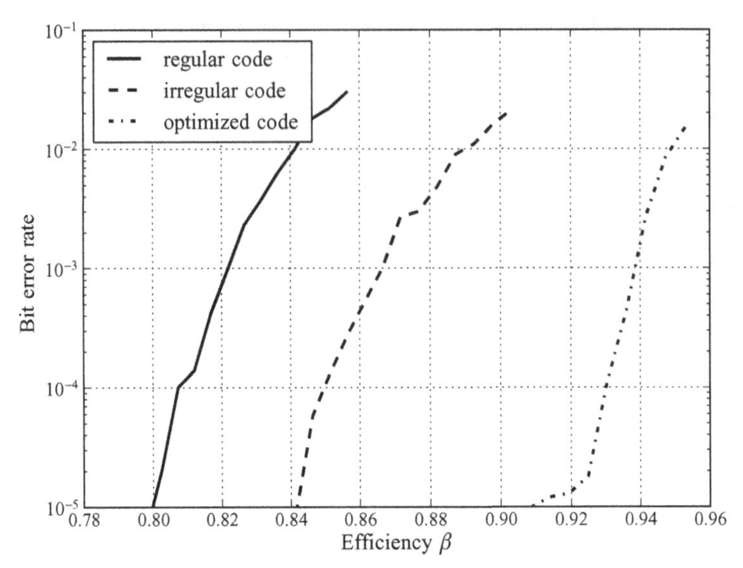

Figure 6.7 Error rate versus efficiency for reconciliation of binary random variables based on LDPC codes. The regular and irregular codes are rate-$\frac{1}{2}$ codes with degree distributions given in Example 6.1 and Example 6.2 and length 10^4. The optimized code is the rate-$\frac{1}{2}$ code of [58] with length 5×10^5.

Remark 6.3. *To ensure the generation of identical keys with privacy amplification, note that the sequence* $\mathbf{x}$ *should be reconstructed* exactly, *and even a bit error rate as low as* 10^{-5} *is not acceptable. The presence of errors is usually well detected by the message-passing decoder, and the key-distillation process could simply be aborted in such cases. However, discarding an entire sequence that contains a small fraction of errors incurs a significant efficiency loss. A more efficient technique consists of concatenating a high-rate outer code, such as a BCH code, to correct the few remaining errors.*

6.5 Reconciliation of general memoryless sources

We now turn our attention to a general memoryless source $(\mathcal{XY}, p_{XY})$, which might be discrete or continuous. If X is a general discrete random variable, Proposition 4.5 applies once again, and reconciliation can be treated as a Slepian–Wolf coding problem. If X is a continuous random variable, its lossless reconstruction would require infinitely many bits, and the traditional approach is to consider the approximate reconstruction of X under a distortion constraint. However, the objective of reconciliation is slightly different because we want to extract a common (binary) sequence from observations of the components X and Y so that privacy amplification can be used later on. Therefore, a pragmatic approach consists of quantizing X into a discrete random variable X′ to revert back to the discrete case. In principle, the source X′ can be compressed at a rate

R arbitrarily close to $\mathbb{H}(X'|Y)$ and the resulting reconciliation efficiency is

$$\beta = \frac{\mathbb{H}(X') - \mathbb{H}(X'|Y)}{\mathbb{I}(X;Y)} = \frac{\mathbb{I}(X';Y)}{\mathbb{I}(X;Y)}. \tag{6.2}$$

In general, $\beta < 1$, but the penalty inflicted by quantization can be made as small as desired by choosing a fine enough quantizer.

Remark 6.4. *A scalar quantizer is sufficient to obtain near-optimal performance as long as there is no rate constraint on public communication. However, for the same amount of information exchanged over the public channel, a vector quantizer will have a better performance than a scalar quantizer.*

6.5.1 Multilevel reconciliation

In this section, we describe a generic protocol for the reconciliation of a memoryless source $(\mathcal{X}\mathcal{Y}, p_{XY})$ for which $|\mathcal{X}| < \infty$. As discussed above, this protocol is also useful for continuous random variables, and we study the case of Gaussian random variables in Section 6.5.2.

In principle, we could design a Slepian–Wolf code that operates on symbols in the alphabet $\mathcal{X}$ directly, but it is more convenient to use binary symbols only. Letting $\ell = \lceil \log|\mathcal{X}| \rceil$, every symbol $x \in \mathcal{X}$ can be assigned a unique ℓ-bit binary label denoted by the vector $(g_1(x) \dots g_\ell(x))^\intercal$, where

$$\forall i \in [\![1, \ell]\!] \qquad g_i : \mathcal{X} \to \text{GF}(2).$$

The source X is then equivalent to a binary vector source $B^\ell \triangleq (g_1(X) \dots g_\ell(X))^\intercal$. For $i \in [\![1, \ell]\!]$, we call the component source $B_i \triangleq g_i(X)$ the ith *binary level*, since it corresponds to the ith "level" in the binary representation of X. If we were to try to encode and decode the ℓ binary levels independently with ideal Slepian–Wolf codes, we would use a compression rate R_i arbitrarily close to $\mathbb{H}(B_i|Y)$ for each level $i \in [\![1, \ell]\!]$, and the overall compression rate would be

$$\sum_{i=1}^{\ell} \mathbb{H}(B_i|Y) \geqslant \sum_{i=1}^{\ell} \mathbb{H}\left(B_i|B^{i-1}Y\right) = \mathbb{H}\left(B^\ell|Y\right) = \mathbb{H}(X|Y),$$

with equality if and only if B_i is independent of B^{i-1} given Y for all $i \in [\![1, \ell]\!]$. Therefore, in general, encoding and decoding the levels separately is suboptimal.

It is possible to achieve better compression by considering the separate encoding/joint-decoding procedure illustrated in Figure 6.8. We start by considering the binary source B_1 and let $\mathcal{E}_1$ be the event representing the occurrence of a decoding error. For any $\gamma, \epsilon > 0$, the Slepian–Wolf theorem guarantees the existence of a code compressing B_1 at rate $R_1 = \mathbb{H}(B_1|Y) + \gamma$ with probability of error $\mathbb{P}[\mathcal{E}_1] \leqslant \epsilon$. Assuming that level B_1 is successfully decoded, we can now treat it as additional side information available to the receiver. Hence, we encode B_2 using a code with compression rate $R_2 = \mathbb{H}(B_2|B_1 Y) + \gamma$ that guarantees a probability of error of $\mathbb{P}\left[\mathcal{E}_2|\mathcal{E}_1^c\right] \leqslant \epsilon$. By repeating this procedure for

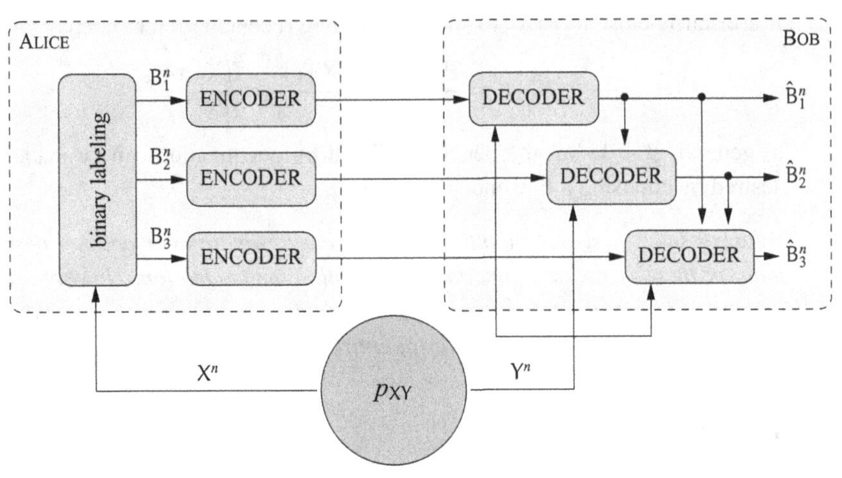

Figure 6.8 Example of a multilevel reconciliation protocol for $\ell = 3$. Each level B_i is decoded using Y^n and previously decoded levels as side information.

each level $i \in [\![1, \ell]\!]$, B_i is encoded *independently* at a rate $R_i = \mathbb{H}(B_i|B^{i-1}Y) + \gamma$ with a code ensuring a probability of decoding error of $\mathbb{P}\left[\mathcal{E}_i \mid \bigcap_{j=1}^{i-1} \mathcal{E}_j^c\right] \leqslant \epsilon$. However, the levels must be decoded *successively*, using the previously decoded levels as side information. The overall probability of error is

$$\begin{aligned}\mathbb{P}\left[\bigcup_{i=1}^{\ell} \mathcal{E}_i\right] &= \mathbb{P}\left[\bigcup_{i=1}^{\ell}\left(\mathcal{E}_i \cup \bigcap_{j=1}^{i-1} \mathcal{E}_i^c\right)\right] \\ &\leqslant \sum_{i=1}^{\ell} \mathbb{P}\left[\mathcal{E}_i \cup \bigcap_{j=1}^{i-1} \mathcal{E}_i^c\right] \\ &\leqslant \sum_{i=1}^{\ell} \mathbb{P}\left[\mathcal{E}_i \mid \bigcap_{j=1}^{i-1} \mathcal{E}_i^c\right] \\ &\leqslant \ell\epsilon,\end{aligned}$$

which can be made as small as desired with a large enough blocklength n since the number of levels ℓ is fixed. In addition, the overall compression rate is then

$$R_{\text{tot}} = \sum_{i=1}^{\ell} \mathbb{H}(B_i|B^{i-1}Y) + \ell\gamma = \mathbb{H}(B^{\ell}|Y) + \ell\gamma = \mathbb{H}(X|Y) + \ell\gamma,$$

which can be made as close as desired to the optimal compression rate $\mathbb{H}(X|Y)$. This optimal encoding/decoding scheme is called *multilevel reconciliation* because of its similarity to multilevel channel coding with multistage decoding.

Remark 6.5. *The optimality of multilevel reconciliation does not depend on the labeling* $\{g_i\}_\ell$ *used to transform* X *into the vector of binary sources* $(B_1 \dots B_\ell)^{\mathsf{T}}$. *However, different*

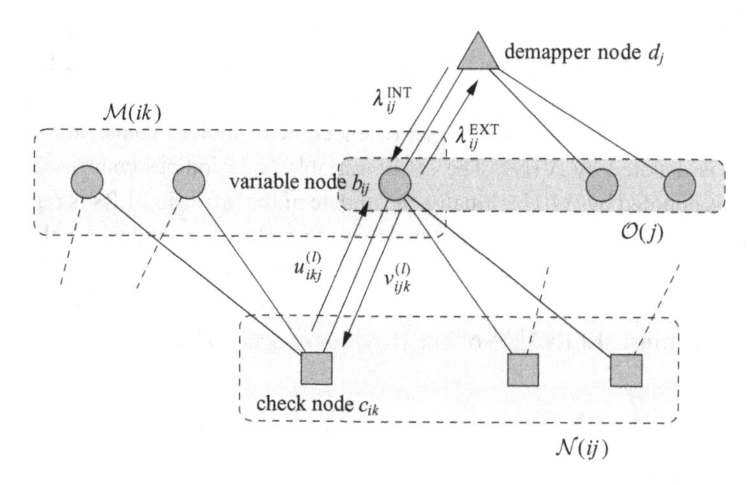

Figure 6.9 Illustration of a message-passing algorithm for multilevel reconciliation. The set of check nodes connected to a variable note b_{ij} is denoted $\mathcal{N}(ij)$. The set of variable nodes connected to a check node c_{ik} is denoted $\mathcal{M}(ik)$. The set of variable nodes connected to a demapper node d_j is denoted $\mathcal{O}(j)$.

choices of labeling result in different values for the compression rates $R_i = \mathbb{H}(\mathrm{B}_i|\mathrm{B}^{i-1}\mathrm{Y})$ *for* $i \in [\![1, \ell]\!]$, *which can in turn have an effect on the complexity of the code design. For instance, it can be quite difficult to design high-rate or low-rate codes with near-optimal performance. We provide heuristic guidelines for the choice of labeling in Section 6.5.2.*

As in Section 6.4, we can implement multilevel reconciliation efficiently using syndrome coding with LDPC codes for each of the ℓ levels. Specifically, given a vector of realizations $\mathbf{x} = (x_1 \dots x_n)$ of the source X, the labeling creates ℓ binary vectors $\mathbf{b}_i = g_i(\mathbf{x})$ for $i \in [\![1, \ell]\!]$, which can be represented in matrix form as

$$\begin{pmatrix} \mathbf{b}_1 \\ \vdots \\ \mathbf{b}_i \\ \vdots \\ \mathbf{b}_\ell \end{pmatrix} = \begin{pmatrix} b_{11} & b_{12} & \dots & b_{1n} \\ \vdots & \vdots & \vdots & \vdots \\ b_{i1} & b_{i2} & \dots & b_{in} \\ \vdots & \vdots & \vdots & \vdots \\ b_{\ell 1} & b_{\ell 2} & \dots & b_{\ell n} \end{pmatrix} \begin{matrix} \text{level } 1 \\ \vdots \\ \text{level } i \\ \vdots \\ \text{level } \ell \end{matrix}$$

For each level $i \in [\![1, \ell]\!]$, the encoder computes the syndromes $\mathbf{s}_i = \mathbf{H}_i\mathbf{b}_i$ with the parity-check matrix $\mathbf{H}_i \in \mathrm{GF}(2)^{k_i \times n}$ of an $(n, n - k_i)$ linear code. Note that, for a fixed $j \in [\![1, n]\!]$, the bits $\{\mathrm{B}_{1j}, \mathrm{B}_{2j}, \dots \mathrm{B}_{\ell j}\}$ are correlated because they stem from the same symbol X_j. Consequently, the intrinsic LLR of every bit b_{ij} with $i \in [\![1, \ell]\!]$ depends not only on the side information y_j but also on the estimation of the bits $\{b_{mj}\}_\ell \setminus \{b_{ij}\}$. Therefore, the message-passing algorithm described in Section 6.4 must be modified to take into account the result of previously decoded levels.

To describe the full message-passing algorithm, we introduce the following notation, which is also illustrated in Figure 6.9 for convenience. For each level $i \in [\![1, \ell]\!]$, the encoder computes the vector $\mathbf{b}_i = g_i(\mathbf{x}) = (b_{i1} \dots b_{in})^\mathsf{T}$, where each b_{ij} corresponds to the ith bit in the binary description of symbol x_j. The receiver, who obtains $\mathbf{y} = (y_1 \dots y_n)^\mathsf{T}$ as side observation, receives the syndrome vector $\mathbf{s}_i = \mathbf{H}_i\mathbf{b}_i$ of k_i bits, whose entries correspond to the values of check node $\{c_{i1}, \dots, c_{ik_i}\}$ in the Tanner graph of the LDPC code for level i. The set of check-node indices connected to a variable node b_{ij} is denoted by $\mathcal{N}(ij)$. The set of variable-node indices connected to a check node c_{ik} is denoted by $\mathcal{M}(ik)$. Finally, the update of the intrinsic LLRs is represented by n *demapper nodes* $\{d_j\}_n$ connecting the Tanner graphs of individual levels. The set of variable nodes $\{b_{1j}, \dots, b_{\ell j}\}$ stemming from the same symbol x_j is denoted by $\mathcal{O}(j)$. One can show that the intrinsic LLR $\lambda_{ij}^{\text{INT}}$ of bit $b_{i,j}$ should be calculated from symbol y_j and the extrinsic LLRs $\lambda_{mj}^{\text{EXT}}$ of bits $\{b_{mj}\}$ for $m \in \mathcal{O}(j) \setminus i$ as

$$\lambda_{ij}^{\text{INT}} = \log \left(\frac{\displaystyle\sum_{x \in \mathcal{X}: g_i(x)=0} p_{\mathrm{XY}}(x, y_j) \exp\left(\sum_{m \in \mathcal{O}(j)\setminus i} (1 - g_m(x))\lambda_{mj}^{\text{EXT}} \right)}{\displaystyle\sum_{x \in \mathcal{X}: g_i(x)=1} p_{\mathrm{XY}}(x, y_j) \exp\left(\sum_{m \in \mathcal{O}(j)\setminus i} (1 - g_m(x))\lambda_{mj}^{\text{EXT}} \right)} \right). \tag{6.3}$$

Note that the extrinsic LLRs $\lambda_{mj}^{\text{EXT}}$ appear as weighting factors in front of the joint probability terms $p_{\mathrm{XY}}(x, y_i)$. As the magnitude of $\lambda_{mj}^{\text{EXT}}$ increases (that is, the estimation of b_{mj} becomes more reliable), some symbols x are given more importance than others in the calculation of $\lambda_{ij}^{\text{INT}}$. For clarity, we denote the operation defined by (6.3) as

$$\lambda_{ij}^{\text{INT}} = \boxplus_{m \in \mathcal{O}(j)\setminus i} \lambda_{mj}^{\text{EXT}}.$$

The entire decoding algorithm is described in Table 6.3.

If the overall graph contains no cycles, the algorithm can be shown to compute exactly the a-posteriori LLRs

$$\log \left(\frac{\mathbb{P}\left[\mathrm{B}_{ij} = 0 | \mathbf{y}, \mathbf{s}\right]}{\mathbb{P}\left[\mathrm{B}_{ij} = 1 | \mathbf{y}, \mathbf{s}\right]} \right).$$

In practice, finite-length LDPC codes contain cycles, but the algorithm still provides reasonable approximations of the real a-posteriori LLRs.

Remark 6.6. *Since the update of λ_{ij}^{INT} is based on extrinsic LLRs, it is possible to modify the scheduling of the previous algorithm and to start decoding level i even if the previous levels $[\![1, i-1]\!]$ have not been entirely decoded. In that case, level i cannot be decoded entirely, but the extrinsic LLRs might be fed back to the message-passing algorithm of the previous levels. Although this feedback is not necessary in principle, it does improve the performance of practical implementations.*

Remark 6.7. *The optimality of multilevel reconciliation requires different compression rates, and therefore different codes, for each level. To simplify the code design, it is possible to use a single code across all levels, albeit with a reduced efficiency. Whereas multilevel reconciliation is similar to multilevel coding, this simplified approach is*

Table 6.3 Belief-propagation algorithm for multilevel reconciliation

---

☐ **Initialization.**

► For each $i \in [\![1, \ell]\!]$, for each $j \in [\![1, n]\!]$ and for each $k \in [\![1, k_i]\!]$

$$\lambda_{ij}^{\text{INT}} = \lambda_{ij}^{\text{EXT}} = v_{ijk}^{(0)} = u_{ikj}^{(0)} = 0.$$

☐ **Iterations across levels.** For each level $i \in [\![1, \ell]\!]$

► For each $j \in [\![1, n]\!]$

$$\lambda_{ij}^{\text{INT}} = \boxplus_{m \in \mathcal{O}(j)\backslash i} \lambda_{mj}^{\text{EXT}}.$$

► Iteration within level. For each $l \in [\![1, l_{\max}]\!]$

▷ For each $j \in [\![1, n]\!]$ and for each $k \in \mathcal{N}(ij)$

$$v_{ijk}^{(l)} = \lambda_{ij}^{\text{INT}} + \sum_{m \in \mathcal{N}(ij)\backslash k} u_{imj}^{(l-1)}.$$

▷ For each $k \in [\![1, k_i]\!]$ and for each $j \in \mathcal{M}(ik)$

$$u_{ikj}^{(l)} = 2 \tanh^{-1}\left((1 - 2s_{ik}) \prod_{m \in \mathcal{M}(ik)\backslash j} \tanh\left(\frac{v_{imk}^{(l)}}{2}\right)\right).$$

► **Extrinsic information.** For each $j \in [\![1, n]\!]$

$$\lambda_{ij}^{\text{EXT}} = \sum_{m \in \mathcal{N}(ij)} u_{imj}^{(l_{\max})}.$$

☐ **Hard decisions.** For each $i \in [\![1, \ell]\!]$ and for each $j \in [\![1, n]\!]$

$$\hat{b}_{ij} = \frac{1}{2}\left(1 - \text{sign}\left(\lambda_{ij}^{\text{INT}} + \lambda_{ij}^{\text{EXT}}\right)\right).$$

---

similar to bit-interleaved coded modulation. The decoding algorithm described in Table 6.3 is easily adapted to syndrome coding with a single LDPC code.

6.5.2 Multilevel reconciliation of Gaussian sources

In this section, we detail the construction of a multilevel reconciliation scheme for a memoryless Gaussian source. We assume

$$\mathsf{X} \sim \mathcal{N}(0, 1) \quad \text{and} \quad \mathsf{Y} = \mathsf{X} + \mathsf{N} \ \text{ with } \ \mathsf{N} \sim \mathcal{N}(0, \sigma^2),$$

and we define the signal-to-noise ratio of this source as $\text{SNR} \triangleq 1/\sigma^2$. Notice that this type of source could be simulated by transmitting Gaussian noise over a Gaussian channel. Before we design a set of LDPC codes for multilevel reconciliation, we first need to construct a scalar quantizer, define a labeling, and compute the compression rate required at each level.

By construction, the joint distribution of X and Y satisfies the property

$$\forall (x, y) \in \mathbb{R}^2 \qquad p_{\mathsf{YX}}(y, x) = p_{\mathsf{YX}}(-y, -x),$$

which it is desirable to preserve when designing the scalar quantizer. Specifically, we restrict ourselves to quantizers $Q : \mathbb{R} \to \mathbb{R}$ such that

$$\forall (x, y)^2 \in \mathbb{R} \qquad p_{\mathsf{YX}}(y, Q(x)) = p_{\mathsf{YX}}(-y, -Q(x)),$$

and we let $\mathsf{X}' = Q(\mathsf{X})$. There is some leeway for choosing the number of quantization intervals, but one should choose them such that $\mathbb{I}(\mathsf{X}'; \mathsf{Y})$ is close to $\mathbb{I}(\mathsf{X}; \mathsf{Y})$ in order to minimize the efficiency penalty in (6.2). Naturally, the higher the SNR is, the more quantization intervals are needed.

Example 6.8. For simplicity, we consider quantizers with equal-width quantization intervals (except for the two intervals on the boundaries). If SNR = 3, the quantizer with 16 intervals maximizing the mutual information $\mathbb{I}(\mathsf{X}'; \mathsf{Y})$ is as given below.

Quantization intervals	
$(-\infty, -2.348]$	$(+0.000, +0.254]$
$(-2.348, -1.808]$	$(+0.254, +0.514]$
$(-1.808 - 1.412]$	$(+0.514, +0.787]$
$(-1.412, -1.081]$	$(+0.787, +1.081]$
$(-1.081, -0.787]$	$(+1.081, +1.412]$
$(-0.787, -0.514]$	$(+1.412, +1.808]$
$(-0.514, -0.254]$	$(+1.808, +2.348]$
$(-0.254 + 0.000]$	$(+2.348, +\infty)$

The quantizer yields $\mathbb{I}(\mathsf{X}'; \mathsf{Y}) \approx 0.98$ bits, compared with $\mathbb{I}(\mathsf{X}; \mathsf{Y}) = 1$ bit. The entropy of the quantized source is $\mathbb{H}(\mathsf{X}') \approx 3.78$ bits.

We assume that the intervals can be labeled with ℓ bits. Once a labeling $\{g_i\}_\ell$ has been defined, we obtain a binary vector source $\mathsf{B}^\ell \triangleq (\mathsf{B}_1, \dots, \mathsf{B}_\ell)$ with $\mathsf{B}_i \triangleq g_i(\mathsf{X}')$ for $i \in [\![1, \ell]\!]$, which is equivalent to the quantized source X'. The optimal compression rates that would be required for ideal codes are then $R_i = \mathbb{H}\left(\mathsf{B}_i | \mathsf{Y}\mathsf{B}^{i-1}\right)$. These rates are easily calculated by writing R_i as

$$R_i = \mathbb{H}\left(\mathsf{B}_i | \mathsf{Y}\mathsf{B}^{i-1}\right) = \mathbb{H}\left(\mathsf{B}^i | \mathsf{Y}\right) - \mathbb{H}\left(\mathsf{B}^{i-1} | \mathsf{Y}\right).$$

The entropy $\mathbb{H}\left(\mathsf{B}^i | \mathsf{Y}\right)$ is given explicitly by

$$-\sum_{\mathbf{b} \in \mathrm{GF}(2)^i} \int_{-\infty}^{\infty} \left(\int_{\mathcal{A}_i(\mathbf{b})} p_{\mathsf{X}|\mathsf{Y}}(x|y)\mathrm{d}x \right) \log \left(\int_{\mathcal{A}_i(\mathbf{b})} p_{\mathsf{XY}}(x, y)\mathrm{d}x \right) \mathrm{d}y,$$

with

$$\mathcal{A}_i(\mathbf{b}) \triangleq \{x : (g_1(Q(x)), \dots, g_i(Q(x))) = \mathbf{b}\}.$$

Note that the optimal compression rates needed at each level depend on the specific choice of labeling considered.

Example 6.9. We consider the quantizer obtained in Example 6.8. The simplest labeling is the *natural labeling* given below, in which the first level consists of the least significant bits.

Natural labeling																
Level 4	0	0	0	0	0	0	0	0	1	1	1	1	1	1	1	1
Level 3	0	0	0	0	1	1	1	1	0	0	0	0	1	1	1	1
Level 2	0	0	1	1	0	0	1	1	0	0	1	1	0	0	1	1
Level 1	0	1	0	1	0	1	0	1	0	1	0	1	0	1	0	1

The optimal rates for SNR = 3 are then the following.

Level	Compression rate	Code rate for syndrome coding
4	0.079	0.921
3	0.741	0.259
2	0.984	0.016
1	0.987	0.013

Another simple labeling is the *reverse natural labeling*, which is similar to natural labeling, but for which the first level consists of the most significant bits.

Reverse natural labeling																
Level 4	0	1	0	1	0	1	0	1	0	1	0	1	0	1	0	1
Level 3	0	0	1	1	0	0	1	1	0	0	1	1	0	0	1	1
Level 2	0	0	0	0	1	1	1	1	0	0	0	0	1	1	1	1
Level 1	0	0	0	0	0	0	0	0	1	1	1	1	1	1	1	1

The optimal rates for SNR = 3 are now the following.

Level	Compression rate	Code rate for syndrome coding
4	0.936	0.064
3	0.801	0.199
2	0.547	0.453
1	0.507	0.493

If we could construct ideal codes at all rates, then no labeling would be better than any other. However, in practice, it is difficult to design efficient codes with compression rates close to unity. Rather than designing efficient codes, it is actually easier not to compress at all and to simply disclose the bits of the entire level. This simplification induces a small penalty, but, for compression rates close to unity, this is more efficient than trying to design a powerful code. For instance, for the natural labeling in Example 6.9, disclosing levels 1 and 2 incurs a negligible efficiency loss. In addition, finite-length LDPC codes do not have the performance of ideal codes; therefore, the code rates at each level must be chosen to be below the ideal ones, which inflicts an efficiency penalty. Nevertheless, one can still achieve high efficiencies, as illustrated by the following example.

Example 6.10. We consider the quantizer of Example 6.8 used in conjunction with natural labeling. Instead of the optimal rates computed in Example 6.9 for SNR $= 3$, we use the following:

Level	Compression rate	Code rate
4	0.14	0.86
3	0.76	0.24
2	1.0	0.0
1	1.0	0.0

Levels 1 and 2 are entirely disclosed so that there are only two codes to design, with rates 0.24 and 0.86, respectively. We choose the rate-0.24 LDPC codes with degree distributions

$$\begin{aligned}\lambda_1(x) &= 0.249\,17x + 0.163\,92x^2 + 0.000\,01x^3 + 0.159\,92x^5 + 0.023\,23x^6\\ &\quad + 0.080\,87x^{12} + 0.019\,58x^{13} + 0.036\,39x^{20} + 0.018\,61x^{26}\\ &\quad + 0.029\,56x^{27} + 0.010\,06x^{30} + 0.048\,15x^{32} + 0.160\,51x^{99},\\ \rho_1(x) &= 0.1x^4 + 0.9x^5,\end{aligned}$$

and the rate-0.86 LDPC code with degree distributions

$$\begin{aligned}\lambda_2(x) &= 0.227\,34x + 0.131\,33x^2 + 0.641\,33x^4,\\ \rho_2(x) &= x^{24}.\end{aligned}$$

Figure 6.10 shows the error-rate versus efficiency performance obtained with the multi-level reconciliation algorithm of Section 6.5.1 and various choices of blocklength. Note that long codes can achieve a probability of error of 10^{-5} at an efficiency close to 90%.

6.6 Secure communication over wiretap channels

Although wiretap codes are not known for general channels, it is nevertheless possible to construct practical codes that are based on sequential key-distillation strategies with one-way reconciliation. In fact, for a WTC $(\mathcal{X}, p_{YZ|X}, \mathcal{Y}, \mathcal{Z})$, one can implement the following four-stage protocol.

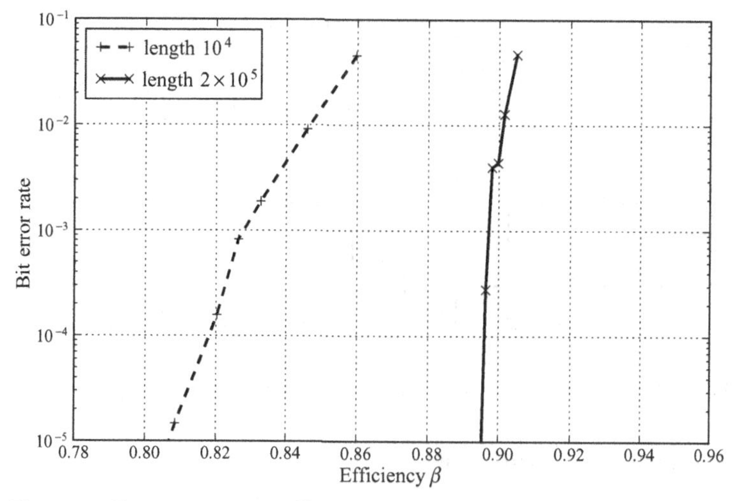

Figure 6.10 Error rate versus efficiency of a finite-length LDPC-based multilevel reconciliation scheme.

1. **Randomness sharing.** Alice generates n realizations of a random variable X with distribution p_X and transmits them through the WTC. Bob and Eve observe n realizations of correlated random variables Y and Z, respectively. The resulting joint distribution factorizes as $p_{XYZ} = p_{YZ|X}p_X$.
2. **Information reconciliation.** Alice computes syndromes using the multilevel reconciliation protocol described in Section 6.5.1, and transmits them over the main channel $(\mathcal{X}, p_{Y|X}, \mathcal{Y})$ using a channel error-correcting code. In principle, the rate of the channel code can be chosen arbitrarily close to the capacity C_m.
3. **Privacy amplification.** Alice chooses a hash function at random in a universal family and transmits this choice over the main channel using a channel error-correcting code. Alice and Bob distill a secret key K.
4. **Secure communication.** Alice uses the key K to encrypt a message M with a one-time pad and transmits the encrypted message over the main channel using a channel error-correcting code.

In general, this procedure is fairly inefficient because many channel uses are wasted to generate a source and distill a secret key instead of for communicating secure messages. The secure rate R_s achieved by this procedure can be estimated as follows. For a given input distribution p_X, about $n\mathbb{H}(X|Y)$ bits are required for information reconciliation, which can be transmitted with $n\mathbb{H}(X|Y)/C_m$ additional channel uses. The choice of a hash function in the family of hash functions in Example 4.6 requires about $n\mathbb{H}(X)$ bits for privacy amplification, which can be transmitted with another $n\mathbb{H}(X)/C_m$ channel uses. Finally, the secret key K distilled by Alice and Bob contains on the order of $n(\mathbb{I}(X;Y) - \mathbb{I}(X;Z))$ bits, which allows the secure transmission of the same number of message bits and requires $n(\mathbb{I}(X;Y) - \mathbb{I}(X;Z))/C_m$ channel uses. Hence, the secure

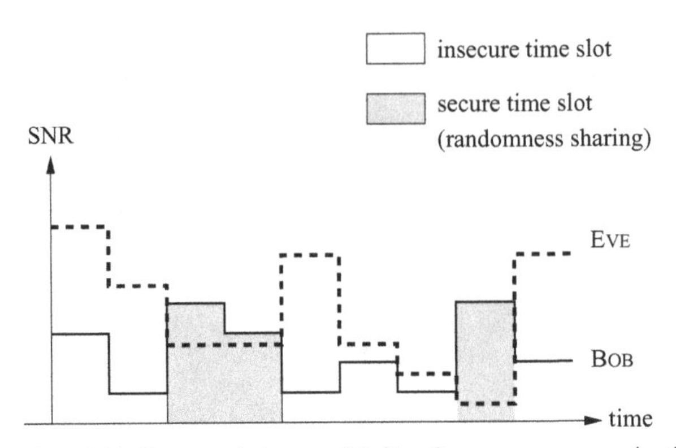

Figure 6.11 Opportunistic use of fading for secure communication.

communication rate is approximately

$$R_s \approx \frac{C_m(\mathbb{I}(X;Y) - \mathbb{I}(X;Z))}{C_m + \mathbb{H}(X|Y) + \mathbb{H}(X) + \mathbb{I}(X;Y) - \mathbb{I}(X;Z)}.$$

Example 6.11. Let us consider a binary WTC in which the main channel and the eavesdropper's channel are both binary symmetric channels with cross-over probabilities p_m and p_e, respectively ($p_e > p_m$). Assume that the procedure described above is used with $X \sim \mathcal{B}(\frac{1}{2})$. Then the secure communication rate is at most

$$R_s = \frac{1 - \mathbb{H}_b(p_m)}{2 + \mathbb{H}_b(p_e) - \mathbb{H}_b(p_m)} C_s.$$

In the extreme case $p_m = 1$ and $p_e = \frac{1}{2} + \epsilon$ for some small $\epsilon > 0$, note that C_s is on the order of 1 bit per channel use but the rate achieved R_s is only on the order of $\frac{1}{3}$ bits per channel use.

Example 6.12. Let us consider the quasi-static fading WTCs discussed in Section 5.2.3 with full channel state information. As illustrated in Figure 6.11, the four-stage protocol can be implemented opportunistically in such a way that randomness sharing is performed only during *secure time slots* for which the legitimate receiver has a better instantaneous SNR than that of the eavesdropper. Reconciliation, privacy amplification, and secure communication are performed during the remaining time slots. This does not affect the secure rate because the key distilled during the protocol is secure against an eavesdropper who obtains the reconciliation and privacy-amplification messages perfectly.

To avoid sharing randomness at a faster rate than that at which it can be processed, it could be necessary to use a fraction of the secure time slots for reconciliation, privacy amplification, or secure communication; however, if the eavesdropper has a much higher SNR on average, all of the communication required for secret-key distillation is performed during insecure time slots. In such a case, the protocol achieves secure rates close to the secrecy capacity of the channel.

Remark 6.8. *Although we have assumed that privacy amplification is performed using the hash functions of Example 4.6, there is not much to be gained by choosing another universal family of hash functions. It can be shown that the minimum number of hash functions in a universal family* $\mathcal{H} = \{h : \mathrm{GF}(2)^n \rightarrow \mathrm{GF}(2)^k\}$ *is* 2^{n-k} *[67]; in practice, no families with fewer than* 2^n *functions are known. We could in principle do better by performing privacy amplification with extractors.*

6.7 Bibliographical notes

LDPC codes were invented by Gallager in 1963 [110], and experienced a renewed interest with the works of MacKay and others. Density evolution and the threshold property of LDPC codes under message-passing decoding were investigated by Richardson and Urbanke [111]. A Gaussian-approximation version of density evolution leading to a linear optimization problem was analyzed by Chung, Forney, Richardson, and Urbanke [112]. This enabled the design of irregular LDPC codes performing extremely close to the Shannon limit [113, 114, 115]. Although it is not known whether there exist sequences of capacity-achieving LDPC codes over arbitrary channels, it is possible to construct sequences of capacity-achieving LDPC codes for the erasure channel. The threshold for the block error probability of LDPC ensembles under belief-propagation decoding was analyzed by Jin and Richardson [109].

The first wiretap-code constructions were proposed for the wiretap channel of type II by Ozarow and Wyner [116]. Thangaraj, Dihidar, Calderbank, McLaughlin, and Merolla generalized these ideas to other wiretap channels [117] and proposed an explicit coset coding scheme based on LDPC codes for the erasure wiretap channel. Suresh, Subramanian, Thangaraj, Bloch, and McLaughlin later proved that the same construction can be used to guarantee strong secrecy, albeit at lower rates [118, 119]. A similar construction based on two-edge-type LDPC codes was proposed by Rathi, Andersson, Thobaben, Kliewer, and Skoglund [120]. For discrete additive noise channels, Cohen and Zemor showed that any random linear code used in conjunction with coset coding is likely to satisfy a strong secrecy requirement [121]. All of the aforementioned constructions are connected to nested codes, as highlighted by Liu, Liang, Poor, and Spasojević [122]. In a slightly different spirit, Verriest and Hellman analyzed the use of convolutional encoding over wiretap channels with noiseless main channel and binary symmetric eavesdropper's channel [123]. For the wiretap channel of type II, Wei established a direct relation between the equivocation guaranteed by a linear code and its generalized Hamming weights [124]. Recently, there has also been a lot of interest in the design of wiretap codes based on polar codes [125, 126, 127, 128]. The polarization and capacity-achieving properties of polar codes over binary-input symmetric-output channels allow one to perform a security analysis that closely follows the equivocation calculation used in Chapter 3.

The challenges posed by the design of wiretap codes ensuring information-theoretic security have fostered the development of alternative secrecy metrics. For instance, Klinc, Ha, McLaughlin, Barros, and Kwak proposed the use of punctured LDPC codes to drive the eavesdropper's bit error probability to an arbitrarily high level over the Gaussian wiretap channel [129]. In the same spirit, Belfiore, Oggier, and Solé proposed the use of lattice codes over the Gaussian wiretap channel [130, 131] and, since the error probability of lattices over Gaussian channels can be related to their theta series, they proposed a secrecy criterion based on the theta series of lattices.

The idea of performing source coding with side information by using the syndromes of good channel codes was proposed by Wyner [132]. Liveris, Xiong, and Georghiades applied this idea to binary memoryless sources with LDPC codes and demonstrated that performance close to the optimal limit could be attained [133]. A threshold analysis of the message-passing algorithm has been proposed by Chen, He, and Jamohan [134].

The reconciliation of binary random variables was first considered by Brassard and Salvail in the context of quantum key distribution with the suboptimal yet computationally efficient algorithm CASCADE [56]. An extensive performance analysis of LDPC-based reconciliation for binary random variables was recently reported by Elkouss, Leverrier, Alléaume, and Boutros [58]. Multilevel reconciliation was first considered by Van Assche, Cardinal, and Cerf [59] for the reconciliation of continuous random variables, and implemented with turbo-codes by Nguyen, Van Assche, and Cerf [60]. Bloch, Thangaraj, McLaughlin, and Merolla reformulated the reconciliation of continuous random variables as a coded modulation problem and used LDPC codes to achieve near-optimal performance [61]. A special case of the general algorithm was developed independently by Ye, Reznik, and Shah [62, 78]. A multidimensional reconciliation scheme for continuous random variables based on the algebraic properties of octonions was also proposed by Leverrier, Alléaume, Boutros, Zémor, and Grangier [135], and was proved to be particularly efficient in the low-SNR regime. Conceptually, multilevel reconciliation is the source-coding counterpart of multilevel coding, a good survey of which can be found in [136] (see also [137] for bit-interleaved coded modulation). LDPC-based multilevel coding has been analyzed quite extensively, see for instance [138].

7 System aspects

At the time of their initial conception, most common network protocols, such as the Transmission Control Protocol (TCP) and the Internet Protocol (IP), were not developed with security concerns in mind. When DARPA launched the first steps towards the packet-switched network that gave birth to the modern Internet, engineering efforts were targeted towards the challenges of guaranteeing reliable communication of information packets across multiple stations from the source to its final destination. The reasons for this are not difficult to identify: the deployed devices were under the control of a few selected institutions, networking and computing technology was not readily available to potential attackers, electronic commerce was a distant goal, and the existing trust among the few users of the primitive network was sufficient to allow all attention to be focused on getting a fully functional computer network up and running.

A few decades later, with the exponential growth in number of users, devices, and connections, issues such as network access, authentication, integrity, and confidentiality became paramount for ensuring that the Internet and, more recently, broadband wireless networks could offer services that are secure and ultimately trusted by users of all ages and professions. By then, however, the layered architecture, in which the fundamental problems of transmission, medium access, routing, reliability, and congestion control are dealt with separately at different layers, was already ingrained in the available network devices and operating systems. To avoid redesigning the entire network architecture subject to the prescribed security guarantees, the solutions adopted essentially resort to adding authentication and encryption to the various layers of the existing protocol stack.

In this chapter, we shall focus our attention on the security of wireless networks as a case study for the implementation of physical-layer security. This class of systems can be deemed an extreme case in that they combine the liabilities inherent to the broadcast property of the wireless medium with the many vulnerabilities of the Internet, with which they share the basic TCP/IP networking architecture and several essential communication protocols. On the basis of a critical overview of existing security mechanisms at each layer of the network architecture, we identify how physical-layer security can be integrated into the system with clear benefits in terms of confidentiality and robustness against active attacks.

7.1 Basic security primitives

Before elaborating on how each networking layer implements the standard security measures against known attacks, we discuss briefly the basic building blocks currently used in almost every security sub-system. For this purpose, we focus on the three main security services: integrity, confidentiality, and authenticity. Other services, such as non-repudiation and access control, often require solutions to be found in the realm of policy rather than among the engineering disciplines.

In most applications of relevance, the ruling paradigm is that of computational security. In other words, the designers of the system implicitly assume that the adversary has limited computing power and is thus unable to break strong ciphers or overcome mathematical problems deemed hard to solve with classical computers. Existing systems are typically secured through a mix of symmetric encryption (confidentiality), hash functions (integrity), and public-key cryptography (secret-key distribution and authentication).

7.1.1 Symmetric encryption

Under the assumption that the legitimate communication partners are in possession of a shared secret key, as illustrated in Figure 1.1, symmetric encryption algorithms (or *ciphers*) are designed to ensure that (a) the plaintext message to be sent can easily be converted into a cryptogram using the secret key, (b) the cryptogram can easily be re-converted into the plaintext with the secret key, and (c) it is very hard, if not impossible, to recover the plaintext from the cryptogram without the key in useful time (a property also referred to as a *one-way trapdoor function*). Useful time means here that an attacker must be unable to break the encryption while there is some advantage in learning the sent message.

Block ciphers achieve this goal by breaking the plaintext into blocks with a fixed number of bits and applying an array of substitutions and permutations that depend on the secret key. State-of-the-art block ciphers employ the principles of *diffusion* and *confusion*. The former ensures that each input plaintext bit influences almost all output ciphertext symbols, whereas the latter implies that it is hard to obtain the key if the ciphertext alone is available for cryptanalysis.

The most prominent examples are the Data Encryption Standard (DES) and the Advanced Encryption Standard (AES), both of which emerged from standardization efforts by the US National Institute of Standards and Technology (NIST). The DES uses a so-called Feistel architecture, which implements diffusion and confusion by repeating the same substitution and permutation operations in multiple identical rounds, albeit with different sub-keys that are derived from the shared secret key. The AES improves over the DES by increasing the block length, implementing substitution and permutation operations that are highly non-linear, and allowing very fast execution. Special *modes of encryption* are used when (a) the size of the plaintext is not a multiple of the defined block length or (b) the same input block appears more than once and the corresponding output ciphertext blocks must be different.

Instead of dividing the input data into blocks, *stream ciphers* produce one encrypted symbol for every input symbol in a continuous fashion. This mode of operation is similar to that of the one-time pad described in Chapter 1. However, rather than mixing each input symbol with perfectly random key symbols, typical stream ciphers use pseudo-random sequences that are functions of the secret key.

Evaluating the security of symmetric encryption mechanisms is widely considered to be a difficult task. In contrast to the case of the one-time pad in information-theoretic security, there are no mathematical proofs for the assured level of secrecy. Standard practices include testing the randomness of the output ciphertext (the closer to perfect randomness the better), counting the number of operations required for a brute-force attack (i.e. trying out all possible keys) and cryptanalysis with known or chosen plaintext–ciphertext pairs. There is also growing consensus that the encryption algorithm should be made public, so that anyone can come up with and publish on efficient attacks against the most widely used ciphers. It follows that the secrecy of the encrypted information must depend solely on the secret key.

7.1.2 Public-key cryptography

The main drawback of symmetric encryption lies in the need for the legitimate communication partners to share a secret key. Under the computational security paradigm, this can be solved in an elegant way by means of *public-key cryptography* (also known as *asymmetric cryptography*). Instead of a secret key, each of the legitimate partners uses a pair of different keys, more specifically a private key, which is not shared, and a public key, which is available to the legitimate receiver and any potential attacker. The encryption algorithm is designed to ensure that the private key and the public key can be used interchangeably, i.e. a cryptogram generated with the public key can be decrypted only with the corresponding private key and vice versa. Thus, if Alice wants to send a confidential message to Bob, she can use his public key to encrypt the message knowing that only he will be able to recover the message, because no one else knows his private key. If the objective is to authenticate the message using a digital signature, Alice can encrypt the message using her private key, so that Bob can verify the sender's authenticity by decrypting with her public key.

The security of public-key cryptography relies on the computational intractability of certain mathematical operations, most notably the prime factorization of large integers or the inversion of the discrete logarithm function. The RSA scheme, named after its creators Rivest, Shamir, and Adleman, puts public-key cryptography into practice by exploiting the properties of exponentiation in a finite field over integers modulo a prime. To generate the public key and the private key, each user must first select two large primes p and q (about 100 digits each) and compute both their product n and its Euler totient function $\phi(n) = (p-1)(q-1)$. The next step is to pick a number e uniformly at random among all $z < \phi(n)$ that are prime relative to $\phi(n)$, and compute also the multiplicative inverse $d < \phi(n)$ of e modulus $\phi(n)$. The public key and the private key are then composed by (e, n) and (d, n), respectively. To encrypt a message block m, satisfying $0 \leqslant m < n$, Alice takes Bob's public key $k_{\text{pu}}^{\text{B}} = (e, n)$ and computes the cryptogram c according to

$$c = m^e \mod n. \tag{7.1}$$

Using his private key $k_{\text{pr}}^{\text{B}} = (d, n)$, Bob can decrypt m by calculating

$$m = c^d \mod n. \tag{7.2}$$

The correctness of the encryption and decryption mechanisms is ensured by the fact that e and d are multiplicative inverses modulus $\phi(n)$. More specifically, it follows from Euler's theorem that

$$ed = 1 + k\phi(n) \tag{7.3}$$

for some k. Hence,

$$c^d = m^{ed} = m^{1+k\phi(n)} = m^1 = m. \tag{7.4}$$

The RSA scheme and similar public-key cryptography schemes typically use large keys (1024 to 2048 bits), which renders a brute-force attack practically unfeasible with classical computers. It is now widely accepted that, if stable quantum computers can be built with a moderate number of quantum bits, then it will be possible to factorize n and obtain the primes p and q in useful time. This would break hard cryptographic primitives such as those used in public-key cryptography. More recently, elliptic curves have emerged as a group upon the basis of which strong public-key encryption can be obtained with smaller key sizes. However, the vulnerabilities with respect to quantum attacks are similar.

7.1.3 Hash functions

The standard way to ensure the integrity or the message authenticity of the encrypted data is to generate a fixed-length *message digest* using a *one-way hash function*. As their names indicate, the digest is much smaller than the original data and the one-way hash function cannot be reversed. More specifically, it is expected that the hash function satisfies the following properties.

- *Pre-image-resistant.* For any hash value or digest it is hard to find a pre-image that generates that hash value.
- *Second-pre-image-resistant or weakly collision-free.* Given a hash value and the message that originated it, finding another message that generates the same value is hard.
- *Collision-resistant or strongly collision-free.* It is hard to find two messages that generate the same hash value or digest.

The digest typically results from one or more mixing and compression operations on message blocks, which are sometimes implemented using a block cipher and a special encryption mode. Flipping only one bit in the original data should already result in a very different digest with practically no correlation with the previous digest. Examples of hash functions currently in use include the SHA family of hash functions standardized by the NIST. Said cryptographic one-way hash functions should not be confused with the

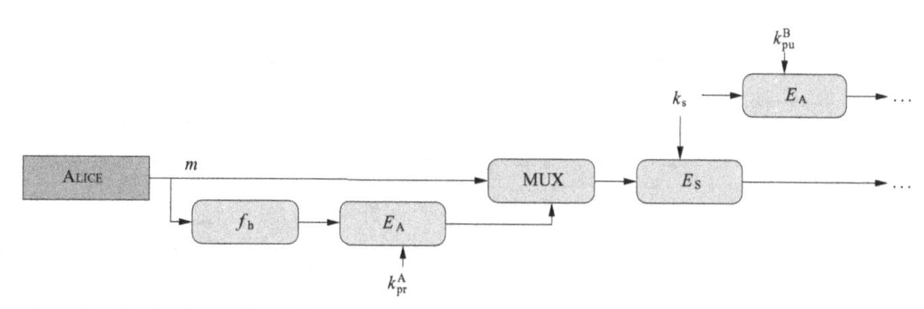

Figure 7.1 Typical security sub-system (sender side).

universal family of hash functions used in information-theoretic secret-key generation, as defined in Chapter 4.

7.1.4 Authentication, integrity, and confidentiality

In principle, the basic mechanisms of public-key cryptography would be sufficient to achieve the fundamental security goals of authentication, integrity, and confidentiality. However, since these mechanisms are very demanding from a computational point of view, most secure systems use a combination of public-key cryptography, symmetric encryption, and one-way hashing.

Figure 7.1 shows a typical solution from the point of view of the sender, Alice. She starts by generating the message digest $f_h(m)$, which she encrypts asymmetrically (E_A) with her private key k_{pr}^{A}. The fact that Alice is the only person in possession of k_{pr}^{A} assures the required sender authentication. Since the message digest is much smaller than the message, the computational overhead that comes with public-key cryptography is not a cause for concern. The authenticated digest, which corresponds to a digital signature, is then multiplexed with the original message. The output is then protected efficiently via symmetric encryption with the session key k_s. To share the session key with Bob securely, Alice uses once again public-key cryptography, this time with Bob's public key k_{pu}^{B}.

On the reception side, illustrated in Figure 7.2, Bob recovers k_s by decrypting it with his private key. He then uses this key to obtain the sent message, to which he applies the known hash function f_h. Using Alice's public key, he can decrypt the sent message digest, and compare it with the result he obtained in the previous step. If the computed hash value is equal to the decrypted message digest, then he can rest assured that the integrity of the message and the authenticity of the sender are guaranteed.

7.1.5 Key-reuse and authentication

Suppose that Alice and Bob share a secret key k. Although in theory it is possible for Alice and Bob to refresh k by generating a new random key k for every message and exchanging k' securely, the overhead incurred in real systems in terms of computation and number of transmissions forces practitioners to compromise and reuse the same key

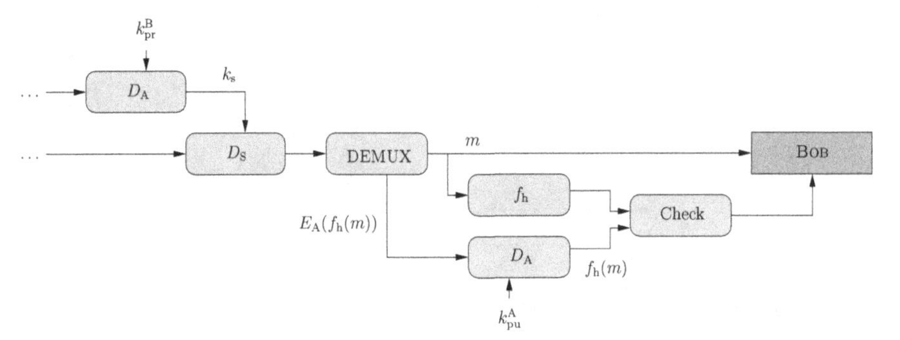

Figure 7.2 Typical security sub-system (reception side).

multiple times. Feasible alternatives include scheduled modifications by means of simple operations (e.g. a shift or an XOR with another key sequence) and re-keying by means of pseudo-random generators, which use the original key as the seed for subsequent keys.

Although changing the key in these ways intuitively makes it harder for the attacker to break the security of the system, from an information-theoretic point of view using functions of the key for encryption does not change the amount of randomness or the uncertainty of the attacker with respect to the protected messages and the secret key.

An active attacker, say Charles, with access to the sent cryptograms will seek not only to break the key but also to impersonate the legitimate sender, either by modifying intercepted messages or by generating new ones using the captured secret key. If we assume that in the worst case the attacker has unlimited computing power, the natural question is then whether Charles can make Bob believe that the faked messages come from Alice.

The scenario described here admits an information-theoretic treatment. Alice wants to send n messages M_i with $i \in [\![1, n]\!]$ in sequence at different points in time. To authenticate the messages, she uses the shared secret key K, and generates one $\mathsf{Y}_i = f(\mathsf{M}_i, \mathsf{K})$ for every M_i to be sent. At time i, Bob must use his knowledge of the key K and all past cryptograms Y_j with $j \in [\![1, i-1]\!]$ to determine whether Y_i is authentic or not. The decision process can be framed as a hypothesis test, in which Y_i was generated either by Alice or by an attacker. A type-I error occurs if Bob rejects the message even though it was actually generated by Alice. On the other hand, Bob incurs a type-II error if he accepts the message when it was generated by the attacker. If we set the probability of type-I errors to zero, i.e. if Bob is expected to accept every authentic message sent by Alice, then the probability P_{SA} that the attack is successful is lower bounded by

$$P_{\text{SA}} \geqslant 2^{-\mathbb{H}(\mathsf{K})/(n+1)}. \tag{7.5}$$

In simple terms, the difficulty in guessing the key is reduced significantly with every re-utilization of the key.

Carter–Wegman universal families of hash functions can be used to ensure unconditional authentication. In other words, it is possible to build digital-signature schemes or authentication tags that cannot be forged or modified even by an attacker with infinite computing power. In precise terms, we say that the scheme is unbreakable with

Table 7.1 The TCP/IP architecture, features, and security mechanisms

Level	Layer	Tasks	Security mechanisms
5	Application	Runs processes and applications	End-to-end cryptography
4	Transport	Reliable communication and congestion control	Secure Sockets Layer (SSL) and Transport Layer Security (TLS)
3	Network	Routing and forwarding	Internet Protocol Security (IPSec)
2	Link	Medium-access control	End-to-end cryptography
1	Physical	Transmission	Spreading against narrow-band jamming

probability p if an attacker who gains access to m and the corresponding authentication tag $f(m)$ has a probability less than or equal to p of guessing the tag of another message m'. To transmit an authenticated message m, the legitimate partners must share a secret key, which consists of the message number i and a function f. The latter is chosen uniformly among a universal family of hash functions that map the message set M to the set of tags T. The fraction of functions in the same class that map m' (different from m) to a particular tag t' is $1/|T|$. Thus, setting $|T| \geqslant 1/p$ ensures that the attacker has a success probability of at most p. To authenticate multiple messages, the sender computes the authentication tag $f(m)$ and then performs a bit-wise XOR of the tag with the message number, which is also known at the receiver. The message number can be used only once. If the authentication tag is k bits long, then the attacker cannot find a message whose tag can be guessed with probability higher than $1/2^k$.

7.2 Security schemes in the layered architecture

The typical architecture of contemporary networks evolved from the Open System Interconnection (OSI) Reference Model to the prevalent five-layer TCP/IP architecture, which is depicted in Table 7.1. The tasks listed for each layer reflect the initial concerns with the reliable communication of information packets from one end of the network to the other. Processes and applications running at the top layer generate data streams, which are segmented and encapsulated into separate packets. Computers and routers then forward the packets across multiple links until they reach the destination. Communication links may exist over different physical media with multiple transmitters, thus requiring specific transmission schemes and channel arbitration among various devices.

Each layer provides the adjacent upper layer with an abstraction of the network. In simple terms, the link layer sees a channel with or without collisions of transmitted packets, the networking layer sees a graph of links with certain rates, the transport layer sees an end-to-end connection with or without packet losses, and the application layer sees a bit-pipe with or without delivery guarantees. It is striking that in their original form none of the layers takes security aspects into consideration.

To fill this gap, the X.800 standard defined a set of fundamental security services based on the original OSI networking model. As summarized in Table 7.2, these services are

Table 7.2 Physical-layer security vis-à-vis the X.800 standard

Category	Service	Objective
Authentication	Peer-entity authentication	Confidence in the identity of communication partners
	Data-origin authentication	Confidence in the source of the received data
Access control	Access control	Prevent unauthorized use of a resource
Data confidentiality	Connection confidentiality	Protect a connection
	Connectionless confidentiality	Protect a single data block
	Selective-field confidentiality	Protect selected fields on a connection or in a data block
	Traffic-flow confidentiality	Prevent information acquisition from traffic observation
Data integrity	Connection integrity with recovery	Detect and correct active attacks on a connection
	Connection integrity without recovery	Detect active attacks on a connection
	Selective-field connection integrity	Detect active attacks on specific fields
	Connectionless integrity	Detect modification and some forms of replay of a single data block
	Selective-field connectionless integrity	Detect active attacks on specific fields of a data block
Non-repudiation	Origin	Prove that a data block was sent by the source
	Destination	Prove that a data block was received by the destination

divided into five categories and fulfill specific security objectives. In the following, we shall demonstrate that some of these services can be enhanced using physical-layer techniques, which at best provide information-theoretic security and at the very least cause significant degradation of the signals observed by the eavesdropper. But first, we set the stage for the integration of physical-layer security into contemporary networks by elaborating on the security mechanisms that have been introduced at each layer to ensure integrity, authentication, and confidentiality.

Application layer

The security mechanisms implemented at the uppermost layer of the protocol stack depend heavily on the application under consideration. A security architecture similar to the one depicted in Figure 7.1 is often implemented to secure email or web-browsing services by means of end-to-end cryptography. Public-key cryptography is used extensively for the purpose of authentication and the sharing of session keys. To ensure the authenticity of public keys, a public-key infrastructure must include trusted third parties, also called certification authorities (CAs). The service they provide consists of managing large directories of registered users and devices, issuing digitally signed certificates of their public keys upon request. These certificates are then used to establish trusted connections among clients and servers.

With the advent of peer-to-peer systems, in which every node is both a client and a server to other nodes in the network, typical network functions such as packet forwarding and flow control are increasingly assigned to so-called overlay networks, which consist of virtual logical links among processes that run in different computing nodes. This form of virtualization simplifies network management and information dissemination, while allowing the construction of topologies that increase the overall robustness of the networked application. The primary concern in this context relates to active attacks by which information flows are compromised through the injection of erroneous packets. The solution here is to enforce integrity checks by means of cryptographic hash functions.

Transport layer

Transport protocols providing connection-oriented or connectionless services are typically implemented in the operating system. Security extensions, most notably the Secure Sockets Layer (SSL) and the Transport Layer Security (TLS) standard of the Internet Engineering Task Force (IETF), provide message encryption and server authentication, with optional support for client authentication. The system stores a list of trusted CAs, from which certificates can be obtained in order to exchange session keys and establish secured connections. The schemes employed are again very similar to the one shown in Figure 7.1, and the transport headers of the transmitted segments are modified to reflect the security primitives employed.

Network layer

Since the main task of network routers is to store packets and forward them towards the right next hop until the destination is reached, their operation is limited to the network layer and below. Consequently, all security mechanisms are implemented at the level of IP datagrams. The IPSec standard allows for the establishment of a unidirectional network-level security association and the introduction of a so-called authentication header, which includes a connection identifier and the digital signature of the sending node. The encapsulating security payload (ESP) protocol is then responsible for ensuring the confidentiality of the data by means of encryption. Beyond the datagrams that carry application data, it is very important to protect also the control traffic, which ensures that the routing tables are correct and up to date. Successful attacks on link advertisements and other control messages can make it impossible for routers to forward the packets correctly, which eventually leads to a serious disruption of network services.

To protect the network against intrusion attacks, it is common to install firewalls at the gateway nodes. Their goal is to filter out packets that are identified as suspicious on the basis of predefined rules that are under the discretion of network administrators. Beyond enforcing some form of access control, firewalls prevent attackers from learning about the available network resources and how they are mapped to different addresses and ports. In addition, they constitute a first line of defense against denial-of-service

(DoS) attacks in which attackers flood the network with fake connection requests or other forms of signaling.

Link layer

While the network layer is agnostic with respect to the underlying channel, the link layer, which governs the access to the channel by multiple devices, depends naturally on the communication medium of choice. It is thus not surprising that security mechanisms are more often found at the networking layer and above, where the same security subsystem can operate irrespective of the physical characteristics of the link over which communication is taking place.

However, whereas security primitives implemented at the network layer address only the vulnerabilities pertaining to end-to-end connectivity, link-layer security is concerned with the direct connections among devices. In contrast with wired networks, where access to the communication link requires tapping a cable or optical link, wireless networks are easy targets for eavesdroppers and intruders. All that is required is a computer with a wireless interface, both of which have now become very affordable commodities. Thus, security solutions at the link layer of wireless communications systems are aimed at ensuring that only authorized devices can communicate over the channel. One such instance is the Extensible Authentication Protocol (EAP), which defines a security framework over which a device can prove its identity and gain access to a wireless network.

7.3 Practical case studies

The previous overview of the security mechanisms that are typically implemented at the various layers above the physical layer underlines the fact that the individual solutions for each layer are tailored to its specific tasks with little relation to the security primitives implemented elsewhere in the system. A device may, for example, be authenticated over a wireless link and operate IPSec at the network layer, while at the application layer a browser is accessing a web site securely using the end-to-end cryptography implemented by TLS. In some cases, this patchwork of security mechanisms can be redundant, requiring for instance parallel authentication steps at different layers or multiple encryptions. At the same time, some vulnerabilities are left open at other layers. This is well illustrated by the fact that end-to-end cryptography at higher layers is unable to prevent traffic-analysis attacks carried out at the lower layers, such that an opponent can learn about the topology of the network and the type of sessions it is carrying.

Owing to the currently existing trust in the security levels assured by cryptographic mechanisms implemented at higher layers, it is fair to say that most available engineering solutions do not exploit the full potential of the physical layer in increasing the overall security of the system. We start by providing a few motivating examples and then proceed with a system-oriented view of how physical-layer security inspired by information-theoretic security can be incorporated into the wireless communication networks of today.

Optical communication networks

Communication networks based on optical fibers were among the first communication infrastructures to embrace physical-layer security. Since the information bearing signals are carried along the cable in the form of light with total internal reflection, typical attacks are physical in nature, thereby diverting the light in order to disrupt the communication service, degrade the quality of service, or extract information about the traffic and its content. Although there is very little radiation from a fiber-optic cable that would allow non-intrusive eavesdropping, an attacker with access to a cable can cut the fiber or bend it in a way that allows light to be captured from or released into the fiber. Sophisticated attacks are capable of intercepting selected wavelengths while keeping others intact. Optical jamming can then take different forms, depending on whether the attacker chooses to inject noise, delayed versions of the captured signal (repeat-back jamming), or some other disrupting signal (correlated jamming). Even if the destination is capable of monitoring subtle fluctuations in received power, many of these attacks can go undetected. Physical-layer security solutions for optical communications include robust signaling schemes, coding, and active limitation of the power and bandwidth of input signals. Since the communication rates are extremely high, coding solutions that require complex processing are hard to implement and therefore very expensive.

Spread spectrum

It is fair to say that, much like other classes of communication networks, wireless systems have not been designed with security requirements as their chief concern and main figure of merit. There is, however, one notable exception: spread-spectrum (SS) systems. Since the modulation schemes in this class were invented first and foremost for military applications, their designers aimed at counteracting the possibility of an enemy attacker detecting and jamming the signals sent by the legitimate transmitter. The key idea is to use pseudo-random sequences to spread the original narrow-band signal over a wide band of frequencies, thus lowering the probability of interception and reducing the overall vulnerability to narrow-band jamming. At the receiver the wide-band signal is de-spread back to its original bandwidth, whereas the jamming signal is spread over a wide band of frequencies. This in turn reduces the power of the interference caused by the jammer in the frequency bandwidth of the original signal.

There are two main SS techniques: direct sequence spreading (DSS) and frequency hopping (FH). DSS modulates a signal s_i onto a train of rectangular pulses $p(t)$ of duration T_c, also called chips, whose values are governed by a pseudo-random spreading sequence known both to the transmitter and to the receiver. The outcome of this process,

$$s(t) = \sum_i s_i p(t - iT_c),$$

is then transmitted to the receiver. To recover the original signal, the receiver must multiply the acquired signal once again by the spreading sequence. The main challenge is to synchronize the pseudo-random sequence used by the receiver with the one used by the transmitter. It follows that spreading sequences must have a strong autocorrelation. Furthermore, sophisticated search and tracking algorithms are needed in order to synchronize and to keep the synchronization between transmitter and receiver. This in turns limits the attainable bandwidth of DSS signals.

As the name indicates, FH spreads the signal by shifting it to different frequencies that are dictated by a pseudo-random sequence. This operation can be expressed in the complex plane by

$$s(t) = \sum_i \exp(\mathrm{j}(2\pi f_i + \phi_i))p(t - iT_\mathrm{h}),$$

where f_i denotes the frequency shift, ϕ_i is a random phase, and T_h is the so-called hop time. Similarly to DSS, the receiver must acquire and track the pseudo-random FH sequence. The modulation is called fast FH if the number of hops per transmitted symbol is equal to or above one. Otherwise, we speak of slow FH modulation, whereby multiple symbols are transmitted in each hop.

Today, spread-spectrum techniques are used extensively in wireless networks both for military and for civilian use. Low-probability-of-interception (LPI) systems hide the transmissions by exploiting the fact that signals modulated according to pseudo-random spreading sequences are hard to distinguish from white noise. In cellular mobile communication networks, spread-spectrum techniques are used primarily for multiple-user channel accessing, whereby all users transmit simultaneously, albeit using different spreading sequences. Since the spreading sequences are orthogonal to each other, the receiver can recover the signal transmitted by each individual user by multiplying the received signal by the corresponding spreading sequence. Privacy can be enhanced by keeping the spreading sequences secret from potential attackers.

Mobile communication systems

The Global System for Mobile Communications (GSM) standard is widely deployed around the world, offering mobile telephony and short-messaging services over wireless channels. Its security architecture is somewhat unconventional in that authentication is dealt with at the application level while confidentiality is implemented at the physical layer. Data are thus encrypted after having been processed by the channel coding and interleaving blocks, as shown in Figure 7.3. The authentication follows a standard challenge–response protocol that relies on the shared secret embedded in the Subscriber Identity Module (SIM). The encryption stage is based on a shared session key and a specific stream cipher, the so-called A5 algorithm. In addition, GSM uses frequency hopping as a means to combat multipath fading and to randomize co-channel interference. Since frequency hopping is a form of spread-spectrum communication, GSM is inherently robust against narrow-band jamming.

The General Packet Radio Service (GPRS) extends the functionality of GSM to enable data communication at rates up to several tens of kbits/s. In contrast to GSM, the system designers of GPRS opted to place the encryption algorithm at the Logical Link Control

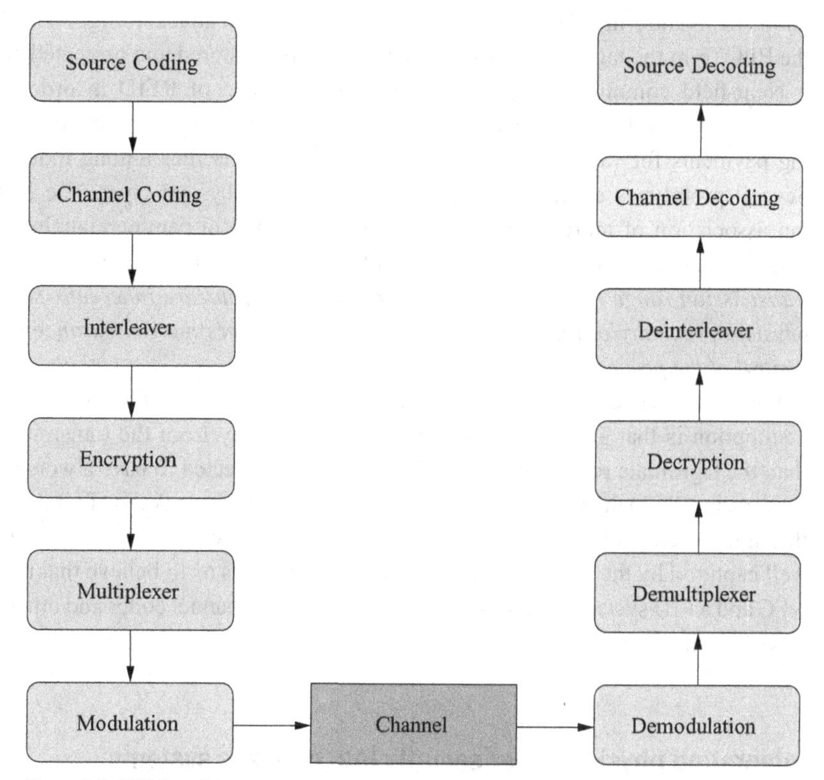

Figure 7.3 GSM architecture.

(LLC) layer instead of at the physical layer. Moreover, the algorithm uses a different stream cipher. More recent standards for mobile broadband communications such as the Universal Mobile Telecommunications System (UMTS) and Long Term Evolution (LTE) also do not employ physical-layer security, except for the aforementioned spread-spectrum techniques. The same is true for the Dedicated Short-Range Communications (DSRCs) for vehicular networks.

RFID and near-field communications

Radio-frequency identification (RFID) tags are emerging as a convenient means to track objects, such as toll cards on highways, passenger luggage in airports, consumer goods in retail, and parcels in shipping businesses. Active, semi-passive, and passive RFID tags all share the same principle of operation: (1) the antenna of the tag receives electromagnetic energy transmitted by the antenna of the RFID reader and (2) the tag sends the EPC back to the reader via the radio channel. The range of transmission can take values up to 100 m. Active tags use a battery to power both their circuitry and their transmissions, whereas passive tags must use some of the power collected by their antenna during reception. Semi-passive tags use battery power for the circuitry only. In a typical system, the tag provides a 96-bit number (also known as an electronic product code, EPC) that points to an entry in a database with restricted access. An attacker might be able to read the EPC from the tag, but the rest of the information is stored and protected elsewhere.

Near-field communication (NFC) builds on the idea of RFID in order to enable high-rate data transfers at very short ranges. The applications envisioned include making payments for various services or sharing multimedia files among mobile phones. Several prototypes exploit inductive coupling at the physical layer. The NFC forum (an association of more than 130 companies interested in commercializing near-field communication technology) makes the following claim on its web site: *Because the transmission range is so short, NFC-enabled transactions are inherently secure. Also, physical proximity of the device to the reader gives users the reassurance of being in control of the process.*

Physical-layer security is so far absent from NFC and RFID technologies. A common assumption is that an eavesdropper will be farther away from the transmitting device than the legitimate reader. The eavesdropper is thus expected to have a worse signal-to-noise ratio (SNR) than that of the reader, which prevents the leakage of vital information through the electromagnetic waves traversing the channel. Clearly, this assumption is well captured by the wiretap-channel model, which leads us to believe that more secure NFC and RFID systems can be developed using secure channel codes and other physical-layer security techniques.

7.4 Integrating physical-layer security into wireless systems

From the previous sections it is clear that physical-layer security cannot be viewed as a panacea for solving all of the security concerns that exist in today's networks. There is evidence, however, that careful design of the physical layer, promoting these security concerns as fundamental performance criteria, will lead to wireless networks that are arguably more secure than those whose security is rooted on cryptographic primitives alone. We shall now investigate how this can be achieved by applying the information-theoretic security principles described in detail in previous chapters.

Impact on the security architecture

To increase the security of a wireless network in a bottom-up fashion, i.e. from the lowest layer to higher layers, the design of the physical layer must go beyond its traditional role of ensuring virtually error-free transmission across imperfect channels by means of powerful error-correction coding and adequate modulation schemes. Figure 7.4 illustrates two new security functions that can be introduced at the physical layer, either as stand-alone features or in combination with the cryptographic mechanisms at higher layers. One function consists of reducing the number of error-free bits that an eavesdropper can extract from the transmitted signal. This can be achieved in several scenarios of practical interest by using a special class of codes, which we call *secure channel codes*. The other function is concerned with exploiting the randomness of the communications channel to generate secret keys at the physical layer. The keys can then be passed on to

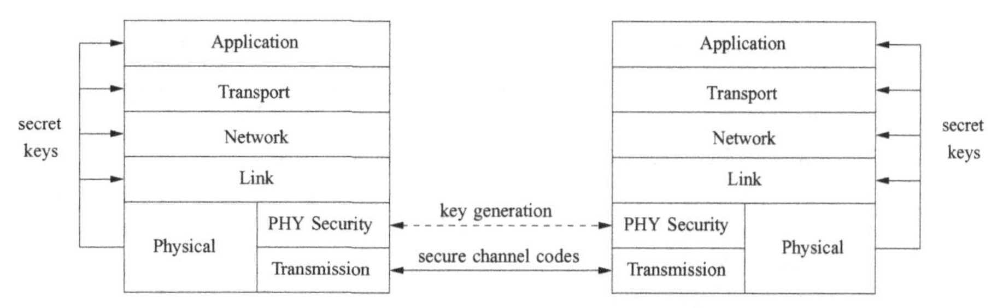

Figure 7.4 Integration of physical-layer security into the network architecture.

higher layers for use with existing cryptographic schemes. The next sections elaborate further on the merits and drawbacks of these two methodologies.

Secure channel codes

The key idea of the first approach presented in the previous section is to substitute typical channel codes and modulation schemes, which ensure reliability but not security, by the class of code constructions described in Chapter 6. These can achieve secrecy and reliable communication at a tolerable price in terms of achievable transmission rate and computational complexity. The codes degrade significantly the amount of information that an eavesdropper is capable of extracting from the observed signals, provided that his SNR is inferior to that of the legitimate receiver over the duration of a transmitted codeword.

While it is fairly straightforward to discover the state of the channel between legitimate communication partners by means of pilot symbols, as explained below, the same is obviously not true for the eavesdropper, who is likely to hide his presence or at the very least conceal his location. We are thus left with the options of (a) transmitting only when the SNR is high (and therefore more likely to exceed the SNR of the eavesdropper), (b) ensuring by physical means that the eavesdropper cannot place an antenna within a certain range, or (c) degrading the eavesdropper's reception by means of jamming the regions where his antenna is assumed to be located, as explained in Chapter 8.

Secure channel codes can be used at the physical layer in a modular fashion, i.e. independently of the security architecture implemented at higher layers. If the sent datagrams are protected by cryptographic primitives, be it through symmetric encryption or public-key cryptography, the use of secure channel codes can strengthen the overall security level significantly by virtue of the fact that the eavesdropper is no longer able to obtain an error-free copy of the cryptogram. Instead, even less sophisticated secure channel codes, which increase the error probability at the eavesdropper rather than the total equivocation (or uncertainty) about the sent information bits, can force the attacker to deal with a large number of erroneous symbols that are virtually indistinguishable from the scrambled bits resulting from cryptographic operations. Therefore, classical tools, such as differential and linear cryptanalysis, no longer apply and must be modified to account for bit errors in the encrypted stream. In some cases, the best that can be achieved is to form a list of possible message sequences. It is to be expected that, in

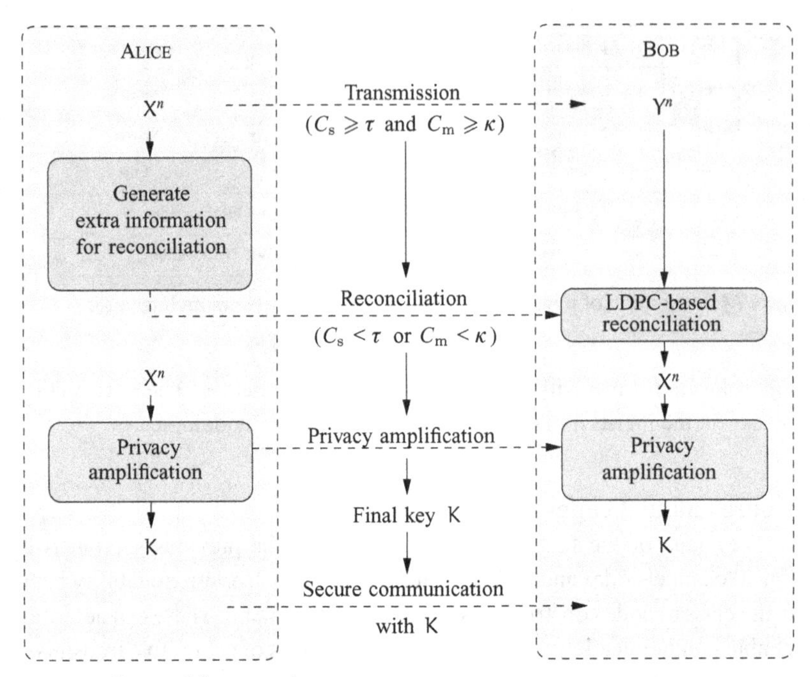

Figure 7.5 Protocol for secret-key agreement.

most cases of interest, the task of the enemy cryptanalyst will become significantly more complex.

When secure channel codes are used in an opportunistic manner depending on current estimates of the SNR, transmissions occur only if the received power is above a certain threshold. Communication rules such as these will obviously affect the transmission schedule and the access to the channel. This suggests joint design of the physical and link layers, as well as an optional adaptation of routing and retransmission at the network and transport layers, respectively.

Insofar as the actual implementation of secure channel codes in real systems is concerned, the main challenges will of course vary from platform to platform. The transmission schemes at the physical layer are typically implemented in a device driver, e.g. for the wireless interface card. With the proliferation of software-defined radios, it is reasonable to assume that security-oriented mechanisms at the physical layer will be easier to integrate into future communication systems.

Secret-key agreement at the physical layer

The second security function that can be implemented at the physical layer is secret-key agreement. As explained in Chapter 4, after sharing some common randomness in the form of correlated symbols, Alice and Bob can agree on a secret key by means of reconciliation and privacy amplification. Figure 7.5 illustrates a wireless security protocol that achieves this purpose. When the estimated secrecy capacity C_s and main channel capacity C_m from Alice to Bob are found to be above the prescribed thresholds,

τ and κ, respectively, Alice transmits Gaussian-shaped symbols X^n to Bob, who observes the channel outputs Y^n. The two sequences of Gaussian symbols are correlated and form the core of common randomness from which Alice and Bob can generate their key. Using a quantizer with a bit mapping, Alice obtains a sequence of bits. To ensure that the two legitimate partners share exactly the same sequence from the observed symbols, Alice sends some side information in the form of a number of low-density parity-check (LDPC) bits (using, for instance, the reconciliation protocol described in Section 6.5.2), which are protected by a powerful channel code. Bob is thus able to receive these bits virtually without error, and feeds them to a belief-propagation decoder for reconciliation. The side information can be sent at any time, even when Eve has a better channel than Alice. To ensure that Eve is unable to extract any information about the key from the noisy symbols and the side information she has access to, Alice and Bob use privacy amplification, thus arriving at a smaller sequence of secret bits, which can be used to communicate securely.

It has been shown that, with a fraction of time dedicated to secret-key generation as small as 1%, such a secret-key agreement protocol can renew a 256-bit encryption key every 25 kbits, i.e. with SNRs of 10 dB and 20 dB for the main and eavesdropper channels, respectively, and, transmitting at an average rate of 2 Mbps, a secret key could be replaced by a new random key every 16 milliseconds. Although these estimates may be optimistic, continuous generation of secret keys at the physical layer thus emerges as a promising solution for the key-reuse problem described.

As in the case of secure channel codes, we are confronted with the need to modify the physical layer, which implies redesigning a wireless interface card and writing a new device driver. Again, software-defined radio may prove valuable in adapting the physical layer for security functionalities. Since common randomness is shared in an opportunistic way, the medium-access control at the link layer, which governs the scheduling of transmissions on the basis of channel state information, will have to be adapted accordingly. This calls for cross-layer design of security protocols, which must ensure also that higher layers are capable of acquiring and using the secret keys generated at the lowest layer. In current systems, this involves altering the communication and security components of the operating system.

The role of channel state information

The objective of physical-layer security is for Alice and Bob to communicate reliability at a certain target rate while leaking the least possible number of information bits to Eve. If the SNRs both of the main channel and of the eavesdropper's channel are known perfectly to Alice, she can compute the secrecy capacity and adjust her transmissions accordingly. When the secrecy capacity turns out to be zero, Alice can decide not to transmit, thus avoiding the likely disclosure of information. On the other hand, when the secrecy capacity is strictly positive, Alice can choose a secure channel code that operates at that rate and achieve information-theoretic security. Alternatively, she can share common randomness with Bob, when the secrecy capacity is positive, and then generate a secret key. Either way, knowledge of the state of both channels is necessary for assuring information-theoretic security.

Estimating the channel between the legitimate users is common practice in the most widespread wireless communication systems. Typically, each burst contains a known training sequence that is transmitted together with the coded bits. In GSM, for example, training bits account for about 20% of the traffic. After detecting the transmitted signals, the receiver can use the training bits and the received signal samples to estimate the channel impulse response. The most common algorithms are the least-squares and linear minimum mean-squared error estimators. Often a feedback channel is available, over which the receiver can inform the transmitter about the state of the channel by sending either the received signal or quantized values of the main channel parameters. In some cases, such as time-division duplex (TDD) systems, in which the coherence time of the channel is sufficiently long, the channels from transmitter to receiver and vice versa are identical. This form of reciprocity obviously simplifies the sharing of channel state information.

In some scenarios, the legitimate communication partners may have partial knowledge of the eavesdropper's channel. This corresponds, for instance, to the situation in which Eve is another active user in the wireless network (e.g. in a TDD environment), so that Alice can estimate the eavesdropper's channel during Eve's transmissions. Naturally, an attacker that aims to intercept or disturb the communication between the legitimate partners cannot be expected to participate in the channel-estimation process or to follow any conventions set by the communication protocol. One possible attack consists of sending low-power signals, which lead Alice and Bob to underestimate the SNR at Eve's receiver. If Eve behaves as a passive attacker, no training signals are available for estimating the state of the eavesdropper's channel. In this case, the best that Alice and Bob can do is to use a conservative estimate based, for example, on the verifiable assumption that Eve is located outside a certain physical perimeter. Studies show that protocols such as the one illustrated in Figure 7.5 are considerably robust against channel-estimation errors.

Authentication requirements

We showed how secure channel codes can be used to increase the levels of confidentiality provided by a wireless network and how secret-key agreement at the physical layer helps solve the key-reuse problem by exploring the randomness of the communications channel. In all cases, it is assumed that the attacker is passive and does not try to impersonate either Alice or Bob. In most scenarios of interest, which do not grant this type of assurance, all communications must be authenticated, otherwise an active attacker is able not only to intercept the datagrams sent over the wireless channel but also to transmit fake signals aimed at confusing Alice and Bob. In the simplest instance, an attacker could disrupt the communications by jamming their transmissions, which can be viewed as a form of DoS. Alternatively, an impersonation attack would allow the attacker to play the role of the man-in-the-middle, for example generating secret keys with Bob while pretending to be Alice, and vice versa, and modifying the exchanged messages without being noticed.

Notice that a small shared secret is sufficient to authenticate the first transmission. Subsequent transmissions can then be authenticated using the secret keys that can be generated at the physical layer through common randomness, reconciliation, and privacy amplification. Despite our ability to authenticate subsequent transmissions, physical-layer security as it stands does not solve the key-distribution problem and we are still left with the question of how to share the initial secret key, which is critical for bootstrapping the security sub-systems of the wireless communication network.

One way is to rely on the traditional approach using public-key cryptography. More specifically, nodes in the network would rely on a public-key infrastructure (PKI) to obtain authenticated public keys from a trusted certification authority and then use these public keys to share the initial secret key, which allows them to authenticate the first transmission and thus establish a secure link. This approach obviously inherits all the flaws of public-key cryptography, most notably the reliance on computational security and the complexity of establishing a PKI.

If Alice (or Bob) can be trusted to be the first one to establish a connection then it is possible to authenticate transmissions at the physical layer using the impulse response of the channel. The key idea is that Alice sends a known probing signal, which Bob can use to estimate the channel's impulse response and infer the channel state, more specifically the instantaneous fading coefficient. Since it has been shown experimentally that a probing signal sent by a different sender transmitting from a different position is going to generate a different impulse response with overwhelming probability, Bob will be able to detect any attempt by Eve to impersonate Alice after the first transmission. All he needs to do is to compare the observed impulse responses for different transmissions.

7.5 Bibliographical notes

The subject of classical cryptography and computational security is well addressed in a number of textbooks. Stallings gives a general overview of security systems as they exist today [139]. The design of ciphers based on the notions of confusion and diffusion was studied by Shannon in [1]. The principles of public-key cryptography were first published by Diffie and Hellman in [140]. Rivest, Shamir, and Adleman proposed the widely used RSA scheme in [141]. A comprehensive and practically oriented treatment of symmetric encryption and public-key cryptography is given by Schneier in [142]. Another essential reference on these matters is the handbook by Menezes, Van Oorschot, and Vanstone [143]. A survey of physical-layer-security aspects of all-optical communications by Médard, Marquis, Barry, and Finn can be found in [144]. Wegman–Carter authentication schemes are discussed in [66]. An information-theoretic treatment of authentication was presented by Maurer in [145]. Several security schemes presented by Li, Xu, Miller, and Trappe in [146] for wireless networks are based on the impulse response of the channel. Bloch, Barros, Rodrigues, and McLaughlin [105] developed the secret-key agreement protocol described in Section 7.4.

Part IV

Other applications of information-theoretic security

8 Secrecy and jamming in multi-user channels

In all of the previous chapters, we discussed the possibility of secure transmissions at the physical layer for communication models involving only two legitimate parties and a single eavesdropper. These results generalize in part to situations with more complex communication schemes, additional legitimate parties, or additional eavesdroppers. Because of the increased complexity of these "multi-user" channel models, the results one can hope to obtain are, in general, not as precise as the ones obtained in earlier chapters. In particular, it becomes seldom possible to obtain a single-letter characterization of the secrecy capacity and one must often resort to the calculation of upper and lower bounds. Nevertheless, the analysis of multi-user communication channels still provides useful insight into the design of secure communication schemes; in particular it highlights several characteristics of secure communications, most notably the importance of *cooperation*, *feedback*, and *interference*. Although these aspects have been studied extensively in the context of reliable communications and are now reasonably well understood, they do not necessarily affect secure communications in the same way as they affect reliable communications. For instance, while it is well known that cooperation among transmitters is beneficial and improves reliability, the fact that interference is also helpful for secrecy is perhaps counter-intuitive.

There are numerous variations of multi-user channel models with secrecy constraints; rather than enumerating them all, we study the problem of secure communication over a two-way Gaussian wiretap channel. This model exemplifies the specific features of most multi-user secure communication systems and its analysis directly leverages the techniques and results presented in previous chapters. We refer the interested reader to the appendix at the end of this chapter and to the monograph of Liang *et al.* [147] for an extensive list of references on multi-user secure communications.

We start this chapter by introducing the two-way Gaussian wiretap channel (Section 8.1). We then discuss in detail three secure communication strategies: cooperative jamming (Section 8.2), coded cooperative jamming (Section 8.3), and key exchange (Section 8.4).

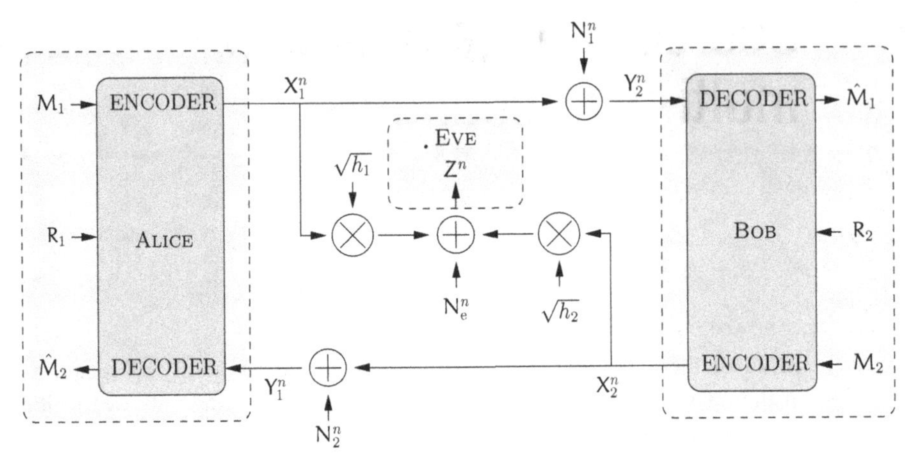

Figure 8.1 Two-way Gaussian wiretap channel with interference at the eavesdropper. M_1 and M_2 represent the messages transmitted by Alice and Bob. R_1 and R_2 represent local randomness used in the encoders.

8.1 Two-way Gaussian wiretap channel

As shown in Corollary 5.1, the secrecy capacity of a Gaussian wiretap channel is given by

$$C_s = \left(\frac{1}{2}\log\left(1+\frac{P}{\sigma_m^2}\right) - \frac{1}{2}\log\left(1+\frac{P}{\sigma_e^2}\right)\right)^+,$$

and the secrecy capacity is zero if $\sigma_m^2 \geqslant \sigma_e^2$, independently of the transmit power P. To improve secure communication rates, one should either increase the signal-to-noise ratio (SNR) of the legitimate receiver or decrease the SNR of the eavesdropper. A natural approach by which to achieve the latter is to introduce interferers into the system. In particular, if the eavesdropper happens to be located closer to the interferers than the legitimate receiver, interferences may have a more detrimental effect on her than on the legitimate receiver, which can result in increased secure communication rates. Notice that this approach implicitly requires knowledge of the locations of all of the transmitters and receivers or some knowledge of all of the instantaneous channel characteristics, so that inteferers do not harm the legitimate receiver unnecessarily. In practice, this knowledge would be obtained via some cooperation mechanism between nodes; therefore, this approach was called *cooperative jamming* by Tekin and Yener to highlight the importance of interfering intelligently.

The concept of cooperative jamming can be applied in many different settings, and we refer the reader to the bibliographical notes for examples. In this chapter, we restrict our attention to the situation in which Alice or Bob plays the role of the interferer. Specifically, we consider the channel model illustrated in Figure 8.1, in which Alice and Bob communicate in full duplex over orthogonal Gaussian channels while their signals interfere at Eve's terminal. We call this model a *two-way Gaussian wiretap channel* (TWWTC for short). At every time instant i, Alice transmits symbol $X_{1,i}$

while receiving $Y_{1,i}$, Bob transmits $X_{2,i}$ while receiving $Y_{2,i}$, and Eve observes Z_i. The relationships among these symbols are given by

$$\begin{aligned} Y_{2,i} &= X_{1,i} + N_{1,i}, \\ Y_{1,i} &= X_{2,i} + N_{2,i}, \\ Z_i &= \sqrt{h_1}X_{1,i} + \sqrt{h_2}X_{2,i} + N_{e,i}. \end{aligned} \tag{8.1}$$

The processes $\{N_{1,i}\}_{i\geqslant 1}$, $\{N_{2,i}\}_{i\geqslant 1}$, and $\{N_{e,i}\}_{i\geqslant 1}$ are i.i.d. and distributed according to $\mathcal{N}(0, 1)$; the real channel gains $h_1 > 0$ and $h_2 > 0$ account for the position of the eavesdropper with respect to Alice and Bob and are assumed known to all parties, including the eavesdropper. The inputs X_1^n and X_2^n to the channel are subject to the average power constraints

$$\frac{1}{n}\sum_{i=1}^{n}\mathbb{E}\left[X_{1,i}^2\right] \leqslant P_1 \quad\text{and}\quad \frac{1}{n}\sum_{i=1}^{n}\mathbb{E}\left[X_{2,i}^2\right] \leqslant P_2.$$

Remark 8.1. *For the sake of generality, we could introduce gains g_1 and g_2 on the forward and backward channels between Alice and Bob and introduce different noise variances:*

$$\begin{aligned} Y_{2,i} &= \sqrt{g_1}X_{1,i} + N_{1,i}, \\ Y_{1,i} &= \sqrt{g_2}X_{2,i} + N_{2,i}, \\ Z_i &= \sqrt{h_1}X_{1,i} + \sqrt{h_2}X_{2,i} + N_{e,i}, \end{aligned}$$

where $\{N_{1,i}\}_{i\geqslant 1}$, $\{N_{2,i}\}_{i\geqslant 1}$, and $\{N_{e,i}\}_{i\geqslant 1}$ are i.i.d. zero-mean Gaussian noises with variances σ_1^2, σ_2^2, and σ_e^2, respectively. Nevertheless, by scaling the signals as $\tilde{Y}_{1,i} = (1/\sigma_1)Y_{1,i}$, $\tilde{X}_{1,i} = \sqrt{g_1/\sigma_1^2}X_{1,i}$, $\tilde{Y}_{2,i} = (1/\sigma_2)Y_{2,i}$, $\tilde{X}_{2,i} = \sqrt{g_2/\sigma_2^2}X_{2,i}$, and $\tilde{Z}_i = (1/\sigma_e)Z_i$, introducing channel gains $\tilde{h}_1 = h_1\sigma_1^2/(g_1\sigma_e^2)$ and $\tilde{h}_2 = h_2\sigma_2^2/(g_2\sigma_e^2)$, and redefining the power constraints as $\tilde{P}_1 = (g_1/\sigma_1^2)P_1$ and $\tilde{P}_2 = (g_2/\sigma_2^2)P_2$, one can check that we can always revert back to the more tractable model given in (8.1).

The orthogonality of the channels between Alice and Bob implicitly relies on the assumption that any self-interference can be perfectly canceled out, and the form of the interference at the eavesdropper's location is valid provided that all signals are synchronized. Realistically, interfering signals are unlikely to be perfectly synchronized when they reach the eavesdropper; nevertheless, the effect of mis-synchronization can be partly included in the magnitudes of the coefficients h_1 and h_2.

The ability to achieve secure communications over the TWWTC relies once more on the use of stochastic encoders. As was done in Chapter 3 and Chapter 4, it is convenient to explicitly introduce the randomness in the encoder and to assume that Alice has access to the realizations of a DMS $(\mathcal{R}_1, p_{R_1})$ while Bob has access to the realization of a DMS $(\mathcal{R}_2, p_{R_2})$; the DMSs are independent of each other and of the noise in the channel. For clarity in the definitions, we also denote the alphabets in which the symbols X_1, X_2, Y_1, and Y_2 take their values by the letters $\mathcal{X}_1$, $\mathcal{X}_2$, $\mathcal{Y}_1$, and $\mathcal{Y}_2$, respectively. A generic code for the two-way wiretap channel is then defined as follows.

Definition 8.1. *A* $(2^{nR_1}, 2^{nR_2}, n,)$ *code* $\mathcal{C}_n$ *for the TWWTC consists of*

- *two message sets,* $\mathcal{M}_1 = [\![1, 2^{nR_1}]\!]$ *and* $\mathcal{M}_2 = [\![1, 2^{nR_2}]\!]$*;*
- *two sources of local randomness,* $(\mathcal{R}_1, p_{R_1})$ *and* $(\mathcal{R}_2, p_{R_2})$*;*
- *two sequences of encoding functions,* $f_{1,i} : \mathcal{M}_1 \times \mathcal{R}_1 \times \mathcal{Y}_1^{i-1} \to \mathcal{X}_1$ *and* $f_{2,i} : \mathcal{M}_2 \times \mathcal{R}_2 \times \mathcal{Y}_2^{i-1} \to \mathcal{X}_2$ *for* $i \in [\![1, n]\!]$*, which generate symbols based on the message to transmit, local randomness, and previous observations;*
- *two decoding functions,* $g_1 : \mathcal{Y}_1^n \times \mathcal{R}_1 \times \mathcal{M}_1 \to \mathcal{M}_2 \cup \{?\}$ *and* $g_2 : \mathcal{Y}_2^n \times \mathcal{R}_2 \times \mathcal{M}_2 \to \mathcal{M}_1 \cup \{?\}$.

Note that the DMSs $(\mathcal{R}_1, p_{R_1})$ and $(\mathcal{R}_2, p_{R_2})$ can be optimized as part of the code design. The $(2^{nR_1}, 2^{nR_2}, n,)$ code $\mathcal{C}_n$ is assumed known by Alice, Bob, and Eve. We also assume that the messages $M_1 \in \mathcal{M}_1$ and $M_2 \in \mathcal{M}_2$ are independent and uniformly distributed in their respective sets. The reliability performance of a $\mathcal{C}_n$ is then measured in terms of the probability of error

$$\mathbf{P}_e(\mathcal{C}_n) \triangleq \mathbb{P}\left[\hat{M}_2 \neq M_2 \text{ or } \hat{M}_1 \neq M_1 \middle| \mathcal{C}_n\right],$$

while its secrecy performance is measured in terms of the leakage

$$\mathbf{L}(\mathcal{C}_n) \triangleq \mathbb{I}(M_1 M_2; Z^n | \mathcal{C}_n h_1 h_2).$$

The conditioning on h_1 and h_2 reflects the fact that the channel gains are known to the eavesdropper; however, we write $\mathbb{I}(M_1 M_2; Z^n | \mathcal{C}_n)$ to simplify the notation.

Definition 8.2. *A rate pair* (R_1, R_2) *is achievable for the TWWTC if there exists a sequence of* $(2^{nR_1}, 2^{nR_2}, n,)$ *codes* $\{\mathcal{C}_n\}_{n \geqslant 1}$ *such that*

$$\lim_{n\to\infty} \mathbf{P}_e(\mathcal{C}_n) = 0 \qquad \textit{(reliability condition)}, \tag{8.2}$$

$$\lim_{n\to\infty} \frac{1}{n}\mathbf{L}(\mathcal{C}_n) = 0 \qquad \textit{(weak secrecy condition)}. \tag{8.3}$$

Note that the secrecy condition requires a vanishing information rate leaked to the eavesdropper for messages M_1 and M_2 *jointly*, which is a stronger requirement than a vanishing information rate for messages M_1 and M_2 individually. In fact, the chain rule of mutual information and the independence of messages M_1 and M_2 guarantee that

$$\begin{aligned}\mathbb{I}(M_1 M_2; Z^n | \mathcal{C}_n) &= \mathbb{I}(M_1; Z^n | \mathcal{C}_n) + \mathbb{I}(M_1; Z^n | M_2 \mathcal{C}_n)\\ &\geqslant \mathbb{I}(M_1; Z^n | \mathcal{C}_n) + \mathbb{I}(M_2; Z^n | \mathcal{C}_n).\end{aligned}$$

Therefore, messages are protected individually if they are protected jointly, but the converse need not be true.

We are interested in characterizing the entire region of achievable rate pairs (R_1, R_2). Unfortunately, it is rather difficult to obtain an exact characterization because, in principle, the coding schemes in Definition 8.1 can exploit both the interference of transmitted signals at the eavesdropper's terminal and feedback. To obtain some insight, we study instead several simpler strategies that partially *decouple* these two effects;

although these strategies are likely to be suboptimal, their analysis is more amenable and will yield achievable rate regions as a function of the channel parameters h_1 and h_2 and the power constraints P_1 and P_2. Specifically, we investigate the following coding schemes.

- **Cooperative jamming:** one of the legitimate parties sacrifices his entire rate to jam the eavesdropper; this strategy has little effect on the eavesdropper's SNR if the channel gains h_1 and h_2 are small, but jamming with noise can be implemented no matter what the values of h_1 and h_2 are, and does not require synchronization between the legitimate parties.
- **Coded cooperative jamming:** both Alice and Bob transmit coded information over the channel; if $h_1 \approx h_2$, codewords interfere with roughly the same strength at Eve's terminal, which allows Alice and Bob to increase their secure communication while communicating messages; however, if h_1 or h_2 is too large, this strategy is likely to be ineffective because the eavesdropper can probably decode the interfering signals.
- **Key-exchange:** one of the legitimate parties sacrifices part of its secure communication rate to exchange a secret key, which is later used by the other party to encrypt messages with a one-time pad; this is perhaps the simplest strategy that exploits feedback, but the key-distillation strategies described in Chapter 4 could also be adapted for the TWWTC.

As a benchmark for achievable secure communication rates, we consider the region achieved with a coding scheme in which Alice and Bob ignore the interference created at the eavesdropper's terminal and do not exploit the feedback allowed by the two-way nature of the channel. This is a special instance of the generic code in Definition 8.1, for which

- Alice has a single encoding function $f_1 : \mathcal{M}_1 \times \mathcal{R}_1 \to \mathcal{X}_1^n$ and a single decoding function $g_1 : \mathcal{Y}_1^n \to \mathcal{M}_2 \cup \{?\}$;
- Bob has a single encoding function $f_2 : \mathcal{M}_2 \times \mathcal{R}_2 \to \mathcal{X}_2^n$ and a single decoding function $g_2 : \mathcal{Y}_2^n \to \mathcal{M}_1 \cup \{?\}$.

To present all subsequent results concisely, we introduce the function

$$C : \mathbb{R}^+ \to \mathbb{R}^+ : x \mapsto \frac{1}{2}\log(1+x),$$

such that $C(x)$ represents the capacity of a Gaussian channel with received SNR x.

Proposition 8.1. *The rate region $\mathcal{R}_0$ defined by*

$$\mathcal{R}_0 \triangleq \left\{(R_1, R_2)\colon \begin{array}{l} 0 \leqslant R_1 < (C(P_1) - C(h_1 P_1))^+ \\ 0 \leqslant R_2 < (C(P_2) - C(h_2 P_2))^+ \end{array}\right\}$$

is achievable with independently designed wiretap codes that ignore feedback and the presence of interference at the eavesdropper's terminal.

Proof. Fix $\epsilon > 0$. We introduce the random variable $Z_{1,i} \triangleq \sqrt{h_1}X_{1,i} + N_{e,i}$, which represents the observation of an eavesdropper who cancels out Bob's interference $X_{2,i}$ perfectly, and we choose a rate R_1 that satisfies

$$0 \leqslant R_1 < (C(P_1) - C(h_1P_1))^+. \tag{8.4}$$

By Corollary 5.1, there exists a $(2^{nR_1}, n)$ code $\mathcal{C}_1$ for communication between Alice and Bob such that

$$\mathbb{P}\left[\hat{M}_1 \neq M_1 | \mathcal{C}_1\right] \leqslant \epsilon \quad \text{and} \quad \frac{1}{n}\mathbb{I}(M_1; Z_1^n | \mathcal{C}_1) \leqslant \epsilon. \tag{8.5}$$

Similarly, we introduce $Z_{2,i} \triangleq \sqrt{h_2}X_{2,i} + N_{e,i}$, which represents the observation of an eavesdropper who cancels out Alice's interference $X_{1,i}$ perfectly, and we choose a rate R_2 that satisfies

$$0 \leqslant R_2 < (C(P_2) - C(h_2P_2))^+. \tag{8.6}$$

Again, Corollary 5.1 ensures the existence of a $(2^{nR_2}, n)$ code $\mathcal{C}_2$ for communication between Bob and Alice such that

$$\mathbb{P}\left[\hat{M}_2 \neq M_2 | \mathcal{C}_2\right] \leqslant \epsilon \quad \text{and} \quad \frac{1}{n}\mathbb{I}(M_2; Z_2^n | \mathcal{C}_2) \leqslant \epsilon. \tag{8.7}$$

The pair of codes $(\mathcal{C}_1, \mathcal{C}_2)$ defines a special instance of a $(2^{nR_1}, 2^{nR_2}, n)$ code $\mathcal{C}_n$ for the TWWTC, which we show achieves the rate pair (R_1, R_2) in the sense of Definition 8.2. By the union bound, (8.5), and (8.7), we obtain

$$\begin{aligned}\mathbb{P}\left[\hat{M}_1 \neq M_1 \text{ or } \hat{M}_2 \neq M_2 | \mathcal{C}_1\mathcal{C}_2\right] &\leqslant \mathbb{P}\left[\hat{M}_1 \neq M_1 | \mathcal{C}_1\right] + \mathbb{P}\left[\hat{M}_2 \neq M_2 | \mathcal{C}_2\right] \\ &\leqslant \delta(\epsilon).\end{aligned}$$

In addition, the information rate leaked to the eavesdropper about messages M_1 and M_2 can be bounded as

$$\begin{aligned}\frac{1}{n}\mathbb{I}(M_1M_2; Z^n | \mathcal{C}_1\mathcal{C}_2) &= \frac{1}{n}\mathbb{I}(M_1; Z^n | \mathcal{C}_1\mathcal{C}_2) + \frac{1}{n}\mathbb{I}(M_2; Z^n | M_1\mathcal{C}_1\mathcal{C}_2) \\ &\leqslant \frac{1}{n}\mathbb{I}(M_1; Z^nX_2^n | \mathcal{C}_1\mathcal{C}_2) + \frac{1}{n}\mathbb{I}(M_2; Z^nX_1^n | M_1\mathcal{C}_1\mathcal{C}_2) \\ &= \frac{1}{n}\mathbb{I}(M_1; Z_1^nX_2^n | \mathcal{C}_1\mathcal{C}_2) + \frac{1}{n}\mathbb{I}(M_2; Z_2^nX_1^n | M_1\mathcal{C}_1\mathcal{C}_2),\end{aligned}$$

where the last equality follows because of the one-to-one mapping between (Z_1^n, X_2^n) and (Z^n, X_2^n) and the one-to-one mapping between (Z_2^n, X_1^n) and (Z^n, X_1^n). Since X_2^n is independent of M_1 and Z_1^n by construction, notice that

$$\frac{1}{n}\mathbb{I}(M_1; Z_1^nX_2^n | \mathcal{C}_1\mathcal{C}_2) = \frac{1}{n}\mathbb{I}(M_1; Z_1^n | \mathcal{C}_1).$$

Similarly, since M_1 and X_1^n are independent of M_2 and Z_2^n by construction,

$$\frac{1}{n}\mathbb{I}(M_2; Z_2^nX_1^n | M_1\mathcal{C}_1\mathcal{C}_2) = \frac{1}{n}\mathbb{I}(M_2; Z_2^n | \mathcal{C}_2).$$

Therefore, by (8.5) and (8.7),

$$\begin{aligned}\frac{1}{n}\mathbb{I}(\mathrm{M}_1\mathrm{M}_2;\mathrm{Z}^n|\mathcal{C}_1\mathcal{C}_2) &\leqslant \frac{1}{n}\mathbb{I}(\mathrm{M}_1;\mathrm{Z}_1^n|\mathcal{C}_1) + \frac{1}{n}\mathbb{I}(\mathrm{M}_2;\mathrm{Z}_2^n|\mathcal{C}_2)\\ &\leqslant \delta(\epsilon).\end{aligned}$$

Since $\epsilon > 0$ is arbitrary, all the rate pairs (R_1, R_2) satisfying (8.4) and (8.6) are achievable over the TWWTC. □

Note that $\mathcal{R}_0$ has a square shape, but the region collapses to a segment as soon as $h_1 \geqslant 1$ or $h_2 \geqslant 1$; that is, as soon as the eavesdropper obtains a better SNR than either that of Alice or that of Bob.

8.2 Cooperative jamming

As a first attempt to increase secure communication rates, we analyze a communication strategy in which Alice and Bob take turns jamming Eve to reduce her SNR. Formally, we call *cooperative jamming code* (cooperative jamming for short) an instance of the generic code in Definition 8.1 such that

- there is only one party (say Alice) that transmits a message without relying on feedback; in other words, we consider a *single* encoding function $f_1 : \mathcal{M}_1 \times \mathcal{R}_1 \to \mathcal{X}_1^n$ and a *single* decoding function $g_2 : \mathcal{R}_2 \times \mathcal{Y}_2^n \to \mathcal{M}_1 \cup \{?\}$;
- the other party (Bob) transmits a jamming signal, which could depend on past channel observations; in other words, we consider a sequence of jamming functions $f_{2,i} : \mathcal{R}_2 \times \mathcal{Y}_2^{i-1} \to \mathcal{X}_2$ for $i \in [\![1, n]\!]$.

The probability of error of such a code reduces to $\mathbb{P}\left[\hat{\mathrm{M}}_1 \neq \mathrm{M}_1|\mathcal{C}_n\right]$. Notice that restricting codes to cooperative jamming is tantamount to considering the simplified channel model illustrated in Figure 8.2, which we refer to as the *cooperative jamming channel model*. Although a cooperative code does not exploit feedback, we emphasize that it is not a trivial code because the jamming signals are allowed to depend on past observation and can be quite sophisticated. Finally, note that the roles of Alice and Bob can be reversed, with Bob transmitting messages while Alice is jamming.

Proposition 8.2 (Tekin and Yener). *The region $\mathcal{R}_{\text{cj}}$ defined as*

$$\mathcal{R}_{\text{cj}} \triangleq \bigcup_{\alpha\in[0,1]} \left\{ (R_1, R_2)\colon \begin{array}{l} 0 \leqslant R_1 < \alpha\left(C(P_1) - C\left(\dfrac{h_1 P_1}{1 + h_2 P_2}\right)\right)^+ \\ 0 \leqslant R_2 < (1-\alpha)\left(C(P_2) - C\left(\dfrac{h_2 P_2}{1 + h_1 P_1}\right)\right)^+ \end{array} \right\}$$

is achievable with cooperative jamming codes.

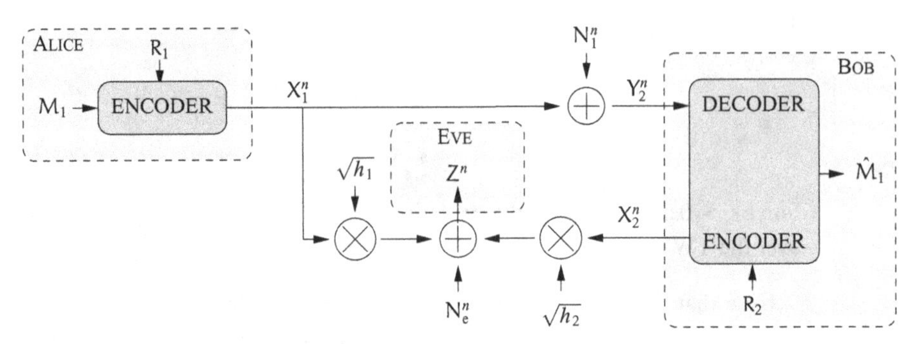

Figure 8.2 Cooperative jamming channel model. Bob helps Alice by jamming the eavesdropper and reducing its SNR.

Proof. Assume that Bob uses his entire power to jam Eve with i.i.d. Gaussian noise with variance[1] P_2. Effectively, this strategy transforms the two-way Gaussian wiretap channel into a one-way Gaussian wiretap channel from Alice to Bob characterized by the input–output relationships

$$\mathsf{Y}_{2,i} = \mathsf{X}_{1,i} + \mathsf{N}_{1,i},$$
$$\mathsf{Z}_i = \mathsf{X}_{1,i} + \mathsf{N}'_{\mathrm{e},i},$$

where $\mathsf{N}'_{\mathrm{e},i}$ is a zero-mean Gaussian random variable with variance $1 + h_2 P_2$. By Corollary 5.1, the secrecy capacity of this wiretap channel is

$$C_1 = \left(C(P_1) - C\left(\frac{h_1 P_1}{1 + h_2 P_2} \right) \right)^+.$$

Therefore, all rate pairs (R_1, R_2) with $R_2 = 0$ and $R_1 < C_1$ are achievable.

Similarly, if Alice uses her power to jam Eve with i.i.d. Gaussian noise with variance P_1, Bob effectively communicates with Alice over a Gaussian wiretap channel with secrecy capacity

$$C_2 = \left(C(P_2) - C\left(\frac{h_2 P_2}{1 + h_1 P_1} \right) \right)^+.$$

Therefore, all rate pairs (R_1, R_2) with $R_1 = 0$ and $R_2 < C_2$ are achievable.

The full region is obtained by time-sharing between these two modes of operation: during a fraction α of the time, Alice transmits with power P_1 while Bob jams with power P_2 and, during the remaining fraction $(1 - \alpha)$ of the time, Alice jams with power P_1 while Bob communicates with power P_2. □

Alice's and Bob's maximum secure communication rates in Proposition 8.2 are always higher than those obtained with the benchmark strategy; however, the jamming terminal ignores the symbols it receives and one can wonder whether adapting the jamming to

[1] Strictly speaking, transmitting i.i.d. Gaussian noise with power P_2 may violate the power constraint; nevertheless, if the variance is set to $P_2 - \epsilon$ for some arbitrary $\epsilon > 0$, then the probability of violating the constraints can be made arbitrarily small for n large enough and the results remain virtually unchanged.

past observations could yield higher secure communication rates. In an idealized case, if Bob had access to the signal X_1^n sent by Alice, he could tremendously reduce the eavesdropper's SNR by performing correlated jamming and partially canceling out X_1^n. In our setting, Bob only has causal and imperfect knowledge of X_1^n, but he might still be able to exploit the structure of the codewords. Perhaps surprisingly, it turns out that jamming Gaussian noise seems close to optimal. We establish this result precisely by deriving an upper bound on the secure communication rates achievable by Alice when Bob performs cooperative jamming.

Proposition 8.3 (Bloch). *The secure rates achieved by Alice with Bob performing cooperative jamming must satisfy*

$$R_1 \leqslant \max \min(\mathbb{I}(X_1; Y_2), \mathbb{I}(X_1; Y_2|ZX_2) + \mathbb{I}(X_2; Z|X_1)),$$

where the maximization is over random variables X_1, X_2, Y_2, *and* Z *with joint distribution* $p_{X_1X_2Y_2Z}$ *such that*

$$\forall (x_1, x_2, y_2, z) \in \mathcal{X}_1 \times \mathcal{X}_2 \times \mathcal{Y}_2 \times \mathcal{Z}$$
$$p_{X_1X_2Y_2Z}(x_1, x_2, y_2, z) = p_{Y_2Z|X_1X_2}(y_2, z|x_1, x_2)p_{X_1X_2}(x_1, x_2)$$

and such that $\mathbb{E}\left[X_1^2\right] \leqslant P_1$, $\mathbb{E}\left[X_2^2\right] \leqslant P_2$.

Proof. Let R_1 be an achievable rate with cooperative jamming, and let $\epsilon > 0$. For n sufficiently large, there exists a $(2^{nR_1}, n)$ code $\mathcal{C}_n$ such that $\mathbf{P}_e(\mathcal{C}_n) \leqslant \delta(\epsilon)$. In the following, we omit the condition on $\mathcal{C}_n$ to simplify the notation. Fano's inequality also ensures that $(1/n)\mathbb{H}(M_1|Y_2^n) \leqslant \delta(\epsilon)$; therefore,

$$\begin{aligned}
R_1 &\leqslant \frac{1}{n}\mathbb{H}(M_1) \\
&\leqslant \frac{1}{n}\mathbb{H}(M_1) - \frac{1}{n}\mathbb{H}(M_1|Y_2^n) + \delta(\epsilon) \\
&= \frac{1}{n}\mathbb{I}(M_1; Y_2^n) + \delta(\epsilon) \\
&\leqslant \frac{1}{n}\mathbb{I}(X_1^n; Y_2^n) + \delta(\epsilon) \\
&\leqslant \frac{1}{n}\sum_{j=1}^{n}\mathbb{I}(X_{1,j}; Y_{2,j}) + \delta(\epsilon).
\end{aligned} \tag{8.8}$$

We now develop a second upper bound that depends on the jamming input X_2 and the eavesdropper's observation Z. We do so by computing an upper bound for the secret-key capacity of the cooperative jamming channel model in Figure 8.2. This approach is motivated by two observations. First, any upper bound for the secret-key capacity is also an upper bound for the secrecy capacity because introducing public discussion cannot reduce achievable secrecy rates. Second, we already know how to obtain a single-letter upper bound for the secret-key capacity of a channel model from Theorem 4.8. The cooperative jamming channel model is slightly different from the channel model of Chapter 4 because not only Alice but also Bob has an input to the channel. Nevertheless, the proof of Theorem 4.8 can be adapted to account for this additional channel input.

A key-distillation strategy for the cooperative jamming channel model is formally defined as follows.

Definition 8.3. *A $(2^{nR}, n)$ key-distillation strategy $\mathcal{S}_n$ for a cooperative jamming channel model consists of*

- *a key alphabet $\mathcal{K} = [\![1, 2^{nR}]\!]$;*
- *an alphabet $\mathcal{A}$ used by Alice to communicate over the public channel;*
- *an alphabet $\mathcal{B}$ used by Bob to communicate over the public channel;*
- *a source of local randomness for Alice $(\mathcal{R}_1, p_{\mathrm{R}_1})$;*
- *a source of local randomness for Bob $(\mathcal{R}_2, p_{\mathrm{R}_2})$;*
- *an integer $r \in \mathbb{N}^*$ that represents the number of rounds of communication;*
- *a set of n distinct integers $\{i_j\} \subseteq [\![1, r]\!]$ that represents the rounds in which Alice and Bob transmit symbols over the channel;*
- *$r - n$ encoding functions $f_i : \mathcal{B}^{i-1} \times \mathcal{R}_1 \to \mathcal{A}$ for $i \in [\![1, r]\!] \setminus \{i_j\}_n$;*
- *$r - n$ encoding functions g_i for $i \in [\![1, r]\!] \setminus \{i_j\}_n$ of the form $g_i : \mathcal{Y}_2^j \times \mathcal{A}^{i-1} \times \mathcal{R}_2 \to \mathcal{B}$ if $i \in [\![i_j + 1, i_{j+1} - 1]\!]$;*
- *n functions $h_j : \mathcal{B}^{i_j-1} \times \mathcal{R}_1 \to \mathcal{X}_1$ for $j \in [\![1, n]\!]$ to generate channel inputs;*
- *n functions $h'_j : \mathcal{A}^{i_j-1} \times \mathcal{R}_2 \times \mathcal{Y}_2^{j-1} \to \mathcal{X}_2$ for $j \in [\![1, n]\!]$ to generate channel inputs;*
- *a key-distillation function $\kappa_{\mathrm{a}} : \mathcal{X}_1^n \times \mathcal{B}^r \times \mathcal{R}_1 \to \mathcal{K}$;*
- *a key-distillation function $\kappa_{\mathrm{b}} : \mathcal{Y}_2^n \times \mathcal{A}^r \times \mathcal{R}_2 \to \mathcal{K}$;*

and operates as follows:

- *Alice generates a realization r_x of her source of local randomness while Bob generates r_y from his;*
- *in round $i \in [\![1, i_1 - 1]\!]$, Alice transmits message $a_i = f_i\left(b^{i-1}, r_1\right)$ and Bob transmits message $b_i = g_i\left(a^{i-1}, r_2\right)$;*
- *in round i_j with $j \in [\![1, n]\!]$, Alice transmits symbol $x_{1,j} = h_j\left(b^{i_j-1}, r_1\right)$ and Bob transmits symbol $x_{2,j} = h'_j\left(a^{i_j-1}, r_2, y_2^{j-1}\right)$ over the channel; Bob and Eve observe the symbols $y_{2,j}$ and z_j, respectively.*
- *in round $i \in [\![i_j + 1, i_{j+1} - 1]\!]$, Alice transmits message $a_i = f_i\left(b^{i-1}, r_1\right)$ and Bob transmits message $b_i = g_i\left(y_2^j, a^{i-1}, r_2\right)$.*
- *after the last round, Alice computes a key $\mathsf{K} = \kappa_{\mathrm{a}}(x^n, b^r, r_1)$ and Bob computes a key $\hat{\mathsf{K}} = \kappa_{\mathrm{b}}(y^n, a^r, r_2)$.*

In addition, the vectors of channel inputs X_1^n and X_2^n should satisfy the power constraints $(1/n)\sum_{i=1}^n \mathbb{E}\left[\mathsf{X}_{1,i}^2\right] \leqslant P_1$ and $(1/n)\sum_{i=1}^n \mathbb{E}\left[\mathsf{X}_{2,i}^2\right] \leqslant P_2$.

By convention, we set $i_{n+1} \triangleq r + 1$, $i_0 = 0$, $\mathcal{A}^0 = 0$, and $\mathcal{B}^0 \triangleq 0$. As in Chapter 4, the indices $\{i_j\}_n$ and the sources of local randomness $(\mathcal{R}_1, p_{\mathrm{R}_1})$ and $(\mathcal{R}_2, p_{\mathrm{R}_2})$ can be optimized as part of the strategy. A rate R is an achievable key rate for the cooperative channel model if the conditions in Definition 4.3 are satisfied. If R is an achievable secret-key rate, we can follow the same steps as those leading to (4.20) and show that

$$
\begin{aligned}
R &\leqslant \mathbb{H}(\mathrm{K}) \\
&\leqslant \frac{1}{n}\mathbb{I}\left(\mathrm{R}_1\mathrm{X}_1^n;\mathrm{R}_2\mathrm{Y}_2^n|\mathrm{A}^r\mathrm{B}^r\mathrm{Z}^n\right)+\delta(\epsilon) \\
&= \frac{1}{n}\mathbb{I}\left(\mathrm{R}_1;\mathrm{R}_2\mathrm{Y}_2^n|\mathrm{A}^r\mathrm{B}^r\mathrm{Z}^n\right)+\delta(\epsilon),
\end{aligned}
$$

where the last equality follows because $\mathrm{X}_j = h_j(\mathrm{B}^{i_j-1}, \mathcal{R}_1)$. As in the proof of Theorem 4.8, we introduce the random variable Λ_j which represents the messages exchanged over the public channel between two successive uses of the channel:

$$
\begin{aligned}
\Lambda_0 &\triangleq \left(\mathrm{A}_1,\dots,\mathrm{A}_{i_1-1},\mathrm{B}_1,\dots,\mathrm{B}_{i_1-1}\right), \\
\Lambda_j &\triangleq \left(\mathrm{A}_{i_j+1}\dots\mathrm{A}_{i_{j+1}-1},\mathrm{B}_{i_j+1}\dots\mathrm{B}_{i_{j+1}-1}\right) \quad \text{for } j\in[\![1,n]\!].
\end{aligned}
$$

We then expand $\mathbb{I}\left(\mathrm{R}_1;\mathrm{R}_2\mathrm{Y}_2^n|\mathrm{A}^r\mathrm{B}^r\mathrm{Z}^n\right)$ as

$$
\begin{aligned}
&\mathbb{I}\left(\mathrm{R}_1;\mathrm{R}_2\mathrm{Y}_2^n|\mathrm{A}^r\mathrm{B}^r\mathrm{Z}^n\right) \\
&\quad = \mathbb{I}(\mathrm{R}_1;\mathrm{R}_2|\Lambda_0)+\sum_{j=1}^n\left[\mathbb{I}\left(\mathrm{R}_1;\Lambda_j|\mathrm{Z}^j\mathrm{Y}_2^j\mathrm{R}_2\Lambda_0\Lambda^{j-1}\right)-\mathbb{I}\left(\mathrm{R}_1;\Lambda_j|\mathrm{Z}^j\Lambda_0\Lambda^{j-1}\right)\right] \\
&\quad\quad +\sum_{j=1}^n\left[\mathbb{I}\left(\mathrm{R}_1;\mathrm{Y}_{2,j}\mathrm{Z}_j|\mathrm{Z}^{j-1}\mathrm{Y}_2^{j-1}\mathrm{R}_2\Lambda_0\Lambda^{j-1}\right)-\mathbb{I}\left(\mathrm{R}_1;\mathrm{Z}_j|\mathrm{Z}^{j-1}\Lambda_0\Lambda^{j-1}\right)\right]. \quad (8.9)
\end{aligned}
$$

As in (4.65), the first term in (8.9) satisfies $\mathbb{I}(\mathrm{R}_1;\mathrm{R}_2|\Lambda_0)=0$ by Lemma 4.2. In addition, the terms in the first sum of (8.9) satisfy

$$
\mathbb{I}\left(\mathrm{R}_1;\Lambda_j|\mathrm{Z}^j\mathrm{Y}_2^j\mathrm{R}_2\Lambda_0\Lambda^{j-1}\right)-\mathbb{I}\left(\mathrm{R}_1;\Lambda_j|\mathrm{Z}^j\Lambda_0\Lambda^{j-1}\right)\leqslant 0,
$$

as has already been proved for Theorem 4.8. The terms in the second sum of (8.9) can be rewritten as

$$
\begin{aligned}
&\mathbb{I}\left(\mathrm{R}_1;\mathrm{Y}_{2,j}\mathrm{Z}_j|\mathrm{Z}^{j-1}\mathrm{Y}_2^{j-1}\mathrm{R}_2\Lambda_0\Lambda^{j-1}\right)-\mathbb{I}\left(\mathrm{R}_1;\mathrm{Z}_j|\mathrm{Z}^{j-1}\Lambda_0\Lambda^{j-1}\right) \\
&\quad = \mathbb{H}\left(\mathrm{Y}_{2,j}\mathrm{Z}_j|\mathrm{Z}^{j-1}\mathrm{Y}_2^{j-1}\mathrm{R}_2\Lambda_0\Lambda^{j-1}\right)-\mathbb{H}\left(\mathrm{Y}_{2,j}\mathrm{Z}_j|\mathrm{Z}^{j-1}\mathrm{Y}_2^{j-1}\mathrm{R}_2\mathrm{R}_1\Lambda_0\Lambda^{j-1}\right) \\
&\quad\quad -\mathbb{H}\left(\mathrm{Z}_j|\mathrm{Z}^{j-1}\Lambda_0\Lambda^{j-1}\right)+\mathbb{H}\left(\mathrm{Z}_j|\mathrm{Z}^{j-1}\Lambda_0\Lambda^{j-1}\mathrm{R}_1\right).
\end{aligned}
$$

We can further simplify this expression by recalling that $\mathrm{X}_{1,j}=h_j(\mathrm{B}^{i_j-1},\mathrm{R}_1)$, $\mathrm{X}_{2,j}=h'_j(\mathrm{A}^{i_j-1},\mathrm{R}_2,\mathrm{Y}_2^{j-1})$, and

$$
\mathrm{R}_1\mathrm{R}_2\mathrm{Y}_2^{j-1}\mathrm{Z}^{j-1}\Lambda_0\Lambda^{j-1}\rightarrow\mathrm{X}_{1,j}\mathrm{X}_{2,j}\rightarrow\mathrm{Y}_{2,j}\mathrm{Z}_j
$$

forms a Markov chain. Using these properties, we obtain

$$
\mathbb{H}\left(\mathrm{Y}_{2,j}\mathrm{Z}_j|\mathrm{Z}^{j-1}\mathrm{Y}_2^{j-1}\mathrm{R}_2\mathrm{R}_1\Lambda_0\Lambda^{j-1}\right)=\mathbb{H}\left(\mathrm{Y}_{2,j}\mathrm{Z}_j|\mathrm{X}_{1,j}\mathrm{X}_{2,j}\right)
$$

and

$$
\mathbb{H}\left(\mathrm{Z}_j|\mathrm{Z}^{j-1}\Lambda_0\Lambda^{j-1}\mathrm{R}_1\right)=\mathbb{H}\left(\mathrm{Z}_j|\mathrm{X}_{1,j}\mathrm{Z}^{j-1}\Lambda_0\Lambda^{j-1}\mathrm{R}_1\right)\leqslant\mathbb{H}\left(\mathrm{Z}_j|\mathrm{X}_{1,j}\right).
$$

Therefore, we can bound the terms in the second sum of (8.9) by

$$\begin{aligned}
&\mathbb{H}\left(\mathrm{Y}_{2,j}\mathrm{Z}_j|\mathrm{Z}^{j-1}\mathrm{Y}_2^{j-1}\mathrm{R}_2\Lambda_0\Lambda^{j-1}\right) - \mathbb{H}(\mathrm{Y}_{2,j}\mathrm{Z}_j|\mathrm{X}_{1,j}\mathrm{X}_{2,j}) - \mathbb{H}(\mathrm{Z}_j|\mathrm{Z}^{j-1}\Lambda_0\Lambda^{j-1}) \\
&\quad + \mathbb{H}(\mathrm{Z}_j|\mathrm{X}_{1,j}) \\
&\quad = \mathbb{H}\left(\mathrm{Y}_{2,j}|\mathrm{Z}^j\mathrm{Y}_2^{j-1}\mathrm{R}_2\Lambda_0\Lambda^{j-1}\right) + \mathbb{H}\left(\mathrm{Z}_j|\mathrm{Z}^{j-1}\mathrm{Y}_2^{j-1}\mathrm{R}_2\Lambda_0\Lambda^{j-1}\right) \\
&\qquad - \mathbb{H}(\mathrm{Y}_{2,j}|\mathrm{Z}_j\mathrm{X}_{1,j}\mathrm{X}_{2,j}) - \mathbb{H}(\mathrm{Z}_j|\mathrm{X}_{1,j}\mathrm{X}_{2,j}) - \mathbb{H}(\mathrm{Z}_j|\mathrm{Z}^{j-1}\Lambda_0\Lambda^{j-1}) + \mathbb{H}(\mathrm{Z}_j|\mathrm{X}_{1,j}) \\
&\quad \leqslant \mathbb{H}(\mathrm{Y}_{2,j}|\mathrm{Z}_j\mathrm{X}_{2,j}) - \mathbb{H}(\mathrm{Y}_{2,j}|\mathrm{Z}_j\mathrm{X}_{1,j}\mathrm{X}_{2,j}) - \mathbb{H}(\mathrm{Z}_j|\mathrm{X}_{1,j}\mathrm{X}_{2,j}) + \mathbb{H}(\mathrm{Z}_j|\mathrm{X}_{1,j}) \\
&\quad = \mathbb{I}(\mathrm{X}_{1,j};\mathrm{Y}_{2,j}|\mathrm{Z}_j\mathrm{X}_{2,j}) + \mathbb{I}(\mathrm{Z}_j;\mathrm{X}_{2,j}|\mathrm{X}_{1,j}).
\end{aligned}$$

All in all, we obtain our second bound:

$$R \leqslant \frac{1}{n}\sum_{j=1}^{n}\left(\mathbb{I}(\mathrm{X}_{1,j};\mathrm{Y}_{2,j}|\mathrm{Z}_j\mathrm{X}_{2,j}) + \mathbb{I}(\mathrm{Z}_j;\mathrm{X}_{2,j}|\mathrm{X}_{1,j})\right) + \delta(\epsilon). \tag{8.10}$$

Finally, we introduce a random variable Q that is uniformly distributed on $[\![1,n]\!]$ and independent of all other random variables, and we define

$$\mathrm{X}_1 \triangleq \mathrm{X}_{1,\mathrm{Q}}, \qquad \mathrm{X}_2 \triangleq \mathrm{X}_{2,\mathrm{Q}}, \qquad \mathrm{Y}_2 \triangleq \mathrm{Y}_{2,\mathrm{Q}}, \qquad \text{and} \qquad \mathrm{Z} \triangleq \mathrm{Z}_\mathrm{Q}.$$

The transition probabilities from $\mathrm{X}_1\mathrm{X}_2$ to $\mathrm{Y}_2\mathrm{Z}$ are the original transition probabilities of the channel $p_{\mathrm{Y}_2\mathrm{Z}|\mathrm{X}_1\mathrm{X}_2}$ and, in addition, X_1 and X_2 should satisfy the power constraints $\mathbb{E}\left[\mathrm{X}_1^2\right] \leqslant P_1$ and $\mathbb{E}\left[\mathrm{X}_2^2\right] \leqslant P_2$. By substituting these random variables into (8.8) and (8.10), and using the fact that $\mathrm{Q} \rightarrow \mathrm{X}_1\mathrm{X}_1 \rightarrow \mathrm{Z}\mathrm{Y}_2$ forms a Markov chain, we obtain

$$R \leqslant \min(\mathbb{I}(\mathrm{X}_1;\mathrm{Y}_2), \mathbb{I}(\mathrm{X}_1;\mathrm{Y}_2|\mathrm{Z}\mathrm{X}_2) + \mathbb{I}(\mathrm{X}_2;\mathrm{Z}|\mathrm{X}_1)) + \delta(\epsilon).$$

Since ϵ can be chosen arbitrarily small and since we can optimize the distribution of inputs $\mathrm{X}_1\mathrm{X}_2$, we obtain the desired result. □

The optimization of the upper bound has to be performed over the random variables $(\mathrm{X}_1, \mathrm{X}_2)$ *jointly*; therefore the terms $\mathbb{I}(\mathrm{X}_1;\mathrm{Y}_2|\mathrm{Z}\mathrm{X}_2)$ and $\mathbb{I}(\mathrm{X}_2;\mathrm{Z}|\mathrm{X}_1)$ are not independent. Nevertheless, the result can still be understood intuitively as follows. The term $\mathbb{I}(\mathrm{X}_1;\mathrm{Y}_2|\mathrm{Z}\mathrm{X}_2)$ represents the secrecy rate achieved in the presence of an eavesdropper who would be able to cancel out the jamming signal X_2 perfectly, whereas the second term, $\mathbb{I}(\mathrm{X}_2;\mathrm{Z}|\mathrm{X}_1)$, represents the information that the eavesdropper has to obtain in order to "identify" the jamming signal and cancel it out. By specializing Proposition 8.3 further, we obtain the following result.

Proposition 8.4 (He and Yener). *The region of rates achievable with cooperative jamming is included in the region $\mathcal{R}_{\mathrm{cj}}^{\mathrm{out}}$ defined as*

$$\mathcal{R}_{\mathrm{cj}}^{\mathrm{out}} = \left\{ (R_1, R_2)\colon \begin{array}{l} 0 \leqslant R_1 \leqslant \min\left(C\left(\dfrac{P_1}{1+h_1P_1}\right) + C(h_2P_2), C(P_1)\right) \\ 0 \leqslant R_2 \leqslant \min\left(C\left(\dfrac{P_2}{1+h_2P_2}\right) + C(h_1P_1), C(P_2)\right) \end{array} \right\}.$$

Proof. We can further bound the result of Proposition 8.3 as

$$R_1 \leqslant \min(\max \mathbb{I}(X_1; Y_2), \max(\mathbb{I}(X_1; Y_2|ZX_2) + \mathbb{I}(X_2; Z|X_1))),$$

where the maximum is taken over all joint distributions $p_{X_1X_2}$ such that $\mathbb{E}\left[X_1^2\right] \leqslant P_1$ and $\mathbb{E}\left[X_2^2\right] \leqslant P_2$. The first term, $\mathbb{I}(X_1; Y_2)$, cannot exceed the capacity of the channel from Alice to Bob; therefore,

$$\max \mathbb{I}(X_1; Y_2) \leqslant C(P_1).$$

To bound the term $\max(\mathbb{I}(X_1; Y_2|ZX_2) + \mathbb{I}(X_2; Z|X_1))$, we introduce the random variables

$$Z_1 \triangleq \sqrt{h_1}X_1 + N_e \qquad \text{and} \qquad Z_2 \triangleq \sqrt{h_2}X_2 + N_e,$$

which represent the observations of an eavesdropper who would be able to cancel out either one of the signals X_1 and X_2. Then,

$$\begin{aligned}
\mathbb{I}(X_1; Y_2|ZX_2) &= \mathbb{h}(Y_2|ZX_2) - \mathbb{h}(Y_2|ZX_1X_2)\\
&= \mathbb{h}(Y_2|Z_1X_2) - \mathbb{h}(N_1)\\
&\leqslant \mathbb{h}(Y_2|Z_1) - \mathbb{h}(N_1)\\
&= \mathbb{h}(Y_2Z_1) - \mathbb{h}(Z_1) - \mathbb{h}(N_1).
\end{aligned} \tag{8.11}$$

Similarly,

$$\begin{aligned}
\mathbb{I}(X_2; Z|X_1) &= \mathbb{h}(Z|X_1) - \mathbb{h}(Z|X_1X_2)\\
&= \mathbb{h}(Z_2|X_1) - \mathbb{h}(N_e)\\
&\leqslant \mathbb{h}(Z_2) - \mathbb{h}(N_e).
\end{aligned} \tag{8.12}$$

We now show that the upper bounds (8.11) and (8.12) are maximized by choosing independent Gaussian random variables for X_1 and X_2. Since (8.11) depends only on X_1 and (8.12) depends only on X_2, the bounds are maximized with independent random variables. In addition, the term $\mathbb{h}(Z_2)$ is maximized if Z_2 is Gaussian, which is achieved if X_2 is Gaussian as well. To show that $\mathbb{h}(Y_2|Z_1)$ is also maximized with X_1 Gaussian, let $\text{LLSE}(Z_1)$ denote the linear least-square estimate of Y_2 based on Z_1 and let λ_{LLSE} denote the corresponding estimation error. Then,

$$\mathbb{h}(Y_2|Z_1) = \mathbb{h}(Y_2 - \text{LLSE}(Z_1)|Z_1) \leqslant \mathbb{h}(Y_2 - \text{LLSE}(Z_1)) \leqslant \frac{1}{2}\log(2\pi e\lambda_{\text{LLSE}}).$$

The inequalities are equalities if Y_2 and Z_1 are Gaussian, which is achieved if X_1 is Gaussian; hence, we can evaluate (8.11) and (8.12) with two independent random variables $X_1 \sim \mathcal{N}(0, P_1)$ and $X_2 \sim \mathcal{N}(0, P_2)$.

On substituting $X_2 \sim \mathcal{N}(0, P_2)$ into (8.12), we obtain directly

$$\mathbb{I}(X_2; Z|X_1) \leqslant C(h_2P_2). \tag{8.13}$$

A little more work is needed to bound (8.11). If $X_1 \sim \mathcal{N}(0, P_1)$, then the vector $(Y_2, Z_1)^T$ defined as

$$\begin{pmatrix} Y_2 \\ Z_1 \end{pmatrix} \triangleq \begin{pmatrix} 1 \\ \sqrt{h_1} \end{pmatrix} X_1 + \begin{pmatrix} 1 \\ 0 \end{pmatrix} N_1 + \begin{pmatrix} 0 \\ 1 \end{pmatrix} N_e$$

is also Gaussian with zero mean. Since X_1, N_1, and N_e are independent, its covariance matrix is

$$\mathbf{K}_{Y_1 Z_2} = \begin{pmatrix} 1 + P_1 & \sqrt{h_1} P_1 \\ \sqrt{h_1} P_1 & 1 + h_1 P_1 \end{pmatrix};$$

therefore,

$$\begin{aligned} \mathbb{I}(X_1; Y_2|ZX_2) &\leqslant \mathbb{h}(Y_2 Z_1) - \mathbb{h}(Z_1) - \mathbb{h}(N_1) \\ &= \log\left(2\pi e \left|\mathbf{K}_{Y_1 Z_2}\right|\right) - \frac{1}{2}\log(2\pi e(1 + h_1 P_1)) - \frac{1}{2}\log(2\pi e) \\ &= C(P_1 + h_1 P_1) - C(h_1 P_1) \\ &= C\left(\frac{P_1}{1 + h_1 P_1}\right). \end{aligned} \tag{8.14}$$

On combining (8.13) and (8.14), we obtain the second part of the upper bound for R_1:

$$\max(\mathbb{I}(X_1; Y_2|ZX_2) + \mathbb{I}(X_2; Z|X_1)) \leqslant C\left(\frac{P_1}{1 + h_1 P_1}\right) + C(h_2 P_2).$$

The bound for R_2 is obtained with identical steps by swapping the roles of Alice and Bob. □

The regions $\mathcal{R}_{cj}^{out}$ and $\mathcal{R}_{cj}$ cannot coincide because $\mathcal{R}_{cj}^{out}$ has a square shape whereas $\mathcal{R}_{cj}$ has a triangle shape (obtained with time-sharing). Nevertheless, Proposition 8.4 still provides a reasonably tight bound for the maximum secure rate achieved by either Alice or Bob with cooperative jamming over a wide range of channel parameters. Figure 8.3 illustrates typical regions $\mathcal{R}_{cj}^{out}$, $\mathcal{R}_{cj}$, and $\mathcal{R}_0$, for which the extremal points of $\mathcal{R}_{cj}$ are within a few tenths of a bit of the outer bound $\mathcal{R}_{cj}^{out}$.

Remark 8.2. *It might be somewhat surprising that an outer bound obtained by analyzing the secret-key capacity of a channel model turns out to be useful. Nevertheless, the reasonable tightness of the bound can be understood intuitively by remarking that the addition of a public channel of unlimited capacity is not a trivial enhancement. In fact, the public channel does not modify the randomness of the channel, which is the fundamental source of secrecy. As a matter of fact, we already know from Corollary 3.1 and Theorem 4.8 that the secrecy capacity and secret-key capacity of a degraded wire-tap channel are equal. In the case of cooperative jamming, the channel also exhibits some degradedness, which partly explains why the approach of Proposition 8.3 is useful.*

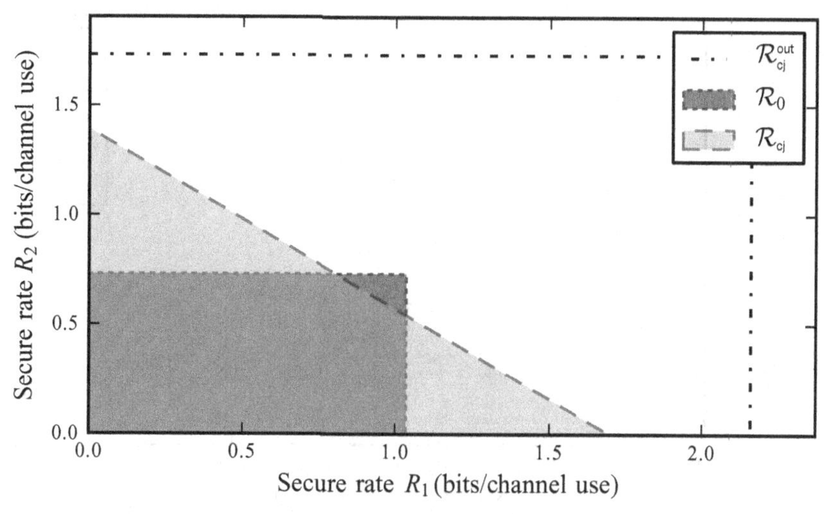

Figure 8.3 Regions $\mathcal{R}_0$, $\mathcal{R}_{\text{cj}}$, and $\mathcal{R}_{\text{cj}}^{\text{out}}$ for $h_1 = 0.2$, $h_2 = 0.3$, $P_1 = 20$, and $P_2 = 10$.

8.3 Coded cooperative jamming

Since cooperative jamming always improves the maximum secure communication rate of either Alice or Bob, cooperative jamming achieves rate pairs that are not achievable with the benchmark strategy and $\mathcal{R}_{\text{cj}} \not\subseteq \mathcal{R}_0$. Unfortunately, cooperative jamming forces either Alice or Bob to stop transmitting information. As a result, if the magnitude of the channel gains h_1 and h_2 is small ($h_1 \ll 1$ and $h_2 \ll 1$), the eavesdropper does not suffer from much interference and Alice and Bob might as well treat their channels as orthogonal wiretap channels; therefore, in general, $\mathcal{R}_0 \not\subseteq \mathcal{R}_{\text{cj}}$ either, and one cannot conclude that cooperative jamming performs strictly better than the benchmark strategy. The numerical example in Figure 8.3 is a specific case of such a situation.

To overcome this limitation, one can wonder whether Alice and Bob could achieve the effect of cooperative jamming while still communicating codewords. The interference of codewords might still have a detrimental effect on the eavesdropper, but without sacrificing the entire rate of either Alice or Bob. To emphasize that the underlying principle is similar to cooperative jamming but that the eavesdropper observes codeword interference, we call such schemes *coded cooperative jamming codes* (coded cooperative jamming for short). Formally, coded cooperative jamming is a specific instance of a code for the two-way wiretap channel such that

- there are two independent encoding functions, $f_1 : \mathcal{M}_1 \times \mathcal{R}_1 \to \mathcal{X}_1^n$ and $f_2 : \mathcal{M}_2 \times \mathcal{R}_2 \to \mathcal{X}_2^n$;
- there are two decoding functions, $g_1 : \mathcal{Y}_1^n \to \mathcal{M}_2 \cup \{?\}$ and $g_2 : \mathcal{Y}_2^n \to \mathcal{M}_1 \cup \{?\}$.

Note that coded cooperative jamming does not exploit feedback, but the codebooks used by Alice and Bob can be optimized jointly prior to transmission to maximize the effectiveness of codeword interference.

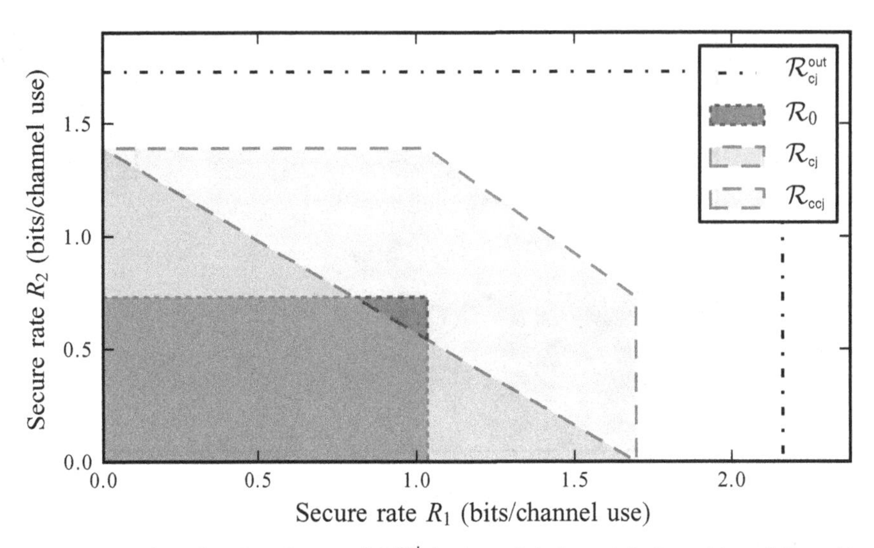

Figure 8.4 Regions $\mathcal{R}_0$, $\mathcal{R}_{\text{cj}}$, $\mathcal{R}_{\text{ccj}}$, and $\mathcal{R}_{\text{cj}}^{\text{out}}$ for $h_1 = 0.2$, $h_2 = 0.3$, $P_1 = 20$, and $P_2 = 10$.

Proposition 8.5 (Tekin and Yener). *The rate region $\mathcal{R}_{\text{ccj}}$ defined as*

$$\mathcal{R}_{\text{ccj}} \triangleq \left\{ (R_1, R_2)\colon \begin{array}{l} 0 \leqslant R_1 < \left(C(P_1) - C\left(\dfrac{h_1 P_1}{1 + h_2 P_2}\right)\right)^+ \\ 0 \leqslant R_2 < \left(C(P_2) - C\left(\dfrac{h_2 P_2}{1 + h_1 P_1}\right)\right)^+ \\ 0 \leqslant R_1 + R_2 < (C(P_1) + C(P_2) - C(h_1 P_1 + h_2 P_2))^+ \end{array} \right\}$$

is achievable with coded cooperative jamming.

Remark 8.3. *The region $\mathcal{R}_{\text{cj}}^{\text{out}}$ in Proposition 8.3 is also an outer bound for $\mathcal{R}_{\text{ccj}}$. In fact, $\mathcal{R}_{\text{cj}}^{\text{out}}$ has been derived for the case of arbitrary jamming signals, which includes codewords as a special case, and, in addition, $\mathcal{R}_{\text{cj}}^{\text{out}}$ has been obtained without requiring the jamming signals to be decoded by the non-jamming terminal. Removing this constraint cannot reduce the secure rates of the non-jamming terminal; therefore, $\mathcal{R}_{\text{cj}}^{\text{out}}$ is an outer bound for $\mathcal{R}_{\text{cj}}$.*

Before we prove Proposition 8.5, it is useful to analyze its implications. As illustrated in Figure 8.4, the shape of $\mathcal{R}_{\text{ccj}}$ is reminiscent of the capacity region of a multiple-access channel. This similarity is not fortuitous because the channel linking Alice and Bob to Eve is indeed a multiple-access channel, and we explicitly use it in the proof. By comparing Proposition 8.1 and Proposition 8.5, we see that the individual rate constraints on R_1 and R_2 are more stringent in $\mathcal{R}_0$ than they are in $\mathcal{R}_{\text{ccj}}$. In fact, if $(R_1, R_2) \in \mathcal{R}_0$ then the sum-rate satisfies

$$R_1 + R_2 < C(P_1) - C(h_1 P_1) + C(P_2) - C(h_2 P_2),$$

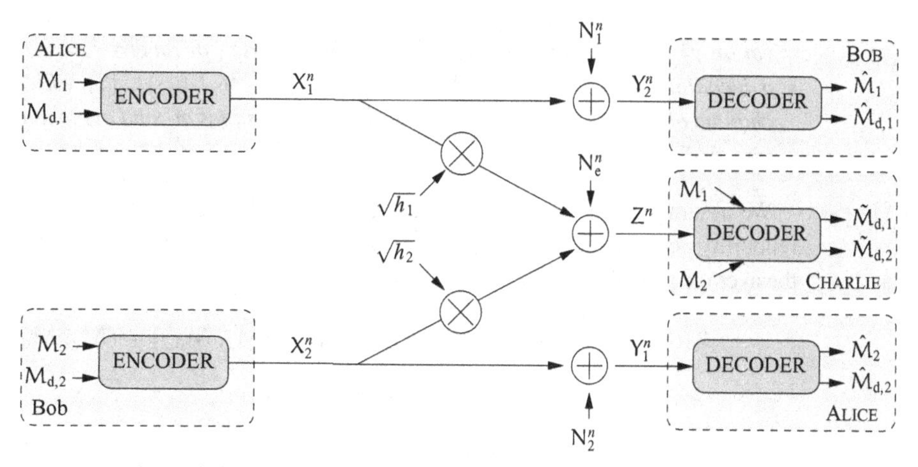

Figure 8.5 Enhanced channel for coded cooperative jamming.

which is more stringent than the sum-rate constraint in $\mathcal{R}_{\text{ccj}}$ because

$$C(P_1)+C(P_2)-C(h_1P_1+h_2P_2)=C(P_1)+C(P_2)-C(h_1P_1)-C\left(\frac{h_2P_2}{1+h_1P_1}\right)$$
$$\geqslant C(P_1)-C(h_1P_1)+C(P_2)-C(h_2P_2).$$

Hence, in contrast with $\mathcal{R}_{\text{cj}}$, we can conclude that $\mathcal{R}_0\subseteq\mathcal{R}_{\text{ccj}}$ for all channel parameters.

We now prove Proposition 8.5. The proof is based on a random-coding argument that combines the wiretap coding technique introduced in Section 3.4.1 with the multiple-access coding technique described in Section 2.3.2. As in Chapter 3, we start by constructing codes for an enhanced channel, which is illustrated in Figure 8.5. This channel enhances the original channel by

- introducing a virtual receiver, hereafter named Charlie, who observes the same output Z^n as Eve and has also access to the messages M_1 and M_2 through an error-free side channel;
- using a message $\mathsf{M}_{\text{d},1}$ with uniform distribution over $\llbracket 1,2^{nR_{\text{d},1}}\rrbracket$ in place of the source of local randomness $(\mathcal{R}_1,p_{\mathsf{R}_1})$ and another message $\mathsf{M}_{\text{d},2}$ with uniform distribution over $\llbracket 1,2^{nR_{\text{d},2}}\rrbracket$ in place of $(\mathcal{R}_2,p_{\mathsf{R}_2})$.

Formally, a code for the enhanced channel is defined as follows.

Definition 8.4. *A $(2^{nR_1},2^{nR_{\text{d},1}},2^{nR_2},2^{nR_{\text{d},2}},n)$ code $\mathcal{C}_n$ for the enhanced channel consists of*

- *four message sets, $\mathcal{M}_1=\llbracket 1,2^{nR_1}\rrbracket$, $\mathcal{M}_{\text{d},1}=\llbracket 1,2^{nR_{\text{d},1}}\rrbracket$, $\mathcal{M}_2=\llbracket 1,2^{nR_2}\rrbracket$, and $\mathcal{M}_{\text{d},2}=\llbracket 1,2^{nR_{\text{d},2}}\rrbracket$;*
- *two encoding functions, $f_1:\mathcal{M}_1\times\mathcal{M}_{\text{d},1}\to\mathcal{X}_1^n$ and $f_2:\mathcal{M}_2\times\mathcal{M}_{\text{d},2}\to\mathcal{X}_2^n$;*
- *a decoding function $g_1:\mathcal{Y}_1^n\to(\mathcal{M}_2\times\mathcal{M}_{2,\text{d}})\cup\{?\}$, which maps each channel observation y_1^n to a message pair $(\hat{m}_2,\hat{m}_{\text{d},2})\in\mathcal{M}_2\times\mathcal{M}_{\text{d},2}$ or an error message ?;*

- *a decoding function* $g_2 : \mathcal{Y}_2^n \to (\mathcal{M}_1 \times \mathcal{M}_{d,1}) \cup \{?\}$, *which maps each channel observation* y_2^n *to a message pair* $(\hat{m}_1, \hat{m}_{d,1}) \in \mathcal{M}_1 \times \mathcal{M}_{d,1}$ *or an error message* ?;
- *a decoding function* $h : \mathcal{Z}^n \times \mathcal{M}_1 \times \mathcal{M}_2 \to (\mathcal{M}_{d,1} \times \mathcal{M}_{d,2}) \cup \{?\}$, *which maps each channel observation* z^n *and the corresponding messages* m_1 *and* m_2 *to a message pair* $(\tilde{m}_1, \tilde{m}_2) \in \mathcal{M}_{d,1} \times \mathcal{M}_{d,2}$ *or an error message* ?.

We assume that all messages are uniformly distributed in their respective sets. The reliability performance of a $(2^{nR_1}, 2^{nR_{d,1}}, 2^{nR_2}, 2^{nR_{d,2}}, n)$ code $\mathcal{C}_n$ is measured in terms of the average probability of error

$$\mathbf{P}_e(\mathcal{C}_n) \triangleq \mathbb{P}\Big[(\hat{M}_1, \hat{M}_{d,1}) \neq (M_1, M_{d,1}) \text{ or } (\hat{M}_2, \hat{M}_{d,2}) \neq (M_2, M_{d,2})$$
$$\text{or } (\tilde{M}_{d,1}, \tilde{M}_{d,2}) \neq (M_{d,1}, M_{d,2}) | \mathcal{C}_n\Big].$$

Since $M_{d,1}$ and $M_{d,2}$ are dummy messages that correspond to a specific choice for sources of local randomness $(\mathcal{R}_1, p_{R_1})$ and $(\mathcal{R}_2, p_{R_2})$, a $(2^{nR_1}, 2^{nR_{d,1}}, 2^{nR_2}, 2^{nR_{d,2}}, n)$ code $\mathcal{C}_n$ for the enhanced channel is also a $(2^{nR_1}, 2^{nR_2}, n)$ coded cooperative jamming code $\mathcal{C}_n$ for the original channel; the probability of error over the original channel does not exceed the probability of error over the enhanced channel since

$$\mathbb{P}\Big[(M_1, M_2) \neq (\hat{M}_1, \hat{M}_2) | \mathcal{C}_n\Big] \leqslant \mathbf{P}_e(\mathcal{C}_n).$$

In addition, by virtue of Fano's inequality,

$$\frac{1}{n}\mathbb{H}(M_{d,1}, M_{d,2} | Z^n M_1 M_2 \mathcal{C}_n) \leqslant \delta(\mathbf{P}_e(\mathcal{C}_n)). \tag{8.15}$$

The above inequality is useful to evaluate the leakage to the eavesdropper guaranteed by $\mathcal{C}_n$ later on.

We begin by choosing two independent probability distributions p_{X_1} on $\mathcal{X}_1$ and p_{X_2} on $\mathcal{X}_2$. Let $0 < \epsilon < \mu_{X_1X_2YZ}$, where

$$\mu_{X_1X_2YZ} \triangleq \min_{(x_1,x_2,y,z)\in\mathcal{X}_1\times\mathcal{X}_2\times\mathcal{Y}\times\mathcal{Z}} p_{X_1}(x_1)p_{X_2}(x_2)p_{YZ|X_1X_2}(y, z|x_1, x_2),$$

and let $n \in \mathbb{N}^*$. Let $R_1 > 0$, $R_{d,1} > 0$, $R_2 > 0$, and $R_{d,2} > 0$ be rates to be specified later. We construct a $(2^{nR_1}, 2^{nR_{d,1}}, 2^{nR_2}, 2^{nR_{d,2}}, n)$ code for the enhanced channel as follows.

- *Codebook construction.* Construct a codebook $\mathcal{C}_1$ with $\lceil 2^{nR_1}\rceil\lceil 2^{nR_{d,1}}\rceil$ codewords labeled $x_1^n(m_1, m_{d,1})$ for $m_1 \in [\![1, 2^{nR_1}]\!]$ and $m_{d,1} \in [\![1, 2^{nR_{d,1}}]\!]$, by generating the symbols $x_{1,i}(m_1, m_{d,1})$ for $i \in [\![1, n]\!]$, $m_1 \in [\![1, 2^{nR_1}]\!]$, and $m_{d,1} \in [\![1, 2^{nR_{d,1}}]\!]$ independently according to p_{X_1}; similarly, construct a codebook $\mathcal{C}_2$ with $\lceil 2^{nR_2}\rceil\lceil 2^{nR_{d,2}}\rceil$ codewords labeled $x_2^n(m_2, m_{d,2})$ for $m_2 \in [\![2, 2^{nR_2}]\!]$ and $m_{d,2} \in [\![2, 2^{nR_{d,2}}]\!]$, by generating the symbols $x_{2,i}(m_2, m_{d,2})$ for $i \in [\![2, n]\!]$, $m_2 \in [\![2, 2^{nR_2}]\!]$, and $m_{d,2} \in [\![2, 2^{nR_{d,2}}]\!]$ independently according to p_{X_2}.
- *Alice's encoder* f_1. Given $(m_1, m_{d,1})$, transmit $x_1^n(m_1, m_{d,1})$.
- *Bob's encoder* f_2. Given $(m_2, m_{d,2})$, transmit $x_2^n(m_2, m_{d,2})$.
- *Alice's decoder* f_1. Given y_1^n, output $(\hat{m}_2, \hat{m}_{d,2})$ if it is the unique message pair such that $\big(x_2^n(\hat{m}_2, \hat{m}_{d,2}), y_1^n\big) \in \mathcal{T}_\epsilon^n(X_2Y_1)$; otherwise, output an error ?.

- *Bob's decoder* f_2. Given y_2^n, output $(\hat{m}_1, \hat{m}_{d,1})$ if it is the unique message pair such that $\left(x_1^n(\hat{m}_1, \hat{m}_{d,1}), y_2^n\right) \in \mathcal{T}_\epsilon^n(X_1Y_2)$; otherwise, output an error ?.
- *Charlie's decoder* f_2. Given z^n, m_1, and m_2, output $(\tilde{m}_{d,1}, \tilde{m}_{d,2})$ if it is the unique message pair such that $\left(x_1^n(m_1, \tilde{m}_{d,1}), x_2^n(m_2, \tilde{m}_{d,2}), z^n\right) \in \mathcal{T}_\epsilon^n(X_1X_2Z)$; otherwise, output an error ?.

The random variable that represents the randomly generated code $\mathcal{C}_n = (\mathcal{C}_1, \mathcal{C}_2)$ is denoted by C_n. By combining the arguments used in Section 2.3.2 for the MAC and in Section 3.4.1 for the WTC, we can show that, if

$$\begin{aligned}
R_1 + R_{d,1} &< \mathbb{I}(X_1; Y_2) - \delta(\epsilon),\\
R_2 + R_{d,2} &< \mathbb{I}(X_2; Y_1) - \delta(\epsilon),\\
R_{d,1} &< \mathbb{I}(X_1; Z|X_2) - \delta(\epsilon),\\
R_{d,2} &< \mathbb{I}(X_2; Z|X_1) - \delta(\epsilon),\\
R_{d,1} + R_{d,2} &< \mathbb{I}(X_1X_2; Z) - \delta(\epsilon),
\end{aligned}$$

then $\mathbb{E}[\mathbf{P}_e(C_n)] \leqslant \delta(\epsilon)$ for n large enough. In particular, for the choice[2] $X_1 \sim \mathcal{N}(0, P_1)$ and $X_2 \sim \mathcal{N}(0, P_2)$, the constraints become

$$\begin{aligned}
R_1 + R_{d,1} &< C(P_1) - \delta(\epsilon),\\
R_2 + R_{d,2} &< C(P_2) - \delta(\epsilon),\\
R_{d,1} &< C(h_1 P_1) - \delta(\epsilon),\\
R_{d,2} &< C(h_2 P_2) - \delta(\epsilon),\\
R_{d,1} + R_{d,2} &< C(h_1 P_1 + h_2 P_2) - \delta(\epsilon).
\end{aligned} \tag{8.16}$$

We now compute an upper bound for $\mathbb{E}[(1/n)\mathbf{L}(C_n)]$ by following steps similar to those in Section 3.4.1:

$$\begin{aligned}
\mathbb{E}\left[\frac{1}{n}\mathbf{L}(C_n)\right] &= \frac{1}{n}\mathbb{I}(M_1M_2; Z^n|C_n)\\
&= \frac{1}{n}\mathbb{I}(M_1M_2M_{d,1}M_{d,2}; Z^n|C_n) - \frac{1}{n}\mathbb{I}(M_{d,1}M_{d,2}; Z^n|M_1M_2C_n)\\
&= \frac{1}{n}\mathbb{I}\left(X_1^nX_2^n; Z^n|C_n\right) - \frac{1}{n}\mathbb{H}(M_{d,1}M_{d,2}|M_1M_2C_n)\\
&\quad + \frac{1}{n}\mathbb{H}(M_{d,1}M_{d,2}|Z^nM_1M_2C_n)\\
&= \frac{1}{n}\mathbb{I}\left(X_1^nX_2^n; Z^n|C_n\right) - \frac{1}{n}\mathbb{H}(M_{d,1}M_{d,2}|C_n)\\
&\quad + \frac{1}{n}\mathbb{H}(M_{d,1}M_{d,2}|Z^nM_1M_2C_n).
\end{aligned} \tag{8.17}$$

[2] See the proof of Theorem 5.1 for a discussion about the transition to continuous channels.

By construction,

$$\frac{1}{n}\mathbb{H}(\mathrm{M}_{\mathrm{d},1}\mathrm{M}_{\mathrm{d},2}|\mathrm{C}_n) \geqslant R_{\mathrm{d},1} + R_{\mathrm{d},2}. \tag{8.18}$$

Next, using (8.15) and $\mathbb{E}[\mathbf{P}_\mathrm{e}(\mathrm{C}_n)] \leqslant \delta(\epsilon)$, we obtain

$$\begin{aligned}\frac{1}{n}\mathbb{H}(\mathrm{M}_{\mathrm{d},1}\mathrm{M}_{\mathrm{d},2}|\mathrm{Z}^n\mathrm{M}_1\mathrm{M}_2\mathrm{C}_n) &= \sum_{\mathcal{C}_n} p_{\mathrm{C}_n}(\mathcal{C}_n)\,\frac{1}{n}\mathbb{H}(\mathrm{M}_{\mathrm{d},1}\mathrm{M}_{\mathrm{d},2}|\mathrm{Z}^n\mathrm{M}_1\mathrm{M}_2\mathcal{C}_n)\\ &\leqslant \delta(n) + \mathbb{E}[\mathbf{P}_\mathrm{e}(\mathrm{C}_n)](R_{\mathrm{d},1} + R_{\mathrm{d},2} + \delta(n))\\ &\leqslant \delta(\epsilon).\end{aligned} \tag{8.19}$$

Finally, note that $\mathrm{C}_n \rightarrow \mathrm{X}_1^n\mathrm{X}_2^n \rightarrow \mathrm{Z}^n$ forms a Markov chain; therefore,

$$\frac{1}{n}\mathbb{I}\left(\mathrm{X}_1^n\mathrm{X}_2^n;\mathrm{Z}^n|\mathrm{C}_n\right) \leqslant \frac{1}{n}\mathbb{I}\left(\mathrm{X}_1^n\mathrm{X}_2^n;\mathrm{Z}^n\right) = \mathbb{I}(\mathrm{X}_1\mathrm{X}_2;\mathrm{Z}). \tag{8.20}$$

On substituting (8.18), (8.19), and (8.20) into (8.17), we obtain

$$\begin{aligned}\mathbb{E}\left[\frac{1}{n}\mathbf{L}(\mathcal{C}_n)\right] &\leqslant \mathbb{I}(\mathrm{X}_1\mathrm{X}_2;\mathrm{Z}) - R_{\mathrm{d},1} - R_{\mathrm{d},2} + \delta(\epsilon)\\ &= C(h_1P_1 + h_2P_2) - R_{\mathrm{d},1} - R_{\mathrm{d},2} + \delta(\epsilon).\end{aligned}$$

Note that, for any rate pair (R_1, R_2) such that

$$\begin{aligned}R_1 &< \left(C(P_1) - C\left(\frac{h_1P_1}{1 + h_2P_2}\right)\right)^+,\\ R_2 &< \left(C(P_2) - C\left(\frac{h_2P_2}{1 + h_1P_1}\right)\right)^+,\\ R_1 + R_2 &< (C(P_1) + C(P_2) - C(h_1P_1 + h_2P_2))^+,\end{aligned} \tag{8.21}$$

we can choose a rate pair $(R_{\mathrm{d},1}, R_{\mathrm{d},2})$ that satisfies

$$\begin{aligned}R_1 + R_{\mathrm{d},1} &< C(P_1) - \delta(\epsilon),\\ R_2 + R_{\mathrm{d},2} &< C(P_2) - \delta(\epsilon),\\ R_{\mathrm{d},1} + R_{\mathrm{d},2} &= C(h_1P_1 + h_2P_2) - \delta(\epsilon),\end{aligned}$$

so that R_1, $R_{\mathrm{d},1}$, R_2, and $R_{\mathrm{d},2}$ satisfy the constraints in (8.16). This choice then guarantees that

$$\mathbb{E}\left[\frac{1}{n}\mathbf{L}(\mathcal{C}_n)\right] \leqslant \delta(\epsilon).$$

By applying the selection lemma to the random variable C_n and the functions $\mathbf{P}_\mathrm{e}$ and $\mathbf{L}$, we conclude that there exists a specific code $\mathcal{C}_n = (\mathcal{C}_1, \mathcal{C}_2)$ such that $\mathbf{P}_\mathrm{e}(\mathcal{C}_n) \leqslant \delta(\epsilon)$ and $\mathbf{L}(\mathcal{C}_n) \leqslant \delta(\epsilon)$; hence, the rate pairs (R_1, R_2) satisfying (8.21) are achievable.

Remark 8.4. *Although the codes $\mathcal{C}_1$ and $\mathcal{C}_2$ are generated according to independent distributions, note that the selection lemma selects the two codes* jointly*; therefore,* *the codes are used independently by Alice and Bob but optimized jointly to guarantee secrecy.*

8.4 Key-exchange

The last scheme we analyze for the TWWTC combines some of the ideas of cooperative jamming with a simple feedback mechanism that allows one user to *transfer* part of its secret rate to the other user. The motivation for this scheme is the situation in which one of the channel gains is high, say $h_1 \gg 1$. On the one hand, Eve observes Alice's signals with a high SNR, which limits Alice's secure rates even if Bob jams. On the other hand, Eve greatly suffers from Alice's jamming, which increases Bob's secure rates. The increase in Bob's secure communication rates can be so great that it becomes advantageous for Alice to jam and help Bob send her a secret key, which she later uses to encrypt her messages with a one-time pad. The strategy we just described leads to the region of achievable rates specified in the following proposition.

Proposition 8.6 (He and Yener). *The rate region $\mathcal{R}_{\text{fb}}$ defined by*

$$\mathcal{R}_{\text{fb}} \triangleq \bigcup_{\alpha\in[0,1]} \left\{(R_1, R_2): \begin{array}{l} 0 \leqslant R_1 < \alpha R_1^* \\ 0 \leqslant R_2 < (1-\alpha)R_2^* \end{array}\right\}$$

with

$$R_1^* = \max_{\beta\in[0,1]} \min\left(\beta C(P_1), \beta\left(C(P_1) - C\left(\frac{h_1P_1}{1+h_2P_2}\right)\right)^+ + (1-\beta)\left(C(P_2) - C\left(\frac{h_2P_2}{1+h_1P_1}\right)\right)^+\right)$$

and

$$R_2^* = \max_{\beta\in[0,1]} \min\left(\beta C(P_2), \beta\left(C(P_2) - C\left(\frac{h_2P_2}{1+h_1P_1}\right)\right)^+ + (1-\beta)\left(C(P_1) - C\left(\frac{h_1P_1}{1+h_2P_2}\right)\right)^+\right)$$

is achievable with cooperative jamming and key exchange.

Proof. We formalize the idea sketched earlier and we analyze a scheme that operates in two phases of $(1-\beta)n$ and βn channel uses, respectively, for some $\beta \in [0, 1]$. During the first $(1-\beta)n$ channel uses, Alice jams with Gaussian noise while Bob uses a wiretap code to transmit a secret key. Proposition 8.2 guarantees that Bob's key-transmission rate per n channel uses can be arbitrarily close to

$$R_{\text{f}} = (1-\beta)\left(C(P_2) - C\left(\frac{h_2P_2}{1+h_1P_1}\right)\right)^+. \tag{8.22}$$

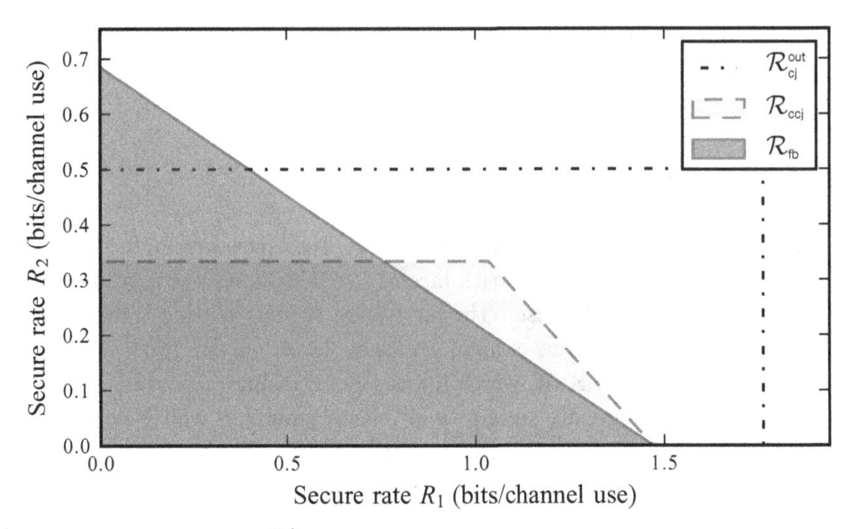

Figure 8.6 Regions $\mathcal{R}_{\text{ccj}}$, $\mathcal{R}_{\text{cj}}^{\text{out}}$, and $\mathcal{R}_{\text{fb}}$ for $h_1 = 0.2$, $h_2 = 1.3$, $P_1 = 20$, and $P_2 = 1$.

During the remaining βn channel uses, Bob jams with Gaussian noise while Alice transmits a secret message by combining a wiretap code and a one-time pad with the secret key sent by Alice as in Section 3.6.2. By Proposition 8.2 and Proposition 3.9, the secure transmission rate can be arbitrarily close to

$$\min\left(\beta C(P_1), \beta\left(C(P_1) - C\left(\frac{h_1 P_1}{1 + h_2 P_2}\right)\right)^+ + R_{\text{f}}\right). \tag{8.23}$$

On substituting (8.22) into (8.23) and optimizing over β we obtain the desired rate R_1^*. The rate R_2^* is obtained by swapping the roles of Alice and Bob, and the full region is obtained by time-sharing between these two modes of operation. □

The choice $\beta = 1$ in Proposition 8.6 eliminates the key-exchange phase of the scheme, which then reduces to the cooperative jamming in Section 8.2. Hence, we can conclude that $\mathcal{R}_{\text{cj}} \subseteq \mathcal{R}_{\text{fb}}$ for all channel parameters. The feedback scheme suffers from the same drawbacks as cooperative jamming, in that it forces either Alice or Bob to stop transmitting. Nevertheless, as illustrated in Figure 8.6, it is possible to find channel parameters for which the feedback scheme achieves rate pairs outside $\mathcal{R}_{\text{cj}}^{\text{out}}$; this result confirms that feedback and interference are fundamentally different mechanisms, thereby calling for coding schemes that combine the two techniques.

Remark 8.5. *The scheme described above is by no means the only possible scheme combining cooperative jamming and feedback. In fact, one can combine the key-exchange mechanism with coded cooperative jamming or adapt some of the key-distillation schemes described in Chapter 4. Unfortunately, it becomes rather difficult to obtain closed-form expressions for achievable rates.*

8.5 Bibliographical notes

Multi-user information-theoretic models with secrecy constraints have been the subject of intensive research. We provide a non-exhaustive but, we hope, representative list of references. Additional information can be found in the monograph of Liang *et al.* [147] and the book edited by Liu and Trappe [148].

The concept of coded cooperative jamming and cooperative jamming was introduced by Tekin and Yener for the two-way Gaussian wiretap channel and the K-user Gaussian multiple-access channel with multiple eavesdroppers [149]. The near-optimality of Gaussian noise for cooperative jamming over the two-way Gaussian wiretap channel was established by He and Yener [150] and by Bloch [151] using two different techniques. The proof developed in this chapter follows the approach of Bloch, and the property that $\mathbb{h}(Y|Z)$ is maximized for X Gaussian is due to Médard [152], an extension of which to MIMO channels was provided by Khisti and Wornell [83]. For the multiple-access channel, cooperative jamming with Gaussian noise and coded cooperative jamming with Gaussian codebooks is suboptimal, as was observed by Tang, Liu, Spasojević, and Poor [153, 154]. This suboptimality was confirmed by He and Yener, who recently showed that structured codebooks based on lattices outperform Gaussian codebooks, in general [155, 156]. The idea of cooperative jamming can be applied in many other settings, such as broadcast channels with cooperative receivers, as studied by Ekrem and Ulukus [157] and Bloch and Thangaraj [158], or from a secrecy-outage perspective in a wireless fading environment, as done by Vilela, Bloch, Barros, and McLaughlin [159]. Comprehensive discussions of cooperative jamming by He and Yener and by Ekrem and Ulukus can be found in [148, Chapter 4 and Chapter 7].

The role of interference in multi-user systems has been analyzed in various settings. For instance, Simeone and Yener [160] and Liang, Somekh-Baruch, Poor, Shamai, and Verdú [161] investigated cognitive channels with secrecy constraints, in which non-causal knowledge of other users' messages allows the legitimate user to interfere intelligently and to gain an advantage over the eavesdropper. In a slightly different setting, Mitrpant, Vinck, and Luo [162] investigated the combination of dirty-paper coding and wiretap coding and showed how knowledge of a non-causal interfering signal can be exploited for secrecy.

The importance of feedback for secure communications was originally highlighted in the context of secret-key agreement, which we discussed extensively in Chapter 4. Nevertheless, the study of secret-key capacity relies on the existence of a public channel with unlimited capacity, and recent works have analyzed the role of feedback without this assumption. For instance, Amariucai and Wei [163], Lai, El Gamal, and Poor [164], and Gündüz, Brown, and Poor [165] analyzed models in which the feedback takes place over a noisy channel but also interferes with the eavesdropper's observations. Ardestanizadeh, Franceschetti, Javidi, and Kim also analyzed a wiretap channel with rate-limited confidential feedback [34]. Note that most of these works can be viewed as special cases of the two-way wiretap channel. The key-exchange strategy for the two-way wiretap channel described in this chapter was proposed by He and Yener [150], and was combined with coded cooperative jamming by El Gamal, Koyluoglu, Youssef, and El Gamal [166]. All these works show that, in general, strategies relying on feedback perform strictly better than do strategies without feedback. The secrecy capacity or secrecy-capacity region of channels with feedback remains elusive, although some headway has been made by He and Yener [167] for the two-way wiretap channel.

The study of feedback is one facet of the more general problem of cooperation for secrecy. For instance, the trade-off between cooperation and security has been studied in the context of relay channels with confidential messages by Oohama [168], Lai and El Gamal [169], Yuksel and Erkip [170], and He and Yener [171, 172, 173]. Among the conclusions that can be drawn from these studies is the fact that relaying improves the end-to-end communication rate between a source and a destination even if the relay is to be kept ignorant of the messages. An overview of "cooperative secrecy" by Ekrem and Ulukus can be found in [148, Chapter 7]

The generalization of the broadcast channel with confidential messages to multiple receivers and multiple eavesdroppers was studied by Khisti, Tchamkerten, and Wornell [174] as well as by Liang, Poor, and Shamai [85]. These results can be treated as special cases of the compound wiretap channels investigated by Liang, Kramer, Poor, and Shamai [175] and Bloch and Laneman [26]. Note that compound channels are relevant in practice because they offer a way of modeling the uncertainty that one might have about the actual channel to the eavesdropper. The generalization of the source models and channel models for secret-key agreement to multiple terminals has been investigated by Csiszár and Narayan [176, 177, 178] as well as by Nitinawarat, Barg, Narayan, Ye, and Reznik [179, 180].

In a slightly different spirit, the impact of security constraints in a network has been investigated by Haenggi [181] and Pinto, Barros, and Win [182] using tools from stochastic geometry. Secure communication in deterministic arbitrary networks has also been investigated by Perron, Diggavi, and Telatar [183].

9 Network-coding security

Many of the applications of classical coding techniques can be found at the physical layer of contemporary communication systems. However, coding ideas have recently found their way into networking research, most strikingly in the form of algebraic codes for networks. The existing body of work on network coding ranges from determinations of the fundamental limits of communication networks to the development of efficient, robust, and secure network-coding protocols. This chapter provides an overview of the field of network coding with particular emphasis on how the unique characteristics of network codes can be exploited to achieve high levels of security with manageable complexity. We survey network-coding vulnerabilities and attacks, and compare them with those of state-of-the-art routing algorithms. Some emphasis will be placed on active attacks, which can lead to severe degradation of network-coded information flows. Then, we show how to leverage the intrinsic properties of network coding for information security and secret-key distribution, in particular how to exploit the fact that nodes observe algebraic combinations of packets instead of the data packets themselves. Although the prevalent design methodology for network protocols views security as something of an add-on to be included after the main communication tasks have been addressed, we shall contend that the special characteristics of network coding warrant a more comprehensive approach, namely one that gives equal importance to security concerns. The commonalities with code constructions for physical-layer security will be highlighted and further investigated.

9.1 Fundamentals of network coding

The main concept behind network coding is that data throughput and network robustness can be considerably improved by allowing the intermediate nodes in a network to mix different data flows through algebraic combinations of multiple datagrams. This key idea, which clearly breaks with the ruling store-and-forward paradigm of current message-routing solutions, is illustrated in Figure 9.1. To exchange messages a and b, nodes A and B must route their packets through node S. Clearly, the traditional scheme shown on top would require four transmissions. However, if S is allowed to perform network coding with simple exclusive-or (XOR) operations, as illustrated in the lower diagram, $a \oplus b$ can be sent in a single broadcast transmission (instead of one transmission with b followed by another one with a). By combining the received data with the stored

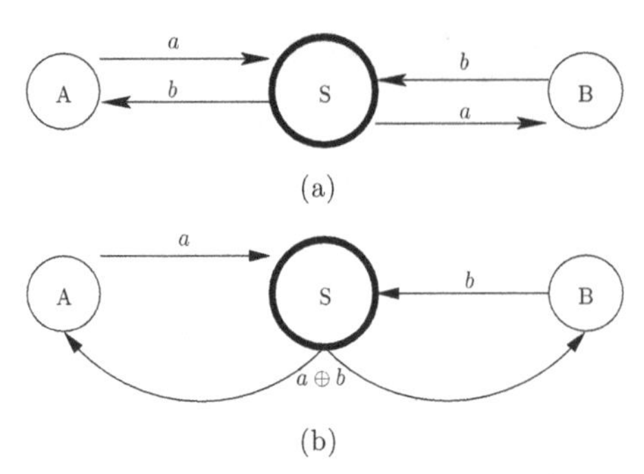

Figure 9.1 A typical wireless network-coding example.

message, A, which possesses a, can recover b, and B can recover a using b. Consequently, in this scenario, network coding saves one transmission (thus saving energy) and one time slot (thus reducing the delay). More sophisticated network-coding protocols view packets as a collection of symbols from a particular finite field and forward linear combinations of these symbols across the network, thus leveraging basic features of linear codes such as erasure-correction capability and well-understood encoding and decoding algorithms.

The resulting techniques seem particularly useful for highly volatile networks, such as mobile ad-hoc networks, sensor networks, and peer-to-peer communications, where stringent constraints due to power restrictions, limited computation capabilities, or unpredictable user dynamics can be countered by broadcasting encoded packets to multiple nodes simultaneously until the destination has enough degrees of freedom to decode and recover the original data, as illustrated by the example in Figure 9.1.

Using information-theoretic reasoning, it is possible to prove that the multicast capacity of a network is equal to the minimum of the maximum flows between the source and any of the individual destinations. Most importantly, routing alone is in general not sufficient to achieve this fundamental limit – intermediate nodes are required to mix the data units they receive from their neighbors using non-trivial coding operations.

The intuition behind this result is well illustrated by the butterfly network shown in Figure 9.2, where each edge is assumed to have unitary capacity. If node 1 wishes to send a multicast flow to sinks 6 and 7 at the max-flow min-cut bound, which in this case is 2, the only way to overcome the bottleneck between nodes 4 and 5 is for node 4 to combine the incoming symbols through an XOR operation. Sinks 6 and 7 can then use the symbols they receive directly from nodes 2 and 3, respectively, in order to reverse this XOR operation and reconstruct the desired multicast flow.

It has also been shown that linear codes are sufficient to achieve the multicast capacity, yielding the algebraic framework for network coding that has since fueled a strong surge in network-coding research.

Although establishing the information-theoretic limits of communication networks with multiple unicast or multicast sessions still seems a distant goal, there is reasonable evidence that network coding allows a trade-off of communication versus computational costs. Furthermore, network coding brings noticeable benefits in terms of throughput,

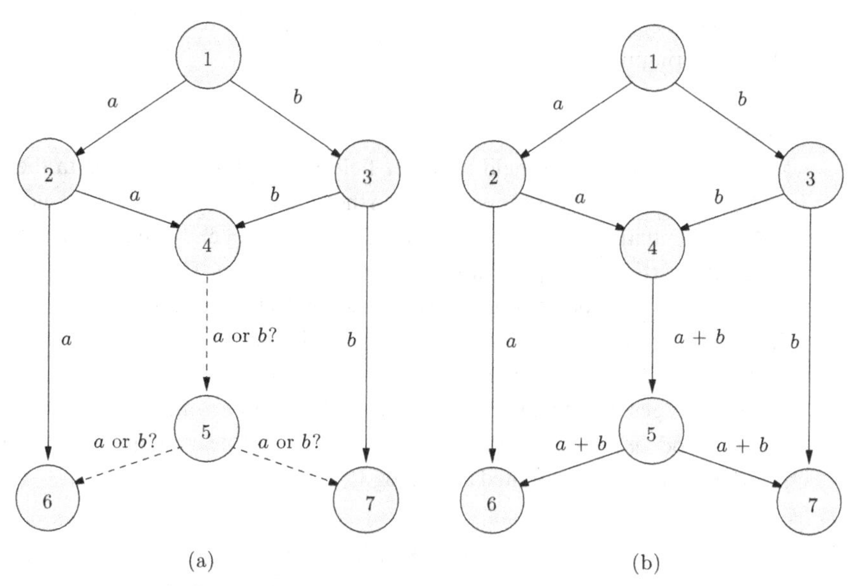

Figure 9.2 Canonical network-coding example.

reliability, and fault tolerance in a variety of relevant networking scenarios beyond the single multicast case, including wireless broadcast and peer-to-peer systems. In particular, random linear network coding (RLNC) provides a fully distributed methodology for network coding, whereby each node in the network selects independently and randomly a set of coefficients and uses them to form linear combinations of the data symbols (or packets) it receives. These linear combinations are then sent over the outgoing links together with the coding coefficients until the receivers are able to decode the original data using Gaussian elimination.

9.2 Network-coding basics

Network coding with XOR operations

In the simplest form of network coding all operations are carried out in the binary field. In more practical terms this implies that packets are mixed (and unmixed) by means of straightforward XOR operations. The basic principle and the underlying advantages of this low-complexity approach are illustrated in Figure 9.1. Suppose that, in a wireless network, nodes A and B must route their packets through node S, in order to exchange messages a and b. S sends $a \oplus b$ in a single broadcast transmission (see Figure 9.1(b)). By combining the received data with the stored message, A, which possesses packet a, can recover b, and B can recover a using b, in both cases with one XOR operation.

Naturally, this idea can be easily extended to scenarios with more than three terminals. A node that has several packets in its buffer can choose judiciously how to combine the packets for the next broadcast transmission. This decision may depend on several factors, including the delay or bandwidth requirements, the available knowledge about the communication channels, the network topology, and feedback about the buffer states of neighboring nodes.

Linear network coding

Moving from the binary field to larger field sizes allows the nodes in the network to perform more sophisticated operations on the incoming packets. In the standard network-coding formulation the network is represented as an acyclic directed graph $G = (V, E)$, where the vertices V represent the terminals, the edges E correspond to the available communication links, and each data unit is represented by a symbol from a predefined finite field F with $q = 2^m$ for some integer m. Suppose now that node v has in its buffer a collection of symbols denoted $X_1, \ldots, X_K$ and receives symbols $Y_1, \ldots, Y_L$ from its incoming edges. In linear network coding, the next data symbol Y_e to be transmitted to a different node u on the outgoing edge $e = (v, u)$ is a linear combination of the available symbols and can be computed according to

$$Y_e = \sum_{i=1}^{K} \alpha_i X_i + \sum_{j=1}^{L} \beta_j Y_j,$$

where $\alpha_i, \beta_j \in F$ are the linear coefficients used to encode the data. Since all of the operations are modulo operations over a finite field, mixing packets through linear network coding does not increase the packet size. To recover the transmitted symbols, the destination node waits until it has enough independent linear combinations (or degrees of freedom) and then performs Gaussian elimination to solve the resulting linear system.

An important property of linear network coding is that for a fixed network topology it is possible to find the optimal network code for a multicast session by means of a polynomial-time algorithm that establishes the exact operations that each node has to carry out. This is in sharp contrast with the multicast routing problem (without coding), which is well known to be NP-hard. An important consequence of the algorithm is that it allows us to bound the alphabet size for the network code or, equivalently, the required bandwidth per link.

Random network coding

If the network is large or highly dynamic, computing the network code in a centralized fashion may quickly become unfeasible. Fortunately, it is possible to perform network coding in a fully distributed way, by allowing the nodes to choose their linear coefficients independently and uniformly at random over all elements of the finite field F. This approach achieves the multicast capacity of the network, provided that the field size is sufficiently large. From a more practical point of view, random linear network coding can be implemented effectively in packet-oriented networks by (1) segmenting the payload into data symbols that can be viewed as elements of F, (2) computing the required linear combinations on a symbol-by-symbol basis using random coefficients, (3) placing the resulting symbols on the payload of the outgoing packet, and (4) updating the header with the random coefficients required for decoding.

Table 9.1 Network coding in the layered architecture

Layer	Network-coding (NC) examples
Application	Random NC in overlay networks, distributed storage, content distribution, peer-to-peer applications
Transport	Flow control with special ACKs, reliable NC multicast with feedback, combined NC and back-pressure algorithms
Network	Deterministic NC, random NC with directed diffusion
Link	Opportunistic NC, wireless multicast with feedback
Physical	Analog NC

Upon receiving a sufficient number of packets, each destination node can obtain the coding matrix from the headers of the incoming packets and recover the original data symbols. Notice that, due to the mixing operations, each packet loses its individuality in the sense that the information of a single original packet is spread over multiple instances. Thus, as long as there are enough degrees of freedom to decode the original symbols, the destination is not particularly concerned with obtaining a specific packet.

9.3 System aspects of network coding

Having covered some of the basic principles of network coding, we are now ready to discuss their application in the layered architecture of the dominant communication networks. With this goal in mind, we shall pursue a top-down approach, from the application down to the physical layer, as summarized in Table 9.1.

The application layer is particularly well suited as a first arena for the development of network-coding protocols. Implementing the required software at the outer edge of the network means that neither the routing and medium-access protocols nor the routers need to be modified or replaced. The routers can be kept exactly as they are. Naturally, the simplified protocol design at the application layer comes at the price of discarding any throughput and robustness benefits that network coding may bring when applied at the lower layers of the protocol stack.

On the other hand, the application layer allows us to define overlay network topologies that are particularly useful in conjunction with random network-coding protocols. One of the real-life applications that exploits these synergies is the Microsoft Secure Content Distribution software, also known as Avalanche. Here, each node wishing to download a large file contacts a number of nodes through a peer-to-peer overlay network and collects linear combinations of file fragments until it is able to recover the entire file. The results indicate that significant reductions in download time and increased robustness against sudden changes in the network are achieved.

Since the random-coding coefficients have to be placed in the header of the packets, random network-coding protocols introduce some extra overhead, which can vary a lot depending on the packet size, the field size, and the number of packets that are combined (also called a *generation*). This is illustrated in Table 9.2. For internet applications,

Table 9.2 Network coding overhead per packet

IP packet size (Bytes)	Generation size	Overhead (%)	
		$q = 2^8$	$q = 2^{16}$
1500	20	1.3	2.7
	50	3.3	6.7
	100	6.7	13.3
	200	13.3	26.7
5000	20	0.4	0.8
	50	1.0	2.0
	100	2.0	4.0
	200	4.0	8.0
8192	20	0.2	0.5
	50	0.6	1.2
	100	1.2	2.4
	200	2.4	4.8

where the IP packet size can be fairly large, the overhead is often negligible. However, in wireless communication networks the size of the generation and resulting overhead must be further restricted because packets are generally smaller. It is also worth mentioning that in error-prone channels the header requires extra protection by means of error-correction codes, which further increase the overhead incurred.

In distributed-storage applications with unreliable storage elements, random network coding increases data persistence by spreading linear combinations in different locations of the network. As long as there exist enough degrees of freedom stored in the network, the data can be decoded even in highly volatile environments.

The facts that the receiver has to wait until it receives enough packets to decode the data and that the decoding algorithm itself has non-negligible complexity ($O(n^3)$ for generation size n) have aroused some skepticism as to whether network coding can be used effectively in real-time applications, e.g. video streaming in peer-to-peer networks. Nevertheless, recent work on combining network coding with a randomized push algorithm shows promising results.

At the transport layer, the Transmission Control Protocol (TCP) typically fulfills two different tasks: (a) it guarantees reliable and in-order delivery of the transmitted data from one end of the network to its final destination by means of retransmissions, and (b) it implements flow and congestion control by monitoring the network state, as well as the buffer levels at the receiver. To fulfill its purpose, TCP relies on the acknowledgments sent by the receiver to decide which packets must be retransmitted and to estimate vital metrics pertaining to the round-trip delay and the congestion level of the network.

In the case of network-coding protocols, the transmitted packets are linear combinations of different packets within one generation, and therefore the acknowledgments sent by the receiving end may provide a very different kind of feedback. More specifically, instead of acknowledging specific packets, each destination node can send back requests for degrees of freedom that increase the dimension of its vector space and allow faster decoding. Upon receiving acknowledgments from the different receivers, the source node minimizes the delay by sending the most innovative linear combination, i.e. the one that is useful to most destination nodes. Without network coding, end-to-end reliability for multicast sessions with delay constraints is generally perceived as a very difficult problem.

Random network coding seems to perform well in combination with back-pressure techniques, which analyze the differences in queue size between the two ends of a link in order to implement distributed flow control and thus maintain the stability of the network.

The single path or route that lies at the core of common routing algorithms is arguably less clear when network coding is involved, because network coding seems most effective when there are multiple paths carrying information from the source to the destination. One way to proceed is to combine network coding with directed diffusion techniques, whereby nodes spread messages of interest in order to route the right data to the right nodes. Even if one opts to use a standard routing algorithm to forward the data, the topology-discovery phase, during which nodes broadcast link-state advertisements to all other nodes, is likely to benefit from the throughput gains of network-coding-based flooding protocols.

At the link layer, opportunistic network coding emerges as a promising technology, which has been implemented successfully in a real wireless mesh network. The main idea is that nodes in a wireless network can learn a lot about the flow of information to neighboring nodes simply by keeping their radios in promiscuous mode during idle times. On the basis of this acquired knowledge and the buffer state information sent by neighboring nodes, it is possible to solve a local multicast problem in the one-hop neighborhood of any given node. The XOR combination of queued packets that is useful to most neighbors can then be broadcast, thus minimizing the number of transmissions and saving valuable bandwidth and power. To reap the largest possible benefits from this approach, the medium-access protocols and scheduling algorithms at the link layer must be redesigned to allow opportunistic transmission.

Network-coding ideas have recently found their way to the physical layer by means of a simple communication scheme whereby two nodes send their signals simultaneously to a common relay over a wireless channel. The transmitted signals interfere in an additive way and the relay amplifies and broadcasts back the signal it received via its own antenna. This form of *analog* network coding, which had already been implemented using software-defined radios, further reduces the number of transmissions required for the nodes in Figure 9.1 to communicate effectively with each other. There is a strong relationship between the resulting schemes and the relay methods that have been proposed in the area of cooperative transmission among multiple receivers (see also Chapter 8).

9.4 Practical network-coding protocols

Although network coding is a fairly recent technology, there exist already protocol proposals regarding the use of network coding for higher throughput or robustness in

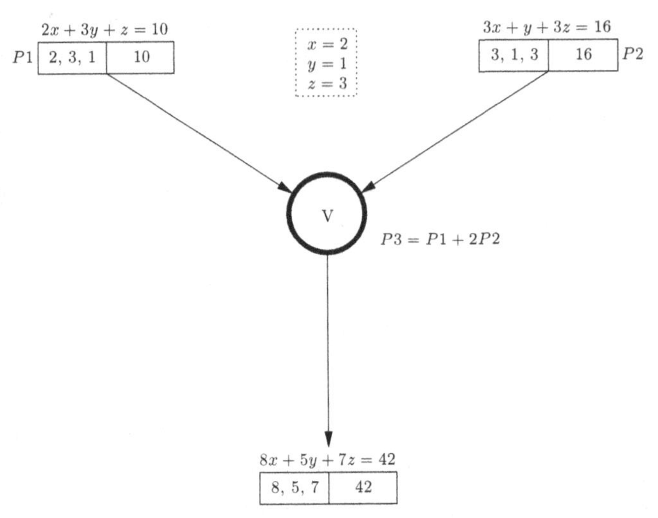

Figure 9.3 An instance of random linear network coding.

various applications and communication networks. From the point of view of network security, it is useful to divide network-coding protocols into two main classes:

(1) *stateless network-coding protocols*, which do not rely on any form of state information to decide when and how to mix different packets in the sender queue;
(2) *state-aware network-coding protocols*, which rely on partial or full network state information (for instance buffer states of neighboring nodes, network topology, or link costs) to compute a network code or determine opportunities to perform network coding in a dynamic fashion.

As will be demonstrated in Section 9.5, the security vulnerabilities of the protocols in the first and second classes are quite different from each other, most notably because the former require state information and node identification to be disseminated in the network and are thus vulnerable to a wide range of impersonation and control-traffic attacks.

Stateless network-coding protocols

As mentioned earlier, random linear network coding (RLNC) is a completely distributed methodology for combining different information flows and therefore leads to stateless protocol design. The basic principle is that each node in the network selects a set of coefficients independently and randomly and then sends linear combinations of the data symbols (or packets) it receives. Figure 9.3 illustrates the linear operations carried out at intermediate node v (using integers for simplicity). The symbols x, y, and z denote the native packets, which convey the information to be obtained at the receivers via Gaussian elimination. $P1$ and $P2$ arrive at intermediate node v in the network through its incoming links. $P3$, which is sent through the only outgoing link of node v, is the result of a random linear combination of $P1$ and $P2$ at node v, with chosen coefficients 1 and 2, respectively.

Recall that the *global encoding vector*, i.e. the matrix of coefficients used to encode the packets, is sent along in the packet header to ensure that the end receivers are capable of decoding the original data. Specifically, it was shown that, if the coefficients are chosen at random from a large enough field, then Gaussian elimination succeeds with overwhelming probability. RLNC can also be used in asynchronous packet networks, and it has been shown that it is capacity-achieving even on lossy packet networks.

A framework for packetized network coding (Practical Network Coding, PNC) leverages RLNC's resilience against disruptions such as packet loss, congestion, and changes of topology, in order to guarantee robust communication over highly dynamic networks with minimal (or no) control information. The framework defines a packet format and a buffering model. The packet format includes in its header the *global encoding vector*, which is the set of linear transformations that the original packet goes through on its path from the source to the destination. The payload of the packets is divided into vectors according to the field size (2^8 or 2^{16}, i.e. each symbol has 8 or 16 bits, respectively). Each of these symbols is then used as a building block for the linear operations performed by the nodes.

The buffering model divides the stream of packets into *generations* of size h, such that packets in the same generation are tagged with a common generation number. Each node sorts the incoming packets in a single buffer according to their generation number. When there is a transmission opportunity at an outgoing edge, the sending node generates a new packet, which contains a random linear combination of all packets in the buffer that belong to the *current* generation. If a packet is *non-innovative*, i.e. if it does not increase the rank of the decoding matrix available at the receiving node, then it is immediately discarded. As soon as the matrix of received packets has full rank, Gaussian elimination is performed at the receivers to recover the original packets.

RLNC seems particularly beneficial in dynamic and unstable networks – that is, networks in which the structure or topology of the network varies within a short time, such as mobile ad-hoc networks and peer-to-peer content-distribution networks. The benefits of RLNC in wireless environments with rare and limited connectivity, either due to mobility or battery scarcity, suggest an algorithm aimed at reducing the overhead of probabilistic routing algorithms with applications in delay-tolerant networks.

The potential impact of RLNC in content-distribution networks can be analyzed as follows. Since each node forwards a random linear combination independently of the information present at other nodes, its operation is completely desynchronized and locally based. Moreover, when collecting a random combination of packets from a randomly chosen node, there is a high probability of obtaining a linearly independent packet in each time slot. Thus, the problem of redundant transmissions, which is typical of traditional flooding approaches, is considerably reduced. In content-distribution applications, there is no need to download one particular fragment. Instead, any linearly independent segment brings innovative information. A possible practical scheme for large files allows nodes to make forwarding decisions solely on the basis of local information. It has been shown that network coding not only improves the expected file-download time, but also improves the overall robustness of the system with respect to frequent changes in the network.

The performance of network coding has been compared with that of traditional coding measures in a distributed-storage setting in which each storage location has only very limited storage space for each file and the objective is that a file-downloader connects to as few storage locations as possible to retrieve a file. It was shown that RLNC performs well without any need for a large amount of additional storage space at a centralized server. There exists also a general graph-theoretic framework for computing lower bounds on the bandwidth required to maintain distributed-storage architectures. It is now known that RLNC achieves these lower bounds.

State-aware network-coding protocols

State-aware network-coding protocols rely on the available state information to optimize the coding operations carried out by each node. The optimization process may target the throughput or the delay, among other performance metrics. Its scope can be *local* or *global*, depending on whether the optimization affects only the operations within the close neighborhood of a node or addresses the end-to-end communication across the entire network. If control traffic is exchanged between neighbors, each node can perform a local optimization step to decide on-the-fly how to mix and transmit the received data packets. One way to implement protocols with global scope is to use the polynomial-time algorithm on a given network graph, which, given a network graph, can be used to determine the optimal network code before starting the communication. By exploiting the broadcast nature of the wireless medium and spreading encoded information in a controlled manner, state-aware protocols promise considerable advantages in terms of throughput, as well as resilience with respect to node failures and packet losses. Efficiency gains come mainly from the fact that nodes make use of every data packet they overhear. In some instances network coding reduces the required amount of control information.

As an example, the COPE protocol inserts a coding layer between the IP and MAC layers for detecting coding opportunities. More specifically, nodes overhear and store packets that are exchanged within their radio range, and then send reception reports to inform their neighbors about which packets they have stored in their buffers. On the basis of these updates, each node computes the optimal XOR mixture of multiple packets in order to reduce the number of transmissions. It has been shown that this approach can lead to strong improvements in terms of throughput and robustness to network dynamics.

9.5 Security vulnerabilities

The aforementioned network-coding protocols expect a well-behaved node to do the following:

- *encode the received packets correctly*, thus contributing to the expected benefits of network coding;
- *forward the encoded packets correctly*, thus enabling the destination nodes to retrieve the intended information;

- *ignore data for which it is not the intended destination*, thus fulfilling basic confidentiality requirements.

In the case of state-aware network-coding protocols, we must add one more rule:

- the node must *participate in the timely dissemination of correct state information*, thus contributing to a sound knowledge base for network-coding decisions.

It follows that an attack on a network-coding-based protocol must result from one or more network nodes breaking one or more of these basic rules. Naturally, the means to achieve a successful attack are, of course, highly dependent on the specific rules of the protocol, and therefore it is reasonable to distinguish between the aforementioned stateless and state-aware classes of network-coding schemes.

If properly applied, stateless network protocols based on RNLC are potentially less subject to some of the typical security issues of traditional routing protocols. First, stateless protocols do not depend on exchange of topology or buffer-state information, which can be faked (e.g. through *link-spoofing* attacks). Second, the impact of *traffic-relay refusal* is reduced, due to the inherent robustness that results from spreading the information by means of network coding. Third, the information retrieval depends solely on the data received and not on the identity of nodes, which ensures some protection against *impersonation* attacks.

In contrast, state-aware network-coding protocols rely on vulnerable control information disseminated among nodes in order to optimize the encodings. On the one hand, this property renders them particularly prone to attacks based on the generation of false control information. On the other hand, control-traffic information can also be used effectively against active attacks, such as the injection of erroneous packets. For protocols with local scope, the negative impact of active attacks is limited to a confined neighborhood, whereas with end-to-end network coding the consequences can be much more devastating. Opportunistic network-coding protocols, which rely on the information overheard by neighboring nodes over the wireless medium, are obviously more amenable to eavesdropping attacks than are their wireline counterparts.

More generally, in comparison with traditional routing, the damage caused by a malicious (Byzantine) node injecting corrupted packets into the information flow is likely to be higher with network coding, irrespective of whether a protocol is stateless or state-aware. Since network coding relies on mixing the content of multiple data packets, a single corrupted packet may very easily corrupt the entire information flow from the sender to the destination at any given time.

9.6 Securing network coding against passive attacks

Having discussed the specific vulnerabilities of network coding, we now turn our attention to the next natural question, namely how to find appropriate mechanisms for securing network-coding protocols. Our main goal here is to show how the specific characteristics of network coding can be leveraged to counter some of the threats posed by

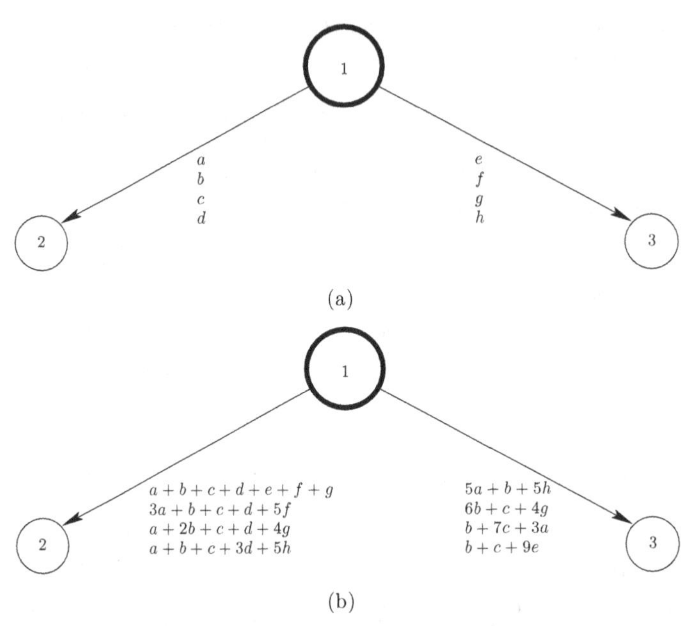

Figure 9.4 Example of algebraic security. The top scheme discloses data to intermediate nodes, whereas the bottom scheme can be deemed algebraically secure.

eavesdroppers and Byzantine attackers. We shall start by presenting countermeasures against passive attacks, with special emphasis on three different scenarios. First, we shall consider nice but curious nodes, which do not break any of the established rules except for not ignoring the data for which they are not the intended receivers. In the second instance, the eavesdropper is able to wiretap a subset of network links. Finally, the third type of attacker is a worst-case eavesdropper who is given full access to all the traffic in the network.

Nice but curious nodes

Consider first a threat model in which the network consists entirely of *nice but curious* nodes, which comply with the communication protocols (in that sense, they are well behaved), but may try to acquire as much information as possible from the data flows that pass through them (in which case, they are potentially ill-intentioned). Under this scenario, stateless protocols that exploit the RLNC scheme described in Section 9.4 possess an intrinsic security feature: depending on the size of the code alphabet and the topology of the network, it is in many instances unlikely that an intermediate node will have enough degrees of freedom to perform Gaussian elimination and gain access to the transmitted data set.

On the basis of this observation, it is possible to characterize the threat level posed by an intermediate node according to an *algebraic security criterion* that takes into account the number of components of the global encoding vector it receives. In the example of Figure 9.4, which uses integers for simplicity, the upper (uncoded) transmission scheme leaves partial data unprotected, whereas in the lower (network-coding) scheme the intermediate nodes 2 and 3 are not able to recover the data symbols.

Wiretappers with access to some links

A different threat model, which is commonly found in the recent literature on secure network coding, assumes that one or more external eavesdroppers (or wiretappers) have access to a subset of the available communication links. The crux of the problem is then the need to find code constructions capable of splitting the data among different links in such a way that reconstruction by the attackers is either very difficult or impossible. Under this assumption, there exist secure linear network codes that achieve perfect information-theoretic secrecy for single-source multicast transmission. The corresponding secure network-coding problem can in fact be cast as a variant of the so-called wiretap channel type-II problem. There, the eavesdropper is able to select and access μ symbols of the n coded symbols transmitted by the legitimate sender. It can be shown that in that case a maximum of $k = n - \mu$ information symbols can be transmitted in perfect secrecy. These results can be generalized to multi-source linear network codes by using the algebraic structure of such codes to derive necessary and sufficient conditions for their security. More specifically, it has been shown that the code constructed maximizes the amount of secure multicast information while minimizing the necessary amount of randomness. Such a coding scheme can achieve a maximum possible rate of $n - \mu$ information-theoretically secure packets, where n is now the number of packets sent from the source to each receiver and μ denotes the number of links that the wiretapper can observe. This can be applied on top of any communication network without requiring any knowledge of the underlying network code and without imposing any coding constraints. The basic idea is to use a "nonlinear" outer code, which is linear over an extension field $\mathbb{F}_{q^m}$, and to exploit the benefits of this extension field. Some contributions propose a different criterion whereby a system is deemed secure if an eavesdropper is unable to get any uncoded or immediately decodable (also called *meaningful*) source data. Other contributions exploit the network topology to ensure that an attacker is unable to get any meaningful information and add a cost function to the secure network-coding problem. The problem then becomes finding a coding scheme that minimizes both the network cost and the probability that the attacker is able to retrieve all the messages of interest.

Since state-aware network-coding protocols with local code optimization (e.g. COPE) expect neighboring nodes to be able to decode all the packets they receive, confidentiality must be ensured by means of end-to-end encryption.

Worst-case eavesdroppers

In this case, the threat model is one in which the attacker has access to all the packets traversing the network but not to the secret keys shared among legitimate communicating parties. Secure Practical Network Coding (abbreviated as SPOC, not SPNC) is a lightweight security scheme for confidentiality in RLNC, which provides a simple yet powerful way to exploit the inherent security of RLNC in order to reduce the number of cryptographic operations required for confidential communication. This is achieved by protecting (or "locking") only the source coefficients required in order to decode the linearly encoded data, while allowing intermediate nodes to run their network-coding operations on substitute "unlocked" coefficients that provably do not compromise the hidden data.

To evaluate the level of security provided by SPOC, one analyzes the mutual information between the encoded data and the two components that can lead to information leakage, namely the matrices of random coefficients and the original data. This analysis, which is independent of any particular cipher used for locking the coefficients, assumes that the encoding matrices are based on variants of RLNC and can be accessed only by the source and sinks. The results, some of which hold even with finite block lengths, prove that information-theoretic security is achievable for any field size without loss in terms of decoding probability. In other words, since correlation attacks based on the encoded data become impossible, protecting the encoding matrix is generally sufficient to ensure the confidentiality of network-coded data.

9.7 Countering Byzantine attacks

Although Byzantine attacks can have a severe impact on the integrity of network-coded information, the specific properties of linear network codes can be used effectively to counteract the impairments caused by traffic-relay refusal or injection of erroneous packets. In particular, RLNC has been shown to be very robust with respect to packet losses induced by node misbehavior. More sophisticated countermeasures, which modify the format of coded packets, can be subdivided into two main categories: (1) end-to-end error correction and (2) misbehavior detection, which can be carried out either packet by packet or in generation-based fashion.

End-to-end error correction

The main advantage of *end-to-end error-correcting codes* is that the burden of applying error-control techniques is left entirely to the source and the destinations, such that intermediate nodes are not required to change their mode of operation. The typical transmission model for end-to-end network coding is well described by a matrix channel $\mathbf{Y} = \mathbf{AX} + \mathbf{Z}$, where $\mathbf{X}$ corresponds to the matrix whose rows are the transmitted packets, $\mathbf{Y}$ is the matrix whose rows are the received packets, $\mathbf{Z}$ denotes the matrix corresponding to the injected error packets after propagation over the network, and $\mathbf{A}$ describes the transfer matrix, which corresponds to the global linear transformation performed on packets as they traverse the network. In terms of performance, error-correction schemes can correct up to the min-cut between the source and the destinations. Rank-metric error-correcting codes in RLNC under this setting appear to work well, including in scenarios in which the channel may supply partial information about erasures and deviations from the sent information flow. Still under the same setting, a probabilistic error model for random network coding provides bounds on capacity and presents a simple coding scheme with polynomial complexity that achieves capacity with an exponentially low probability of failure with respect to both the packet length and the field size. Bounds on the maximum achievable rate in an adversarial setting can be obtained from generalizations of the Hamming and Gilbert–Varshamov bounds.

A somewhat different approach to network error correction consists of robust network codes that have polynomial-time complexity and attain optimal rates in the presence of active attacks. The basic idea is to regard the packets injected by an adversarial node as a second source of information and add enough redundancy to allow the destination to distinguish between relevant and erroneous packets. The capacity achieved depends on the rate at which the attacker can obtain information, as well as on the existence of a shared secret between the source and the sinks.

Misbehavior detection

Generation-based detection schemes generally offer similar advantages to those obtained with network error-correcting codes in that the often computationally expensive task of detecting the modifications introduced by Byzantine attackers is carried out by the destination nodes. The main disadvantage of generation-based detection schemes is that only nodes with enough packets from a generation are able to detect malicious modifications, and thus usage of such detection schemes can result in large end-to-end delays. The underlying assumption is that the attacker cannot see the full rank of the packets in the network. It has been shown that a hash scheme with polynomial complexity can be used without the need for secret-key distribution. However, the use of a block code forces an a-priori decision on the coding rate.

The key idea of *packet-based detection schemes* is that some of the intermediate nodes in the network can detect polluted data on-the-fly and drop the corresponding packets, thus retransmitting only valid data. However, packet-based detection schemes require active participation of intermediate nodes and are dependent on hash functions, which are generally computationally expensive. Alternatively, this type of attack can be mitigated by signature schemes based on homomorphic hash functions. The use of homomorphic hash functions is specifically tailored for network-coding schemes, since the hash of a coded packet can easily be derived from the hashes of previously encoded packets, thus enabling intermediate nodes to verify the validity of encoded packets prior to mixing them algebraically. Unfortunately, homomorphic hash functions are also computationally expensive.

There exists a homomorphic signature scheme for network coding that is based on Weil pairing in elliptic-curve cryptography. Homomorphic hash functions have also been considered in the context of peer-to-peer content distribution with rateless erasure codes for multicast transfers. With the goal of preventing both the waste of large amounts of bandwidth and the pollution of download caches of network clients, each file is compressed to a smaller hash value, with which receivers can check the integrity of downloaded blocks. Beyond its independence from the coding rate, the main advantage of this process is that it is less computationally expensive for large files than are traditional forward error-correction codes (such as Reed–Solomon codes).

A cooperative security scheme can be used for on-the-fly detection of malicious blocks injected in network coding-based peer-to-peer networks. In order to reduce the cost of verifying information on-the-fly while efficiently preventing the propagation of malicious blocks, the authors propose a distributed mechanism whereby every node performs block checks with a certain probability and alerts its neighbors when a suspicious block is found. Techniques to prevent denial-of-service attacks due to the dissemination of alarms are available.

The basic idea is to take advantage of the fact that in linear network coding any valid packet transmitted belongs to the subspace spanned by the original set of vectors. A signature scheme is thus used to check that a given packet belongs to the original subspace. Generating a signature that is not in the subspace yet passes the check has been shown to be hard.

A comparison of the bandwidth overhead required by Byzantine error-correction and -detection schemes can be carried out as follows. The intermediate nodes are divided into regular nodes and trusted nodes, and only the latter are given access to the public key of the Byzantine detection scheme in use. Under these assumptions, it is shown that packet-based detection is most competitive when the probability of attack is high, whereas a generation-based approach is more bandwidth-efficient when the probability of attack is low.

Key-distribution schemes

The ability to distribute secret keys in a secure manner is an obvious fundamental requirement towards assuring cryptographic security. In the case of highly constrained mobile ad-hoc and sensor networks, key pre-distribution schemes emerge as a strong candidate, mainly because they require considerably less computation and communication resources than do trusted-party schemes or public-key infrastructures. The main caveat is that secure connectivity can be achieved only in probabilistic terms, i.e. if each node is loaded with a sufficiently large number of keys drawn at random from a fixed pool, then with high probability it will share at least one key with each neighboring node.

It has been shown that network coding can be an effective tool for establishing secure connections between low-complexity sensor nodes. In contrast with pure key-pre-distribution schemes, it is assumed that a mobile node, e.g. a hand-held device or a laptop computer, is available for activating the network and for helping to establish secure connections between nodes. By exploiting the benefits of network coding, it is possible to design a secret-key-distribution scheme that requires only a small number of pre-stored keys, yet ensures that shared-key connectivity is established with a probability of unity and that the mobile node is provably oblivious to the distributed keys.

The basic idea of the protocol, which is illustrated in Figure 9.5, can be summarized in the following tasks:

(a) prior to sensor-node deployment:
  (1) a large pool P of N keys and their N identifiers are generated off-line;
  (2) a different subset of L keys drawn randomly from P and the corresponding L identifiers are loaded into the memory of each sensor node;
  (3) a table is constructed with the N key identifiers and N sequences that result from performing an XOR of each key with a common protection sequence X;
  (4) the table is stored in the memory of the mobile node;

(b) after sensor-node deployment:
  (1) the mobile node broadcasts HELLO messages that are received by any sensor node within wireless transmission range;

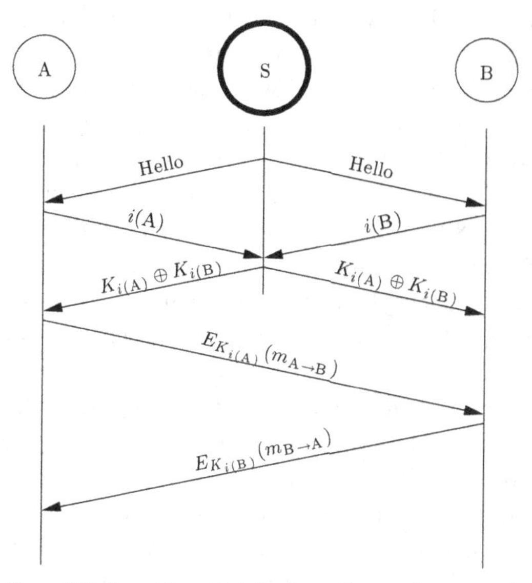

Figure 9.5 Secret-key-distribution scheme. Sensor nodes A and B want to exchange two keys via a mobile node S. The process is initiated by a HELLO message broadcast by S. Upon receiving this message, each sensor node sends back a key identifier $i(\cdot)$ corresponding to one of its keys $K_{i(\cdot)}$. Node S then broadcasts the result of the XOR of the two keys, $K_{i(\mathrm{A})} \oplus K_{i(\mathrm{B})}$. Once this process is concluded, sensor nodes A and B can communicate using the two keys $K_{i(\mathrm{A})}$ and $K_{i(\mathrm{B})}$ (one in each direction). Here, $E_{K_{i(\mathrm{A})}}(m_{\mathrm{A}\to\mathrm{B}})$ denotes a message sent by A to B, encrypted with $K_{i(\mathrm{A})}$, and $E_{K_{i(\mathrm{B})}}(m_{\mathrm{B}\to\mathrm{A}})$ corresponds to a message sent by B to A, encrypted with $K_{i(\mathrm{B})}$.

(2) each sensor node replies with a key identifier;
(3) on the basis of the received key identifiers the mobile node locates the corresponding sequences protected by X and combines them through an XOR network-coding operation, thus canceling out X and obtaining the XOR of the corresponding keys;
(4) the mobile node broadcasts the resulting XOR sequence;
(5) by combining the received XOR sequence with its own key, each node can easily recover the key of its neighbor, thus sharing a pair of keys that is kept secret from the mobile node.

Although the use of network coding hereby presented is limited to XOR operations, more powerful secret-key-distribution schemes are likely to result from using linear combinations of the stored keys.

9.8 Bibliographical notes

The field of network coding emerged from the seminal work of Ahlswede, Cai, Li, and Yeung [184], who proved that the multicast capacity of a general network can be achieved only if intermediate nodes are allowed to encode incoming symbols. Li, Yeung, and Cai proved that linear codes [185] are sufficient to achieve the aforementioned multicast capacity. The algebraic framework for network coding developed by Koetter and Médard in [186] and the development of random linear network coding by Ho *et al.* in [187] led to practical applications in which the nodes in the network generate linear combinations of information symbols using random coefficients. A practical approach to random linear network coding was proposed and tested by Chou, Wu, and Jain in [188]. Microsoft presented the first application of network coding for content distribution in a paper by Gkantsidis and Rodriguez [189]. A system implementation of XOR-based network coding for wireless networks (the COPE protocol) was presented by Katti *et al.* in [190]. Secure network coding with wiretapping limited to a subset of the network links was first addressed by Cai and Yeung in [191]. The connection between secure network coding and the wiretap channel of type II of Ozarow and Wyner [192] was investigated by Rouayheb and Soljanin in [193]. A weak criterion for the algebraic security provided by network coding was presented by Bhattad and Narayanan in [194]. The security potential of the algebraic structure of network coding in large-scale networks was analyzed by Lima, Médard, and Barros in [195]. Vilela, Lima and Barros provided in [196] a lightweight security solution for network coding, whereby only the coding coefficients must be encrypted. The proposed scheme found its first application in wireless video [197]. Solutions for Byzantine attacks were presented by Jaggi *et al.* in [198] and by Gkantsidis and Rodriguez in [199]. An information-theoretic treatment of network error correction was provided by Cai and Yeung in [200]. Koetter and Kschischang later presented code constructions for network error correction in [201]. Kim *et al.* provided a unified treatment of robust network coding for peer-to-peer networks under Byzantine attacks in [202]. Oliveira, Costa, and Barros investigated the properties of wireless network coding for secret-key distribution in [203]. Kim, Barros, Médard, and Koetter showed how to detect misbehaving nodes in wireless networks with network coding using an algebraic watchdog in [204]. A detailed account of the many facets of network coding can be found in the books by Ho and Lun [205] and by Fragouli and Soljanin [206, 207].

References

[1] C. E. Shannon, "Communication theory of secrecy systems," *Bell System Technical Journal*, vol. 28, no. 4, pp. 656–715, April 1949.

[2] R. G. Gallager, *Information Theory and Reliable Communication*. Wiley, 1968.

[3] T. M. Cover and J. A. Thomas, *Elements of Information Theory*, 2nd edn. Wiley-Interscience, 2006.

[4] R. W. Yeung, *A First Course in Information Theory*. Springer, 2002.

[5] I. Csiszár and J. Körner, *Information Theory: Coding Theorems for Discrete Memoryless Systems*. Akadémiai Kiadó, 1981.

[6] G. Kramer, *Topics in Multi-User Information Theory*. NOW Publishers, 2008.

[7] C. E. Shannon, "A mathematical theory of communication," *Bell System Technical Journal*, vol. 27, nos. 7/10, pp. 379–423/623–656, July/October 1948.

[8] G. Kramer, "Capacity results for the discrete memoryless networks," *IEEE Transactions on Information Theory*, vol. 49, no. 1, pp. 4–21, January 2003.

[9] D. Slepian and J. K. Wolf, "Noiseless coding of correlated information sources," *IEEE Transactions on Information Theory*, vol. 19, no. 4, pp. 471–480, July 1973.

[10] M. H. M. Costa, "Writing on dirty paper," *IEEE Transactions on Information Theory*, vol. 29, no. 3, pp. 439–441, May 1983.

[11] A. Wyner and J. Ziv, "The rate-distortion function for source coding with side information at the decoder," *IEEE Transactions on Information Theory*, vol. 22, no. 1, pp. 1–10, January 1976.

[12] R. Ahlswede, "Multi-way communication channels," in *Proc. International Symposium on Information Theory*, Thakadsor, Armenian SSR, USSR, September 1971, pp. 23–52.

[13] H. Liao, "Multiple access channels," Ph.D. dissertation, University of Hawaii, 1972.

[14] T. M. Cover, "Broadcast channels," *IEEE Transactions on Information Theory*, vol. 18, no. 1, pp. 2–14, January 1972.

[15] P. Bergmans, "Random coding theorem for broadcast channels with degraded components," *IEEE Transactions on Information Theory*, vol. 19, no. 2, pp. 197–207, March 1973.

[16] R. G. Gallager, "Capacity and coding for degraded broadcast channels," *Problemy Peredachi Informatsii*, vol. 10, no. 3, pp. 3–14, 1974.

[17] T. M. Cover, "Comments on broadcast channels," *IEEE Transactions on Information Theory*, vol. 44, no. 6, pp. 2524–2530, October 1998.

[18] I. Csiszár and J. Körner, "Broadcast channels with confidential messages," *IEEE Transactions on Information Theory*, vol. 24, no. 3, pp. 339–348, May 1978.

[19] G. S. Vernam, "Cipher printing telegraph systems for secret wire and radio telegraphic communications," *Transactions of the American Institute of Electrical Engineers*, vol. 45, no. 1, pp. 295–301, January 1926.

[20] G. D. Forney, Jr., "On the role of MMSE estimation in approaching the information-theoretic limits of linear Gaussian channels: Shannon meets Wiener," in *Proc. 41st Annual Allerton Conference on Communication, Control, and Computing*, Monticello, IL, USA, October 2003, pp. 430–439.

[21] A. D. Wyner, "The wire-tap channel," *Bell System Technical Journal*, vol. 54, no. 8, pp. 1355–1367, October 1975.

[22] S. Leung-Yan-Cheong, "On a special class of wiretap channels," *IEEE Transactions on Information Theory*, vol. 23, no. 5, pp. 625–627, September 1977.

[23] J. L. Massey, "A simplified treatment of Wyner's wire-tap channels," in *Proc. 21st Allerton Conference on Communication, Control, and Computing*, Monticello, IL, USA, October 1983, pp. 268–276.

[24] I. Csiszár, "Almost independence and secrecy capacity," *Problemy Peredachi Informatsii*, vol. 32, no. 1, pp. 40–47, January–March 1996.

[25] M. Hayashi, "General nonasymptotic and asymptotic formulas in channel resolvability and identification capacity and their application to the wiretap channels," *IEEE Transactions on Information Theory*, vol. 52, no. 4, pp. 1562–1575, April 2006.

[26] M. Bloch and J. N. Laneman, "On the secrecy capacity of arbitrary wiretap channels," in *Proc. 46th Allerton Conference on Communication, Control, and Computing*, Monticello, IL, USA, September 2008, pp. 818–825.

[27] B. P. Dunn, M. Bloch, and J. N. Laneman, "Secure bits through queues," in *Proc. IEEE Information Theory Workshop on Networking and Information Theory*, Volos, Greece, June 2009, pp. 37–41.

[28] J. Körner and K. Marton, "Comparison of two noisy channels," in *Proc. Topics in Information Theory*, Keszthely, Hungary, 1977, pp. 411–423.

[29] M. van Dijk, "On a special class of broadcast channels with confidential messages," *IEEE Transactions on Information Theory*, vol. 43, no. 2, pp. 712–714, March 1997.

[30] C. Nair, "Capacity regions of two new classes of 2-receiver broadcast channels," in *Proc. IEEE International Symposium on Information Theory*, Seoul, South Korea, July 2009, pp. 1839–1843.

[31] H. Yamamoto, "Rate-distortion theory for the Shannon cipher systems," *IEEE Transactions on Information Theory*, vol. 43, no. 3, pp. 827–835, May 1997.

[32] S. K. Leung-Yan-Cheong, "Multi-user and wiretap channels including feedback," Ph.D. dissertation, Stanford University, 1976.

[33] R. Ahlswede and N. Cai, "Transmission, identification and common randomness capacities for wire-tap channels with secure feedback from the decoder," in *General Theory of Information Transfer and Combinatorics*. Springer-Verlag, 2006, pp. 258–275.

[34] E. Ardestanizadeh, M. Franceschetti, T. Javidi, and Y.-H. Kim, "Wiretap channel with secure rate-limited feedback," *IEEE Transactions on Information Theory*, vol. 55, no. 12, pp. 5353–5361, December 2009.

[35] M. Feder and N. Merhav, "Relations between entropy and error probability," *IEEE Transactions on Information Theory*, vol. 40, no. 1, pp. 259–266, January 1994.

[36] K. Yasui, T. Suko, and T. Matsushima, "An algorithm for computing the secrecy capacity of broadcast channels with confidential messages," in *Proc. IEEE International Symposium on Information Theory*, Nice, France, July 2007, pp. 936–940.

[37] K. R. Gowtham and A. Thangaraj, "Computation of secrecy capacity for more-capable channel pairs," in *Proc. IEEE International Symposium on Information Theory*, Toronto, Canada, July 2008, pp. 529–533.

[38] G. Van Assche, *Quantum Cryptography and Secret-Key Distillation*. Cambridge University Press, 2006.

[39] M. A. Nielsen and I. L. Chuang, *Quantum Computation and Quantum Information*. Cambridge University Press, 2000.

[40] N. Gisin, G. Ribordy, W. Tittel, and H. Zbinden, "Quantum cryptography," *Reviews of Modern Physics*, vol. 74, no. 1, pp. 145–195, January 2002.

[41] U. Maurer, R. Renner, and S. Wolf, "Unbreakable keys from random noise," in *Security with Noisy Data*, P. Tuyls, B. Skoric, and T. Kevenaar, Eds. Springer-Verlag, 2007, pp. 21–44.

[42] U. Maurer, "Secret key agreement by public discussion from common information," *IEEE Transactions on Information Theory*, vol. 39, no. 3, pp. 733–742, May 1993.

[43] R. Ahlswede and I. Csiszár, "Common randomness in information theory and cryptography. I. Secret sharing," *IEEE Transactions on Information Theory*, vol. 39, no. 4, pp. 1121–1132, July 1993.

[44] A. A. Gohari and V. Anantharam, "Information-theoretic key agreement of multiple terminals – part I," *IEEE Transactions on Information Theory*, vol. 56, no. 8, pp. 3973–3996, August 2010.

[45] A. A. Gohari and V. Anatharam, "Information-theoretic key agreement of multiple terminals – part II: Channel models," *IEEE Transactions on Information Theory*, vol. 56, no. 8, pp. 3997–4010, August 2010.

[46] U. M. Maurer and S. Wolf, "Unconditionally secure key agreement and intrinsic conditional information," *IEEE Transactions on Information Theory*, vol. 45, no. 2, pp. 499–514, March 1999.

[47] R. Renner and S. Wolf, "New bounds in secret-key agreement: The gap between formation and secrecy extraction," in *Proc. EUROCRYPT*, 2003, pp. 652–577.

[48] I. Csiszár and P. Narayan, "Common randomness and secret key generation with a helper," *IEEE Transactions on Information Theory*, vol. 46, no. 2, pp. 344–366, March 2000.

[49] A. Khisti, S. N. Diggavi, and G. Wornell, "Secret-key generation with correlated sources and noisy channels," in *Proc. IEEE International Symposium on Information Theory*, Toronto, Canada, July 2008, pp. 1005–1009.

[50] V. Prabhakaran and K. Ramchandran, "A separation result for secure communication," in *Proc. 45th Allerton Conference on Communications, Control and Computing*, Monticello, IL, USA, September 2007, pp. 34–41.

[51] S. Watanabe and Y. Oohama, "Secret key agreement from vector Gaussian sources by rate limited public communications," in *Proc. IEEE International Symposium on Information Theory*, Austin, TX, USA, June 2010, pp. 2597–2601; see also arXiv:1001.3705v1.

[52] M. J. Gander and U. M. Maurer, "On the secret-key rate of binary random variables," in *Proc. IEEE International Symposium on Information Theory*, Trondheim, Norway, June 1994, p. 351.

[53] S. Liu, H. C. A. Van Tilborg, and M. Van Dijk, "A practical protocol for advantage distillation and information reconciliation," *Designs, Codes and Cryptography*, vol. 30, no. 1, pp. 39–62, August 2003.

[54] M. Naito, S. Watanabe, R. Matsumoto, and T. Uyematsu, "Secret key agreement by reliability information of signals in Gaussian Maurer's models," in *Proc. IEEE International Symposium on Information Theory*, Toronto, Canada, July 2008, pp. 727–731.

[55] J. Muramatsu, K. Yoshimura, and P. Davis, "Secret key capacity and advantage distillation capacity," in *Proc. IEEE International Symposium on Information Theory*, Seattle, WA, USA, July 2006, pp. 2598–2602.

[56] G. Brassard and L. Salvail, "Secret-key reconciliation by public discussion," in *Proc. Advances in Cryptology – Eurocrypt*, T. Helleseth, Ed. Springer-Verlag, May 1993, pp. 411–423.

[57] W. T. Buttler, S. K. Lamoreaux, J. R. Torgerson, G. H. Nickel, C. H. Donahue, and C. G. Peterson, "Fast, efficient error reconciliation for quantum cryptography," *Physical Review A*, vol. 67, no. 5, pp. 052 303/1–8, May 2003.

[58] D. Elkouss, A. Leverrier, R. Alléaume, and J. J. Boutros, "Efficient reconciliation protocol for discrete-variable quantum key distribution," in *Proc. IEEE International Symposium on Information Theory*, Seoul, South Korea, July 2009, pp. 1879–1883.

[59] G. Van Assche, J. Cardinal, and N. J. Cerf, "Reconciliation of a quantum-distributed Gaussian key," *IEEE Transactions on Information Theory*, vol. 50, no. 2, pp. 394–400, February 2004.

[60] K.-C. Nguyen, G. Van Assche, and N. J. Cerf, "Side-information coding with turbo codes and its application to quantum key distribution," in *Proc. International Symposium on Information Theory and its Applications*, Parma, Italy, October 2004, pp. 1274–1279.

[61] M. Bloch, A. Thangaraj, S. W. McLaughlin, and J.-M. Merolla, "LDPC-based Gaussian key reconciliation," in *Proc. IEEE Information Theory Workshop*, Punta del Este, Uruguay, March 2006, pp. 116–120; extended report available at arXiv:cs.IT/0509041.

[62] C. Ye, A. Reznik, and Y. Shah, "Extracting secrecy from jointly Gaussian random variables," in *Proc. IEEE International Symposium on Information Theory*, Seattle, WA, USA, July 2006, pp. 2593–2597.

[63] A. Rényi, "On measures of entropy and information," in *Proc. Fourth Berkeley Symposium on Mathematical Statistics and Probability*, vol. 1, Berkeley, CA, USA, 1961, pp. 547–561.

[64] C. H. Bennett, G. Brassard, C. Crépeau, and U. Maurer, "Generalized privacy amplification," *IEEE Transactions on Information Theory*, vol. 41, no. 6, pp. 1915–1923, November 1995.

[65] J. L. Carter and M. N. Wegman, "Universal classes of hash functions," *Journal of Computer and System Sciences*, vol. 18, no. 2, pp. 143–154, April 1979.

[66] M. N. Wegman and J. Carter, "New hash functions and their use in authentication and set equality," *Journal of Computer Sciences and Systems*, vol. 22, no. 3, pp. 265–279, June 1981.

[67] D. R. Stinson, "Universal hashing and authentication codes," *Designs, Codes and Cryptography*, vol. 4, no. 4, pp. 369–380, October 1994.

[68] U. M. Maurer and S. Wolf, "Information-theoretic key agreement: From weak to strong secrecy for free," in *Advances in Cryptology – Eurocrypt 2000*, B. Preneel, Ed. Springer-Verlag, 2000, p. 351.

[69] S. Vadhan, "Extracting all the randomness from a weakly random source," Massachusetts Institute of Technology, Cambridge, MA, USA, Technical Report, 1998.

[70] C. Cachin, "Entropy measures and unconditional security in cryptography." Ph.D. dissertation, ETH Zürich, Zurich, Switzerland, 1997.

[71] C. Cachin and U. M. Maurer, "Linking information reconciliation and privacy amplification," *Journal of Cryptology*, vol. 10, no. 2, pp. 97–110, March 1997.

[72] J. Muramatsu, "Secret key agreement from correlated source outputs using low density parity check matrices," *IEICE Transactions on Fundamentals of Electronics, Communications and Computer Sciences*, vol. E89-A, no. 7, pp. 2036–2046, July 2006.

[73] S. Nitinawarat, "Secret key generation for correlated Gaussian sources," in *Proc. 45th Allerton Conference on Communications, Control and Computing*, Monticello, IL, USA, September 2007, pp. 1054–1058.

[74] U. Maurer and S. Wolf, "Secret-key agreement over unauthenticated public channels – part I. Definitions and a completeness result," *IEEE Transactions on Information Theory*, vol. 49, no. 4, pp. 822–831, April 2003.

[75] U. Maurer and S. Wolf, "Secret-key agreement over unauthenticated public channels – part II. The simulatability condition," *IEEE Transactions on Information Theory*, vol. 49, no. 4, pp. 832–838, April 2003.

[76] U. Maurer and S. Wolf, "Secret-key agreement over unauthenticated public channels – part III. Privacy amplification," *IEEE Transactions on Information Theory*, vol. 49, no. 4, pp. 839–851, April 2003.

[77] H. Imai, K. Kobara, and K. Morozov, "On the possibility of key agreement using variable directional antenna," in *Proc. 1st Joint Workshop on Information Security*, Seoul, South Korea, September 2006, pp. 153–167.

[78] C. Ye, S. Mathur, A. Reznik, Y. Shah, W. Trappe, and N. B. Mandayam, "Information-theoretically secret key generation for fading wireless channels," *IEEE Transactions on Information Forensics and Security*, vol. 5, no. 2, pp. 240–254, June 2010.

[79] C. Chen and M. A. Jensen, "Secret key establishment using temporally and spatially correlated wireless channel coefficients," *IEEE Transactions on Mobile Computing*, vol. 10, no. 2, pp. 205–215, February 2011.

[80] F. Grosshans, G. Van Assche, J. Wenger, R. Brouri, N. J. Cerf, and P. Grangier, "Quantum key distribution using Gaussian-modulated coherent states," *Letters to Nature*, vol. 421, no. 6920, pp. 238–241, January 2003.

[81] J. Lodewyck, M. Bloch, R. García-Patrón, S. Fossier, E. Karpov, E. Diamanti, T. Debuisschert, N. J. Cerf, R. Tualle-Brouri, S. W. McLaughlin, and P. Grangier, "Quantum key distribution over 25 km with an all-fiber continuous-variable system," *Physical Review A*, vol. 76, pp. 042 305/1–10, October 2007.

[82] H. Chabanne and G. Fumaroli, "Noisy cryptographic protocols for low-cost RFID tags," *IEEE Transactions on Information Theory*, vol. 52, no. 8, pp. 3562–3566, August 2006.

[83] A. Khisti and G. Wornell, "The MIMOME channel," in *Proc. 45th Allerton Conference on Communication, Control and Computing*, Monticello, IL, USA, September 2007, pp. 625–632; available online: http://allegro.mit.edu/pubs/posted/journal/2008-khisti-wornell-it.pdf.

[84] A. Khisti and G. Wornell, "Secure transmission with multiple antennas – II: The MIMOME wiretap channels," *IEEE Transactions on Information Theory*, vol. 56, no. 11, pp. 5515–5532, November 2010.

[85] Y. Liang, H. V. Poor, and S. Shamai (Shitz), "Secure communication over fading channels," *IEEE Transactions on Information Theory*, vol. 54, no. 6, pp. 2470–2492, June 2008.

[86] P. K. Gopala, L. Lai, and H. El Gamal, "On the secrecy capacity of fading channels," *IEEE Transactions on Information Theory*, vol. 54, no. 10, pp. 4687–4698, October 2008.

[87] S. K. Leung-Yan-Cheong and M. E. Hellman, "The Gaussian wire-tap channels," *IEEE Transactions on Information Theory*, vol. 24, no. 4, pp. 451–456, July 1978.

[88] R. Bustin, R. Liu, H. V. Poor, and S. Shamai, "An MMSE approach to the secrecy capacity of the MIMO Gaussian wiretap channels," *EURASIP Journal on Wireless Communications and Networking*, vol. 2009, pp. 370 970/1–8, 2009.

[89] F. Oggier and B. Hassibi, "The secrecy capacity of the MIMO wiretap channels," in *Proc. IEEE International Symposium on Information Theory*, Toronto, Canada, July 2008, pp. 524–528.

[90] T. Liu and S. Shamai, "A note on the secrecy capacity of the multiple-antenna wiretap channel," *IEEE Transactions on Information Theory*, vol. 55, no. 6, pp. 2547–2553, June 2009.

[91] S. Shafiee, N. Liu, and S. Ulukus, "Towards the secrecy capacity of the Gaussian MIMO wire-tap channel: The 2–2–1 channels," *IEEE Transactions on Information Theory*, vol. 55, no. 9, pp. 4033–4039, September 2009.

[92] M. Gursoy, "Secure communication in the low-SNR regime: A characterization of the energy–secrecy tradeoff," in *Proc. IEEE International Symposium on Information Theory*, Seoul, South Korea, July 2009, pp. 2291–2295.

[93] R. Negi and S. Goel, "Secret communication using artificial noise," in *62nd Vehicular Technology Conference (VTC)*, vol. 3, Dallas, TX, USA, September 2005, pp. 1906–1910.

[94] S. Goel and R. Negi, "Guaranteeing secrecy using artificial noise," *IEEE Transactions on Wireless Communications*, vol. 7, no. 6, pp. 2180–2189, June 2008.

[95] A. Khisti, G. Wornell, A. Wiesel, and Y. Eldar, "On the Gaussian MIMO wiretap channels," in *Proc. IEEE International Symposium on Information Theory*, Nice, France, July 2007, pp. 2471–2475.

[96] J. Barros and M. R. D. Rodrigues, "Secrecy capacity of wireless channels," in *Proc. IEEE International Symposium on Information Theory*, Seattle, WA, USA, July 2006, pp. 356–360.

[97] Z. Li, R. Yates, and W. Trappe, "Secret communication with a fading eavesdropper channel," in *Proc. IEEE International Symposium on Information Theory*, Nice, France, June 2007, pp. 1296–1300.

[98] H. Jeon, N. Kim, M. Kim, H. Lee, and J. Ha, "Secrecy capacity over correlated ergodic fading channels," in *Proc. IEEE Military Communications Conference*, San Diego, CA, USA, November 2008, pp. 1–7.

[99] O. O. Koyluoglu, H. El Gamal, L. Lai, and H. V. Poor, "On the secure degrees of freedom in the k-user Gaussian interference channels," in *Proc. IEEE International Symposium on Information Theory*, Toronto, Canada, July 2008, pp. 384–388.

[100] M. Kobayashi, M. Debbah, and S. S. (Shitz), "Secured communication over frequency-selective fading channels: A practical Vandermonde precoding," *EURASIP Journal on Wireless Communications and Networking*, vol. 2009, pp. 386 547/1–19, 2009.

[101] M. Bloch and J. N. Laneman, "Information-spectrum methods for information-theoretic security," in *Proc. Information Theory and Applications Workshop*, San Diego, CA, USA, February 2009, pp. 23–28.

[102] X. Tang, R. Liu, P. Spasojević, and H. V. Poor, "On the throughput of secure hybrid-ARQ protocols for Gaussian block-fading channels," *IEEE Transactions on Information Theory*, vol. 55, no. 4, pp. 1575–1591, April 2009.

[103] X. Tang, H. Poor, R. Liu, and P. Spasojević, "Secret-key sharing based on layered broadcast coding over fading channels," in *Proc. IEEE International Symposium on Information Theory*, Seoul, South Korea, July 2009, pp. 2762–2766.

[104] Y. Liang, L. Lai, H. Poor, and S. Shamai, "The broadcast approach over fading Gaussian wiretap channels," in *Proc. Information Theory Workshop*, Taormina, Italy, October 2009, pp. 1–5.

[105] M. Bloch, J. Barros, M. R. D. Rodrigues, and S. W. McLaughlin, "Wireless information-theoretic security," *IEEE Transactions on Information Theory*, vol. 54, no. 6, pp. 2515–2534, June 2008.

[106] T. F. Wong, M. Bloch, and J. M. Shea, "Secret sharing over fast-fading MIMO wiretap channels," *EURASIP Journal on Wireless Communications and Networking*, vol. 2009, pp. 506 973/1–17, 2009.

[107] R. Wilson, D. Tse, and R. A. Scholtz, "Channel identification: Secret sharing using reciprocity in ultrawideband channels," *IEEE Transactions on Information Forensics and Security*, vol. 2, no. 3, pp. 364–375, September 2007.

[108] T. Richardson and R. Urbanke, *Modern Coding Theory*. Cambridge University Press, 2008.

[109] H. Jin and T. Richardson, "Block error iterative decoding capacity for LDPC codes," in *Proc. International Symposium on Information Theory*, Adelaide, Australia, 2005, pp. 52–56.

[110] R. G. Gallager, "Low density parity check codes," Ph.D. dissertation, Massachusetts Institute of Technology, Cambridge, MA, USA, 1963.

[111] T. J. Richardson and R. L. Urbanke, "The capacity of low-density parity-check codes under message-passing decoding," *IEEE Transactions on Information Theory*, vol. 47, no. 2, pp. 599–618, February 2001.

[112] S.-Y. Chung, T. J. Richardson, and R. L. Urbanke, "Analysis of sum–product decoding of low-density parity-check codes using a Gaussian approximation," *IEEE Transactions on Information Theory*, vol. 47, no. 2, pp. 657–670, February 2001.

[113] S.-Y. Chung, J. G. D. Forney, T. J. Richardson, and R. Urbanke, "On the design of low-density parity-check codes within 0.0045 dB of the Shannon limit," *IEEE Communications Letters*, vol. 5, no. 2, pp. 58–60, February 2001.

[114] T. J. Richardson, M. A. Shokrollahi, and R. L. Urbanke, "Design of capacity-approaching irregular low-density parity-check codes," *IEEE Transactions on Information Theory*, vol. 47, no. 2, pp. 619–637, February 2001.

[115] A. Shokrollahi and R. Storn, "Design of efficient erasure codes with differential evolution," in *Proc. IEEE International Symposium on Information Theory*, Sorrento, Italy, June 2000.

[116] L. H. Ozarow and A. D. Wyner, "Wire tap channel II," *AT&T Bell Laboratories Technical Journal*, vol. 63, no. 10, pp. 2135–2157, December 1984.

[117] A. Thangaraj, S. Dihidar, A. R. Calderbank, S. W. McLaughlin, and J.-M. Merolla, "Applications of LDPC codes to the wiretap channels," *IEEE Transactions on Information Theory*, vol. 53, no. 8, pp. 2933–2945, August 2007.

[118] A. T. Suresh, A. Subramanian, A. Thangaraj, M. Bloch, and S. McLaughlin, "Strong secrecy for erasure wiretap channels," in *Proc. IEEE Information Theory Workshop*, Dublin, Ireland, 2010, pp. 1–5.

[119] A. Subramanian, A. Thangaraj, M. Bloch, and S. W. McLaughlin, "Strong secrecy on the binary erasure wiretap channel using large-girth LDPC codes," submitted to *IEEE Transactions on Information Forensics and Security*, September 2010. Available online: arXiv:1009.3130.

[120] V. Rathi, M. Andersson, R. Thobaben, J. Kliewer, and M. Skoglund, "Two edge type LDPC codes for the wiretap channels," in *Proc. 43rd Asilomar Conference on Signals, Systems and Computers*, Pacific Grove, CA, USA, November 2009, pp. 834–838.

[121] G. Cohen and G. Zemor, "Syndrome-coding for the wiretap channel revisited," in *Proc. IEEE Information Theory Workshop*, Chengdu, China, October 2006, pp. 33–36.

[122] R. Liu, Y. Liang, H. V. Poor, and P. Spasojević, "Secure nested codes for type II wiretap channels," in *Proc. IEEE Information Theory Workshop*, Lake Tahoe, CA, USA, September 2007, pp. 337–342.

[123] E. Verriest and M. Hellman, “Convolutional encoding for Wyner’s wiretap channels,” *IEEE Transactions on Information Theory*, vol. 25, no. 2, pp. 234–236, March 1979.

[124] V. Wei, “Generalized Hamming weights for linear codes,” *IEEE Transactions on Information Theory*, vol. 37, no. 5, pp. 1412–1418, September 1991.

[125] H. Mahdavifar and A. Vardy, “Achieving the secrecy capacity of wiretap channels using polar codes,” in *Proc. IEEE International Symposium on Information Theory*, Austin, TX, USA, June 2010, pp. 913–917. Available online: arXiv:1001.0210v1.

[126] M. Andersson, V. Rathi, R. Thobaben, J. Kliewer, and M. Skoglund, “Nested polar codes for wiretap and relay channels,” *IEEE Communications Letters*, vol. 14, no. 4, pp. 752–754, June 2010.

[127] O. O. Koyluoglu and H. E. Gamal, “Polar coding for secure transmission and key agreement,” in *Proc. IEEE International Symposium on Personal Indoor and Mobile Radio Communications*, Istanbul, Turkey, 2010, pp. 2698–2703.

[128] E. Hof and S. Shamai, “Secrecy-achieving polar-coding for binary-input memoryless symmetric wire-tap channels,” in *Proc. IEEE Information Theory Workshop*, Dublin, Ireland, 2010, pp. 1–5.

[129] D. Klinc, J. Ha, S. McLaughlin, J. Barros, and B.-J. Kwak, “LDPC codes for the Gaussian wiretap channels,” in *Proc. IEEE Information Theory Workshop*, Taormina, Sicily, October 2009, pp. 95–99.

[130] J.-C. Belfiore and P. Solé, “Unimodular lattices for the Gaussian wiretap channels,” in *Proc. IEEE Information Theory Workshop*, Dublin, Ireland, September 2010, pp. 1–5.

[131] J.-C. Belfiore and F. Oggier, “Secrecy gain: A wiretap lattice code design,” in *Proc. International Symposium on Information Theory and its Applications*, Taichung, Taiwan, October 2010.

[132] A. Wyner, “Recent results in the Shannon theory,” *IEEE Transactions on Information Theory*, vol. 20, no. 1, pp. 2–10, January 1974.

[133] A. D. Liveris, Z. Xiong, and C. N. Georghiades, “Compression of binary sources with side information at the decoder using LDPC codes,” *IEEE Communications Letters*, vol. 6, no. 10, pp. 440–442, October 2002.

[134] J. Chen, D. He, and A. Jagmohan, “Slepian–Wolf code design via source-channel correspondence,” in *Proc. IEEE International Symposium on Information Theory*, Seattle, WA, USA, July 2006, pp. 2433–2437.

[135] A. Leverrier, R. Alléaume, J. Boutros, G. Zémor, and P. Grangier, “Multidimensional reconciliation for a continuous-variable quantum key distribution,” *Physical Review A*, vol. 77, no. 4, p. 042325, April 2008.

[136] U. Wachsmann, R. F. H. Fischer, and J. B. Huber, “Multilevel codes: Theoretical concepts and practical design rules,” *IEEE Transactions on Information Theory*, vol. 45, no. 5, pp. 1361–1391, July 1999.

[137] G. Caire, G. Taricco, and E. Biglieri, “Bit-interleaved coded modulation,” *IEEE Transactions on Information Theory*, vol. 44, no. 3, pp. 927–946, May 1998.

[138] J. Hou, P. H. Siegel, L. B. Milstein, and H. D. Pfister, “Capacity-approaching bandwidth-efficient coded modulation schemes based on low-density parity-check codes,” *IEEE Transactions on Information Theory*, vol. 49, no. 9, pp. 2141–2155, September 2003.

[139] W. Stallings, *Cryptography and Network Security: Principles and Practice*. Prentice Hall, 2010.

[140] W. Diffie and M. Hellman, "New directions in cryptography," *IEEE Transactions on Information Theory*, vol. 22, no. 6, pp. 644–654, November 1976.

[141] R. Rivest, A. Shamir, and L. Adleman, "A method for obtaining digital signatures and public-key cryptosystems," *Communications of the ACM*, vol. 21, no. 2, pp. 120–126, February 1978.

[142] B. Schneier, *Applied Cryptography: Protocols, Algorithms, and Source Code in C*, 2nd edn. Wiley, 1996.

[143] A. Menezes, P. Van Oorschot, and S. Vanstone, *Handbook of Applied Cryptography*, 5th edn. CRC Press, 2001.

[144] M. Médard, D. Marquis, R. Barry, and S. Finn, "Security issues in all-optical networks," *IEEE Network*, vol. 11, no. 3, pp. 42–48, May–June 1997.

[145] U. M. Maurer, "Authentication theory and hypothesis testing," *IEEE Transactions on Information Theory*, vol. 46, no. 4, pp. 1350–1356, July 2000.

[146] Z. Li, W. Xu, R. Miller, and W. Trappe, "Securing wireless systems via lower layer enforcements," in *Proc. 5th ACM Workshop on Wireless Security*, Los Angeles, CA, USA, September 2006, pp. 33–42.

[147] Y. Liang, H. V. Poor, and S. S. (Shitz), *Information-Theoretic Security*. Now Publishers, 2009.

[148] R. Liu and W. Trappe, Eds., *Securing Wireless Communications at the Physical Layer*. Springer-Verlag, 2010.

[149] E. Tekin and A. Yener, "The general Gaussian multiple-access and two-way wiretap channels: Achievable rates and cooperative jamming," *IEEE Transactions on Information Theory*, vol. 54, no. 6, pp. 2735–2751, June 2008.

[150] X. He and A. Yener, "On the role of feedback in two-way secure communications," in *Proc. 42nd Asilomar Conference on Signals, Systems and Computers*, Pacific Grove, CA, USA, October 2008, pp. 1093–1097.

[151] M. Bloch, "Channel scrambling for secrecy," in *Proc. of IEEE International Symposium on Information Theory*, Seoul, South Korea, July 2009, pp. 2452–2456.

[152] M. Médard, "Capacity of correlated jamming channels," in *Proc. Allerton Conference on Communications, Computing and Control*, Monticello, IL, USA, October 1997.

[153] X. Tang, R. Liu, P. Spasojević, and H. V. Poor, "Interference-assisted secret communications," in *Proc. IEEE Information Theory Workshop*, Porto, Portugal, May 2008, pp. 164–168.

[154] X. Tang, R. Liu, P. Spasojević, and H. V. Poor, "The Gaussian wiretap channel with a helping interferer," in *Proc. IEEE International Symposium on Information Theory*, Toronto, Canada, July 2008, pp. 389–393.

[155] X. He and A. Yener, "Providing secrecy with lattice codes," in *Proc. 46th Annual Allerton Conference on Communication, Control, and Computing*, Monticello, IL, USA, September 2008, pp. 1199–1206.

[156] X. He and A. Yener, "Secure degrees of freedom for Gaussian channels with interference: Structured codes outperform Gaussian signaling," in *Proc. IEEE Global Telecommunications Conference*, Honolulu, HI, USA, December 2009, pp. 1–6.

[157] E. Ekrem and S. Ulukus, "Secrecy in cooperative relay broadcast channels," in *Proc. IEEE International Symposium on Information Theory*, Toronto, Canada, July 2008, pp. 2217–2221.

[158] M. Bloch and A. Thangaraj, "Confidential messages to a cooperative relay," in *Proc. IEEE Information Theory Workshop*, Porto, Portugal, May 2008, pp. 154–158.
[159] J. P. Vilela, M. Bloch, J. Barros, and S. W. McLaughlin, "Friendly jamming for wireless secrecy," in *Proc. IEEE International Conference on Communications*, Cape Town, South Africa, May 2010, pp. 1550–3607.
[160] O. Simeone and A. Yener, "The cognitive multiple access wire-tap channel," in *Proc. 43rd Annual Conference on Information Sciences and Systems*, Baltimore, MD, USA, March 2009, pp. 158–163.
[161] Y. Liang, A. Somekh-Baruch, H. V. Poor, S. Shamai, and S. Verdú, "Capacity of cognitive interference channels with and without secrecy," *IEEE Transactions on Information Theory*, vol. 55, no. 2, pp. 604–619, February 2009.
[162] C. Mitrpant, A. J. H. Vinck, and Y. Luo, "An achievable region for the Gaussian wiretap channel with side information," *IEEE Transactions on Information Theory*, vol. 52, no. 5, pp. 2181–2190, May 2006.
[163] G. T. Amariucai and S. Wei, "Secrecy rates of binary wiretapper channels using feedback schemes," in *Proc. 42nd Annual Conference on Information Sciences and Systems*, Princeton, NJ, USA, March 2008, pp. 624–629.
[164] L. Lai, H. El Gamal, and H. V. Poor, "The wiretap channel with feedback: Encryption over the channel," *IEEE Transactions on Information Theory*, vol. 54, no. 11, pp. 5059–5067, November 2008.
[165] D. Gündüz, D. R. Brown, and H. V. Poor, "Secret communication with feedback," in *Proc. International Symposium on Information Theory and Its Applications*, Auckland, New Zealand, May 2008, pp. 1–6.
[166] A. El Gamal, O. O. Koyluoglu, M. Youssef, and H. El Gamal, "New achievable secrecy rate regions for the two way wiretap channels," in *Proc. IEEE Information Theory Workshop*, Cairo, Egypt, January 2010, pp. 1–5.
[167] X. He and A. Yener, "A new outer bound for the secrecy capacity region of the Gaussian two-way wiretap channels," in *Proc. IEEE International Conference on Communications*, Cape Town, South Africa, May 2010, pp. 1–5.
[168] Y. Oohama, "Coding for relay channels with confidential messages," in *Proc. IEEE Information Theory Workshop*, Cairns, Australia, September 2001, pp. 87–89.
[169] L. Lai and H. El Gamal, "The relay-eavesdropper channel: Cooperation for secrecy," *IEEE Transactions on Information Theory*, vol. 54, no. 9, pp. 4005–4019, September 2008.
[170] M. Yuksel and E. Erkip, "Secure communication with a relay helping the wire-tapper," in *Proc. IEEE Information Theory Workshop*, Lake Tahoe, CA, USA, September 2007, pp. 595–600.
[171] X. He and A. Yener, "Secure communication with a Byzantine relay," in *Proc. IEEE International Symposium on Information Theory*, Seoul, South Korea, July 2009, pp. 2096–2100.
[172] X. He and A. Yener, "Two-hop secure communication using an untrusted relay," *EURASIP Journal on Wireless Communication and Networking*, vol. 2009, pp. 305 146/1–13, 2009.
[173] X. He and A. Yener, "Cooperation with an untrusted relay: A secrecy perspective," *IEEE Transactions on Information Theory*, vol. 56, no. 8, pp. 3807–3827, August 2010.
[174] A. Khisti, A. Tchamkerten, and G. W. Wornell, "Secure broadcasting over fading channels," *IEEE Transactions on Information Theory*, vol. 54, no. 6, pp. 2453–2469, June 2008.
[175] Y. Liang, G. Kramer, H. V. Poor, and S. S. (Shitz), "Compound wiretap channels," *EURASIP Journal on Wireless Communications and Networking*, vol. 2009, pp. 142 374/1–12, 2009.

[176] I. Csiszár and P. Narayan, "Secrecy capacities for multiple terminals," *IEEE Transactions on Information Theory*, vol. 50, no. 12, pp. 3047–3061, December 2004.

[177] I. Csiszár and P. Narayan, "Secrecy capacities for multiterminal channel models," *IEEE Transactions on Information Theory*, vol. 54, no. 6, pp. 2437–2452, June 2008.

[178] I. Csiszár and P. Narayan, "Secrecy generation for multiple input multiple output channel models," in *Proc. IEEE International Symosium on Information Theory*, Seoul, South Korea, July 2009, pp. 2447–2451.

[179] S. Nitinawarat, A. Barg, P. Narayan, C. Ye, and A. Reznik, "Perfect secrecy, perfect omniscience and Steiner tree packing," in *Proc. IEEE International Symposium on Information Theory*, Seoul, South Korea, July 2009, pp. 1288–1292.

[180] S. Nitinawarat and P. Narayan, "Perfect secrecy and combinatorial tree packing," in *Proc. IEEE International Symposium on Information Theory*, Austin, TX, USA, June 2010, pp. 2622–2626.

[181] M. Haenggi, "The secrecy graph and some of its properties," in *Proc. IEEE International Symposium on Information Theory*, Toronto, Canada, July 2008, pp. 539–543.

[182] P. C. Pinto, J. Barros, and M. Z. Win, "Physical-layer security in stochastic wireless networks," in *Proc. 11th IEEE Singapore International Conference on Communication Systems*, Guangzhou, Singapore, November 2008, pp. 974–979.

[183] E. Perron, S. Diggavi, and E. Telatar, "On noise insertion strategies for wireless network secrecy," in *Proc. Information Theory and Applications Workshop*, San Diego, CA, USA, February 2009, pp. 77–84.

[184] R. Ahlswede, N. Cai, S. Li, and R. Yeung, "Network information flow," *IEEE Transactions on Information Theory*, vol. 46, no. 4, pp. 1204–1216, July 2000.

[185] S. Li, R. Yeung, and N. Cai, "Linear network coding," *IEEE Transactions on Information Theory*, vol. 49, no. 2, pp. 371–381, February 2003.

[186] R. Koetter and M. Médard, "An algebraic approach to network coding," *IEEE/ACM Transactions on Networking*, vol. 11, no. 5, pp. 782–795, October 2003.

[187] T. Ho, M. Médard, R. Koetter, D. Karger, M. Effros, J. Shi, and B. Leong, "A random linear network coding approach to multicast," *IEEE Transactions on Information Theory*, vol. 52, no. 10, pp. 4413–4430, October 2006.

[188] P. Chou, Y. Wu, and K. Jain, "Practical network coding," in *Proc. 41st Allerton Conference on Communication, Control, and Computing*, Monticello, IL, USA, October 2003.

[189] C. Gkantsidis and P. Rodriguez, "Network coding for large scale content distribution," in *Proc. 24th Annual Joint Conference of the IEEE Computer and Communications Societies (INFOCOM)*, Miami, FL, USA, March 2005, pp. 2235–2245.

[190] S. Katti, H. Rahul, W. Hu, D. Katabi, M. Médard, and J. Crowcroft, "XORs in the air: practical wireless network coding," in *Proc. Conference on Applications, Technologies, Architectures, and Protocols for Computer Communications*, Pisa, Italy, August 2006, pp. 243–254.

[191] N. Cai and R. Yeung, "Secure network coding," in *Proc. IEEE International Symposium on Information Theory*, Lausanne, Switzerland, July 2002, p. 323.

[192] L. Ozarow and A. Wyner, "Wire-tap channel II," *AT&T Bell Labs Technical Journal*, vol. 63, no. 10, pp. 2135–2157, December 1984.

[193] S. Rouayheb and E. Soljanin, "On wiretap networks II," in *Proc. IEEE International Symposium on Information Theory*, Nice, France, June 2007, pp. 551–555.

[194] K. Bhattad and K. Narayanan, "Weakly secure network coding," in *Proc. First Workshop on Network Coding, Theory, and Applications (NetCod)*, Riva del Garda, Italy, February 2005.

[195] L. Lima, M. Médard, and J. Barros, "Random linear network coding: A free cipher?" in *Proc. IEEE International Symposium on Information Theory*, Nice, France, June 2007, pp. 546–550.

[196] J. Vilela, L. Lima, and J. Barros, "Lightweight security for network coding," in *Proc. IEEE International Conference on Communications*, Beijing, China, May 2008, pp. 1750–1754.

[197] L. Lima, S. Gheorghiu, J. Barros, M. Médard, and A. Toledo, "Secure network coding for multi-resolution wireless video streaming," *IEEE Journal on Selected Areas in Communications*, vol. 28, no. 3, pp. 377–388, April 2010.

[198] S. Jaggi, M. Langberg, S. Katti, T. Ho, D. Katabi, and M. Médard, "Resilient network coding in the presence of Byzantine adversaries," in *IEEE INFOCOM*, May 2007.

[199] C. Gkantsidis and P. Rodriguez, "Cooperative security for network coding file distribution," in *IEEE INFOCOM*, Barcelona, Spain, April 2006.

[200] N. Cai and R. Yeung, "Network error correction," in *Proc. IEEE International Symposium on Information Theory*, Kanagawa, Japan, July 2003, p. 101.

[201] R. Koetter and F. Kschischang, "Coding for errors and erasures in random network coding," *IEEE Transactions on Information Theory*, vol. 54, no. 8, pp. 3579–3591, August 2008.

[202] M. Kim, L. Lima, F. Zhao, J. Barros, M. Médard, R. Koetter, T. Kalker, and K. Han, "On counteracting Byzantine attacks in network coded peer-to-peer networks," *IEEE Journal on Selected Areas in Communications*, vol. 28, no. 5, pp. 692–702, June 2010.

[203] P. F. Oliveira, R. A. Costa, and J. Barros, "Mobile secret key distribution with network coding," in *Proc. International Conference on Security and Cryptography*, Barcelona, Spain, July 2007, pp. 1–4.

[204] M. J. Kim, J. Barros, M. Médard, and R. Koetter, "An algebraic watchdog for wireless network coding," in *Proc. IEEE International Symposium on Information Theory*, Seoul, South Korea, July 2009, pp. 1159–1163.

[205] T. Ho and D. Lun, *Network Coding: An Introduction*. Cambridge University Press, 2008.

[206] C. Fragouli and E. Soljanin, "Network coding fundamentals," *Foundations and Trends in Networking*, vol. 2, no. 1, pp. 1–133, 2007.

[207] C. Fragouli and E. Soljanin, "Network coding applications," *Foundations and Trends in Networking*, vol. 2, no. 2, pp. 135–269, 2007.

Author index

Subject index

Note: Page numbers in bold refer to figures and tables.

Printed in the United States
By Bookmasters